Engaging Countries

Global Environmental Accord: Strategies for Sustainability and Institutional Innovation
Nazli Choucri, editor

Global Accord: Environmental Challenges and International Responses
Nazli Choucri, editor

Institutions for the Earth: Sources of Effective International Environmental Protection
Peter M. Haas, Robert O. Keohane, and Marc A. Levy, editors

Intentional Oil Pollution at Sea: Environmental Policy and Treaty Compliance
Ronald B. Mitchell

Institutions for Environmental Aid: Pitfalls and Promise
Robert O. Keohane and Marc A. Levy, editors

Global Governance: Drawing Insights from the Environmental Experience
Oran R. Young, editor

The Struggle for Accountability: The World Bank, NGOs, and Grassroots Movements
Jonathan A. Fox and L. David Brown

The Implementation and Effectiveness of International Environmental Commitments: Theory and Practice
David G. Victor, Kal Raustiala, and Eugene Skolnikoff, editors

Global Environmental Diplomacy: Negotiating Environmental Agreements for the World, 1973–1992
Mostafa K. Tolba with Iwona Rummel-Bulska

Engaging Countries

Strengthening Compliance with International Environmental Accords

edited by Edith Brown Weiss and Harold K. Jacobson

The MIT Press
Cambridge, Massachusetts
London, England

First MIT Press paperback edition, 2000

This book was set in Sabon on the Monotype "Prism Plus" PostScript Imagesetter by Asco Trade Typesetting Ltd., Hong Kong.

Printed and bound in the United States of America.

Library of Congress Cataloging-in-Publication Data

Engaging countries : strengthening compliance with international environmental accords / edited by Edith Brown Weiss and Harold K. Jacobson.
p. cm. — (Global environmental accord)
Includes bibliographical references and index.
ISBN 0-262-23198-0 (hardcover : alk. paper), 0-262-73132-0 (pb)
1. Environmental law, International. 2. Environmental policy.
3. Treaties. 4. International organization. I. Weiss, Edith Brown, 1942– .
II. Jacobson, Harold Karan. III. Series.
K3585.4.E545 1998
341.7'62—dc21 98-17993
CIP

Series Foreword

A new recognition of profound interconnections between social and natural systems is challenging conventional intellectual constructs as well as the policy predispositions informed by them. Our current intellectual challenge is to develop the analytical and theoretical underpinnings crucial to our understanding of the relationships between the two systems. Our policy challenge is to identify and implement effective decision-making approaches to managing the global environment.

The Series on Global Environmental Accord adopts an integrated perspective on national, international, cross-border, and cross-jurisdictional problems, priorities, and purposes. It examines the sources and consequences of social transactions as these relate to environmental conditions and concerns. Our goal is to make a contribution to both the intellectual and the policy endeavors.

Tables

Contents

Figures

Preface

Books, like countries and international institutions, have histories. These histories result from small steps taken incrementally and sequentially. There is an interplay of events, ideas, personalities, and conversations. This book certainly has a history. We recount it so that readers can understand the process that produced the words before them, and so that others who might pursue similar multidisciplinary, multinational research projects can keep our experiences in mind as they choose their paths. Recounting the history of the work also provides an occasion to acknowledge our many debts to those who helped us along the way.

The history of the book begins with the formation by the Social Science Research Council (SSRC) of the Committee for Research on Global Environmental Change in 1989. Edith Brown Weiss was appointed chair, and Harold Jacobson, a member of the SSRC committee. At its early meetings the committee considered topics that would be important in understanding the human dimensions of global environmental change. The present and potential role of national and international law and institutions figured prominently in these discussions. It was clear that international accords—treaties and other international agreements—would play a vital role in efforts to mitigate global change. It was also clear that while much was known about the creation of treaties, sometimes called regime formation, little was known about what happened after treaties were signed and came into force. The more the discussions progressed, the greater the agreement became among the members of the committee that compliance with international environmental accords was a prominent candidate for research. It was a topic for which a variety of social science theories were relevant and a topic that could yield findings that would have important applications in policy.

It was also a topic that interested both of us and was related to work that we had done over the years. But it was so vast that even two dedicated individuals could not realistically investigate it empirically. This was also true of other topics that the SSRC committee identified: land-use/land-cover change; the role of property rights in shaping land-use change; and social learning to deal with global environmental change. The committee decided that it should promote research consortia to deal with such topics. These consortia would be led by members of the committee and would include as many people as would be necessary to accomplish the research. We agreed to lead a research consortium on national compliance with international environmental accords.

Then came the tasks of creating a research design, recruiting a research team, finding funding to support the research, and conducting the research. The SSRC agreed to bring together a group of scholars in the spring of 1990 to craft a research design. We decided that the topic demanded that the group be multidisciplinary; that it had to include lawyers, political scientists, economists, and sociologists; and that it had to have as members some who could approach the problem of compliance generally and others who had specialized knowledge about particular countries. We created an initial research design at the meeting, and we began the process of recruiting the research consortium. Most of the participants in the first meeting stayed with the project and became coauthors.

Funding was next. In 1992 the Ford Foundation decided to make an award to the project. The foundation recommended that our consortium include at least one national of each of the countries that we decided to study, and agreed to provide the funding that would be necessary for this. Later that year, the John D. and Catherine T. MacArthur Foundation made an award to the project, and its grant provided funds that enabled us to do much more to include young scholars in the activities of the research consortium. Both opportunities enormously enhanced the project. The fact that each chapter on a country was prepared by a team consisting of American or British scholars and scholars from the country concerned led to wonderful insights and helped to avoid ethnocentric interpretations. Younger scholars gave the consortium energy, vitality, and fresh insights, and the project introduced them to the complicated world of multidisciplinary and multinational research. In 1993, the United States National Science Foundation (NSF) also made an award to the project. Thus, within three years of the creation of the research design, basic funding for the project was in place.

Research started even before we had all of the funding that we needed. As the research progressed, we met periodically as a consortium and pursued empirical research both individually and in small groups. We sought to develop a common approach so that analyses of compliance in the countries included in our sample would be comparable and so that the country and crosscutting analyses would fit together.

Becoming a unified research team was a major challenge and accomplishment. Social science lacks a strong tradition of group research. Indeed, often a premium is placed on individualism, perhaps even extreme individualism. We all were deeply immersed in the traditions and jargons of our own disciplines. Working together required developing an understanding of the ways others approached problems.

Even ensuring that communications among us were fully and properly understood required effort, and we had to develop our own specialized meanings of terms, particularly the core terms "implementation," "compliance," and "effectiveness."

We brought to the research different methodological preferences. Some of us were deeply committed to behavioral approaches and quantitative analyses; others were equally deeply committed to more humanistic approaches. Yet we had to have common understandings about methodology. Without this it would have been impossible for us to conclude that some countries were doing a better job than others of complying with the international environmental accords, or that the extent of compliance of all of the countries in our study had changed over time. The consensus that we reached on measurement fell short of the precision that some of us sought and went beyond what others initially considered reasonable. We all learned from the debate and are all content, though in varying degrees, with the final result.

Meetings were crucial to becoming a unified research team. In addition to the initial session in Bermuda, we met at the University of Michigan's Institute for Social Research in Ann Arbor (twice); the Graduate Institute of International Studies in Geneva, Switzerland; and the East–West Center in Honolulu, Hawaii. We thank the hosts of each meeting. A subset of authors of the country chapters met in May 1996 at the 1763 Inn, Loudoun County, Virginia, to harmonize their work.

Only through prolonged talking could we have reached the understandings that we did. No amount of exchanges of E-mail, paper documents, or telephone calls could have accomplished this. The understandings were, above all, the product of the goodwill and commitment to scholarship of the members of the consortium. All had at one time or another to subordinate individual preferences to group decisions, and all willingly did so. As consortium leaders and editors, we are deeply grateful for the forbearance of the members of the consortium.

Beyond agreement on methodology, the meetings produced an important shift in the focus of the research. Initially, we were primarily concerned about comparisons across countries. It soon became apparent, however, that the results of these comparisons were sometimes rather obvious and not terribly interesting. We found that, generally, rich and democratic countries had a better record of compliance than those that were less well off and less democratic. The reasons for this were not hard to fathom. What became much more interesting to us was why countries did a better job of complying with some treaties than with others (often but not always the same treaties), and why the compliance of all countries changed over time. Change over time became the mantra of the project. We should not take full credit for this shift in focus. Though our own discussions were leading us this way, peer review reports amassed in the NSF review process gave us a strong nudge toward a longitudinal analysis. Each of the major sources of funds for the project had an important influence in shaping its direction.

From the outset, we wanted the project to yield results that could have practical application. The fact that the United Nations Conference on Environment and Development (UNCED), which was held at Rio de Janeiro in June 1992, and which in Working Group Three focused on the enforcement of environmental law, occurred during the formative stages of the project surely helped make this goal salient. We knew that new international environmental accords would result from UNCED, some—particularly the United Nations Framework Convention on Climate Change (FCCC)—being much more far-reaching than any previously negotiated. We hoped that the results of our project would contribute to the design of future international agreements and help the process of implementing the FCCC.

To link our project with the real world of dealing with environmental issues, we held several symposia that included individuals from governments, international governmental organizations, nongovernmental organizations, and industry, as well as scholars, during the course of the project. We wanted to learn from these individuals and to check our preliminary findings with them to see if they were reasonable and had apparent validity. In 1994 the entire consortium met at the Graduate Institute of International Studies with officials from the secretariats responsible for the administration of four of the treaties that we were studying and of one other environmental treaty. Several individuals who had been involved in treaty negotiations were invited to address us during this meeting, and they participated in the discussions.

Closer to the completion of the project, we presented our preliminary findings at symposia organized and sponsored by the United Nations Environment Programme (UNEP) at its headquarters in Nairobi and at the Georgetown University Law Center, and at seminars and workshops at the Woodrow Wilson International Center for Scholars in Washington, D.C., the Environmental Studies Unit of the Graduate Institute for International Studies in Geneva, Switzerland, the headquarters of the United Nations University in Tokyo, Japan and the Council on Foreign Relations in New York. The findings of the country chapter on India and the general conclusions were presented by Ronald Herring and Erach Bharucha at the Center for Environmental Law in the building of the WorldWide Fund for Nature, India, in New Delhi. Each of these encounters gave us new insights and helped us hone and refine our findings.

We also benefited from the comments we received when we made presentations to our colleagues on the SSRC Committee for Research on Global Environmental Change; at meetings of the American Association for the Advancement of Science, the American Political Science Association, the American Society of International Law, the International Studies Association, the International Political Science Association, and the Swedish Council for Research; at our own and other universities both in the United States and abroad. The SSRC committee provided especially useful oversight and advice. The criticisms and suggestions that we received in all of these discussions helped us in many, many ways. To those engaged in complex projects, we strongly recommend regularly talking to others in more or less formal settings about the course of the work. The benefits are enormous.

Several people played special roles and made particular contributions during the course of the project. Richard Rockwell was the SSRC program officer who created the Committee for Research on Global Environmental Change. He recruited us to the committee, and he helped us in more ways than we can describe. He helped us get the funding for the initial meeting and played a crucial role in obtaining the grants from the Ford and MacArthur foundations. He identified several of the people who became members of the consortium. We owe him special thanks. David Major replaced Richard Rockwell as the SSRC program officer responsible for the committee and its activities, and he and his assistant Sarah Gordon helped us with the innumerable details that arise in arranging meetings and overseeing accounts. Ellen Schaeffer, the international law librarian at the Georgetown University Law Center, aided enormously in guiding us to and finding essential resources. Andrea Lachenmayr assisted us in creating tables and figures. Barbara Opal helped arrange meetings, and she and Julie Weatherbee miraculously turned our manuscripts, produced in diverse formats and styles, into a uniform manuscript. Madeline Sunley, Enza Vescera, and Melissa Vaughn of MIT Press have gracefully and cheerfully helped us navigate the complex passage of turning a project into a book.

The members of the research consortium include more than those who ultimately became authors of the chapters in this book. In addition to ourselves, they included Danae Aitchison, Murillo de Aragão, Anthony Balinga, Laszlo Bencze, Erach Bharucha, Piers Blaikie, Stephen Bunker, Abram Chayes, Antonia Handler Chayes, James Clem, Ellen Comisso, Pamela Cothran, Liz Economy, Fang Xiaoming, James Vincent Feinerman, John Flynn, Koichiro Fujikura, Michael Glennon, Saul Halfon, Peter Hardi, Allison Hayward, Ronald Herring, Philipp Hildebrand, Takesada Iwahashi, Jasmin Jagada, Sheila Jasanoff, Tim Kessler, Ron Mitchell, Elena Nikitina, Kenneth da Nobrega, Michel Oksenberg, Neema Pathak, Jonathan Richards, Gideon Rottem, Alberta Sbragia, Thomas Schelling, Cheryl Shanks, John Mope Simo, Anjili Soni, Alison Stewart, David Vogel, Nadia Wetzler, Wu Zijin, Zhang Shuqing, Andreas Ziegler, and William Zimmerman. All of these individuals made important contributions to the project. We are deeply grateful to our collaborators.

We appreciate the helpful comments on the manuscript of the reviewers for MIT Press. We are especially grateful to our friends who helped us along the way, particularly Winfried Lang, who graciously read and commented on various drafts, and Sun Lin, former director of the Environmental Law Unit at UNEP, who assisted us in many ways.

We and all members of the consortium are indebted to the many people whom we interviewed as part of the research and to all those who provided data for the project. We could not have done the research without their cooperation. They are too numerous to mention. There is no way that we can repay them; we can only hope that this book will contribute to helping them attain their goals.

We are deeply grateful for the grants from the Ford Foundation, the MacArthur Foundation, the U.S. National Science Foundation (grant SBR 9223158), the Social Science Research Council, and the East–West Center (Hawaii) that made the research possible. The research was done under the broad auspices of the Committee for Research on Global Environmental Change of the Social Science Research Council.

We of course have overall responsibility for the book and whatever flaws it may contain.

Contributors

Murillo de Aragão
University of Brasília
Brasília, D.F.
Brazil

Erach Bharucha
Saken Valentina Society
Poona, India

Piers Blaikie
School of Development Studies
University of East Anglia
Norwich, United Kingdom

Edith Brown Weiss
Georgetown University Law Center
Georgetown University
Washington, D.C.

Stephen Bunker
Department of Sociology
University of Wisconsin
Madison, Wisconsin

Abram Chayes
Harvard University Law School
Harvard University
Cambridge, Massachusetts

Antonia Handler Chayes
Conflict Management Group
Cambridge, Massachusetts

James Clem
Russian Research Center
Harvard University
Cambridge, Massachusetts

Ellen Comisso
Department of Political Science
University of California, San Diego
San Diego, California

Elizabeth Economy
Council on Foreign Relations
New York, New York

James V. Feinerman
Georgetown University Law School
Georgetown University
Washington, D.C.

Koichiro Fujikura
Waseda University Law School
Waseda University
Tokyo, Japan

Michael J. Glennon
School of Law
University of California, Davis
Davis, California

Peter Hardi
International Institute for Sustainable Development
Winnipeg, Manitoba, Canada

Ronald Herring
Department of Government
Cornell University
Ithaca, New York

Philipp M. Hildebrand
World Economic Forum
Geneva, Switzerland

Harold K. Jacobson
Center for Political Studies
University of Michigan
Ann Arbor, Michigan

Sheila Jasanoff
Department of Science and Technology Studies
Cornell University
Ithaca, New York

Timothy Kessler
San Diego, California

Ronald B. Mitchell
University of Oregon
Eugene, Oregon

Elena Nikitina
Institute for World Economy and
International Relations
Academy of Sciences
Moscow, Russian Federation

Michel Oksenberg
Stanford University
Stanford, California

John Mope Simo
World Wildlife Fund Representation
Yaouandé, Cameroon

Alberta M. Sbragia
Center for West European Studies
University of Pittsburgh,
Pittsburgh, Pennsylvania

Alison L. Stewart, Esq.
Los Angeles, California

David Vogel
Walter A. Haas School of Business
University of California, Berkeley
Berkeley, California

William Zimmerman
Center for Political Studies
University of Michigan
Ann Arbor, Michigan

1
A Framework for Analysis

Harold K. Jacobson and Edith Brown Weiss

In 1972, countries gathered in Stockholm at the United Nations Conference on the Human Environment to launch a global effort to protect, preserve, and enhance the environment. International environmental accords—treaties and other binding international agreements—became central components of the strategy that was adopted. International agreements orient and coordinate the behavior of states and ultimately of enterprises, nongovernmental organizations (NGOs), and individuals, steering behavior away from activities that are environmentally destructive and toward those that are environmentally benign. At the time of the Stockholm conference, there were only a few dozen multilateral treaties dealing with environmental issues.

By 1992, when countries gathered again to deal with the global environment at the United Nations Conference on Environment and Development at Rio de Janeiro, there were more than 900 international legal instruments (mostly binding) that were either fully directed to environmental protection or had more than one important provision addressing the issue.[1] The substantive and procedural duties contained in the accords had become more stringent, detailed, and comprehensive, and the range of issues subject to such accords had expanded (Brown Weiss 1993). Two major treaties were signed at Rio, one on climate change and the other on biodiversity, together with several important nonbinding legal instruments. The negotiation of international treaties and other international legal instruments to protect the global environment has continued.

Yet we know very little about national implementation and compliance with the treaties and other international legal instruments that have been negotiated, despite their importance and growing number. Even if no more accords were to be negotiated, it would be essential to make those that are in force work effectively.

International accords are only as effective as the parties make them. Effectiveness is the result not only of how governments implement accords (the formal legislation or regulations that countries adopt to comply with the accord) but also of compliance (the observance of these regulations and the commitments contained in the international accord). Weak legislation can produce weak compliance; unenforced strong legislation can have the same effect. One cannot simply read domestic legislation to determine whether countries are complying. While some claim that most states comply with most international treaties most of the time, there are reasons to believe that national implementation of and compliance with international accords is not only imperfect but often inadequate, and that such implementation as takes place varies significantly among countries.

It is not known to what extent international environmental accords have or have not evoked compliance, or whether the same factors that presumably motivate compliance with arms control, trade, or human rights agreements motivate compliance with environmental accords. There have been only a few systematic studies of factors affecting compliance with international environmental accords into which countries have *already* entered (Mitchell 1994). Our study takes a major step toward drawing empirically from the experience of existing international environmental law those lessons that might instruct us how better to proceed in the future.

The Stylized View of Compliance and Reality

A traditional, stylized view of international law might maintain that (1) countries accept treaties only when their governments have concluded that the treaties are in their interest; (2) because of this, countries generally comply with treaties; and (3) when countries do not comply with treaties, sanctions are employed both to punish offenders and to serve as deterrents designed to encourage first-order compliance.

Reality with respect to many types of treaties, particularly environmental accords, is quite different. While countries may join only treaties that they regard as in their interest, there are various reasons that countries find them in their interest, and these reasons affect their willingness and ability to comply with the treaties. Governments may choose to accept a treaty because of a desire to climb onto an international bandwagon, or because of pressures from other governments—both inducements and threats—with leverage over them. Or there may be domestic interests that force the issue. In some cases, countries may enter treaties without intending to modify their behavior significantly so as to comply fully. Even if they intend to comply, some countries may find it difficult or impossible, because they lack the capacity to do so.

Scattered evidence suggests that implementation of and compliance with international environmental accords are often haphazard and ragged. Parties rarely resort to adjudication of violations or employ significant sanctions against noncomplying parties. While blandishments may be used to encourage compliance, these are rarely of major proportions. Nevertheless, as the experience with some human rights treaties and labor agreements illustrates, over time many countries gradually do more to implement treaties and improve compliance.

This less elegant reality of imperfect, varied, and changing implementation and compliance is the starting point for this study. The purpose of this study is to discover factors that lead to improved implementation of and compliance with treaties that cover environmental issues. We assume that cost–benefit calculations are murky, military sanctions are out of the question, and economic sanctions are exceptional and may violate international trading arrangements. We also assume that the propensity of various countries to comply with different treaties will vary and will change over time. Our task is to understand the factors that shape the variation and propel the change.

Compliance takes place in an international system that is in the process of change, and our analyses will highlight several of these changes. The traditional view of the international system as hierarchical and focused almost exclusively on states has evolved into one that is nonhierarchical. Effective power is increasingly being organized in a nonhierarchical manner. While sovereign states continue as principal actors, and are the only ones that can levy taxes, and conscript and raise armies, these functions have declined in importance relative to newly important issues, such as environmental protection and sustainable development. There are now many relevant actors in addition to states: international governmental organizations (IGOs), nongovernmental organizations, enterprises, other nonstate actors, and individuals. The *Yearbook of International Organizations* records in its 1997/1998 edition 6,115 intergovernmental organizations and 40,306 nongovernmental organizations (UIA 1997). Nonstate entities are performing increasingly complex tasks, especially in the newer issue areas.

New information technologies empower nonstate actors to participate in developing, implementing, and complying with international law. Interest groups can form almost instantaneously on the Internet to press for actions in a country, a situation that can both help and hinder compliance. The technologies facilitate linkages among nonstate actors and communications among nonstate actors, IGOs, and treaty secretariats.

International law is also changing in form. The sharp lines between public and private international law are blurring, the sharp divide between international and domestic law is fading, and the difference between the effectiveness of legally binding and nonbinding international instruments is being questioned (Brown Weiss 1996). These developments are reflected in our analysis.

Our research focused on compliance with specific international treaties and did not directly address issues of compliance with customary international law. Assessing compliance with these rules presents additional problems. It is difficult to identify rules of customary international law and, once they are identified, to determine their precise boundaries, unless the rules are codified in international legal instruments. Moreover, it is difficult to identify failures to implement the rules or to comply with them, since there are no formal parties to monitor compliance with them. However, it is reasonable to expect that the framework of analysis and the model developed later in this study would apply to national compliance with customary international law, although it would be more difficult to acquire data needed for the analysis.

Assessing Implementation, Compliance, and Effectiveness

An essential first step in the analysis is to have clear definitions of implementation, compliance, and effectiveness.

Implementation refers to measures that states take to make international accords effective in their domestic law. Some accords are self-executing; that is, they do not require national legislation to become effective. But most international accords require national legislation or regulations to become effective. Countries adopt different approaches, ranging from accounting procedures to incentives such as tax relief to induce compliance, and from public admonishment to sanctions for noncompliance. This study seeks to identify systematically the various methods that are employed for implementing international accords and to analyze which strategies are used with what degree of effectiveness in securing compliance.

Compliance goes beyond implementation. It refers to whether countries in fact adhere to the provisions of the accord and to the implementing measures that they have instituted. The answer cannot be taken as given even if laws and regulations are in place. Measuring compliance is more difficult than measuring implementation. It involves assessing the extent to which governments follow through on the steps that they have taken to implement international accords. In the end, assessing the extent of compliance is a matter of judgment.

Compliance has several dimensions. Treaties contain specific obligations, some of which are procedural, such as the requirement to report, and others that are substantive, such as the obligation to cease or control an activity. In addition, preambles or initial articles in treaties place these specific obligations in a broad normative framework, which we refer to as the spirit of the treaty.

Compliance is probably never perfect; substantial compliance is what is sought by those who advocate treaties and agreements. We seek to assess the extent to which substantial compliance is achieved with the procedural and substantive obligations contained in treaties, as well as with the spirit or broad norm involved in the treaties, and to compare the extent of success within and among political units and over time.

In principle, the compliance of countries with their obligations under international environmental accords can be measured, but we found precise measurement elusive. In examining procedural obligations such as the requirement to submit reports, it is relatively straightforward to determine whether a report has or has not been submitted on time, though it is more difficult to assess whether the report is sufficiently complete or is accurate. Similarly, one can tell if the production of a prohibited substance, such as chlorofluorocarbons, has been discontinued. But measuring emissions resulting from changing refrigerants in automobile air conditioners is much more problematic. And how should an occasional emission be weighted in relation to the overall cessation of production? In the end, we have had to rely on our own expert judgments to assess compliance. Our assessments are not made at precise intervals but, rather, in rough orders of magnitude—substantial compliance, moderate compliance, weak compliance, or no action taken toward compliance. But even these rough ordinal measures allow us to detect differences among countries and treaties, as well as change over time.

Effectiveness is related to, but is not identical with, compliance. Countries may be in compliance with a treaty, but the treaty may nevertheless be ineffective in attaining its objectives. And even treaties that are effective in attaining their stated objectives may not be effective in addressing the problems that they were intended to address. To illustrate the point, compliance with a treaty may result in the cessation of an activity that contributed to pollution, but it may also lead to an overall increase of pollution by encouraging other activities as substitutes whose consequences are even worse. Or a treaty prohibiting international trade in a certain species of monkey could effectively stop the trade but have little impact on the decimation of the monkey populations, which may be consumed within the national borders where they are located, rather than abroad, after being exported.

Table 1.1 shows the several dimensions of implementation, compliance, and effectiveness. Our project is particularly concerned with assessing implementation and compliance. Effectiveness is crucially important, but until implementation and

Table 1.1
Implementation, compliance, and effectiveness

I. Implementation
II. Compliance
 A. Compliance with the specific obligations of the treaty
  1. Procedural obligations
  2. Substantive obligations
 B. Compliance with the spirit of the treaty
III. Effectiveness
 A. In achieving the stated objectives of the treaty
 B. In addressing the problems that led to the treaty

compliance are better understood, the contribution of treaties to solving international environmental problems cannot be known. Learning about implementation and compliance is an essential first step to learning about effectiveness.

Many factors may affect a country's implementation of and compliance with international accords. We are interested in how several interrelated factors affect the extent to which, and the way in which, countries have met their commitments. These factors include characteristics of the activity; characteristics of the accord; the international environment; and factors involving the country.

Characteristics of the Activity

Environmental accords are about human activities— activities that extract resources, produce pollutants or other emissions, change ecosystems, or reduce biodiversity. Some substances or activities have little economic importance, while others have consequences for entire economies. Some also have little intrinsic economic value, but the process of compliance can disrupt economic activities in many other areas. Some are easy to monitor, while others can be detected only through very intrusive measures. The costs and benefits of regulating substances and activities, and their distribution among various social classes and geographical regions, can also be important. The organization of the activities—whether they are conducted by large multinational organizations or small firms or individuals—could also make a difference.

Characteristics of the Accord

The characteristics of the treaty or agreement are an important factor. Some issues relate to the process by which the accord was negotiated. By whom and how was the process initiated? What form did the negotiations take? What were the extent and the depth of agreement? The substantive characteristics of the accord also raise important issues. What is the nature of the obligations contained in the accord? Are the duties general or precise? Are they binding or hortatory? What implementation and compliance mechanisms are contained in the accords? How does the agreement treat countries that do not join?

The Montreal Protocol on Substances That Deplete the Ozone Layer and the Convention on International Trade in Endangered Species obligate parties not to trade controlled substances with countries that are not parties to the agreement. How effective is this provision in inducing countries to join and comply? What benefits accrue to countries that are parties to the accords? What special dilemmas do the accords produce, such as the problem of how an item, once placed on the World Heritage Convention's list of protected sites, ever gets taken off that list?

The International Environment

The actions of other states in implementing and complying with the accord can also affect a state's compliance with an agreement. To what extent have other countries' noncompliance or compliance with the accord affected the willingness of countries to abide by the accord? How does the answer to this question vary with the subject and obligations of the international accord? To what extent can a state be a free-loader under the accord?

International governmental organizations have important roles in promoting the implementation of and compliance with international accords. We investigate how countries relate to the IGOs that have responsibilities for these accords. What importance, if any, was attached to involvement by international organizations, such as the United Nations, the World Bank, and regional development banks? International conferences, such as those held at Stockholm and Rio, could also play a role. International nongovernmental organizations and the worldwide media could also be important.

Factors Involving the Country

The social, cultural, political, and economic characteristics of the countries clearly influence implementation and compliance. We assess the relative importance of its broad political culture, the level of its economic development, and the trajectory and pace of its economic growth or decline in shaping a country's actions. Are there cultural traditions that influence how a country complies? What difference does it make whether the country has a market or a planned economy, or if it is mixed? Does it make a difference in which sector the substance or activity is included?

What are the effects of the characteristics of the political system? Is the country democratic? How strong and effective is the bureaucracy? How effective is coordination among the relevant ministries and between the national and provincial governments? What is the nature of the legal system? What procedures are required to adopt the regulations or other strategies necessary to implement the agreement? How strong and independent is the country's judicial system?

A country's policy history regarding the substance or activity being regulated is another basic factor. What was the country doing about the substance or activity before adhering to the international accord? Had the country already recognized the existence of an environmental problem? What role did the country play in the negotiation of the accord?

It is clear that people make a difference. Some leaders are more committed to and effective in promoting compliance with international environmental accords. Some countries have drawn leadership on an issue from the scientific community; others have not had such communities on which to draw. What are the consequences of changes in and differences among leaders?

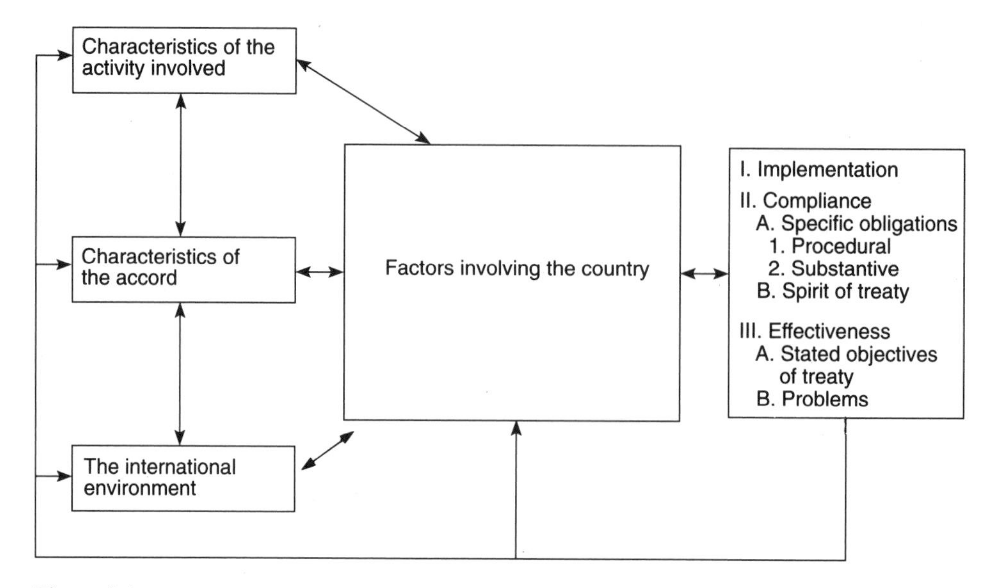

Figure 1.1
A model of factors that affect implementation, compliance, and effectiveness

Nongovernmental organizations can play important roles at the national level, just as they can at the international level. What is the strength of nongovernmental groups, including those engaged in lobbying and domestic and international agenda-setting?

It is broadly assumed that the more information there is about an environmental issue and the clearer the understanding of the issue, the more effective implementation and compliance will be. This assumption impels much of the work of international organizations. We want to assess how the availability of information about and the extent of understanding of the environmental issues covered in the treaties affects national implementation and compliance. Another important issue is whether a culture of compliance exists within the country.

Figure 1.1 presents a graphic representation of the interaction of these factors with a state's implementation of and compliance with an accord and the effectiveness of the accord. This graphic representation constitutes a wiring diagram or a rudimentary depiction of a model. It includes feedback effects from implementation, compliance, and effectiveness. The fact that many of the arrows go in both directions indicates that we expect there to be interactive effects.

In examining these factors, we want to explore the hypotheses that are nested within the questions posed in the paragraphs above. It would be pretentious to claim that we could test the hypotheses. We have too many variables and too few cases, and several of our crucial variables are, at best, measured inexactly.

Some of these hypotheses have been deduced from rational choice assumptions, others have been derived from the existing literature on international law and institutions, and still others have been derived from preliminary analyses of empirical data that we gathered. Many of the hypotheses simply state common sense. Some relationships are obvious. Some have been identified in assessments by the U.S. General Accounting Office and the secretary-general of the United Nations Conference on Environment and Development (USGAO 1992b; UN 1992).

Based on the realist and rational choice assumptions that undergird game theory, one would expect that the smaller the costs and the greater the benefits associated with an accord, however difficult they may be to calculate, the greater the probability of implementation and compliance. The likelihood of significant sanctions would be included in the prospective cost element of this hypothesis. Since implementation and compliance require monetary and bureaucratic resources, it would seem logical that the larger a country's gross national product (GNP) and the higher its per capita GNP, the greater the probability of implementation and compliance. Because costly measures can be accommodated with minimal or no redistribution in a period of rapid economic growth, the higher the rate of a country's economic growth, the greater the probability of implementation and compliance.

Since pressure from domestic groups and the mass public comprises an important mechanism for promoting implementation and compliance with treaty obligations, many scholars and policy makers have assumed that the more a country adheres to democratic norms concerning political and civil rights and political participation, the greater the probability of implementation and compliance. Several have also assumed that decentralization would promote more effective compliance.

These hypotheses are so strongly grounded in common sense that from the outset of our work, we expected them to be supported by the evidence produced by our analyses, to bound other findings. We expected that the variables involved in these hypotheses would explain the largest share of the variance among countries and treaties, but we also were very conscious that it would not be very surprising or particularly helpful to those interested in improving compliance for us to discover that rich countries are more likely to comply with treaties than are poor countries, or that countries are more likely to comply with treaties that impose few burdens. Countries cannot be made rich overnight, nor can the burdens that compliance might entail always be eliminated.

The relationships that hold greatest fascination for us are those that can be manipulated in the relatively short run through policy interventions. They include the characteristics of the accords, the international environment, and some factors involving the countries.

Many of these, particularly those involving domestic and international processes, have been identified in broad general studies by Abram and Antonia Handler

Chayes (1993, 1995), Roger Fisher (1981), Peter M. Haas (1990, 1992), Robert O. Keohane (1984, 1986, 1988), and Elinor Ostrum (1990). Factors that one or more of them have stressed are (1) creating communities of interested parties, especially scientists and specialists in the topic, or what have been termed "epistemic communities"; (2) increasing the amount, quality, and availability of information about the issues involved, so that they can be understood; (3) involving domestic officials and bureaucracies, so that their personal interests and reputations become issues at stake; and (4) generating international momentum toward compliance, which increases the benefits of compliance, and the costs and consequences of noncompliance, for adhering countries.

Hypotheses concerning these factors are straightforward. Since epistemic communities are deeply committed to the goals of particular accords because of their knowledge and professional interest, the greater the size, strength, and activism of epistemic communities, the greater the probability of implementation and compliance.

With respect to information, the hypothesis is: The greater the flow of scientific and technical information about targeted activities, in a form that is understood by governments and public pressure groups, the greater the likelihood of implementation and compliance. Particularly with environmental accords, these actors must rely on scientific and technical information flows to identify and assess risks, address targeted activities, and identify available technical options to enhance compliance.

Hypotheses concerning domestic processes are straightforward. Two are very important. Because repeated encounters and associations with counterparts and concern for reputations have a powerful impact on behavior, the more involved a country's domestic officials and bureaucracies are in the preparation, implementation, and oversight of an accord, the greater the probability of implementation and compliance.

With respect to international momentum, the most direct hypothesis is: The greater the number of countries that have ratified an accord and the greater the extent of their implementation and compliance, the greater the probability of implementation and compliance by any individual adhering country. Countries have a deep and abiding interest in creating and maintaining a relatively stable and predictable international political environment. The more stable and predictable an environment is, the higher the costs of disrupting it, and thus the greater the probability of strong implementation and compliance. International momentum also has broader aspects, such as how committed international public opinion is to the issue. International nongovernmental organizations (INGOs) capture and articulate important sections of international public opinion; hence, the stronger and more active the INGOs are in the area of the treaty, the greater the probability of strong implementation and compliance.

We are also interested in other process factors, such as leadership and the extent of transparency surrounding the activity covered by the accord. We assume that the more committed a country's leaders are to the goals of an accord, the greater the probability of implementation and compliance. Leaders are chosen for many reasons. To some extent, in terms of the focus of this research, a country's choice of leaders is a stochastic process. We study how leaders, whatever their initial inclinations, become more engaged in ensuring compliance with the accord.

Transparency may promote compliance because it makes noncompliance more apparent and makes it easier for international and domestic actors to take actions to encourage and enforce accountability. Transparency is closely linked to the character of the issue covered by the treaty and democratic norms. Cultural factors may affect the acceptability of transparency internally, and hence its effectiveness in inducing compliance. We test the hypothesis that transparency promotes implementation and compliance, and seek to identify the conditions that bound the hypothesis. We also analyze the extent to which transparency can be promoted.

Finally, since administrative and bureaucratic capacity is essential for implementing accords, we explore the extent to which such capacity has been, and could be, increased independent from broader economic, political, and social development. The greater the capacity of the political unit to implement the accord, the more likely it is that it will comply. Administrative and bureaucratic capacity depends upon economic resources, but it also involves education, technical training and skills, and attitudes.

Our research efforts focus on these factors because previous research indicates that they are important. The evidence for them, however, is largely anecdotal. We examine them empirically on a multicountry basis. Since the factors are subject to directed and purposeful modification, policies could be adopted that target them in order to increase implementation and compliance with international environmental accords.

Stating the hypotheses that have guided our research as starkly as we have may make the investigation appear overly mechanistic. The analyses could not have been conducted in a mechanical and simplistic way. Context and institutions are extremely important. The hypotheses are important because they provide a framework that guides and directs the investigation, useful both as a way of structuring our analyses and guiding comparisons, and also as a way to link our study with others at a basic, practical level as well as at a theoretical one.

In varying degrees, the factors that are involved in the hypotheses that are of most concern to us are subject to deliberate manipulation by those who prepare accords and are responsible for overseeing their implementation and compliance with them. The research seeks to develop a basis for reasoned speculation about how manipulation of these factors might improve implementation and compliance.

The Countries and the Treaties

To investigate these hypotheses and issues, we chose to focus on eight countries and the European Union, and five international treaties covering different areas of environmental concern. We chose the nine political units and the five treaties in the hope that the study will yield knowledge that will have worldwide utility and will pertain to most kinds of environmental accords concluded in the future.

Political Units of Great Importance and Widely Differing Character

The eight countries are Brazil, Cameroon, China, Hungary, India, Japan, the Soviet Union/Russian Federation, and the United States. We also include a group of countries, the European Union (EU). We chose these countries because they are crucially important to the effective implementation of broad international environmental accords. They include those that have contributed most to the anthropogenic effects that bring about global change (Japan, Russia, the United States, and the European Union and its members), and those that have the potential of making major contributions to anthropogenic effects (Brazil, India, and China).

We included Cameroon and Hungary to illustrate the problems and processes of implementation and compliance in smaller countries that, although their total contribution to global environmental problems may be individually small, comprise by far the largest number of states in the global political system.

We chose the European Union because it is increasingly behaving as a state actor through directives and regulations that are applicable in all member states.[2] It represents a new form of governmental organization, one that conceivably could be duplicated elsewhere, such as among the states that constituted the former Soviet Union. It merits study for that reason, as well as for the reason that the EU will be a major political and economic actor in future negotiations. Although the EU was a party to only two of the treaties, its member states (who join treaties as individual states) were generally parties to most of them.

In the 1990s these countries included about three-fifths of the world's population, their gross national products constituted about four-fifths of the world product, and they contributed more than two-thirds of the global greenhouse emissions. Table 1.2 presents data that show the importance of the nine political units for environmental issues. They spanned the globe and encompassed a range of forms of political organization and culture. Furthermore, they included low-, middle-, and high-income countries, some with mixed-market economies and others that were

Table 1.2
The political units and their characteristics

Political unit	Population (millions)	GNP per capita (U.S. dollars, 1990)	GNP per capita growth rate, 1965–1990	CO_2 emissions per capita, 1989 (tons of carbon)	CFC net annual atmospheric increase, 1986 (thousands of metric tons)	Percentage change in forest and woodland, 1989
Brazil	150	2,680	3.3	0.38	9	−0.4
Cameroon	12	960	3.0	0.14	0	−0.4
China	1,134	370	5.8	0.59	18	0.0
EU[1]	343	17,058	2.5	2.20	228	1.1[2]
Hungary	11	2,780	—	1.65	1	0.6
India	850	350	1.9	.21	0	−0.2
Japan	124	25,430	4.1	2.31	58	0.0
Russia	148.7	3,220[3]	—	3.60	—	—
United States	250	21,790	1.7	5.37	197	−0.1
Average	—	8,293	2.8	1.8	63.88	0.08
World average	—	4,964	1.6	1.09	5.4	−0.13
Total	3,022.7	—	—	—	511	—
World total	5,222	—	—	—	659	—

Source: World Bank, *The Environment Data Book: A Guide to Statistics on the Environment and Development* (Washington, DC: World Bank, 1993), 10–13.
[1] Not including Luxembourg.
[2] Not including Belgium.
[3] 1991.

restructuring centrally planned economies. Some of these countries could be particularly affected by global change.

Five International Environmental Treaties

The five international treaties were chosen so as to maximize the knowledge that could be gained about ways of managing global environmental change. We deliberately selected both agreements that control pollution (the brown issues) and that protect natural resources (the green issues). There is considerable variety among these treaties in what they address and how they deal with it. We selected only treaties to which there is a significant number of parties and for which there is already some experience with implementation and compliance. A study of proposed accords that have not yet been implemented would tell us little about what makes for crafting a successful agreement. The five treaties we chose include three that deal with the management of natural resources and two that are aimed at controlling pollution.

Natural Resource Treaties

1. The United Nations Educational, Scientific and Cultural Organization Convention for the Protection of the World Cultural and Natural Heritage, November 16, 1972, 27 U.S.T. 37, T.I.A.S. no. 8226 (referred to as the World Heritage Convention).

2. The Washington Convention on International Trade in Endangered Species of Wild Fauna and Flora, March 3, 1973, 27 U.S.T. 1087, T.I.A.S. no. 8249 (referred to as CITES).

3. The International Tropical Timber Agreement, November 18, 1983, U.N. Doc. TD/Timber/11/Rev. 1 (1984) (referred to as the International Tropical Timber Agreement). The 1994 International Tropical Timber Agreement replaced this agreement as of January 1, 1997, when it entered into force.

Pollution Control Treaties

4. The International Maritime Convention on the Prevention of Marine Pollution by Dumping of Wastes and Other Matter, December 29, 1972, 26 U.S.T. 2403, T.I.A.S. no. 8165 (referred to as the London Convention of 1972 and formerly referred to as the London Ocean Dumping Convention). The 1996 Protocol will replace this agreement when ratified by the requisite number of countries.

5. The Montreal Protocol on Substances That Deplete the Ozone Layer, September 6, 1987, 26 I.L.M. 1550 (referred to as the Montreal Protocol), together with

5a. The Vienna Convention for the Protection of the Ozone Layer, March 22, 1985, 26 I.L.M. 1529 (referred to as the Vienna Convention). This is the framework treaty under which the Montreal Protocol was negotiated. We look at this treaty only insofar as it relates to the Montreal Protocol.

These treaties have been chosen because (1) they address both pollution and natural resource problems; (2) they involve several key environmental issues con-

Table 1.3
Adherence of the political units to the treaties (year joined)

Political unit	World Heritage	CITES	Tropical Timber	London Convention	Vienna Convention	Montreal Protocol
Brazil	1977	1975	1985	1982	1990	1990
Cameroon	1982	1981	1985		1989	1989
China	1985	1981	1986	1985	1989	1991
European Union			1985		1988	1989
Hungary	1985	1985		1976	1988	1989
India	1977	1976	1986		1991	1992
Japan	1992	1980	1985	1980	1988	1989
Russia	1988	1992	1986	1976	1988	1989
United States	1973	1975	1990	1975	1988	1989

nected with global change; (3) they contain a range of types of obligations and use various techniques to promote implementation and compliance; (4) they involve different jurisdictional frameworks and issues that occur primarily within states' borders, that cross borders, and that are inherently global in nature; (5) they have been in effect long enough that there is an adequate database to analyze implementation and compliance; and (6) all of the countries are parties to at least three of the treaties, and most are parties to all of them.

Table 1.3 shows the years in which the nine political units joined the respective treaties.

Five of the political units were parties to all five treaties, and three were parties to four. The European Union was a party to only two of the treaties, but its member states were parties to all of them. These nine political units began joining the treaties in 1973 and continued through 1992, when Japan became a party to the World Heritage Convention and India a party to the Montreal Protocol.

Chapter 5 describes the treaties in detail: the obligations they impose, and the procedures and institutions they create. More important, it analyzes how they came into being and how they have evolved over time. It presents a living history of the treaties.

Methodological Issues

The research tasks involved in this project are inherently difficult in several respects. Practicality required that we limit the number of countries and treaties in the study. We recognize that countries that are very important for several of the treaties are not included, such as Australia for the World Heritage Convention, and Malaysia and Indonesia for the International Tropical Timber Agreement. We know that a number of interacting external and internal factors affect the extent to which countries implement and comply with treaties, as well as the pace and character of the actions that they take. Measurement of the extent of implementation and compliance, and of the factors that affect countries' actions concerning implementation and compliance, is imprecise at best. Moreover, some of the data essential to the analyses could be gained only through interviews.

One aspect of our research design consists of structured focused comparisons of the experience of diverse political units in implementing these specific international treaties. The comparisons follow the methodology set forth by Alexander L. George and Richard Smoke (1974:95–103). Comparison is necessary because there are considerable differences among the ways in which countries, even at similar levels of economic development, frame and implement policies designed to protect the environment (Jasanoff 1986, 1991). Since we are interested in the effects of different types of problems and legal arrangements, we are interested in comparing the records of individual countries with respect to different treaties as well as comparing the records of different countries. International and global regulatory efforts must take account of these practices.

A second and perhaps more important aspect of the research design consists of comparisons over time. We are interested in whether countries' compliance improves or declines over time. For those interested in using international environmental accords to protect the global environment, the relevant issue is how to strengthen all countries' compliance over time.

In some respects we have a larger number of observations than it may appear at first glance. Because we are interested in changes in political units' implementation of and compliance with their treaty obligations over time, our number of observations is the number of political units involved in the study, times the number of treaties they have ratified, times the number of years each unit has been a party to each treaty. Calculated in this manner, our number of possible observations is 502. We also observe the behavior of the political units in the period before negotiations that led to the treaties. Including this, we could have 1,485 observations. Unfortunately, the obligations imposed by the five treaties are comparable only in limited ways, so in several analyses we have been forced to keep our comparisons within a single treaty.

Measurement of implementation and compliance poses serious difficulties, as was noted above. Some measures, for instance, on the time lag between a state's becoming a party to a treaty and its adoption of implementing legislation and regulations, can be precise and other issues, such as whether reports were filed and violations were noted and reported, can be coded "yes" or "no," although it is more difficult to capture the extent to which the reports are complete and accurate.

Resources devoted to implementation and compliance in principle could be measured precisely. Yet there are difficulties, and in practice the data are murky. Many national and international officials who are involved in implementation and compliance efforts spend only part of their time on these tasks, and find it difficult to say exactly how much time they do spend. The same problem occurs with respect to budgets. Finally, there are many issues where we must rely on our own expert judgments. Whether a country is in substantial compliance with a treaty is in the end essentially a matter of judgment. Thus the study contains a mixture of quantitative and qualitative analytical techniques.

Since we have several poorly measured variables, many causes, and numerous interactions among the variables, we know that our study cannot yield clearly specified causal propositions. It can and will, however, yield interesting and useful findings even though they may be imprecise.

The Plan of the Book

The analysis that follows can be divided into sections. The first section provides a context for the analysis. The first chapter in this section, chapter 2, reviews the experience within countries in securing compliance with domestic environmental legislation and regulations. The analysis focuses primarily upon the United States and the European Union. Compliance is less than perfect in both units. This shows that even in political systems with hierarchical systems of authority, subordinate units, for a variety of reasons, do not always execute the directions of superior authorities. Sometimes we expect far too much from international law because we expect perfect compliance. We forget that even within countries, compliance with national laws is almost never perfect. Looking at the domestic experience within the United States and the European Union helps to frame the subsequent analyses by showing how complex and difficult compliance is, even when regulations can be enforced through the coercive power of the state.

Chapter 3 reviews the broad experience in a variety of fields, including arms control, labor, and trade, in securing compliance with international accords. Compliance with international environmental accords poses special problems, but there are also many generic problems that apply to any international accord, regardless of its subject matter.

Chapter 4 deals with the nature of scientific knowledge. Often in debate about environmental matters there is a simple assumption that science can produce clear answers that include unambiguous prescriptions for individual and common action. This chapter shows how problematic this assumption is.

The second section, chapter 5, reviews the development of the five treaties, how their terms have changed through amendment and interpretative documents such as guidelines, and how their implementation has evolved. It compares the evolution of the agreements.

The next section, chapters 6 through 14, comprises analyses of what the nine political units in our sample have done to implement and comply with treaty obligations. The first three of these chapters, 6 through 8, deal with the United States, the European Union, and Japan, all developed democratic countries. The next three chapters, 9 through 11, include analyses of the Soviet Union/the Russian Federation, Hungary, and China, countries in varying stages of transition from centrally planned to market economies. The final three chapters, 12 through 14, address India and Cameroon, low-income countries, and Brazil, a middle-income country. The country chapters are arranged in this order to facilitate comparisons among countries in somewhat similar conditions.

Chapter 15, the conclusion, analyzes compliance across the countries and across time, and presents a theoretical explanatory model. It offers insights and policy prescriptions to policy makers, scholars, and citizens who are concerned with strengthening treaty implementation and compliance.

Notes

1. This number is derived from a comprehensive collection of legal instruments that includes agreements with important provisions relating to the environment (Weiss et al. 1992). The commonly cited figure of about 150 multilateral environmental agreements is taken from UNEP data and includes only multilateral agreements totally directed to environmental issues (UNEP 1991b).

2. Our analysis focused primarily on the period when the European Union had twelve members: Belgium, Denmark, the Federal Republic of Germany, France, Greece, Ireland, Italy, Luxembourg, the Netherlands, Portugal, Spain, and the United Kingdom.

2

How Compliance Happens and Doesn't Happen Domestically

David Vogel and Timothy Kessler

This chapter reviews the literature on compliance with environmental legislation in the United States and the European Community (EC) by both governments and the private sector. It identifies the main factors that have contributed to compliance (and noncompliance) with national environmental laws and, where applicable, relates these factors to the country case studies in this volume.

Both the United States and the EC are characterized by a division of labor between the institutions that make environmental policy and those that are responsible for implementing and enforcing it. In the United States, federal environmental statutes are the responsibility of the Environmental Protection Agency (EPA). However, the actual implementation of federal regulations is often left to state or local governments. In the EC, the government of each member state must pass legislation implementing Community directives before they become legally binding. These laws must then be enforced by national, regional, or local bureaucracies. At each stage of the implementation process, there are substantial opportunities for noncompliance. And in fact a persistent gap has emerged between the standards established by national regulations and the behavior of governmental institutions.

For example, in the United States, despite a significant increase in the political salience of environmental protection as well as of resources devoted to it, through 1990 the EPA had managed to meet only 14 percent of the deadlines for environmental improvement mandated by Congress (Lazarus 1991:324). Eleven years after the passage of the Clean Air Act Amendments of 1970, 87 percent of the nation's integrated iron and steel facilities, 54 percent of its primary smelters, and 19 percent of its petroleum refineries did not meet federal and state emissions limits. A decade after the passage of the 1972 Clean Water Act, only half of the nation's lakes and streams had met the water quality standards established by this legislation (Vogel 1986:165–166).

In the late 1980s the National Wildlife Federation identified more than 100,000 violations of the Safe Drinking Water Act per year (Adler and Lord 1991:789). Although the 1976 Toxic Substances Control Act required the EPA to review over 50,000 chemicals, as well as each of the 1,000 chemicals introduced each year, to determine if they "may present an unreasonable risk of injury to health or the environment," by 1985 fewer than 100 had been reviewed (Lazarus 1991:327).

Noncompliance with regulations governing hazardous wastes has been particularly serious. According to a 1989 General Accounting Office (GAO) study, over 40 percent of industries that release toxic waste into sewers were in violation of discharge regulations (Adler and Lord 1991:789). A 1988 GAO study revealed that even for landfills, EPA's highest enforcement priority, compliance rates were only about 50 percent (Humphrey and Paddock 1990:34).

A substantial compliance gap also has emerged within the European Community. "Only a few of the EC's environmental laws have been sufficiently implemented and enforced by Member States" (Crockett and Schultz 1991:181). In February 1990, the EC commissioner in charge of environmental protection publicly complained about member states' unsatisfactory implementation of environmental directives (Wagerbaum 1990:465). In June 1990, in its Declaration of the Environmental Imperative, the EC Council of Ministers acknowledged the extent of the compliance gap (Wagerbaum 1990:455). The next month, 130 members of the European Parliament proposed the creation of a committee on the application of EC environmental legislation.

More than two years later, the situation had hardly improved. In September 1992, *the Official Journal of the European Communities* published a report arguing that although EC environmental directives represent the most important means of prevention, member states are slow to implement them and "there is a significant gap between the set of rules in force and their actual application" (quoted in "EC Directives ..." 1992:642).

Administrative Capacity

A critical factor that affects compliance is the government's administrative capacity. Regulatory bodies need to have sufficient administrative, scientific, and legal resources to issue rules and regulations and to monitor their enforcement. As the number of environmental laws has grown, the capacity of governmental agencies to administer them often has not kept pace. In many cases, both the Council of Ministers and the United States Congress have issued directives and laws without taking into account the capacity of regulatory bureaucracies to implement them. The result has frequently been administrative overload.

For example, Congress, responding to the sudden upsurge of public support for environmental regulations between 1969 and 1972, approved eight major pieces of environmental legislation; many of these laws required detailed administrative rulemaking before they could be enforced. This requirement proved to be beyond the

EPA's administrative capacities. It took the agency until 1976 to establish an administrative framework for implementing the provisions of the 1972 Water Quality Act, which required all emitters to employ the best practicable technology—by 1977. As of 1980, the EPA had yet to act on 643 proposed changes in state-established controls on existing sources of air pollution, and new-source performance standards had not been issued for a number of major industrial sources of air pollution (Vogel 1986:165–166).

Congress gave the EPA too much to do in too little time. Trying to eliminate as many environmental hazards as possible, and acting in great haste, the legislators on Capitol Hill instructed the EPA to set standards for *all* major air pollutants (1970) and water pollutants (1972), to regulate all pesticides (1972), to control solid-waste disposal (1976), and to eliminate the toxic substances among thousands of industrial chemicals (1977) (Crandall 1987:71). Moreover, Congress specifically required the EPA to "eliminate water pollution, and all risk from air pollution, prevent hazardous waste from reaching ground water, establish standards for all toxic drinking water contaminants, and register all pesticides" (Lazarus 1991:324).

Former EPA head William Ruckelshaus notes that "Congress has established levels of perfection for the EPA such that it is doomed from the start. When people are presented with the impossible, they either freeze or study a situation to death. The EPA has done both" (quoted in Ramirez 1989:140). According to another EPA head, William Reilly, "Congress and the courts had imposed 800 deadlines on the agency through 1989" (Lazarus 1991:323).

Consequently, "EPA officials are compelled to spend as much as 90 percent of their time defending their actions in court or in congressional hearings" (Lazarus 1991:355).

Moreover, the federal nature of the American political system adds an additional hurdle to administrative implementation. "The federal government, while it makes policy, does not implement much of it. And since the implementors, the states and localities, do not participate in federal decision-making, questions of implementation are not paid much heed" (Sbragia 1992:63). Many states lack the administrative capacities to implement federal requirements. For example, a study of United States air quality management concludes that among the most important reasons for the failure of states to achieve regulatory compliance were administrative shortcomings, including "inadequate permit review, infrequent and cursory inspections, announced inspections, poorly trained inspectors, high turnover of inspection staff, and inadequate monitoring techniques or equipment" (Brady and Bower 1983:44).

There also is no clear pattern for the distribution of responsibilities to either state or federal authorities in the enforcement of federal environmental regulations.

"Instead, the allocation of responsibility appears to be haphazard, responding to short-term problems rather than to any consistent theory of the appropriate long-term roles of various levels of government" (Humphrey and Paddock 1990:31).

The EC Council of Ministers also has imposed substantial demands on the member states. Whereas between 1983 and 1985 the Council approved a total of forty environmental directives, between 1989 and 1991 it approved more environmental legislation than during the previous twenty years. The EC now has more than 500 directives in effect, and is adding new ones at the rate of about 100 per year.

However, the administrative capacities of the then twelve member states to implement these directives vary considerably. Some member states have highly developed regulatory bureaucracies, and others have very little administrative expertise in the area of environmental regulation. In particular, the administrative capacities of the Greek, Italian, Spanish, and Portuguese governments to comply with EC environmental directives are substantially less than those of Germany, Denmark, and France, with the other states falling in between.

A study of the implementation of the EC's Seveso Directive, whose purpose was to reduce the occurrence and severity of industrial accidents through a system of safety reports and emergency plans, concluded that compliance was best among member states that already had a basic control system in place. In countries where none existed, "traditional practices had to be replaced by a sophisticated new control regime [leading] to delays in implementing the directive of up to 5 years" (Bennett 1993:25).

Moreover, even those states with strong institutional capacity in other areas are relatively weak in the environmental area. Environmental ministries have typically been established relatively recently; "even the Federal Republic [of Germany] ... has only had a federal Ministry of the Environment since mid-1986. And that ministry does not have effective jurisdiction" (Sbragia 1992:81). The administrative structure of the member states further complicates compliance: in some nations, the responsibility for implementation rests with central governmental units, whereas in others it is handled by provincial or local authorities. And, as in the United States, the administrative capacities of the latter tend to vary widely.

Administrative capacity is closely (though by no means perfectly) associated with level of national income. Accordingly, the pattern of compliance with EC directives has been affected by the substantial variations in the levels of economic development among the member states. Not surprisingly, the poorest nations of Europe have had the least success in implementing EC environmental directives.

However, differing levels of economic development are not the only factor affecting levels of national compliance. For example, while the per capita incomes of

Germany, the Netherlands, and Denmark are comparable, Denmark's record of compliance with EC directives is substantially better. Ironically, this appears to be due to the advanced levels of national environmental regulation already in effect in Germany and the Netherlands. "In both countries, a well developed and sophisticated system of legislation and administration relating to environmental protection results in a lack of motivation to adapt national measures fully to new Community requirements" (Collins and Earnshaw 1992:219–220).

In fact, some member states with markedly inferior institutions for environmental protection have had fewer Court of Justice rulings against them. Ironically, this appears to be due to the undeveloped state of environmental protection administration in these countries; they had less need to overcome the "bureaucratic inertia" of experienced administrators who resented having to modify the enforcement of existing regulations. This phenomenon also explains some of the divergences in compliance within nation-states: it is often easier for many governments to implement Community directives in areas in which they have not previously legislated.

The levels of compliance by American states demonstrates a similar pattern. A study of state-level implementation of the Resource Conservation and Recovery Act found that the states with the most highly developed regulatory bureaucracies were *not* the best implementers (nor the worst). For such states, labeled "strategic delayers" because of their failure to implement promptly, federal, "national guidelines may not represent the best available solution to the hazardous waste problem. Strategic delayers are more likely to forge their own solutions rather than rely on the federal government to do it for them ..." (Lester and Bowman 1989:750).

The country studies in this volume clearly reveal the critical role of administrative capacity in affecting compliance with international environmental agreements. In general, compliance tends to be greatest in nations whose government bureaucracies are the most competent and powerful. Thus, on balance, the nations with the best overall compliance record are Japan and the United States. By contrast, the record of compliance has been poorer in developing countries such as Cameroon, Brazil, Russia, and India, because the latter's administrative capabilities are relatively weak.

Moreover, the latter three countries, as well as China, suffer from an additional handicap: the federal nature of their political systems. Federal systems typically leave policing and enforcement to the regional authorities, many of whose administrative capacities are extremely limited and must be divided among a number of tasks. In India, for example, guerrilla insurgents keep their bases in some of the largest tropical forests and wildlife sanctuaries, and have developed close relations with poachers. Regional authorities may have an interest in maintaining compliance with international treaties, but this goal takes a distant back seat to the problem of fundamental political instability.

Monitoring Enforcement

To enforce environmental laws adequately, a regulatory authority must be able to monitor compliance. Two major factors affect this ability. First, a bureaucracy requires adequate feedback mechanisms. These can assume a variety of forms, including on-site monitoring by inspectors, reporting requirements, the use of complaint mechanisms, and close working relationships with nongovernmental organizations. Second, the effectiveness of these mechanisms is significantly affected by the number and size of the potential violators whose conduct the government is responsible for monitoring.

For example, a major factor contributing to noncompliance in the United States is the sheer number of pollution sources. The EPA's 1987 Toxics Release Inventory covers 19,000 factories and 328 toxic pollutants (Ramirez 1989:139). Moreover, between 20,000 and 40,000 major stationary sources of air pollution and more than 68,000 point sources of water pollution fall within the agency's jurisdiction (Lazarus 1991:324, 327). An even more challenging problem, according to the Office of Technology Assessment, is that there are more than 600,000 active or former solid-waste disposal facilities in the United States, all of which, in principle, pose a potential threat to either public health or the environment (Lazarus 1991:327, 328).

During the second half of the 1980s, the number of regulated entities at the state level in the United States grew dramatically. For example, by 1990 the EPA had given the state of Minnesota responsibility for regulating more than 15,000 small-quantity hazardous-waste generators and 33,000 underground storage tanks. Reporting requirements under the Emergency Planning Act cover more than 10,000 facilities, and the Medical Waste Tracking Act "introduce[d] thousands of previously unregulated facilities into the environmental enforcement systems in participating states (Humphrey and Paddock 1990:32).

As the number of regulated pollution sources multiplied, their relative size shrank, leading to new questions about the best way to deploy scarce resources. Evaluating regulatory compliance is labor-intensive, and personal trips to thousands of small companies or millions of consumers, whose potential regulatory sins are relatively small, have less impact than similar inspections of large producers. As small firms and individuals learn that it is unlikely they will be evaluated, they will place less emphasis on compliance.

The significance of this variable can be illustrated by examining the pattern of compliance with federal automobile emission standards. Compliance with automotive emission standards by newly produced cars is virtually 100 percent, since all that is required for effective enforcement is the monitoring of the production of a handful of automobile companies. By contrast, monitoring the emissions of cars after they are produced has proven much more difficult, since millions of automobile owners must have their cars inspected—at tens of thousands of locations. And these locations in turn must be inspected.

As a general rule, it has been harder to secure compliance with environmental regulations by small firms than by large ones, since the smaller number and high visibility of the latter make their behavior easier to monitor. For example, in the United States it has been estimated that small firms account for more than half of all highly hazardous-waste disposal; they also are more likely than large toxic-waste producers to dispose of their material illegally (Schwartz et al. 1989:282). Robert Kagan quips, "Regulating elephants is different from regulating foxes. It is harder for elephants to hide" (Kagan 1994:397).

A large firm such as DuPont employs numerous scientists, engineers, and, perhaps most important, attorneys who are paid to understand and observe environmental regulations. But it is much less likely that the manager of a neighborhood automotive repair shop is even aware of committing an environmental crime when motor oil from the shop is allowed to run off into the street. Significantly, one of the causes of illegal disposal of hazardous waste by small businesses is inadequate knowledge of regulations (Schwartz et al. 1989:293).

In Europe, the difficulty of administering a uniform system of waste disposal has been increased significantly by the large disparities in the institutional structures underlying most national waste management programs, ranging from a few large, centralized public companies in Denmark, to dozens of small private firms in Great Britain (Brand 1993:242–244). Not surprisingly, the latter have proven to be much more difficult to monitor. Likewise, in both Europe and the United States, government officials have been relatively successful in improving water quality by monitoring the emissions of large firms and upgrading municipal sewage treatment plants. But controlling the water pollution generated by runoffs from farmland has proven much more difficult because of the large number of farms. India has had the same difficulty in regulating tropical forest use by tens of thousands of aboriginal forest people.

Much of the variance in national patterns of compliance with the treaties examined in this volume can be attributed to the number of sources that require monitoring. Thus there appears to be a relatively high level of compliance with the Montreal Protocol and the World Heritage Convention, in part because these agreements place manageable limits on the number of sources or sites that require monitoring. In contrast, compliance with the Convention on International Trade in Endangered Species appears to be much more problematic, in part because the number of potential violators is large. The other agreements fall in between.

Enforcement

The experience of the European Community clearly reveals the importance of adequate enforcement mechanisms to ensure compliance with environmental agreements. Because the Community lacks adequate mechanisms to enforce the directives it approves, "The 1980s has witnessed a dramatic increase in the volume of infringement proceedings brought by the [EC] Commission for member states' non-implementation of EC environmental legislation" (Collins and Earnshaw 1992:216). The EC has no institutional authority with which to threaten, coerce, or otherwise influence recalcitrant members, and each member state is aware that all other members are similarly free from central discipline.

An important reason for unsatisfactory compliance in the EC is that the member states are responsible for interpreting and implementing Community directives: "each Member State tends to use different criteria when implementing an environmental directive, thus frequently producing inconsistent results" (Reitzes 1992:10524). The current structure of EC regulation relies heavily on the member states for self-regulation and information (Reitzes 1992:10525). For example, in the area of integrated waste management, member states have undermined EC directives by failing to provide the appropriate data for analysis, thereby making "the task of formulating an integrated waste management policy for Europe ... hard, if not impossible to accomplish" (Brand 1993:246).

By contrast, the United States government has considerably more enforcement capacity. Indeed, in recent years both the penalties for noncompliance and the willingness of federal enforcement officials to impose them on industry have increased significantly. While the fines for violations have grown larger, numerous violations of environmental laws have been reclassified as felonies and many now carry prison sentences.

"Twenty years ago, none of the major environmental laws in effect contained significant criminal enforcement provisions. Today all do" (Thornburgh 1991:776–777). These changes began in the mid-1980s. For example, the Sentencing Reform Act of 1984 increased the sanctions for first-time white-collar criminals; the Criminal Fine Improvements Act, enacted three years later, raised the level of fines that may be imposed on an organization from $100,000 to $200,000 per offense. The 1988 reauthorization of the Resource Conservation and Recovery Act (RCRA) granted prosecutors the discretion to consider any infraction of the legislation's numerous rules governing the generation, storage, treatment, transport, and disposal of chemicals as a felony.

Under the Clean Air Act Amendments of 1990, "any person who knowingly violates a state implementation plan, or an emission standard for a hazardous air pollutant ... may be imprisoned for up to five years" (Thornburgh 1991:777). That same year, Congress approved the Pollution Prosecution Act, which significantly expanded the size of the Department of Justice's Environmental Crimes Unit, which had been established at the beginning of the 1980s.

Government prosecutors have made much greater use of penalties. According to former U.S. Attorney General Dick Thornburgh, "Criminal enforcement of environmental laws is not merely a goal, it is a national priority" (1991:775–776). Between 1982 and 1990, the federal government obtained 517 convictions, most of them since 1988. In 1990, the Department of Justice's indictment rate increased 33 percent over the previous year, and its conviction rate reached a record high of 95 percent (Thornburgh 1991:778).

The average corporate fine under both RCRA and the Clean Water Act tripled between 1983 and 1990 (Adler and Lord 1991:802). In 1990, the average corporate fine was more than $181,000 (Thornburgh 1991:778). Exxon was assessed a fine of $100 million in connection with the *Exxon Valdez* oil spill, and under a civil settlement agreed to pay the United States and the state of Alaska $1.1 billion over ten years for past cleanup and future restoration costs (Adler and Lord 1991:783–784). As of 1992, cumulative fines and penalties for breaking various environmental laws totaled $332 million (Brunner 1992:1327). This increase in penalties has certainly made American corporations, especially large and highly profitable ones, much more wary of violating federal environmental statutes.

American courts have become more willing to send environmental law violators to jail. In 1982, only seven out of twenty civil indictment referrals to the Department of Justice resulted in convictions, and no jail sentences were imposed. By contrast, in 1989, sixty civil referrals resulted in forty-three convictions: executives were sentenced to total prison time of twenty-seven years (Rasmussen 1992:338). Between 1985 and 1990, the total amount of prison time for breaking environmental laws increased from 6 years to 72 years (Spencer 1992:101). In 1990, 55 percent of the individuals convicted of environmental crimes were given prison sentences; 84 percent actually served time, the average time served being 1.8 years (Adler and Lord 1991:779).

These developments reflect not only a strengthening of environmental enforcement in the United States but also a particular approach to it. There are two basic styles of regulatory enforcement: one emphasizes the formal use of legal penalties to enforce the letter of the law, and the other relies upon flexible enforcement and informal negotiations between the regulated and the regulators. The United States has made increasing use of the former. This "enforced compliance" model relies on "formal, precise, and specific rules; the literal interpretation of the[se] rules; ... the advice of legal technicians (attorneys) and an adversarial orientation toward the regulated" (Hunter and Waterman 1992:404). By contrast, the "negotiated compliance" model, which is far more prevalent in Europe, "assumes greater discretion in the enforcement of the law ... reflect[ing] a dominant orientation toward obtaining compliance with the spirit of the law through the use of general flexible guidelines [and] the discretionary interpretation of rules." It also encourages "bargaining between agency personnel and industry ..." (quoted in Hunter and Waterman 1992:404).

In marked contrast to America's adversarial approach to environmental regulation, Great Britain has leaned far more toward a cooperative or negotiated approach to enforcement. A 1970 White Paper on Protection of the Environment notes that British authorities do not employ penalties as their main deterrent; rather, they are persuaded that public relations disgrace counts more than the fine (Richardson et al. 1983:61–63). This strategy emphasizes mutual problem-solving rather than arm's-length enforcement and punishment. Thus British authorities employ their technical expertise not primarily to prove that a violation has occurred but, rather, to assist firms in reducing or eliminating the problem. According to one veteran inspector, after the initial warning that companies exceed pollution limits, "They ask what they should do. This means we are helping them when we first go down there. We come in as consultants rather than as stick-wavers" (quoted in Richardson et al. 1983: 128–129). As a British pollution control inspector put it: "We look upon our jobs as educating industry, persuading it, cajoling it. We achieve far more this way" (Vogel 1989:89).

The style of enforcement in most other countries leans toward the British model. When "viewed in comparative perspective, the process by which environmental policy is made and implemented has been far more conflictual and adversarial in the United States than in Europe" (Sbragia 1992:57). Nonetheless, the use of criminal sanctions against business violators of environmental regulations has increased in Europe. In 1981, the German government criminalized several environmental violations: over the next decade the number of environmental criminal charges increased from 5,150 to 9,805 (DiMento 1993:137). Sweden, Austria, and Spain also impose criminal sanctions. On the other hand, the British, French, and Italians make little or no use of the criminal law for violations of environmental regulations.

Criticisms of the enforced compliance model are legion (e.g., Bardach and Kagan 1982). Although enforcement does improve compliance, it does so at an extremely high cost. To begin with, relatively few offenders are likely to be prosecuted, because prosecuting firms is very time-consuming. Moreover, the antagonizing atmosphere may reduce the likelihood of voluntary compliance. In addition, the rigid application of environmental standards does not always result in the optimal regulatory outcomes for a particular firm, industry, or problem (Schneider 1993:A1). Finally, regulators are frequently deprived of useful information from industry that may improve the effectiveness or efficiency of particular regulatory standards.

While the threat of imprisonment is likely to encourage managers to monitor compliance by their companies more aggressively, the threat of negative publicity is equally significant. "That a corporate executive has actually served time for an environmental violation is a communication that citizens and the press treat seriously" (DiMento 1993:138). On the other hand, knowing that their company's records may serve as evidence against them may actually discourage managers from finding out more about company compliance. "In fact, 16 percent of the general counsel [surveyed] say they actually have altered their procedures for conducting environmental self-audits because of concern with whether the violations they find could be used against them" (Lavelle 1992:3).

American officials have been criticized for focusing too much of their enforcement efforts on large firms. The logic of this enforcement targeting is clear enough: as noted above, large companies are more visible than small ones, more likely to have created a major problem than a minor one, and more capable of paying for the cleanup.

This in turn has sparked criticism from large firms. "The EPA, they charge, fails to force small contributors to hazardous site contamination to bear their share of cleanup costs" (Barnett 1993:131). Indeed, there does appear to be evidence that large companies suffer greater monetary penalties (Cohen 1992:1092).

In addition, the aggressive use of legal sanctions against business violators of environmental laws has led to accusations of unfairness. For example, United Technologies was prosecuted for illegal hazardous-waste disposal after it had both cleaned it up and paid a $3 million fine—even though EPA found no evidence of deliberate noncompliance. Nonetheless, the government prosecuted "because it felt that, in general, the company didn't put enough resources or care into compliance. A dangerous precedent: Where but in an Orwellian world can a bad attitude be a crime .. ?" (Spencer 1992:102).

It is also important to note that civil litigation offers several advantages over the imposition of criminal penalties. First, although civil proceedings may indeed be costly, they are generally much less expensive, cumbersome, and drawn-out than criminal prosecution. Second, civil procedures and standards of proof are less strict. But perhaps the most important advantage concerns the increased incentives of business to negotiate, if not capitulate, in civil cases, which often result in the payment of heavy fines: "the threat of civil sanctions typically evokes a less defensive response on the part of the targeted polluter, allowing a potentially greater chance for government and polluter to negotiate toward an acceptable resolution of the problem" (DiMento 1993:142).

As a result of such criticisms, a number of students of United States regulation have urged the adoption of a regulatory approach that relies less on legal compulsion. Applying well-known game-theoretical concepts from international relations to environmental enforcement, one political scientist suggests that agencies employ a "tit-for-tat" strategy with regulated firms. Scholtz argues that "net social benefits are generally far higher if enforcement and compliance costs can be minimized through cooperation between agency and firm" (Scholtz 1984:385–386). This approach, which "depends on prompt punishment of any betrayal of trust ... is most feasible when regulator and regulated firm interact frequently ..." (Kagan 1994:396). Another study of regulation suggests that optimum "efficiency" can be achieved by fining organizations, rather than individuals within them, because the latter involves much higher transaction costs. This approach in turn hinges on the firm's ability to sanction its employees internally for actions that contribute to environmental risk (Segerson and Tietenberg 1992:197).

Thanks to more aggressive enforcement in recent years, the resources devoted to environmental improvement by American firms have undoubtedly increased. But at the same time legal expenditures by both government and industry also have increased: an adversarial approach to regulation may be relatively effective, but it is also apt to be highly inefficient. On the other hand, nations with more cooperative approaches to environmental regulation, such as Germany, Japan, and Great Britain, have made substantial progress in improving the quality of their environment.

Notwithstanding the benefits of cooperation, however, the experience of the European Community suggests that some enforcement mechanisms are critical for compliance. Especially in the case of developing nations, in which respect for public authority tends to be low, the use of prosecution, accompanied by severe penalties for noncompliance, may well be salutary. From this perspective, improved compliance with international environmental treaties appears to be linked to changes in the overall policing capacity of both national and local authorities.

The Political Environment

A fourth critical factor affecting compliance with environmental policy is politics. "Policy implementation ... is a fundamentally political process" (Collins and Earnshaw 1992:233). The enforcement of environmental regulations is significantly affected by three political factors: the extent of public support for environmental regulation, the preferences and priorities of politicians, and the political strength of environmental organizations and their access to the policy process.

Public Opinion

In the United States, environmental policy suddenly and dramatically became a major focus of public concern during the late 1960s and early 1970s. A survey conducted during this period reported that "alarm about the environment sprang from nowhere to major proportions in a few short years" (quoted in Jones 1973:23).

Between 1969 and 1970, air and water pollution jumped from tenth to fifth place among the American public's most pressing concerns. On April 22, 1970, as many as 20 million Americans participated in Earth Day. In December 1970, a Harris poll revealed that Americans considered pollution to be "the most serious problem" facing their communities (Vogel 1989:65).

Public support for environmentalism increased rapidly in Europe during the second half of the 1980s. Stimulated in part by the Soviets' Chernobyl disaster and a massive spill of chemical toxins into the Rhine River, environmental issues moved rapidly to the forefront of the EC political agenda. The *Washington Post* observed: "Dead seals in the North Sea, a chemical fire in the Loire, killer algae off the coast of Sweden, contaminated drinking water in Cornwall [England]. A drumbeat of emergencies has intensified the environmental debate this year in Europe ..." (Herman 1988:19). A poll taken in December 1986 reported that 54 percent of the German electorate regarded environmental quality as the most important issue facing their nation (Kirkland 1988:21). An EC publication observed in 1990 that "global problems like ozone depletion and the greenhouse effect, and quality of life issues such as drinking water and air pollution have all contributed in recent years to a 'greening' of European public opinion ..." (*Environmental Policy* 1990:19).

However, while there is little doubt that public support for environmental protection has increased, it is important to distinguish between the "strength" and the "salience" of an issue: "Strength refers to the degree to which people regard an issue as a matter of national or personal concern and want to improve the situation.... Salience, in contrast, is the amount of *immediate* personal interest people have in the issue" (Mitchell 1990:83). It is clear that environmental regulation has become a permanent part of the political agenda in both the United States and western Europe. In this sense, support for environmental protection remains "strong"; most citizens perceive the quality of their physical environment—locally as well as globally—as inadequate, and they expect both government and business to help improve it.

The "salience" of environmental issues has exhibited considerably more variation: the intensity of public concern in both Europe and the United States has fluctuated over time. The environment constantly competes with other policy objectives, whose salience also varies from year to year. Among the most important factors affecting the salience of environmental regulation in Europe and the United States have been economic ones: the *relative* importance of environmental issues has tended to increase when growth is robust, and to decline when the economy stagnates (Vogel 1989). Thus the salience of environmental regulation declined in the United States during the recession of the mid-1970s and increased during the recovery of the 1980s. Similarly, public interest in environmental issues has recently declined somewhat in Europe, due to the persistence of high unemployment in much of the EC.

There is also a second factor that affects the salience of environmental regulation: disasters. Given the "strength" of public support for environmental regulation, salience increases dramatically as a response to highly visible evidence of environmental damage. The public perception of a "crisis" can transform the political agenda, leading to stricter regulations and significantly improved enforcement. Obviously, some dimensions of local, national, or international environmental issues or regulations are more likely to lend themselves to such "crises" than are others: salient issues tend to be headline-sensitive. For example, the jump in the percentage of respondents claiming that the environment "was one of the two most important issues facing the United States today" from 2 percent in 1987 to 16 percent by mid-1989 was due at least in part to the *Exxon Valdez* disaster (Mitchell 1990:84).

Preferences of Elected Politicians

While compliance with particular environmental regulations is likely to be enhanced if support for regulation is both strong and salient, the impact of public attitudes on compliance with environmental regulations is also mediated by the preferences of powerful politicians.

Politicians exercise considerable discretion over the enforcement of environmental policy. One major reason is that they are unlikely to be elected to their position specifically on regulatory issues. With certain exceptions, the enforcement of environmental regulations tends not to be politically salient. Elected politicians play an important role in environmental enforcement through their appointments to regulatory bodies, their instructions to officials responsible for enforcement, and, more important, their influence over the budgetary process.

The latter is apt to be particularly important. Regulatory agencies require funds to hire the personnel to develop rules, keep records, monitor enforcement, and prosecute violators. When these funds are cut back, as they were by the Reagan administration in the early 1980s, compliance diminishes. For example, in the area of toxic-waste disposal, the EPA, constrained by inadequate enforcement staff resources during the early 1980s, failed to issue unilateral orders when negotiations stalled and did not attempt to recover costs once a fund-financed action was completed (Barnett 1993:128).

Nongovernmental Organizations

The impact of public opinion on enforcement is also mediated by a second critical factor: the degree to which environmental groups are both well-organized and able to participate in the policy process. This varies considerably across nations. The United States is notable on both dimensions. American environmental organizations are highly organized: they have large memberships and considerable financial and technical resources. Equally important, the constitutional and political system of the

United States provides them with considerable opportunities to participate in the policy process: they can testify at administrative hearings, lobby the Congress, and file lawsuits to force regulatory agencies to comply with congressional statutes and to order compliance by companies.

Consequently, in the United States, environmental groups have been an important complement to the government's enforcement efforts. They serve as "watchdogs" reporting violations to the appropriate authorities, and they often possess specialized information that government officials can use in both monitoring and prosecuting violations. In this sense, environmental groups function as extensions of government regulators. In addition, environmental organizations are able to pressure bureaucracies to increase their implementation and enforcement efforts.

The American constitutional separation of powers has significantly expanded the opportunities for environmental organizations to affect environmental enforcement. In particular, environmental litigation by public-interest groups has played a critical role in enabling regulatory agencies to resist pressures from business and, on occasion, the president to weaken regulatory enforcement.

According to a study of state implementation of the RCRA in the United States, "in the case of citizen participation, it appears that expanded opportunities for citizen input facilitates [*sic*] implementation.... Pressure by citizens perhaps encourages more rapid implementation" (Lester and Bowman 1989:748). Significantly, civil enforcement of environmental laws increased during the early 1980s, when the states' enforcement efforts were reduced. In the United States, "lax public enforcement appears to have played a role in the rise of citizen suits" (Naysnerski and Tietenberg 1992:45). More generally, civil enforcement "allows the public sector greater flexibility in targeting its limited enforcement resources.... Without private enforcement the government would be forced to spread its limited resources much more thinly over the vast territory regulated by government policy" (Tietenberg 1991:9).

However, the use of citizen suits is politically determined: civil suits by private parties are most likely to be filed when large damages can be imposed and industrial violators are required to reimburse attorney fees (Naysnerski and Tietenberg 1992:45). High damages increase the likelihood that lawsuits will have a deterrent effect, and legal reimbursement significantly reduces the costs of litigation to public-interest law firms. "The availability of attorney fees enables citizen groups ... to undertake more enforcement activity because their litigation resources could be stretched further" (Tietenberg 1991:9).

While the access of environmental organizations to the courts is unique to the United States, both individual citizens and environmental organizations have come to play an important role in improving compliance within the European Community. Because the European Commission lacks any means of assessing the actual implementation of its directives by the member states, "it is almost entirely dependent on EC citizens to bring to its attention the alleged failure of member states to implement in practice Community environmental legislation" (Collins and Earnshaw 1992:231). One environmental lawyer notes that the European Commission relies most extensively on information and complaints submitted by individuals and firms "that are dissatisfied with the measures their countries have taken.... In practice, adequate and timely implementation of the environmental directives has depended on the climate of opinion in the Community, which is itself news-driven" (Reitzes 1992:10525).

In fact, the number of complaints filed with the European Commission alleging national noncompliance with EC directives increased dramatically during the 1980s, in tandem with increased public awareness of and support for EC environmental policy. "The tendency of citizens to use the Community as an avenue of complaint [in EC environment policy] has apparently kept pace with the Community's growing involvement in national political life" (Collins and Earnshaw 1992:231–233).

As with enforcement styles, the ways that environmental organizations participate in the policy process vary considerably among countries. In a number of European countries, environmentalists have formed political parties. "Green" parties are represented in a number of national European legislatures, and in the 1989 elections to the European Parliament, they doubled their total representation to thirty-four seats. In some European nations, environmental organizations constitute effective pressure groups, actively campaigning for stricter legislation and effective enforcement. In other cases, they work closely with government ministries to help implement policies and programs. In still other countries, they are effectively excluded from policy-making and implementation.

Does the evidence suggest that more democracy leads to enhanced environmental compliance? Not necessarily. In the former Soviet Union, democratization reduced national compliance with some international environmental agreements. What democratization does offer citizens is a greater voice in shaping a wider array of public policies, including environmental regulation. If the citizens of democratic or relatively democratic nations strongly support the objectives of the environmental treaties their governments have signed, as appears to be the case in the United States and northern Europe, compliance is likely to be enhanced. On the other hand, if they value other objectives, such as profit from the sales of endangered species or inexpensive refrigeration, then compliance is likely to prove more difficult.

Accordingly, public attitudes play a key role in affecting the preferences of politicians responsible for the implementation of environmental policy. The lack of public support for the preservation of endangered species among large numbers of citizens of India, Japan, and Italy has made it more difficult for their governments to fulfill their obligations under CITES. Similarly, the extent of India's long-term compliance with the provisions of the Montreal Protocol is likely to be significantly affected by the political attitudes and consumer behavior of its growing middle class.

The Economic Environment

Economics also plays a critical role in compliance. Most obviously, enforcement is likely to be affected by the relationship between the costs of compliance and a country's wealth or level of prosperity. Affluent countries experiencing relatively high growth rates are likely to be more willing to comply with relatively costly environmental regulations than are poorer nations or nations whose economies are growing slowly or not at all. Similarly, if the costs of compliance are relatively modest, or if compliance produces a net economic gain, a nation's level of economic well-being may be less critical. In sum, what matters is not so much the absolute costs of compliance as the relationship between these costs and the capacity of a nation's economy to absorb them.

The relative financial burdens of compliance on the private and public sectors is also important. To the extent that the costs of compliance are borne by the private sector, governments are more willing to impose them. Indeed, from the perspective of politicians, such regulatory costs are quite attractive, because indirect or hidden taxes are relatively easy to impose. By contrast, if the costs of compliance must come directly from public funds, politicians may be more reluctant to impose them because they will be required either to raise taxes or to shift funds from other purposes.

In some cases, the costs of compliance with environmental regulations must be borne directly by the government. This is especially true in the case of water pollution: the improvement of water quality usually requires substantial public expenditures, especially by local governments, for the construction of sewers and water treatment facilities. The same is true of the cleanup of hazardous waste, a significant portion of which is produced by the public sector. Since the costs of compliance are substantially greater than the costs of monitoring and enforcement, noncompliance is more likely to occur when the government must comply with its own regulations.

The impact of regulations on individuals is critical. Many environmental regulations affect consumers either indirectly, by raising the prices of particular commodities, or directly, by requiring changes in their consumption patterns, such as by depriving them of particular products or requiring them to use them in a particular way. What matters is not simply the abstract commitment of citizens to environmental protection, but to what extent regulations require them to change some aspect of their lifestyle. The more regulations are or appear to be burdensome to individuals, the less likely people are to comply with them.

However, most environmental regulations primarily require changes in the behavior of firms. And since firms enjoy substantial political influence in all capitalist nations, their response to regulations significantly affects the ability of governments to secure compliance. This is especially true at the local level, where particular firms or industrial sectors may be major employers.

However, what matters to business is not simply (or even primarily) the absolute costs of compliance, but the *relative* costs, or competitive impact. Regulations are never neutral in their effect on different firms and industries: they invariably make some better off and some worse off (Leone 1986). Some regulations can actually improve the competitive position of particular firms or industries, if their compliance costs are less than those of their competitors or if the regulation provides the company with expanded market opportunities. For example, firms producing low-sulfur coal supported laws to reduce acid rain—which is partly the result of burning high-sulfur fuels ("Regulate Us, Please" 1994:69).

In an increasingly integrated global economy, regulations affect international as well as domestic competition. For example, both international and national regulations may impose greater costs on firms in some countries than in others. Alternatively, they may serve as nontariff trade barriers, effectively excluding or raising the costs of imported goods (Vogel 1995). Thus a number of developed countries have imposed restrictions on imports of tropical woods in order to save the world's rain forests; this in turn has served to protect their domestic lumbering and wood-processing industries.

The relative costs and benefits of compliance have clearly played a decisive role in affecting national rates of compliance. Most obviously, national compliance with the Montreal Protocol and the World Heritage Convention by a number of developing nations has been enhanced by the fact that these treaties have been accompanied by net transfers of wealth to their national governments. The support of the United States for the Montreal Protocol and the Vienna Convention was significantly facilitated by the fact that the major domestic producer of chlorofluorocarbons (CFCs), Du Pont, stood to gain a competitive advantage by the phasing out of production of this chemical outside the United States as well. Similarly, it is far easier for a nation to comply with treaties affecting ocean dumping or tropical forests if it does not receive significant benefits from these activities. By the same token, the relative lack of compliance with CITES in India, Russia, and China is closely linked to the large and extremely lucrative black market for products from endangered species.

Finally, although in principle the increased resources resulting from economic development may ultimately lead to a higher rate of compliance with international environmental accords, increased economic growth rates in developing nations, which stimulate consumer demand and expectations, may actually result in reduced compliance rates.

Conclusion

Numerous factors affect the likelihood of compliance with environmental regulations. We have discussed these factors in terms of five broad categories. But although these categories are useful for analytical purposes, their significance cannot be understood in isolation from each other. Compliance is a complex and dynamic process. And the relative importance of any particular factor, or set of factors, is likely to vary from nation to nation, and from regulation to regulation. Moreover, the significance of these factors must be understood dynamically. None is static: all have evolved and are likely to continue to vary over time, leading to changes in the extent of compliance.

3

Managing Compliance: A Comparative Perspective

Abram Chayes, Antonia Handler Chayes, and Ronald B. Mitchell

What analytic framework should we use to evaluate treaty compliance?[1] This chapter draws on our research into compliance with multilateral, bilateral, and regional treaties addressing the principal regulatory agreements dealing with economic, political, environmental, and social problems that require cooperative action among states over time. Specifically, we examine how treaty provisions and the operation of international institutions help make treaties work. The chapter highlights the error of conceptualizing most compliance problems as being due to intentional violations that can, and should, be responded to through enforcement and sanctions. We then argue that strengthening regulatory regimes requires a strategy of integrated, active management of compliance that addresses the real sources of noncompliance, without necessarily expecting to achieve perfect implementation and compliance.

What We Mean by Compliance

We use "compliance" to describe those instances when an actor's behavior conforms to an explicit rule of a treaty (Mitchell 1994). This implies evaluation of compliance with individual provisions rather than with the treaty as a whole. Compliance goes beyond implementation. A government may fulfill requirements to implement the treaty in national law, but not provide required copies of the law to the secretariat of the treaty organization. Nonreporting is often small in itself, but may prove to be indicative of more significant forms of noncompliance. Often the governing rule takes the form of a general standard or principle, so that measuring performance against its requirements is not a simple mathematical exercise. Even when the obligation is embodied in a more explicit rule, the normative content of the language is often far from clear, either to the state when it is taking action or to subsequent critics and observers. Although an international law treaty's obligations apply only to states, environmental treaties seek to influence the behavior of subsidiary governmental units and private actors, and our definition recognizes that the behavior of such actors can also be described in terms of its compliance with a treaty provision. We believe it is rarely the case that noncompliance results from calculated governmental decision, as in the traditional realist assumption. The sources of noncompliance are much more complex and varied. We can ask several questions about treaty-related behavior to improve our evaluation of treaty effectiveness.

Questions Relating to the Behavior Itself

First, does the observed behavior conform to that required by the treaty provision? Behaviors can be compliant, noncompliant, or ambiguous, because some treaty provisions clearly distinguish compliance from noncompliance whereas others do not. Second, what is the degree of compliance? Even under a clear treaty standard, some actors will barely meet the standard while others will substantially exceed it. Actors may be "overcompliers," going beyond what is required; "good faith noncompliers," striving for compliance but falling short; or "intentional violators," who intentionally fail to comply. Third, was the compliance or noncompliance significant? Arms-control debates have frequently focused on the "military significance" of alleged violations. For example, unlike clandestine weapons deployments, accidental releases from nuclear weapons tests have been viewed as noncontroversial, technical noncompliance because they were less important to core treaty goals.

Questions About the Sources of, and Factors Influencing, the Behavior

Each type of behavior—compliant, noncompliant, ambiguous—raises different questions. In cases of compliance, was the behavior "treaty-induced" or "coincidental"? Many countries' compliance with the Nuclear Non-proliferation Treaty's proscriptions reflect their lack of desire or ability to acquire nuclear weapons, rather than regime pressures. Often, behaviors reflect exogenous "coincidental" forces—economic factors, domestic politics, diplomatic pressures, the impacts of the treaty rules and regime.

In cases of noncompliance, was the behavior deliberate? Was a good-faith effort made to comply? Did behavior change? Was some progress made toward the established standard? Noncompliance may be intentional but may also reflect inadvertence, incapacity, or a failure to understand treaty provisions. Even a developed government desiring to fulfill treaty commitments may inadvertently fail to do so because of the inherent uncertainty of its chosen policy instruments. It may fail to meet an emission reduction target because of inaccurate estimates of the necessary enforcement or taxation level. As several country studies demonstrate, governments often fail to comply because they lack the financial, administrative, informational, or regulatory capacities. The problem can be especially acute when the treaty targets private and individual behavior not directly under a government's control, even when the government has strong incentives to comply, as is evident in violations of nuclear export-control agreements. Between incapacity and intentional noncompliance are cases where states expend resources on higher-priority goals to provide their citizens with other forms of improved social welfare and, as a result, compliance with the treaty slips. We believe deliberate treaty violations are the dramatic, but rare, exceptions rather than the rule.

Approaches to Compliance: "Enforcement" or "Management"

What types of treaty provisions ex ante and compliance management strategies ex post induce compliance in a context where noncompliance stems from the range of sources just discussed? Our research indicates that, in the face of noncompliance, coercive sanctions are not only ineffective but inherently unsuitable.

Although an enforcement model of compliance rests on the availability and use of sanctions to deter violations, systemic features of international society severely constrain the use of sanctions. Only two treaties, the United Nations Charter and the Organization of American States Charter, authorize the use of concerted military or economic measures, and these have been invoked in only a dozen or so cases. More frequently, economic sanctions have been used unilaterally to advance particular foreign policy goals, not to enforce treaty obligations. Only powerful states or coalitions of states can use them, and only against weaker states.

States face high political and economic costs at home, collective-action-type difficulties in building international coalitions, and a strong empirical record that their efforts will be ineffective (Martin 1992). What is surprising (and hard to explain) is the persistent predilection of scholars and practitioners for sanctions as a routine way of enforcing regulatory treaties. In our view, efforts to negotiate sanction clauses into treaties and to invoke unilateral sanctions for violations are largely a waste of time.

In contrast, a managerial model of compliance suggests that regimes usually keep noncompliance at acceptable levels by an iterative process of discourse among the parties, the treaty organization, and the wider public. Most states enter agreements intending to comply. Compliance often serves the state interests that led to, and were shaped by, the negotiation process. Treaties legally bind the member states and carry a presumptive obligation to comply. Moreover, compliance reduces decision costs and conforms to bureaucratic modes of action.

States, like other actors, call on each other to justify behavior that departs from agreed-upon norms. The ensuing discourse progressively elaborates the meaning of relevant obligations through cooperative processes of consultation, analysis, and persuasion, rather than coercive punishment. Even in the rare treaty that adopts formal dispute-resolution processes, as did the World Trade Organization (WTO) agreement, parallel processes of review and assessment have been introduced that are more cooperative (Agreement Establishing the Multilateral Trade Organization 1993). Skillful and imaginative treaty organizations and institutions devise ex ante and ex post measures, building on the deep economic and political interdependence of modern states to enhance compliance.

We believe that effective compliance management requires establishing and maintaining a transparent information system and a response system. The information system must produce adequate and accurate information about actors' behaviors under the treaty. The managerial response system must then produce discriminating responses to different types of noncompliance, using both multilateral, treaty-based and unilateral actions to induce behavioral change.

Norms: The Foundation for Compliance

Norms provide the foundation for this compliance process. Here we use "norms" to refer generally to "prescriptions for action in situations of choice, carrying a sense of obligation, a sense that they ought to be followed" (Chayes and Chayes 1995: 113). Treaties embody either a previously established or a recently created set of norms that reflects the relative power and interests of the negotiating states, including their interests in creating and maintaining certain norms.

If a normative consensus on an issue area exists, then much initial compliance may be motivated by this consensus rather than by treaty compliance mechanisms. The international norm of *pacta sunt servanda*, "treaties must be obeyed," and the voluntariness of treaty signature provide further pressures for compliance. Even with the deep inequities in the distribution of power, countries accept agreements that discriminate in favor of powerful states as legitimate, when they have participated in the negotiation, because those agreements may be preferable to available alternatives. For example, the Nuclear Non-proliferation Treaty is viewed as binding on nonnuclear signatory states even though it allows five major powers to remain nuclear "haves" while prohibiting the "have-nots" from attaining such status.

Treaties do not simply reflect norms, however, but can create and strengthen norms that did not previously exist. Treaties banning deployment of nuclear weapons on the seabed, in outer space, and in Antarctica created new international norms. It is easy to imagine that arms races would have developed in these areas had the treaties not generated expectations that others would abide by the norm. Treaties also can widen a norm's scope, as is evident in the human-rights arena, where nations may initially sign a treaty because of public and diplomatic pressures, but over time internalize the norms embodied in the treaty. The almost universal signature of the Chemical Weapons Convention and the United Nations Conference on Environment and Development agreements suggests powerful public international pressures to sign treaties that establish new norms, with the costs of compliance playing a minor role in most countries' calculus regarding signature. Nevertheless, the treaty norms are likely to constrain future behavior.

Even a hegemon may feel constrained by the norms of the regimes it creates. The United States regularly accepted General Agreement on Tariffs and Trade (GATT) decisions that went against its position (Meyer 1978; Lipson 1982). Negotiating and signing a treaty creates a standard, deviation from which demands an explanation to other states, publics, and nongovernmental organizations (NGOs). The seemingly endless discussions in international organizations of the scope and meaning of norms enhance their authoritative character. Actors regularly appeal to legal norms in their justification of behavior, thus legitimizing those norms and reinforcing expectations that will constrain their own future behavior.

The treaty and treaty secretariat need not rely exclusively on norms to alter behavior, however. Several more active means are available to them. The two key components of successful efforts at eliciting compliance are a transparent information system and a managerial strategy of responses to induce compliance.

Developing a Transparent Information System

To manage compliance effectively, the treaty regime must have a transparent information system. We use "transparency" to mean the adequacy, accuracy, availability, and accessibility of knowledge and information about the policies and activities of parties to the treaty, and of the central organizations established by it on matters relevant to compliance and effectiveness, and about the operation of the norms, rules, and procedures established by the treaty.

An information system's transparency can be evaluated against several standards. First, does the system collect a wide range of relevant information on compliance and effectiveness? Second, is the available information perceived as accurate, reliable, and legitimate? Third, does information available to the secretariat get analyzed and processed effectively? Fourth, is the information available to the secretariat made available to industry, NGOs, and publics as well as governments?

The Functions of Transparency

Transparency fosters compliance by permitting actors to coordinate their behavior, reassuring actors who desire to cooperate but fear being "suckered," and deterring actors contemplating noncompliance. In many instances, the actor's independent responses to these forces will assure compliance. Where strategic interaction is insufficient, transparency allows other parties to observe deviations from prescribed conduct and to require that those deviations be accounted for and justified.

Coordination. In simple cases where actors care more that a single rule govern the activity, than which rule governs it, treaties facilitate cooperation by creating and publicizing an agreed-upon rule. Some rules are literally rules of the road, such as rules established for air transport, marine navigation, and satellite communication allocation. Once the parties understand the rules, no actors have incentives to violate them. In other cases, regimes produce collective information that participating states would find it impossible, or prohibitively expensive, to assemble on their own. Various commodity agreements and the International Monetary Fund (IMF) have produced industrial, financial, and economic databases that facilitate numerous loans and private economic transactions. Periodic reporting on SO_2 emissions under the Long-Range Trans-Boundary Air Pollution (LRTAP) Convention led most parties to limit their emissions by some 30 percent unilaterally; only subsequently was a protocol to that effect negotiated. The credible, integrated database that was essential to common scientific judgment and coordinated action would not have emerged in the absence of the regime (Ausubel and Victor 1992).

Reassurance. Transparency also reassures parties that others are meeting their obligations; and if they are not, it permits a timely response. Reassurance is needed when actors otherwise inclined to comply are concerned that they will be placed at a disadvantage if their compliance is not matched by others. To preserve a common pool resource, actors must adopt a "contingent strategy" of committing to follow the rules as long as others do so, but such actors must have "information about the rates of rule conformance adopted by others" (Ostrom 1990: 187). Treaties banning deployment of nuclear weapons in certain environments worked by reassuring each side: since each side had incentives to place weapons in these areas only if the other side did, ordinary surveillance reassured each side of the other's compliance.

Regimes for confidence-building measures in Europe required states to give notice of military maneuvers to assure all Europeans that neither the North Atlantic Treaty Organization nor the Warsaw Pact was preparing a surprise attack. If the conditions of an assurance problem hold—that is, if the benefits to a state exceed its costs of contribution so long as others also comply—transparency supplies the reassurance needed for parties to make safe, advantageous, and credible commitments to follow the rules.

Deterrence. Deterrence is the obverse of reassurance. A party disposed to comply needs reassurance. A party contemplating violation needs to be deterred. Transparency supplies both. The probability that conduct departing from treaty requirements will be discovered operates to reassure the first and to deter the second, and that probability increases with the transparency of the treaty regime. Deterrence succeeds if discovery entails penalties that increase the costs of defection enough to exceed the expected gains. Penalties can involve loss of the anticipated benefits of the regime. Even when direct retaliation seems unlikely, exposure alone can cause behavior to change.

When the United States raised questions about activation of surface-to-air missile (SAM) radars at test ranges in violation of Article VI of the Anti-Ballistic Missile (ABM) Treaty, the Soviets admitted nothing, but the practice stopped a few weeks later. Transparency often can be used to prevent a defector from achieving the benefits of defection. For instance, the ABM regime prohibits precursor activities that the opposite side can readily, quickly, and accurately detect, and to which it can readily respond in ways that would erase any potential gain from the original defection.

In a multilateral setting, the delinquent may suffer more diffuse negative reactions from states and other groups with a stake in the treaty regime. Some of the country studies demonstrate that even fear of negative reputational impacts and diffuse reciprocity may be adequate to deter (Charny 1990; Keohane 1986).

Assembling the Database

Creating a successful transparent information system crucially depends on the types of behavior the treaty seeks to regulate, the way the treaty defines those behaviors, and the available means for identifying whether a proscribed action has occurred.

The framing of the treaty's rules has at least three implications for regime transparency. First, it determines the number of actors whose behavior must be verified. For example, treaties regulating habitat destruction seek to restrain relatively large numbers of actors whose behaviors face little if any regulation, whereas those regulating chlorofluorocarbon (CFC) manufacturing need monitor only a few chemical plants that already face considerable regulation of their activities for other reasons. Second, information regarding certain actors that is already available through other regulatory infrastructures reinforces and facilitates transparency. The International Labor Organization's (ILO) Committee of Experts uses a variety of existing informational sources to identify violations of the many conventions it administers. Third, the type and level of standards must match the capacities of existing monitoring technologies. Transparency regarding compliance increased dramatically in treaties regulating intentional oil pollution when rules limiting discharges at sea were replaced with rules requiring installation of specific equipment to prevent such discharges (Mitchell 1994).

Once the rules are established, transparency requires developing data on the behavior of the parties with respect to the principal treaty norms. Independent data collection by a central organization is rare, and most regulatory treaties contain provisions for some kind of voluntary reporting by the parties, exchange of information, or joint research and data collection (Ausubel and Victor 1992; Chayes and Chayes 1995). Indeed, establishing a database is often the primary initial objective of a framework-type agreement.

Self-reporting

Self-reporting on measures taken to implement the treaty is often central to efforts to create a transparent information system and assure compliance. Examples of requirements for self-reporting are ubiquitous in international regimes of all stripes, from early agreements on slavery to almost all recent environmental accords. However, such requirements do not equate with actual reporting (General Accounting Office 1992b; Mitchell 1994). Reporting is one of the few provisions common to all five treaties in the study, a fact that permits analysis of the extent to which differences in reporting provisions and systems influence the level of actual reporting. Requirements that parties report on planned programs and policies designed to bring them into compliance with a treaty can become the basis for the iterated management process of policy review and assessment, described below. The wide reliance on self-reporting raises two principal issues: failure to report and inaccuracy of reporting.

The principal problems with respect to self-reporting seem to be less the deliberate flouting of reporting requirements than limitations of capacity and of the bureaucratic setting in which reports are generated. Occasionally a country will skip its report to avoid revealing a serious violation, as was apparently the case with Panama and the whaling convention (Birnie 1985). An ILO working group concluded that reporting failures usually stem from administrative and technical difficulties or personnel changes rather than from deliberate refusal. The country studies in this volume provide further support for this conclusion.

The ILO has strong management procedures of blacklisting countries that fail to report, because it views reporting as essential to the compliance process (International Labor Conference 1980). As a result, the ILO has received reports from over 80 percent of its members in every year since its inception, despite long and burdensome requirements (International Labor Conference 1992). Likewise, enforcement data collected in and agreement among fourteen industrialized members of the International Maritime Organization (IMO) permit effective performance of inspections, which is a high priority of the port agencies, and has close to 100 percent reporting rates (Mitchell 1994). In contrast, compliance with the various reporting requirements of the International Convention for the Prevention of Pollution from Ships (MARPOL) is quite low, because the IMO secretariat does not facilitate reporting, makes little use of the information it does receive, and does not censure failures to report (Mitchell 1994).

Why would a state report information that shows it to be out of compliance? We assume that countries usually prepare reports to represent the facts as known, although some nations face difficulties in collecting and analyzing complete and accurate data sets. For most functional agreements, middle- or lower-level officials within relevant ministries prepare reports that are reviewed by officials concerned about their international "audience." This bureaucratic setting provides some insulation against deliberate misreporting. Day-to-day policy-making and administration require reliable statistical data. Indeed, nations often report to international secretariats only data that they already collect for other reasons. This circumstance undoubtedly means that much information frequently remains unreported, but it also makes it somewhat more difficult to "cook the books."

On the other hand, incentives exist to make performance look good. The Soviet Union systematically exceeded international quotas on important whale species, and deliberately misreported kills to the International Whaling Commission. In more open societies, bureaucrats will generally face more checks, and even direct challenges to the accuracy and completeness of their reports. In the human-rights area, NGOs even in closed societies have regularly produced information to challenge reports filed by parties to human-rights treaties, and efforts have recently been made to create such procedures in environmental treaties (Navid 1979; Greenpeace 1990; Chayes and Chayes 1995). The country studies provide some evidence of both the political pressures from domestic legislatures, citizens, and NGOs, and pressure from international groups to make reports available and accurate.

Independent Reporting and Verification

Although environmental treaties usually require only national self-reporting, ways of skirting the "self-incrimination" problems inherent in such systems are increasingly being recognized and put to use. Agreements regulating trade in coffee, endangered species, and weapons require both the exporting and the importing party to report on each transaction (Chayes and Chayes 1994; Trexler 1989). NGOs commonly help verify human-rights treaties. Even in the security-shrouded world of arms control, advocacy NGOs such as the Federation of American Scientists, as well as research organizations such as the Stockholm Institute of Peace Research, Jane's, and the International Institute for Strategic Studies, have developed impressive credentials as independent sources of authoritative information (Laurance et al. 1993). Environmental NGOs have made it their business to collect information on treaty-related behavior and to sponsor scientific measurements of atmospheric conditions, ozone depletion, and species populations, thus providing information independent from that provided by their governments.

The Commission on Sustainable Development explicitly created a legitimate channel for NGOs to provide reports to secretariats in order to facilitate evaluation of compliance and noncompliance. Even industry may provide independent information on compliance. The International Chamber of Shipping, a private consortium of shipping companies, has regularly identified ports that have not provided reception facilities as required by MARPOL, and much CFC production information is provided directly by industry (Mitchell 1994).

The availability of other sources for the same data sometimes facilitates verification of national data. Data from one country can be compared with those from other countries and validated against information available on highly correlated independent statistics. The LRTAP Secretariat develops, and to an extent verifies, emissions reports by comparing them with fuel consumption statistics converted into sulfur emission estimates (Levy 1993). A similar procedure may prove useful for verifying parts of the Framework Convention on Climate Change. Scientific monitoring devices are available and of increasing utility in measuring emissions both directly and indirectly (Ausubel and Victor 1992).

Finally, a secretariat can provide independent verification by direct inquiries. Almost all arms-control treaties signed since the path-breaking Intermediate-Range Nuclear Forces (INF) Treaty have intrusive inspection procedures. Human-rights committees often appoint a rapporteur for a particular problem, with a mandate to gather evidence and information from all available sources, and often to make country visits. The 1971 Convention on Wetlands of International Importance (Ramsar Convention) established on-site monitoring procedures that have been used dozens of times to verify noncompliance and assist countries in identifying strategies to encourage compliance (Ramsar Convention Bureau 1990).

Analysis, Evaluation, and Dissemination

Collected data contribute to compliance management only if the regime provides analysis, evaluation, interpretation, and dissemination of the information acquired. Some regimes make extensive use of the data they collect. As already noted, the IMF compiles, analyzes, and publishes data in formats not otherwise available; the European marine enforcement regime creates a useful real-time database, and commodity agreements produce valuable industry and sector data. The Convention on International Trade in Endangered Species (CITES) makes extensive use of the reports of TRAFFIC, an NGO, to attempt to identify trends in wildlife trade.

However, several factors inhibit secretariats' analysis of the data available to them. First, they lack resources. Secretariats, particularly of environmental treaties, often have huge demands placed on their limited personnel and financial resources. Although all concerned may view analyses of compliance and effectiveness as essential, the need to prepare for the next conference of the parties or to type the transcripts of the last conference often takes priority. Independent analyses, by NGOs or academics, occur episodically at best, often reflect the analyst's agenda, and lack the "impartiality" and "legitimacy" that a secretariat analysis would carry. Indeed, they may fail to fully understand the real sources of the problem.

Second, the secretariat may lack the incentives to conduct certain types of analyses, and may even be directed by member states not to do so. For example, although the European marine enforcement regime requires each state to inspect 25 percent of the ships entering its ports, the annual reports do not analyze inspections on a country-by-country basis because of a desire by signatories to avoid being subject to "shaming" for failing to meet the requirements. Even when such analyses are conducted, they may not be disseminated beyond the governments of the member states because of diplomatic deference and each party's willingness not to publicize another's noncompliance in exchange for not having its own noncompliance identified.

A Managerial Model of Compliance

Most regulatory regimes should be regarded as instruments to manage an issue area over time, rather than as sets of prohibitory rules. Just as private companies and public bureaucracies commonly produce and review information about past performance in order to measure and manage their own progress, so creating transparency in international treaties generates information for assessing compliance of individual parties as well as evaluating overall regime effectiveness.

Essentially, the tasks of managing compliance are threefold:

1. Reviewing and assessing the performance of the parties in order to identify problems with the regime itself and to distinguish intentional violations from other types of noncompliance.
2. Ensuring that appropriate responses to noncompliance and violations produce and maintain a level of compliance "acceptable" to the regime parties.
3. Adjusting the rules to improve regime performance.

As in other managerial settings, the approach is not primarily accusatory or adversarial. In fact, regime management frequently starts with education and building a public constituency and awareness. Although effective regime management requires distinguishing willful violation from unintentional noncompliance, the process starts with the assumption that all regime members are engaged in a common enterprise. Initially, assessments seek to discover how to improve individual and system performance. Secretariats and other parties give states ample—sometimes, it seems, excessive—opportunities to explain and justify their conduct. Technical or financial assistance may be provided. Promises of improvement contain increasingly concrete, detailed, and measurable undertakings. If resistance persists, however, states and the secretariat may take more confrontational stances and intensify pressures for compliance. This process creates pressures to correct suspect conduct attributable to inadvertence, misunderstanding, or inattention while identifying, exposing, and isolating deliberate offenders.

Regime management usually involves three different levels of treaty evaluation. The first, review and assessment, evaluates each member state's performance and seeks to improve it while holding the regime rules relatively constant. The second, dispute resolution and interpretation, helps to clarify those areas of regulation and behavior that, at least initially, pose ambiguities. The third, adaptation and revision, entails the less frequent reappraisal of whether alternative rules would prove more effective at inducing compliance and achieving treaty goals.

Review and Assessment

Review and assessment involves evaluating the performance of the parties with an eye to improving compliance with the existing regime rules and structure. Treaties as diverse as the International Monetary Fund, the INF agreement, human-rights conventions, the Uruguay Round trade agreements, and the recently adopted Framework Convention on Climate Change have adopted review and compliance as central to their compliance systems. These systems involve evaluation of data from self-reporting, assessment of information collected from various sources, and analysis of member performance with respect to treaty requirements. In a well-managed treaty, steps may be taken, if performance is found wanting, to bring about improvement. These steps range from technical and administrative assistance to public exposure and "blacklisting." Together they comprise a powerful set of tools.

The ILO well exemplifies such review and assessment processes. Its Committee of Experts reviews government responses, comments on them, and reports findings to the conference of the parties. Another committee takes up cases of noncompliance, calling on noncomplying countries to explain and defend their behavior. A state may request "direct contact," a site visit by ILO staff to try to work the problem out on the ground. This is similar to the Ramsar Convention monitoring procedure noted above. Finally, persistent violation within defined categories can cause blacklisting, a form of "shaming." This final stage is preceded by a warning phase of a "special paragraph," which puts pressure on a member to come into compliance. Such procedures seek to distinguish noncompliance from intentional violation while simultaneously providing both the ability and the incentives necessary to increase compliance.

In IMF surveillance of national monetary policies, staff teams cooperate closely with local officials in detailed periodic reviews of members' economic performance. The Organization for Economic Cooperation and Development (OECD) elicits compliance with its Codes of Liberalization of Capital Movements and Invisible Transactions through a ratcheting process of assessment and review. States frequently take reservations to obligations under these codes. Those that do, are called upon by the relevant OECD committee to justify existing restrictions and to indicate their plans for removing such reservations in periodic reports. The OECD secretariat's review of the reports provides the foundation for discussion at quarterly meetings of the committee. Because of the pressures, states often remove reservations in the course of this process or on a faster schedule than initially proposed. The OECD has begun to conduct similar systematic reviews of the environmental policies of its members, evaluating them in terms of international commitments, OECD guidelines, and national declaratory policies. A similar policy review and assessment approach was adopted by the GATT in its Trade Policy Review Mechanism, and is now incorporated by treaty in the WTO.

Determining Acceptable Compliance Levels

International treaties, like other legal rules, can withstand significant noncompliance so long as it does not threaten basic objectives (Chayes and Chayes 1994). Not all violations, even persistent ones, threaten the life of the regime. At some point, the benefits of obtaining improved compliance fall below the costs of efforts needed to induce compliance, and the parties will accept the situation as it is. On the other hand, actors view certain defections as so serious that leaving them unaddressed would undermine the regime. For example, although as a general rule CITES tolerates a certain amount of noncompliance, most states viewed the defection of Japan, a major importer, from the ban on ivory trade as a threat to the credibility of the regime as a whole. The treaty legally allowed Japan to opt out. But pressures brought against it, including threats to change the plans to hold the next conference of the parties at Kyoto, induced Japan not to do so.

Levels of acceptable compliance depend upon such factors as the urgency of the issue and the degree of reliance placed by the parties upon the performance of others. The compliance of certain members may prove more crucial to effective treaty operation than that of others, whose noncompliance may be ignored or downplayed. Time matters, too—acts of noncompliance that are unremarked early in a treaty's history may receive quick and harsh responses later on. Treaties may even go through cycles of low-compliance "legislation stages"—during which states negotiate rules whose value stems from the distance between the "good" rules and the "bad" current behavior—and high-compliance "implementation stages"—during which activist states exert strong pressures to bring behavior into line with existing regulations.

Under most circumstances, strict and immediate compliance with every provision of an agreement is neither necessary nor feasible, as the record of domestic regulatory programs makes clear. Nor is there an invariant standard less than 100 percent. When member states perceive the current level of compliance as below the "acceptable" level, they often seek to improve compliance by some of the managerial techniques described here.

Capacity-Building

As traditionally conceived, treaties govern the actions of states. Treaty compliance involves making state behavior conform with treaty rules. However, environmental treaties also seek to change the actions of private actors. The problem of incapacity presents itself at several steps in such a compliance process. (On the importance of capacity-building in improving environmental treaty effectiveness, see Levy, Keohane, and Haas 1993.) First, compliance may require a state to enact legislation regulating the conduct of its corporate and individual citizens in accordance with the stipulations of the treaty. That capability may be deficient in some states, particularly those in transition to democracy. Technical assistance and advice from other states can help to develop workable and appropriate legislation, as shown in experience with the ILO, IMF, and CITES agreements (International Labor Conference 1982; Strange 1974). (The International Union for the Conservation of Nature's Environmental Law Center in Bonn, Germany, catalogs existing legislation and promotes the adoption of model legislation for CITES. [Burhenne-Guilmin 1992].)

The next step is harder. The state must mobilize an effective administrative and political effort to translate the legislation on the books into the reality of changing the behavior of private parties in accordance with treaty norms. Environmental treaties implicate the capacity of the state to govern—to enforce its own rules in significant ways. These problems can arise in developed as well as developing countries and in a wide range of subjects. Pakistan's nuclear weapons program relied heavily on suppliers in industrialized states that were signatories to the Nuclear Non-proliferation Treaty and that had stringent export-control laws. Even on such a high-salience issue, the United States, Germany, France, Great Britain, and other developed countries with interests in and programs for controlling nuclear-related exports failed to control the actions of their private corporate citizenry. Many advanced industrialized countries fail to secure full compliance with domestic clean-air and other pollution regulations, and adoption of a treaty seems unlikely to cure such problems.

Economic instruments, such as taxes and charges, will place new strains on existing infrastructures. Tax collection requires public discipline and a well-functioning bureaucracy. Despite the widely touted efficiency of taxes as a regulatory instrument, economists cannot accurately forecast the exact magnitude of behavioral response to a tax, a situation creating the possibility that a well-planned tax may fail to achieve an emissions target (Epstein and Gupta 1990).

Similarly, compliance with CITES requires customs officers to make fine distinctions among species while simultaneously preventing imports of drugs and other contraband, and moving legitimate shipments rapidly through the customs process.

Even customs officials in countries strongly committed to CITES, like the United States, may lack the necessary abilities and training.

In developing countries, the problem of enforcement capacity is often particularly acute. Inadequacies may exist in the administrative structures, available manpower, procedures for statistical record-keeping, the priority given to enforcement, and financial resources. To remedy such problems, the World Health Organization (WHO), the Food and Agriculture Organization (FAO), and the World Meteorological Organization (WMO) provide technical assistance as a main programmatic activity. The International Atomic Energy Agency spends half of its budget on technical assistance to developing countries in order to promote peaceful uses of nuclear energy. IMF surveillance procedures enhance the technical and professional capacities of finance ministries and central banks. Even when such assistance comes without explicit conditionality clauses, the organizational commitment to the treaty pushes recipients to conform to treaty goals.

Environmental treaties increasingly make assistance conditional on improving compliance. The Montreal Protocol multilateral fund, the Climate Fund of the Framework Convention on Climate Change, and the Global Environmental Facility proceed on the premise that developing countries need technical and financial assistance to facilitate compliance. All these mechanisms are designed to finance the "incremental costs" of compliance, including not only operating projects but also education, training of national enforcement officials, improvement of scientific facilities, assistance to planning departments, enhancement of data systems, and the like.

Reliance on project funding to remedy noncompliance may be misplaced, however. The supply of such funding is subject to classic "public goods" problems: although each treaty party wants other parties to comply, none wants to pay for another party's compliance. Mandatory contribution provisions themselves present compliance problems. Whether the mechanisms established, or those proposed, will maintain the flow of funds necessary for significant increase of treaty compliance remains to be seen.

Finally, the capacity to elicit compliance from domestic actors depends on the nature of treaty rules. Revising treaty rules to match existing monitoring and enforcement capabilities can sometimes prove to be the easiest and cheapest means of "increasing" capacity. MARPOL's discharge requirements proved difficult to enforce because, although aerial observation could detect oil slicks that violated treaty provisions, authorities often could not make the links to particular ships that are necessary for prosecution. Later agreement to require construction of tankers with separate tanks for oil and water ballast allowed much easier verification of compliance through inspection in port. The easing of the enforcement burden has boosted compliance to figures approaching 100 percent (Mitchell 1994).

Similarly, an on-line data collection system for customs documents aids German enforcement of export controls by simplifying processing and analysis of documentation (Reinicke 1994). Peter Haas recounts that in the Mediterranean Action Plan, France first sought pollution measurements that were beyond the range of laboratories and monitoring stations in developing countries on the Mediterranean's south shore. The policy harmonization process revealed that effective pollution control did not require such fine measurements, and produced new rules that allowed the developing countries to monitor their own emissions (Haas 1990).

The Ad Hoc Group of Experts on ozone reporting identified many problems common to other treaty reporting systems that arise from the reporting provisions rather than from national incapacity: reports were unnecessarily complicated, requested useless information, and the like. At its first meeting, the Group concluded that countries lacked the knowledge and technical expertise necessary to collect or provide the relevant data, and made a detailed series of recommendations for addressing the problem (see UNEP 1992; Peet 1994).

Dispute Resolution and Interpretation

Many analysts regard dispute settlement as a side track to, rather than an integral part of, a compliance strategy. But, as we have stated, treaties, like most other legal instruments, are rarely self-defining. When differences over meaning arise in the concrete circumstances of a particular case, whether or not they take the form of a "legal dispute," their resolution serves a dual function. It clarifies the meaning of the norm for all parties, and specifies the performance required of the disputants in the particular circumstances.

As in all legal systems, parties settle most treaty disputes by negotiation without recourse to available formal processes. The question is, What happens when negotiation fails? The United Nations Charter rehearses a familiar sequence of settlement methods: negotiation, inquiry, mediation, conciliation, arbitration, and judicial settlement. Yet the Charter does not require states to invoke any of these, apart from a generalized obligation of peaceful settlement. International lawyers make claims for the value of binding adjudication, either in the International Court of Justice or through a specialized tribunal or arbitral panel. Despite their alleged virtues, the Court and binding arbitration have played a minor role in treaty compliance to date, and seem unlikely to do more in the future. Besides being costly, contentious, cumbersome, and slow—the usual defects of litigation—they have the additional unattractive features of raising the political visibility of the problem and failing to be subject to party control.

Most treaty regimes turn to a variety of relatively informal mediative processes if the disputants cannot resolve the issues themselves. In multilateral regimes outside the security context, institutionalized processes provide scope for the secretariat or uninvolved parties to play an intermediary role. For example, the Convention on International Civil Aviation commits decisions on disagreements over interpretation or application to its Council (Convention on International Civil Aviation 1944). Many bilateral air traffic agreements provide for dispute settlement by the same Council. Since the thirty-three-member body is of an awkward size for carrying out judicial functions, the predominant mode of settlement is informal conciliation. Thomas Buergenthal concludes that "in dealing with disputes arising under the Chicago Acts, the ... Council has been guided by a policy that favors settlement by political and diplomatic rather than judicial means" (Buergenthal 1969: 195).

Based on the experience reviewed, we conclude that the legal stature of the form that dispute settlement takes or the decisions that come out of such efforts makes little difference, so long as parties treat the outcome as authoritative. Nevertheless, recent trends suggest a reversion to more compulsory and binding methods. The most important instance is the WTO, which, after almost two decades of incremental tinkering in GATT, adopted a new procedure in the Uruguay Round that amounts to binding adjudication. Panel opinions are automatically adopted, subject to an appeal on questions of law to a special panel of legal experts. Compulsory adjudication for some issues is also stipulated in the Canadian–United States Free Trade Agreement and in the North American Free Trade Agreement.

An important middle ground is emerging in the form of compulsory conciliation, culminating in a nonbinding recommendation from the conciliators on the issues in dispute, if the parties fail to agree. It avoids the adversarial quality of more formal adjudicatory procedures while at the same time assuring that the regime will be able to address the entire range of disputes without being blocked by a stonewalling respondent. Although the conciliators cannot pronounce a binding decision, their publicly reported view is likely to carry considerable weight both with the disputants and with the parties in general. Such a procedure preserves the niceties of sovereignty and avoids forcing the parties to accept a decision.

In regimes managed by international organizations, the preferred alternative for the resolution of disputes involving legal issues is authoritative or semiauthoritative interpretation by a designated body of the organization, often the secretariat or a legal committee. This provides a far less contentious method for dealing with disputes about the meaning of treaty provisions, and also may help prevent disputes by stemming potentially noncompliant behavior before a party has committed itself to activities that clash with regime goals. A state will tend not to disregard the answer to a question it has submitted, especially if such a nonadversarial context encourages working out differences or misunderstandings. In addition, the interpretative process can provide the ongoing clarification and elaboration of the governing legal rules that courts and administrative agencies perform in domestic legal systems. At the extreme, "interpretation" can provide a means for adapting the norms to significantly changed circumstances.

Of the 125 treaties reviewed by Chayes and Chayes, over half have some sort of provision for nonjudicial interpretation. Beyond those with explicit provisions, many implicitly grant the power of interpretation to a treaty body. Even an unsystematic examination of the practices of some of the governing organizations suggests that, unlike adjudication, member states do avail themselves of procedures for treaty interpretation. Experience under many regulatory regimes indicates that interpretation is a valuable tool for managing treaty implementation, and can avoid some of the contentiousness of traditional dispute-resolution mechanisms.

Adaptation and Revision

A less frequent part of treaty maintenance involves the long-term attempt by the parties to identify ways to improve treaty provisions so as to induce higher levels of compliance. Treaties do not remain static. To endure, they must adapt to inevitable economic, technological, social, and political changes. Traditionally, this required formal treaty amendment, but many recent environmental agreements have adopted a "framework and protocol" approach. The LRTAP regime has adopted protocols on sulfur dioxide, nitrogen oxides, and volatile organic compounds. The 1985 Vienna Convention on the Protection of the Ozone Layer provided only that the parties will cooperate in research and will exchange legal, technical, and scientific information on matters concerning the ozone layer (Vienna Convention for the Protection of the Ozone Layer 1985). Only two years later did the Montreal Protocol provide for cutbacks in consumption of CFCs. In 1990, nations amended the protocol to extend the controlled substances list and to speed up the phaseout. A similar framework-and-protocol approach is being considered in proposals for regimes to govern the size and character of conventional military establishments (Chayes and Chayes 1992).

Because protocols face the same ratification process as the original treaty, dissatisfied parties can block their entry into force. To skirt this problem, some treaties provide for rule adaptation without such formal procedures. The simplest device involves vesting the power to "interpret" the agreement in an organ established by the treaty. The United States Constitution has kept up with the times not primarily through amendments but through the Supreme Court's interpretation of its broad clauses. The IMF Agreement gives similar power to the Governing Board, and numerous key questions, including whether drawings against the Fund's resources could be made conditional on the economic performance of the drawing member, have been resolved by this means (International Monetary Fund 1945, 1952).

Experience under the ABM Treaty illustrates how initial hopes to avoid disagreements by careful treaty phrasing can fail. Article VI of the treaty prohibited "testing in an ABM mode." If the Soviet Union could upgrade its SAM systems for ABM use, the numerous and widely distributed SAM installations would constitute the nationwide ABM system that the treaty prohibited. The United States appended a unilateral statement to the treaty declaring that it would interpret colocation of SAMs at intercontinental ballistic missile (ICBM) test sites and testing of SAM radars concurrently with ICBM reentry as prohibited testing, since these practices were necessary to designing ABM upgrades for SAMs. After the United States observed the Soviet Union continuing these practices, lengthy negotiations in the Standing Consultative Commission in 1978 produced agreement by the Soviets with the American interpretation of Article VI. The two states further refined this interpretation in 1985, after the United States continued to observe concurrent operations of SAM radars and missile tests.

A second mechanism for adaptation used under several treaties is the adoption of "technical" regulations by vote of the parties (usually by a special majority of more than 51 percent), which then bind all parties that do not choose to opt out. The International Civil Aeronautics Organization has such power with respect to operational and safety matters in international air transport (Convention on International Civil Aviation 1944). CITES and the International Convention for the Regulation of Whaling make use of such technical appendices to ensure that changes to species listings and whale quotas, respectively, can take effect without long ratification delays. IMO treaties contain "tacit acceptance" provisions whereby certain amendments adopted by the relevant committee enter into force automatically within sixteen months for all parties that do not explicitly object. In many regulatory treaties, "technical" matters may be relegated to an annex that can be altered by vote of the parties (see 1987 Montreal Protocol on Substances That Deplete the Ozone Layer, art. 2(9); 1990 London Amendments to Montreal Protocol). Even recent United States–Russian arms-control treaties have authorized modifications regarding "technical matters" by executive agreement without reference to the legislative bodies (Koplow 1992).

Treaties characteristically contain self-adjusting mechanisms by which, over a significant range, they can, and in practice commonly do, adapt to changes in the interests of the parties. The need for, and direction of, such adaptation will depend on evaluations of the level of compliance. Compliance deemed adequate to achieve existing goals of the major parties can lead to parties' viewing the treaty as a success. This may then lead them to agree on new treaty goals and rules that establish higher standards for behavior. The adoption of new protocols for nitrogen oxides and volatile organic compounds under LRTAP, and the adoption of earlier phaseout dates for CFCs under the London Amendments to the Montreal Protocol reflect efforts to build on evidence of treaty success.

If the treaty information system reveals high levels of noncompliance due to inadvertence, ambiguity, incapacity, and other unintentional factors, the parties may adopt new rules and procedures that attempt to address these problems. Effective treaty alteration requires sufficient analysis to ensure that the new rules address the true sources of current noncompliance. For example, if the real reason for noncompliance is lack of financial resources, then adopting more specific language is unlikely to improve compliance.

Information may show, however, that noncompliance actually stems from intentional violation. But even in such situations, "deliberate" failure to comply may reflect nothing more than a lack of interest in the goals of the treaty (Lyster 1985). If powerful parties maintain an interest in the treaty goals, however, these parties may succeed in replacing ineffective treaty rules with rules that facilitate the monitoring and enforcement necessary to bring coercive pressures to bear on the violating parties. MARPOL's replacement of discharge standards with equipment standards reflected pressures from the United States and other countries for rules that provided greater transparency and were designed to elicit greater compliance.

New rules may not always improve compliance. They may exacerbate tensions between those seeking effective compliance and those who prefer a more lax treaty regime. The International Whaling Commission adopted an International Observer Scheme in the early 1970s, after years of debate and concern over inaccurate reporting; however, whaling nations' resistance led to a watered-down provision involving voluntary inspections by the personnel of one whaling nation of the catch of another whaling nation (Birnie 1985).

The Role of International Organizations and Nongovernmental Organizations

The activities described above do not arise and operate spontaneously. Their effectiveness depends heavily on the organizational setting in which they are deployed. The parties to treaties, acting on their own, can sometimes activate these instruments. NGOs often can heighten and intensify such efforts. Such a diffuse model has certain appealing features in an era of extreme skepticism about the capacities of government and bureaucratic institutions. However, it is no coincidence that the regimes with the most impressive compliance experience—ILO, IMF, OECD, GATT—depend upon substantial, well-staffed, and well-functioning international organizations. In the contemporary international system, both nongovernmental and intergovernmental organizations play essential roles in the management of compliance.

NGOs began to become involved in international affairs with the emergence of economic and social issues on the international agenda after World War I. They now perform parallel and supplementary functions at almost every step of the regime management process we have described. They provide independent information and data. They verify party reporting, and evaluate and assess party performance. They often provide technical assistance to enable developing countries to participate in negotiating treaties and to comply with reporting and substantive requirements. When noncompliance occurs, they prove crucial to exposure, shaming, and motivating public responses. They enhance pressures on governments to comply both from within and from without. Because they are outside governmental control and have their own goals and definitions of compliance, their actions are not always appreciated.

NGOs within countries can use international treaties to appeal for enforcement of treaty norms. In 1987, for example, two Nigerian lawyers founded the Civil Liberties Organization to represent common prisoners held without charges or trial for extended periods. It soon was bringing class actions on behalf of groups of detainees and publishing reports on individual prisons. Its 1992 report on conditions in Nigerian prisons, *Behind the Wall*, documented the degree to which the country's prisons failed to meet the provisions of numerous human-rights agreements. Shortly thereafter, the government granted amnesty to 5,300 prisoners, doubled prison budgets, and made other policy changes.

A year after the 1973 coup in Chile, documentation by Amnesty International and the International Commission of Jurists of arbitrary detention and arrest, torture, and disappearances provided the basis for charges by the Soviets and others in 1974 before the United Nations Human Rights Commission. By the next year, the Commission gave NGOs the right to appear before it in their own name, rather than as information suppliers to governments, and set up an ad hoc working group to investigate the Chilean abuses. Although the progress was slow and halting, human-rights NGOs and their counterparts within Chile played important roles in providing both the information and the motivation for the progress that has been made.

In whaling, in the 1970s, NGOs conducted an extensive membership drive among anti-whaling states that created the three-quarters majority needed to pass the moratorium on commercial whaling in 1982. American NGOs also consistently pressed for application of United States domestic legislation authorizing sanctions against actions that "diminished the effectiveness" of the whaling convention. Most of these cases did not involve treaty violations, and many involved attempts to pressure nations to join the International Whaling Commission. Various NGOs have regularly undertaken letter-writing and advertising campaigns, consumer boycotts, and sometimes more drastic measures to urge governments and whaling companies to halt whaling.

In all these cases, NGO activities made a difference, although it is sometimes an open question whether their contribution was positive. The development and elaboration of norms and states' compliance with existing norms cannot be explained without reference to NGO actions. Their impact stems from an ability to influence the domestic policy process in many states, as well as an ability to appeal directly to international organizations and the international community over the heads of governments.

International organizations are arenas for almost continuous interactions among the members, their representatives, and the staff. This process involves persuasion and an important element of exchange. They generate a continuous stream of transactions that serve as counters in an unending game of political bargaining and diffuse reciprocity (Keohane 1986). Lisa Martin argues that the involvement of an international institution facilitates a state's efforts to mobilize international support for economic sanctions it seeks to impose against another state, as happened in Britain's use of the European Community to build support for sanctions against Argentina in the Falklands/Malvinas conflict (Martin 1992).

The secretariats of international organizations wield considerable power through their control of the agenda. In the international context, Dr. Mostafa Tolba, the executive director of the United Nations Environment Programme, defined much of the environmental treaty-making agenda in the 1970s and 1980s despite his small budget and the absence of any formal power. Director General Hans Blix took the disclosures of Iraqi nuclear sites as an opportunity to strengthen the IAEA safeguard system. Simply proposing such strengthening to the Board of Governors in that climate forced important progress. The bureaucrats of the European Commission play this role regularly in the politics of the European Union.

International organizations can also influence the policies of their members more directly. During the debates of the 1970s over population growth and family planning, WHO officials successfully resisted the effort to characterize population issues as broad social and economic matters by mobilizing its formidable constituency of health ministries, medical groups, and NGOs to persuade governments and the international community to keep family planning and population issues under the control of the health sector (Finkle and Crane 1976). Bureaucratic alliances between international organizations and the relevant ministries can exert considerable influence on domestic governmental policies. The IMF and WHO use their extensive network of loyal contacts within domestic bureaucracies to promote their policies, and provide information and support to actors pushing for greater compliance.

The industrial states that provide much of the financial support have become increasingly skeptical of international organizations, and this is reflected in their unwillingness to provide funds or establish structures in new treaty regimes capable of developing the treaty mandate. Part of the reason is the well-known ills of bureaucracies that appear in heightened form in the international arena. But part is also that a strong secretariat can attain considerable autonomy from the control of member states. Thus, often, the consent granted by states to regulation by an international regime has not been accompanied by the delegation of authority to a central body with sufficient staff and resources to manage the implementation of the obligations undertaken. In most contemporary environmental treaties, the operational arm of the regime is the conference of parties, not only nominally but in practice. Special committees or working groups staffed by country representatives do much of the preparatory and staff work. The secretariats, consisting of a few officials, are too small to be capable of much policy initiative.

This design is already leading to concerns about "institutional overload" (Levy, Keohane, and Haas 1993). The demand for intensive and extensive party participation strains the resources of even the largest and most dedicated foreign policy establishments. Without modification of this stance, the increasingly complicated and complex tasks facing organizations in many international issue areas will likely go unfulfilled. Implementation, compliance, and enforcement are the quintessential tasks of bureaucracies in all organizations. Reducing the resources available for these purposes will not improve performance. Creating lean, effective, and politically responsive organizations should not be beyond the capacities of the international community at the end of the twentieth century. It will be essential to increasing compliance with international treaties.

Conclusion: Toward an Active, Integrated Management Strategy

The elements of management just discussed are powerful, but to date they have not been perceived as part of a coherent or comprehensive strategy for managing compliance. Few, if any, regulatory agreements display all of them. Particular treaties have developed some of these features quite fully, but others, such as capacity-building, remain rudimentary. Despite some impressive recent developments, current efforts do not yet constitute a comprehensive strategy of active compliance management. For the most part, the authority and resources to create the forms of issue management traditionally found in national governments and in the corporate world have not been made available. More important, the notion that treaty regimes can actively manage compliance through a comprehensive strategy remains neither widely understood nor widely accepted. It begins to emerge only by piecing together disparate efforts to manage the treaty implementation process. We believe this conceptual failure needs correcting. The elements we have described are not merely discrete useful practices; they need to be integrated into a comprehensive management strategy. Ideally, such a management strategy would include implementation with the support of a strong and effective international organization.

The foregoing discussion reflects a view of noncompliance as expected rather than deviant, and as inherent rather than deliberate. This in turn leads to deemphasis on enforcement measures or coercive sanctions, whether formal or informal, except in the most egregious cases. It shifts attention to sources of noncompliance that routine international political processes can manage. Thus, improved dispute-resolution procedures address problems of ambiguity; technical and financial assistance can mitigate, if not eliminate, capacity problems; and transparency and review processes increase the likelihood that national policies are brought progressively into line with agreed international standards.

These approaches merge in a process of "jawboning"—an effort to persuade a state to change its ways that is the characteristic form for eliciting international compliance. Jawboning exploits the de facto necessity for alleged violators to explain and justify suspect conduct. These justifications are evaluated in many forums, both public and private, formal and informal, domestic and international. This process distinguishes justifiable or inadvertent noncompliance—those instances that comport with a good-faith compliance standard—from those relatively rare cases of willful violation. Most compliance problems yield to this process. For those that do not, the process confronts the offending state with the stark choice between conforming to the rule as defined and applied in the particular case, or openly and explicitly flouting its obligation. The discomfort of such a position proves sufficient in most circumstances to get the transgressor to bring its behavior in line with its obligations.

Our analysis leads away from the search for better enforcement measures—"treaties with teeth"—and toward better management of compliance problems. It requires focusing on and improving the mundane, day-to-day interactions and discussions that persuade actors to comply, rather than dramatic episodes of sanctions as a response to clear violations. Treaties will elicit greater compliance when they look for ways to improve the former processes than when they demand the latter.

Notes

1. The framework and much of the material and arguments in this chapter have been developed more fully in Chayes and Chayes (1995).

4

Contingent Knowledge: Implications for Implementation and Compliance

Sheila Jasanoff

International environmental agreements differ from most other types of international regimes in their multifaceted reliance on science and technology. The production, diffusion, and institutionalization of scientific knowledge are rightly seen as essential contributors to environmental standard-setting and enforcement (Wynne 1987; Salter 1988; Jasanoff 1990; Graham 1991). Scientific findings underpin the recognition that an environmental problem of transboundary proportions exists and that it merits attention from the international community. Scientific deliberations help to identify environmental resources that may be at risk, and to define the levels and kinds of protection that are appropriate. Technology, similarly, provides a baseline against which states can assess the feasibility of proposed environmental measures. Once agreements are in place, scientific and technical capacity is required to carry out necessary tasks of monitoring, analysis, and evaluation.

Science, moreover, functions as an important political resource in the formation and implementation of international environmental regimes. There is a general perception, at least in the Western world, that recourse to science enables reason and objectivity to prevail over emotion, ideology, and special interests. Particularly in the international arena, where value commitments are highly disparate and consensus can easily unravel, the possibility of defining environmental obligations in scientific or technical terms offers an inviting way out of conflict and possible political stalemate.

One cannot, then, afford to ignore the role of science and technology in any comprehensive attempt to examine the implementation of environmental agreements. Yet, the effort to integrate science into the kind of legal and political analysis undertaken in this volume presents unusual difficulties. It requires, to begin with, bridge-building between academic literatures that are not ordinarily in close communication. Models for such work are scarce, and different disciplinary traditions, investigative agendas, and analytic discourses pose formidable barriers to integration. Nevertheless, there is a body of scholarship in the social studies of science and technology that carries unavoidable implications for the implementation and enforcement of environmental treaties. Spelling out these implications explicitly is the objective of this chapter.

Such a project necessarily requires the charting of unconventional theoretical waters, but three significant gains are foreseen. First, the social analysis of science and technology promises to deepen our understanding of key concepts related to the implementation of environmental accords, including, foremost, the fundamental notions of "environmental standard" and of "standardization." Second, awareness of the socially constructed character of environmental knowledge allows us to evaluate more informedly the mechanisms for linking knowledge to action in different treaty regimes. Third, such analysis opens the door to a set of normative questions that can be asked only if contemporary debates about the nature of scientific knowledge are factored into discussions of environmental agreements and their implementation (Jasanoff 1996).

Science and Politics: Transparency or Black Box?

Social scientists have not, for the most part, regarded the internationalization of environmental science and its application in different cultural contexts as presenting any novel analytical problems. Science, by virtue of its commitment to truth-seeking and its capacity to uncover reproducible facts about nature, has traditionally been conceived as standing apart from politics, and thus as not amenable to normal social and political investigation. Although much recent scholarship acknowledges that science is neither free from subjective judgment nor immune to controversy, political analysts have nevertheless assumed, unproblematically for the most part, that scientists can resolve their disagreements internally, without reference to worlds outside science. On the international scene, scientific institutions tend to be seen as more authoritative than political ones, and technical accords, such as agreements on standards, instrumentation, and monitoring or control regimes, are thought to promote convergent behavior among states.

The perception of science as neutral territory is consistent with a widely accepted model of decision-making that regards science mainly as a source of objective knowledge for clarifying and rationalizing policy choices. Famously summed up in the phrase "speaking truth to power" (Jasanoff 1992:196, n. 8; Carnegie Commission 1992), this image of science as lacking a politics of its own persists, although decades of research on science-based policy point to the need for a more nuanced characterization. The parochial character of much of the literature on science and policy is partly to blame. Several limitations are noteworthy for our purposes: studies in this area have tended to focus on national rather than international issues; on controversies rather than routine administration; on policy formulation rather than policy implementation; and, above all, on case studies rather than more general theoretical principles. To date, little of this work has taken account of developments in the history and sociology of science that are demanding a rethinking of the relationship between knowledge, power, and political action (exceptions include Wynne 1987; Jasanoff 1990; Bimber and Guston 1995; Cozzens and Woodhouse 1995).

The notion of science as an impartial adjunct to policy rests on several assumptions that have deservedly been called into question: first, it attributes a philosophically debatable objectivity to all of science (science speaks truth); second, it posits a linear, unidirectional process of policy formation (science speaks, and politics accepts, the truth); third, and most important, it conceives of science as an autonomous institution whose ability to define the truth is independent of the "powers" that turn to it for guidance. The picture that has emerged instead from a growing body of recent research shows science as a deeply social institution whose claims are situated within, rather than outside, politics and culture. What implications does this newly socialized picture of scientific activity hold for compliance with international environmental agreements?

Questions about the international system's faith in technical solutions begin to arise as soon as the picture of "autonomous" science and technology is replaced by one that is more contingent and context-dependent (Winner 1977, 1986). Science and technology as they operate in the world are plausibly represented as heterogeneous, sometimes fragile, conglomerates of users, interpreters, claims, texts, instruments, and practices (see Bijker et al. 1987). The components of these belief systems or technological networks may be more or less diverse, extensive, and tightly coupled (Perrow 1984). When the pieces of the system function harmoniously together, knowledge coheres and technology works as intended. Stability, however, is neither automatic nor guaranteed: it requires constant fine-tuning to keep complex scientific and technological networks from breaking apart. Moreover, even the most passive-seeming of technologies—a tall smokestack or a catalytic converter—is not apolitical. Shaped and sustained by social forces, technological processes and artifacts incorporate historically and culturally sanctioned methods of social control (Noble 1978; Winner 1986; Sclove 1995). Far from imparting uniformity to regimes, these socially contingent networks can become sites where political differences crystallize or controversies erupt about what is known or knowable in the physical world, what is within and what is beyond the human capacity to predict and control.

The more one recognizes the socially embedded character of science and technology, the less plausible it becomes that the political problems of noncompliance can be solved by purely technical means. At the most pragmatic level, technical standards in environmental regimes always run the risk of being implemented in different ways, both across countries and by different actors within the same country. Variations in knowledge, training, resources, and skills among responsible authorities, lack of public and media attention, or differential supervision of implementing bodies by nongovernmental organizations (NGOs)—as, for example, in the protection of endangered species under the Convention on International Trade in Endangered Species—may give rise to seriously divergent regulatory practices.

At a more theoretical level, acknowledging the sociopolitical dimensions of science and technology raises two additional sets of questions. One has to do with the meaning, and therefore the assessment, of cross-national uniformity in the implementation of technically grounded standards. If expert knowledge (both scientific and technical) is even partly contingent on local circumstances, then one would expect differences in the institutionalization of expertise within governments and in the larger political culture to lead to important cross-national differences in policy implementation. The problem is not simply that some implementing bodies may fall short of the agreed optimal outcome through lack of capacity or political will. Rather, even when countries claim in good faith to be complying with the *same* standard, their actual practices may well be dissimilar because of the "interpretive flexibility" of technical standards in their varied social and cultural contexts (Bloor 1976; Bijker et al. 1987). Judgments about what constitutes adequate enforcement and compliance will inevitably reflect deep-seated normative understandings about fairness, efficiency, and the appropriate limits of governmental power (Hawkins 1984; Wynne 1987; see also Jasanoff 1986; Putnam 1993; Porter 1995).

It is, of course, commonly recognized that legal standards defining norms of social behavior (e.g., due care, recklessness, "deliberate speed") possess interpretive flexibility. Indeed, in democratic societies the awareness that too much flexibility is a bad thing, because it allows for uncontrolled discretion in enforcement, has prompted corrective action; this, for instance, is the basis for the "void for vagueness" doctrine used to invalidate anti-vagrancy statutes in American constitutional and criminal law. The assumption questioned in this chapter is that the interpretive flexibility of legal norms is adequately tamed, or even that it disappears entirely, when standards are fixed through use of technical measures, artifacts, or practices, and are monitored by international expert agencies. The contingency of knowledge about complex environmental phenomena and of science-based technology leads us to inquire whether compliance with standards can ever be internationally uniform, or whether the social fact of compliance is always locally and flexibly constructed. Compliance, at any rate, emerges as an analytic category that requires unpacking and investigation.

A second set of theoretical concerns grows out of a paradox that is defined by the contrasting metaphors of the "black box" in the social studies of science and of "transparency" in international relations and international law. Sociologists of science and technology use the concept of black-boxing to describe the consolidation of scientific theories and claims, as well as technological networks, into entities that resist being seen through or pulled apart. The "facts" of science and the products of technological systems become resistant in this way when they are backed by sufficiently strong coalitions of actors, institutions, norms, practices, and artifacts (Callon and Law 1982; Callon 1986; Latour 1987). The means by which facts were produced or artifacts took on particular shapes, however contested they may once have been, are eventually boxed up and made impervious to deconstruction in everyday social interaction.

The concept of transparency, by contrast, refers to the properties of international regimes—most importantly, the accessibility of information and implementing institutions or personnel—that make them available for review and monitoring by state as well as nonstate actors. Transparency is conducive to treaty-making, harmonization, and progressive learning because it satisfies the curiosity of skeptics, builds trust, and facilitates enforcement. It is regarded by many students of international regimes as an essential prerequisite for effective implementation (see chapter 3 in this volume).

The juxtaposition of these two metaphors and what they imply about the legitimacy of technical decisions raises a number of interesting questions for students of international environmental agreements. Can black-boxing, which conceals conflict and uncertainty, and transparency, which makes them visible, coexist meaningfully in regimes that depend heavily on science and technology? Environmental treaties, as noted earlier, are particularly dependent on goals and protocols based on inputs from science. The implementation of standards likewise entails intensive use of technology. The more thoroughly black-boxed these inputs are at the time of regime formation, the more we can expect them to withstand deconstruction in national contexts of implementation. In other words, once an international agreement based on science and technology has been formalized, uniformity across countries will be easiest to secure if its components are closed off to criticism and renegotiation. Are the centralization and muted levels of criticism needed to preserve the authority of international standards compatible with maintaining the administrative and political transparency of regimes?

Answers to these practical and theoretical questions may provisionally be sought in four bodies of literature at the intersection of science policy and social studies of science and technology. The first consists of research on the production of scientific knowledge for policy purposes, and shows how stable institutional and procedural arrangements (for example, "national styles" of regulation; see chapter 2 in this volume) can reduce uncertainty about scientific claims and methods, as well as about technological monitoring and verification systems. A second group of studies, located generally in comparative politics, draws attention to the divergent ways in which political culture influences the interpretation of policy-relevant science, leading to cross-national differences in assessments of risk and safety. The third body of work, centered in international relations, seeks to explain how professional networks known as epistemic communities disseminate scientific ideas across political boundaries and link science to policy in international regimes. Fourth, and finally, there is an emerging body of work that assesses the role of international institutions in stabilizing science-based norms across countries and in resolving scientific disputes.

Three themes of great relevance to compliance will emerge from a review of this extended literature: (1) the contingency of science can be overcome through the convergent interpretive practices (standardization) and cognitive evolution of transnational agencies and actors; (2) the harmonization of technical standards does not necessarily wipe out variant local practices or guarantee uniform implementation; and (3) institutions that enable the internationalization of scientific and technical norms do not necessarily ensure meaningful participation by all interested governmental and nongovernmental actors.

Contingent Knowledge, Negotiable Technology

Since the late 1970s, work in the history, philosophy, sociology, and ethnography of science has shown that scientific knowledge claims cannot be understood simply as unmediated truths about the natural world. Pioneering studies by David Bloor and Barry Barnes (1982), H. M. Collins (1985), Bruno Latour (1987), and Steven Shapin and Simon Schaffer (1985), among others, showed that the credibility of scientific claims, even in the "hard" physical sciences, is often best understood as the product of complex and protracted social negotiations. In his enunciation of the "strong programme" for the sociology of science, Bloor (1976) argued that to grasp the nature of scientific knowledge in full, one has to study its social derivation (see Geertz 1983 for a concurring view from anthropology). Elaborating on this theme, Collins (1985) demonstrated that what counts as proper experimental replication, a mainstay of the scientific enterprise, generally cannot be settled through direct appeals to nature but must, rather, be worked out in micronegotiations among scientists. Sometimes these negotiations fail, and scientists then are forced into a never-ending search for "tests of tests of tests" that Collins termed "experimenters' regress."

In a related vein, a path-breaking study by Latour and Woolgar (1979) of scientists at work in the laboratory traced the everyday construction of scientific facts from observed laboratory phenomena through a paper trail of "inscriptions"—texts such as tables, graphs, and charts. Facts became progressively more difficult to refute as they were codified into abstract and visual forms. Equally important from the standpoint of implementation and compliance are studies that have shown how technologies acquire stable, standardized forms through social mediation (see particularly Bijker et al. 1987).

The constructivist turn in social studies of science and technology draws attention to issues that, until recently, were not regarded as problems for policy implementation. If compliance depends on information provision, for example, then it becomes important to ask not merely whether the facts are "right," but also who provided the information and on what authority, how it was validated, whether it embeds possibly undisclosed normative presuppositions, and what room it leaves for conflicting interpretations. If compliance is secured through technological controls, we are alerted to look also at the social and technical practices of the deployers of the technology. If instrument readings or other data-reporting schemes form the basis of compliance, what steps were taken to standardize the instruments, the reporting protocols, or the cultures of monitoring and enforcement across compliance regions? Did the expert bodies entrusted with measuring compliance apply the same criteria to observations in different locations and at different times? And how were divergent conceptualizations of nature accommodated, or possibly excluded, in the processes of standardization and policy-making?

Differences in the interpenetration of the social and technological worlds are not hard to find in accounts of environmental regimes. A mechanical malfunction or deviation from codified practice that rings instant alarm bells in one cultural setting may be ignored in another until disaster strikes (Wynne 1988; Jasanoff 1994). Monitoring technologies may incorporate tacit assumptions about the way in which natural phenomena are caused or distributed and should be measured (Zehr 1994). We will return to examples like these in more detail below. The point to note for now is this: we cannot meaningfully evaluate the effectiveness of technical guarantees of compliance without a detailed understanding of the social context within which they were produced and implemented. Filling in this contextual information is one of the most significant contributions of the country chapters in this book.

Scientific Uncertainty and Social Trust

The socially constructed aspects of scientific knowledge are most apparent when science is used to support and rationalize high-visibility political decisions. Thus, in American technological controversies, political stakeholders routinely participate in producing and interpreting science to suit their policy preferences (see, e.g., Nelkin 1992). Whether the issue is the siting of a nuclear power plant or a hazardous-waste facility, the marketing of a bioengineered pesticide, or the building of a supersonic aircraft, the advocates of technology consistently evaluate the evidence of risk, safety, and efficacy differently than do its opponents. Controversies reveal the unavoidable interpretive flexibility of scientific claims, dispelling the idea of "pure" or "objective" scientific analysis in the political realm. The scientific conclusions adopted by policy makers are almost invariably negotiated products, developed in liberal democracies through formal processes of rule-making and adjudication.

Social organization, institutions, and culture therefore need to be considered in explaining how scientific claims gain authority under conditions of uncertainty. In one example, anthropologists researching expert judgments about deforestation in the Himalayas found radical differences with respect to seemingly objective measures of "per capita fuelwood consumption" (Thompson et al. 1986). The researchers concluded that the "facts" the experts had tried to discover about the Himalayas were so variable and context-bound that they could not be ascertained through a single, universally accepted model of scientific inquiry. What the experts offered were not descriptions of reality but, rather, varying glosses on their own uncertainty. Their estimates were shaped by culturally induced perceptions of what the problem was, and reinforced by their informants' localized experiences and locally contingent worldviews.

Certainty about the truth of a knowledge claim or the dependability of a technology reflects in many instances a judgment about the trustworthiness of the underlying institutional and political arrangements (Irwin and Wynne 1996). Technical controversies dissipate when policy disagreements are resolved and relations of trust are established between policy makers and affected social interests. Thus, in Donald MacKenzie's (1990) fascinating history of American nuclear missile guidance, the meaning of "accuracy" was continually contested as long as huge political consequences hinged on military authorities' being right about the efficacy of particular weapons systems. Disputes about the accuracy of intercontinental ballistic missiles (ICBMs) arose in the late 1970s, when the performance of missiles had to be evaluated. A perfectly dispositive test of accuracy was a contradiction in terms, an invitation to experimenters' regress (Collins 1985). Were tests of components equivalent to tests of the whole weapon? Could trajectories be known with sufficient accuracy in the absence of land-based tracking? Were multiple, correctable, test-range firings an adequate stand-in for the one-shot firing of war? And was enough known about the earth's gravitational field to assess the operational accuracy of inertial guidance across the missile's entire trajectory?

Supporters and critics of ICBMs therefore divided on the second-order question of whether the tests being conducted were *sufficiently like* the expected real-world conditions to justify the going assumptions about missile accuracy. Such similarity determinations must continually be made both in producing reliable scientific knowledge (Barnes 1982) and in testing technology (Pinch 1993). In the ICBM case, uncertainty about accuracy allowed the parties to argue, in turn, "for manned bombers, for the cancellation of MX and a more 'dovish' defense policy, for radio guidance, and for large missile warheads!" (MacKenzie 1990:363). The controversy eventually ended, MacKenzie notes, not because either side was proved scientifically right or wrong, but because declining media attention and a change of policy under President Reagan shifted the weapons debate to other arenas of disagreement.

In environmental and public-health controversies, a decrease in the plasticity of policy options similarly coincides with a reduction in perceived scientific and technical uncertainty. Quarrels about natural phenomena may continue unabated unless parties agree on a restricted set of possible responses. Narrowing the range of decisions causes parties to rein in their fears of the unknown. The lengthy battle of experts over the Storm King power plant on the Hudson River in New York provides an instructive example. Here, years of inconclusive dueling between competing estimates of biological impacts on fish populations ended only when participants agreed to focus on short-term impacts without attempting to predict and control longer-term effects (Barnthouse et al. 1984). With the problem reframed in this fashion, biologists for the two sides found mutually acceptable methods of counting the fish and analyzing the costs and benefits of different control options.

Deeper links between technical uncertainty and social distrust are borne out by work on the connections between regulation and political culture. The impulse toward precise and formal regulatory controls in modern industrial democracies correlates inversely with trust in and deference toward ruling institutions. Objectivity is at one level firmly tied to democratization and politically sanctioned questioning of authority. Neutral, quantitative methods of policy assessment are most valued where the sovereign is most exposed to public criticism. In his ambitious study of science, technology, and the rise of the liberal democratic state, Yaron Ezrahi (1990:197–216) calls attention to the intimate correspondences in America between a decentralized, transparent political culture and the state's reliance on authoritative public facts produced through science.

Comparative case studies of environmental policy-making also bear out the existence of systematic differences between America's scientistic, impersonal style of policy justification and the more qualitative European approaches, warranted by elites who enjoy higher levels of trust (Brickman et al. 1985; Vogel 1986; Jasanoff 1991; Porter 1995). There are intriguing suggestions that policy makers in relatively less open and adversarial European political systems are less inclined to admit uncertainty about science than are their more exposed American counterparts (Wynne and Mayer 1993). In the United States, uncertainty is both more readily acknowledged and more rapidly translated into impersonal scientific discourse (Jasanoff 1987).

Comparisons within Europe also show how divergent, culturally rooted attitudes toward implementation and compliance are translated into substantially different frameworks of technical and social controls. In a comparative study of hazardous-waste management, Brian Wynne (1987) noted that definitions of "hazardous waste" were shaped not by any "intrinsic natural meaning" but by the contingencies of national politics. Precise, inflexible, technical standards were adopted in the fragmented and relatively adversarial Dutch political setting, whereas Britain, with a long-established tradition of informal regulatory negotiation and institutional trust, could get by with a much less transparent system of inexplicit, ad hoc, and inconsistent rules. "Technical standards," Wynne concluded, "are socially constructed, as social languages reflecting their institutional setting" (1989:403). In a suggestive counterpoint, Putnam (1993) has argued that in Italy a civic culture of social cohesion, associationism, and trust has correlated well with the state's power to innovate policy in such habitually contested and uncertain areas of environmental protection as strip mining, fisheries promotion, wildlife protection, and air and water pollution control.

The Meaning of Standards

Standardization is the process by which policy-relevant knowledge is most commonly stabilized across divergent political and cultural spaces. Standards reduce the range of interpretation associated with the observation of environmental phenomena. They usefully focus on the routine operations and practices that become the basis for securing and monitoring compliance with environmental treaties. Agreement on standards lowers the incentives for opposing interests to exploit the contingency of science. Theodore Porter acknowledges the power of quantification, in particular, to assuage public skepticism and provide the appearance of impartial regularity:

> But where [the objectivity of quantification] has been applied more or less routinely, where its contexts of application as well as its methods are at least partly standardized, it can permit administrative decisions to be made quietly, discouraging public activism. In a suspicious democratic order, even truth claims depend on the appearance of objectivity in the sense of impartiality. (1992:47)

Through standardized techniques such as risk assessment or cost-benefit analysis, administrative agencies can extend their control through wide reaches of time and space (Porter 1995).

The downside of quantification, however, is that it may strip away complex social meanings and prematurely flatten important contextual differences. In studying ozone depletion and global warming, for instance, some Western experts sought at first to assess the detrimental properties of gases on the basis of standard measures of "ozone depletion potential" (ODP) or "global warming potential" (GWP). These universally applicable quantitative measures offered enormous advantages for implementation: in particular, they made it possible for harmful emissions to be compared according to "objective" criteria, thereby laying the groundwork for transparent and equitable emissions trading schemes.

Yet, conflicts that arose in 1991 between experts at the United States–based World Resources Institute (WRI) and other scientists showed that the simplifying choices made in GWP calculations incorporated implicit—and inadequately questioned—normative judgments about responsibility for climate change. In particular, WRI's decision to focus on current emissions ranked developing countries higher on the scale of responsibility for the perceived problem than did an alternative ranking based on cumulative emissions (Hammond et al. 1991; Subak 1991). In a similar vein, a report from India's Centre for Science and Environment (CSE), a respected environmental NGO, stressed that all carbon releases could not be treated as the *same* from an ethical standpoint (Agarwal and Narain 1991). According to CSE, WRI's methodology erred in disregarding per capita consumption as a factor in the production of greenhouse gases and in equating the essential agricultural emissions of the world's poor ("subsistence emissions") with nonessential emissions from the consumption patterns of the world's wealthy ("luxury emissions").

Further, as Latour (1987, 1990) recognizes in theorizing the work of scientific representation, and Benedict Anderson (1991) asserts in explicating the coherence of national identities, power relationships are involved in disseminating standardized understandings of the natural and political worlds. Standardization is often achieved through visual means, such as statistical charts, questionnaires, surveys, or maps. Latour felicitously names such representations "immutable mobiles": fixed in form, they can nevertheless be transported around the world, potentially carrying the same meanings (and triggering the same obligations) wherever they are put to use. Administrative agencies with adequate technical capacity to create immutable mobiles become "centers of calculation" (Latour 1990:59); such centers exert power by gathering together in one place, and then redistributing, a wealth of spatially and temporally disconnected information about the world. In effect, such centers participate not merely in representing the world neutrally but in *making* the world according to their preferred cognitive and political specifications.

Anderson is more concerned with the power of representation to forge political alliances. He uses the term "logoization" to describe how the simple mapping of geographical space can exert the imaginative pull that makes a government in Jakarta feel a distant, if painful, kinship with the rebellious populations of West New Guinea (Irian Jaya) while brutally suppressing the dissenting "outsiders" of East Timor (Anderson 1991:176–178). The exploitative force of symbolic representations is central to Anderson's account. His description of the appropriation of Angkor by successive Cambodian political regimes offers a poignant, and for us a highly relevant, example of a world heritage site that changed its status and meaning in the course of violent ideological change:

> [O]n 9 November 1968, as part of the celebrations commemorating the 15th anniversary of Cambodia's independence, Norodom Sihanouk had a large wood and papier-mâché replica of the great Bayon temple of Angkor displayed in the national sports stadium in Phnom Penh. The replica was exceptionally coarse and crude, but it served its purpose—instant recognizability via a history of colonial-era logoization. "Ah, our Bayon"—but with the memory of French colonial restorers wholly banished. French-reconstructed Angkor Wat, again in "jigsaw" form, became ... the central symbol of the successive flags of Sihanouk's royalist, Lon Nol's militarist, and Pol Pot's Jacobin regimes. (Anderson 1991:183)

"Coarse and crude," the replica of Angkor instrumentally employed by Cambodia's political masters to foment nationalism distorted, or even negated, the more complex historical and cultural identity that the World Heritage Convention seeks to preserve through the force of international consensus.

Standardized assessment techniques and instruments can be produced by entities other than sovereign states, and can be deployed for purposes other than shielding the sovereign's discretionary actions. Work such as Chung Lin Kwa's (1987) on the origins of ecosystems' ecology underscores the importance of shared metaphors, images, and discursive practices, as well as of state support, in bringing together far-flung communities of technical and social actors around shared conceptions of environmental problems. Liora Salter (1993) emphasizes the political economy of such work. Standards used in international trade, she suggests, have changed in two important ways in recent years: they are developed increasingly in advance of the technologies they are meant to regulate, so that they participate in shaping the actual forms of technology; and they represent the work of international, regional, or even bilateral standards organizations without substantial political constituencies of their own.

While these private efforts at standardization may open up new opportunities for participation by consumers of technology, the openings may be illusory, with the first and most expert players determining the gradient of the playing field for all subsequent entrants. In the arena of international environmental standards, for example, agreements about what to measure (e.g., the GWP of greenhouse gases) and how to measure it can entrench themselves before underlying ethical and distributive questions have been addressed by the international community at large. Once clothed in technical language, such decisions lose their transparency and acquire a look of impartial credibility that resists criticism by actors lacking the necessary expertise.

Countering the unifying power of standards, of course, are more relativizing realities of international politics and practice. Manuals and protocols offer only skeletal guidance to the human actors who must carry out their dictates within the constraints and contingencies of their local environments. The messiness of the real world creates discretionary space for the individual's (or the institution's) tacit knowledge and moral or aesthetic sensibilities. A superficial similarity in technological standards can paper over a world of divergent practices. Keith Hawkins (1984:24–27) observed, for instance, that field officers enforcing water pollution standards in Great Britain were required to negotiate with dischargers when the latter's expectations of fairness came into conflict with the efficacy principle built into science-based standards. Farmers and industrialists wanted similar dischargers to be treated alike, with similarly stringent controls; by contrast, the scientific enforcement preferred by water authorities pegged the stringency of controls to the receiving waterways, with the aim of achieving like ambient quality through unlike discharge standards.

Tinkering, adaptation, and even outright avoidance may become the rule when incompletely negotiated and poorly understood standards are imposed on excluded or marginalized groups. When safety technologies malfunctioned at Union Carbide's pesticide plant in Bhopal, India, untrained and inexperienced workers relied on their senses to alert them to releases of toxic, and eventually deadly, methyl isocyanate (Jasanoff 1994). Even instrument readings and sophisticated scientific inscriptions may not generate uniform applications in varying contexts. These phenomena offer some comfort from the standpoint of democratic politics, for they create openings for local action and mitigate the disempowering pressure of global standards and standardizing discourse. But they also point to potential difficulties in securing compliance. Without specific provision for evaluation, feedback, and institutional learning, international regimes may lack the capacity to interpret the meaning of local resistance or to draw from it positive insights into the redesign of standards.

The Comparative Politics of Knowledge

Some fifteen years of research on public health and environmental policy-making have undermined the illusion that scientific discoveries in and of themselves are sufficient to produce convergent policy outcomes across states. In this section, we ask what happens when expert understandings that have already gained some stability—that have, to some degree, been black-boxed—are distributed across disparate policy contexts. To paraphrase Latour, how securely does science itself (or, more properly, a real or notional "universal" community of scientists) serve as a "center of calculation," maintaining its hold on the authoritative interpretation of knowledge? The literature here has focused mainly on cross-national comparisons of scientific assessments in industrial countries. It is therefore both conceptually and methodologically limited for our purposes, but it nevertheless offers suggestive lessons for a study of compliance.

Scientific studies deemed reliable and persuasive in one country are frequently dismissed as inadequate for policy guidance in another, although regulators in both are purportedly responding to the same social, political, and economic demands.

Knowledge claims produced by scientists of one's own country command greater respect than research from abroad, belying the proposition that policy makers regard scientific knowledge as unproblematically universal. Active harmonization is needed to make scientists from different nations accept each other's processes of expert certification, such as testing, replication, peer review, and publication. The embrace of nuclear power in France and its rejection in the United States offers but one example of the fact that informed citizens in one democratic society will discern insupportable risks in a technology assessed as safe by their equally informed and equally democratic counterparts in another (Gillespie et al. 1979; McCrea and Markle 1984; Brickman et al. 1985; Jasanoff 1986, 1987; Hoberg 1990; Abraham 1993).

Some of the differences among countries appear to reflect established and recurrent patterns of linkage between public perceptions of risk and other political and institutional parameters, such as trust, cooperation, and social cohesion. We have already noted the turn toward objective analytic methods—particularly quantification—in countries where access to policy makers is relatively open and relations between citizens and the state are relatively adversarial. Other striking national preferences include Britain's reluctance to use animal data as a basis for assessing human health risks (Germany and the United States offer contrasting cases) and the extraordinary reliance of United States decision makers on mathematical modeling as a basis for environmental standard-setting.

Without detailed historical investigation, one can at best speculate about the reasons for such marked variation in governmental preferences for differing types and quantities of scientific proof. Is it far-fetched, for instance, to see in the modern British predilection for empirical proofs the continuing influence of a shared gentlemanly culture whose members have learned to trust each other's methods of experimental demonstration (Shapin and Schaffer 1985; Shapin 1994)? Does mathematical modeling in America respond to a skeptical democratic culture's need for visual displays of the state's efficacy (Ezrahi 1990), even in the face of irreducibly complex social problems that cannot be modeled through physical experimental systems? The important point is not to resolve these questions but to note that scientific "facts" bearing on policy never take root in a neutral interpretive field; they are always dropped into contexts that have already been structured to respond in culturally distinctive ways to professions of expert knowledge.

Cross-national studies of policy responses to environmental risk likewise threaten any simplistic assumption that a fixed set of social variables—openness, adversarial procedures, bureaucratic or hierarchical organization—completely determines the assessment of science and technology. Studies of policy implementation challenge the sharply dualistic thinking of conventional political analysis, in which *either* policy, by virtue of its technical underpinnings, conditions a special alignment of politics, *or* general features of politics always determine the shape of policy. The boundary between politics and technically grounded policy domains is never so precisely drawn in practice; one cannot readily identify where policy begins and politics ends, or how to orient the connecting arrows of influence. A view more in line with current work on risk, regulation, and political culture would regard policy debates rather as sites where a society's dominant visions of social order, including its faith in scientific institutions and their expertise, are continually under negotiation and reassessment.

Scientific and technical discourse remains in any event one of the most powerful resources available to participants in policy debates. In Latour's (1993) terms, to label something as science or technology is to "purify" it of associations with human agency, to relegate it altogether to the physical or material world, and hence to render it socially untouchable (that is, nonnegotiable). Scientific advisers to governments seem to know this instinctively, because they go about constructing boundaries that protect their spheres of influence from the pressures of ordinary interest-group politics (Jasanoff 1990). But the very domain that is constructed as "scientific" by policy actors varies in shape and scope from one political context to another. A problem that in one country is framed as scientific, and hence is formally investigated and characterized by experts, may be dealt with in another country through the informal, negotiating approaches of politics.

Recent national responses to the environmental risks of biotechnology offer a persuasive study in the comparative politics of knowledge. Debates about the need for new controls occurred at roughly the same time in the United States, Britain, and Germany, but political discourse was structured differently in each country, reflecting different historical experiences and patterns of regulatory behavior (Wright 1994; Gottweis 1995; Jasanoff 1995). The ultimate look of policy instruments and institutions differed across the three countries, with associated differences in national responses to specific genetically engineered products. In the United States, once a policy leader on environmental issues, policy makers generally denied the novelty of risks from genetic engineering and dealt with them through incremental administrative rule-making under existing product safety laws. In Britain, policy makers were more reluctant to accept the safety of genetically modified organisms without further research on their safety and physical containability. Their concerns about the unknown, however, were successfully accommodated within an advisory committee structure that made unprecedented concessions to transparency and participation. Finally, in Germany, a massive outpouring of public concern forced the parliament to consider and adopt comprehensive legislation on genetic engineering; once the law was in place, biotechnology regulation largely reverted to the traditional German mode of rule application by invisible bureaucracies.

Such observations from the comparative politics of risk caution that international regimes may incorporate framings of environmental problems that cannot be taken for granted as universally valid. Continual reevaluation is needed to ensure that regimes do not narrowly replicate the epistemic commitments of particular expert or political cultures.

Epistemic Communities

Although domestic institutions and culture can induce divergences even in highly technical regulatory policies, some scholars in international relations have argued that transnational knowledge-based networks, called *epistemic communities*, can exert countervailing pressure toward policy coordination among states (Haas 1992). According to one current definition, epistemic communities are groups of professionals, not necessarily scientists, who have the following in common:

(1) a shared set of normative and principled beliefs, which provide a value-based rationale for the social action of community members;

(2) shared causal beliefs, which are derived from their analysis of practices leading to or contributing to a central set of problems in their domain . . . ;

(3) shared notions of validity—that is, intersubjective, internally derived criteria for weighing and validating knowledge in the domain of their expertise; and

(4) a common policy enterprise. (Haas 1992:3)

The contribution of epistemic communities has been examined in connection with international cooperation on economics, arms control, and, above all, regional and global environmental issues.

The epistemic communities approach enriches starkly oversimplified rational actor models by taking account of the ideas and beliefs of international actors in addition to their calculation of economic utilities. It recognizes the specially persuasive role of scientific knowledge in regime formation, and it helps explain some cases in which states have apparently chosen to participate in cooperative agreements whose costs exceeded their projected benefits (Haas 1990:186–188). Looking at the behavior of epistemic communities introduces a dynamic element into studies of regimes, enabling one to theorize the social acceptance of new knowledge and the possibilities that this entails for institutional learning and regime change. This literature also provides a vehicle for thinking about the role of nonstate actors in international cooperation, especially in an era when technology makes possible various forms of identity-building and collective action that can overcome the physical and political boundaries of states (Anderson 1994; but see Skolnikoff 1993).

If epistemic communities are a force to reckon with in policy formation, then how should we conceive their role in securing compliance with international regimes? Emanuel Adler suggests that it may be unwise to treat epistemic communities simply as a *cause* of convergence—as just another interest group pressuring for policy coordination (Adler 1991; see also Jasanoff 1996). Instead, Adler suggests that epistemic communities should be studied as agents of "cognitive evolution," both forming and formed by international learning and scientific progress. This view of epistemic communities accords well, as suggested in preceding sections, with constructivist work on science and politics that connects the production of particular epistemic ideas about the environment (e.g., cause–effect relationships) to particular forms of environmental politics. This approach suggests that in international regimes as well as in national politics, some institutional features, such as transparency, may be more conducive than others to the formation of effective epistemic communities.

As vehicles for translating knowledge and values across national boundaries, epistemic communities may be well positioned to correct for institutional deficits across countries, to adapt uniform norms or standards to local settings, and to take advantage of local capacity and local opportunity structures. A good example of such constructive role-playing occurs in the chapter on Russia in this volume (chapter 9). Greenpeace, an ingenious and activist environmental NGO, aroused international opinion and prodded Russian officialdom to action through the quasi-scientific strategy of mapping and disclosing the extent of Russia's burial of radioactive wastes in the northern seas. The Greenpeace report was eventually taken up almost unchanged as an official document by President Yeltsin's environmental counselor.

Epistemic communities, however, may serve a less benign function when they extend the work of an unaccountable central directorate with an unexamined scientific worldview, thus helping to perpetuate a hegemonic system. India's implementation of the World Heritage Convention provides an example of this form of interaction (see chapter 12 in this volume). At the Bharatpur WHC site, a well-known bird sanctuary, a narrowly biological understanding of preservation, applied without thought to local circumstances, dictated that grazing domestic buffaloes should be eliminated because they were not part of the area's "natural" ecology. This application of a taken-for-granted scientific norm (domestic buffaloes are not part of nature) not only led to conflicts between park authorities and local graziers but also undercut the WHC mission by changing the now ungrazed wetland into a grassland unsuitable as a habitat for wildfowl.

Indian officials also have criticized the WHC for promoting a Eurocentric bias in the designation of cultural sites, which are selected for their global rather than their local significance. Given its opaqueness and attempted universalism, it is perhaps not coincidental that the WHC is among the least transparent and most static of the regimes studied in this project. We will return to the significance of these points at the end of the chapter.

International Institutions

Interest in international institutions as producers, custodians, and disseminators of scientific information has grown along with concerns about the sustainability of the global environment. In their 1993 survey, Peter Haas, Robert Keohane, and Marc Levy urged an expanded role for environmental institutions in knowledge creation and the "universal circulation of information"—an idea that resonates interestingly with Latour's "centers of calculation." Science emanating from such bodies, the authors argue, could "be nonpartisan and untainted by national concerns, to offset suspicions that monitoring activities constitute political control by another means, or are a disingenuous way to promote the economic advantages of selected groups" (1993:411). The prestigious Carnegie Commission on Science, Technology, and Government (1992) proposed the creation of an international Consultative Group for Research on the Environment to identify and foster opportunities for collaborative, cross-national environmental research. The internationalization of scientific assessment under the aegis of specially constituted expert bodies has emerged as a favored implementation technique under a variety of recent accords, from the Montreal Protocol evaluated in this volume (see chapter 5) to the 1992 Rio conventions.

Persuasion and Participation

Although international organizations have been producing scientific knowledge for many years, there is relatively little systematic literature on the sources of their cognitive authority or their relationships with domestic centers of knowledge. One way for these bodies to produce authoritative knowledge would be to involve and engage all relevant parties, without regard to formal technical qualifications. There are indications, however, that the processes used by supranational bodies to stabilize or black-box knowledge may be quite as opaque and closed to participation or critique as the processes by which national expert institutions ("centers of calculation") often achieve and maintain epistemic dominance.

The strategy of public spectacle or miracle, which Ezrahi (1990) associates with premodern and predemocratic "celebratory" political cultures, can be employed to good effect at the international as well as the national level. The World Health Organization, for example, gained enormous prestige from its successful efforts in the 1960s and 1970s to eradicate smallpox worldwide, and this extraordinary demonstration of efficacy is still routinely invoked to underwrite its more controversial ventures into global public-health policy. Charismatic leadership, similarly, accounts for some of the success achieved by the United Nations Environment

Programme (UNEP) in the knowledge production that laid the basis for the Montreal Protocol on ozone depletion. Mostafa Tolba, then executive director of UNEP, played a central role in forging a transnational consensus of knowledge and values. Richard Benedick, the chief United States negotiator, describes (1990:72) how Tolba's championship of a scientific consensus on chlorofluorocarbons (CFCs) (for example, Tolba's remark at the 1987 Geneva meeting, that "no longer can those who oppose action to regulate CFC releases hide behind scientific dissent") coincided with his careful, behind-the-scenes building of political consensus through closed meetings of key delegation heads. The diplomatic achievement was gained, in part, through exclusion of potentially difficult players. Benedick observes, without apparent irony, "The absence of any developing nation symbolized the South's lack of interest in the details of the control measures; Tolba himself served, in effect, as representative of the developing world."

In the case of the Mediterranean Action Plan (Med Plan), Haas (1990:377–403) suggests that the UNEP built a convergence of scientific worldviews through top-down and unaccountable bureaucratic techniques: training a cadre of marine pollution experts committed to an ecological paradigm; distributing research funds; and producing, by rhetorical and other means, a vision of holistic environmental management.

Haas's "thick description" (Geertz 1983) of the Med Plan centers on the elaborate work that UNEP had to do to hold its ecological epistemic community to a common policy enterprise. It was an unruly group that UNEP needed to pull together: "Many of the individual marine scientists and officials of specific organizations had different views about the nature of the problem of Mediterranean pollution and the appropriate remedies, reflecting their various backgrounds and expertise in disciplines such as marine biology, marine chemistry, marine geology, oceanography, microbiology, public health, and civil engineering." Only by seeking broad umbrella definitions of key terms like "pollution" could UNEP "[blur] the distinctions between otherwise incompatible views" and cater to this varied constituency. Overall policy agreement was maintained in this case by leaving room for locally variant interpretations of centrally articulated, but flexible, scientific concepts.

Knowledge Beyond Borders

Opportunities for negotiation and local adjustment are severely curtailed when the interpretation of scientific knowledge is committed to authoritative, and usually insulated (Wirth 1994), international expert bodies—as, for example, under article XX of the General Agreement on Tariffs and Trade (GATT). The 1989 United States–Canada salmon and herring controversy displayed such an adjudicatory mechanism in operation. The case exemplifies how international expert committees perform complex tasks of sociopolitical legitimation and reassurance under the rubric of evaluating science. They behave in this respect very much like similar bodies constituted within states (Jasanoff 1990). In the international context, however, opportunities to question the assumptions of such institutions are generally more limited than in states with strong traditions of public participation (Hurrell and Kingsbury 1992). The GATT prototype of expert decision-making, in particular, represents a top-down approach to securing cognitive authority that contrasts sharply with the Med Plan's more dispersed model of knowledge-creation and consensus-building.

The fisheries dispute that concerns us here arose under article XX(g) of the GATT, as incorporated by reference into the Canada–United States Free Trade Agreement (FTA). Article XX (General Exceptions) sets out two possible environmental justifications for violating GATT by not accepting international standards under the Agreement on Technical Barriers to Trade (Standards Code):

> Subject to the requirement that such measures are not applied in a manner which would constitute a means of arbitrary or unjustifiable discrimination between countries where the same conditions prevail, or a disguised restriction on international trade, nothing in this Agreement shall be construed to prevent the adoption or enforcement by any contracting party of measures:
>
> . . .
>
> (b) necessary to protect human, animal or plant life or health;
>
> . . .
>
> (g) relating to the conservation of exhaustible natural resources if such measures are made effective in conjunction with restrictions on domestic production or consumption.

In April 1989, Canada introduced a new regulation under its Fisheries Act requiring that salmon and roe herring caught in Canadian waters had to be landed, for sampling purposes, either at a licensed fish landing station in British Columbia or onto a vessel bound for such a destination. The United States argued that this regulation was an illegal export restriction that was not justified under the article XX(g) exception because it did not serve primarily as a conservation measure. Canada argued, to the contrary, that the landing requirement did serve the valid conservation purpose of obtaining high-quality data on the commercially important Pacific salmon and herring fisheries. Like United States courts conducting substantive judicial review of administrative regulations (Jasanoff 1990), the bilateral panel appointed to resolve the dispute looked behind Canada's stated reasons for adopting the landing requirement and made an independent assessment of its validity. The panel's task was to construct an authoritative, independent reading of the evidence, which in this instance conclusively supported the United States position.

The science under scrutiny by the FTA panel was developed and carried out within the framework of international politics. Novel issues that the panel confronted were the following (panel's rulings are indicated in parentheses): Could Canada reasonably be obliged to rely on United States scientific cooperation for data collection (no); were there less restrictive measures for meeting Canada's data needs (yes); was it necessary to monitor 100 percent of the catch in order to have data of adequate quality (no); was there reason to believe that, in the absence of the landing requirement, an overly large or unrepresentative part of the fish populations would escape sampling, thereby yielding misleading results (no, based on existing analysis)?

Objective determinations of what constituted reliable data and valid conservation methods for fisheries management could not be disentangled in the FTA decision-making process from underlying political and institutional issues. Thus, the panel could not rule simply on whether it was scientifically necessary to sample 100 percent of the catch. It had to decide whether this sampling regime made sense in light of the background political assumption that sovereign states cannot be asked to rely wholly on each other's data collection systems. In ratifying the scientific sampling method proposed by the United States, the panel was therefore invisibly shoring up the political authority of this particular sovereign state to do science in its own ways—in effect, to produce nationally contextualized knowledge without having to bow to transnational "peer review" by Canadian authorities. Lines of scientific autonomy and political power were simultaneously reaffirmed in a proceeding that was cast as a straightforward exercise in expert judgment. Yet, because the panel was securely established as an expert body under the FTA's ground rules, it could adjudicate a technopolitical debate without disclosing its underlying political judgments or endangering its cognitive authority.

Competing Meanings, Dissenting Voices

The disputing nations in the salmon–herring controversy were united by longstanding cultural and political ties, including a shared scientific discourse, a common language, and an inherited legal tradition. Convergent economic interests had bound them to a long history of bilateral environmental negotiation. These background conditions help to account for the FTA panel's success in reaching closure on scientific issues with a potentially explosive political dimension. Because the matter before the panel looked sufficiently scientific, the question of relative power, which was, and remains, an important issue in United States–Canadian relations, could be bypassed in silence. In situations where cultural gaps loom larger and power is still less evenly divided, as in the treaty regimes examined in this volume, competing technopolitical understandings may not be so easily mediated. Consensus, if achieved, may mean no more than the adoption of a reductionist or hegemonic cultural position in the guise of applying "universal" science.

Recent writing by historians and sociologists of science, as well as by contemporary social theorists concerned with modernity (Latour 1993; Giddens 1990; Beck 1992) confirms that science offers an especially powerful discursive and institutional framework for creating globally convergent understandings about environmental problems. At the same time, as we have seen above, the ways in which science draws boundaries between nature and culture—their separateness and their interconnections—inevitably reflect a series of prior stabilizations of both the epistemological and the social domains.

In her influential essays on primatology, for example the cultural theorist Donna Haraway (1989) charts the confluence of gender, race, and economic ideology in constructing nineteenth-century museum representations of primate behavior. But it is not just the expected "bad guys" of Haraway's narrative—male scientists, market entrepreneurs, and Western notions of progress—that participate in the creation and projection of dominant cultural paradigms. Environmental groups, for instance, have now joined in the discourse of science to globalize their private, normative understandings about the environment (Yearley 1992, 1996). The biases and worldviews of powerful NGOs such as Greenpeace, no less than those of powerful states, can affect the production and implementation of international scientific and technical norms. To write an adequate history of the evolution of international regimes, such as the narrowing of CITES to an excessive absorption with "charismatic megafauna" (see chapter 5 in this volume), one must therefore interweave, as this volume does, the classical narratives of economic and political dominance with less extensively documented themes of local resistance, indigenous conflict, and transnational social movements (see chapters 12 and 13 in this volume).

It should not surprise us, finally, if attempts to characterize, and prevent, what human beings see as threats to their environment evoke the most densely layered and contradictory meanings that can be devised by the interpretive genius of human societies. There has never been uniformity, after all, in the things that societies are prepared to write off as being outside culture, morally worthless, economically wasteful, and dangerous to established order (Douglas 1966). Even the "varmints" of the western environment (Worster 1977:258–290) are products of historically and culturally situated knowledges and understandings. International norm-building, we may conclude, is likely to proceed most fruitfully when diverse rationales for environmental action can be accommodated under a common umbrella—for instance, under the Med Plan's flexible definition of pollution or the definition of hazardous wastes under European laws. Such arrangements permit treaties to become ongoing sites for negotiation and, in Adler's terms, "cognitive evolution"—where epistemic communities may gradually come into being as seedlings of a wider civil society.

Institutional Responses to Contingency

Importing the literature of science and technology studies into a discussion of implementation and compliance directs our attention with renewed intensity toward the institutional and procedural dimensions of international regimes, to the places where words and ideas are translated into action. Stability in knowledge or its technological applications, we learn, is a condition that does not happen of itself, through nature's unmediated agency; human communities must work hard over time to achieve and maintain it. Both science and technology are constructs, in this sense, of collective human activity. Knowledge claims and artifacts not only are contingent upon particular ways of seeing and knowing, upon culture and history, but also are capable of imparting a misleading veneer of homogeneity to disparate understandings of nature, artifacts, and society. As constitutive elements of environmental accords, terms like "sustainability," "carrying capacity," "precautionary principle," "pollution," and "unreasonable risk" convey to many a reassuring sense that regimes are built on the impersonal and culture-blind authority of science. We have seen, to the contrary, that at their best such concepts serve to create focal points for negotiation and cognitive convergence; at their worst, they remove the power of decision-making to remote "centers of calculation" and promote, among those who are marginalized, resistance, apathy, or a pervasive distrust (Wynne 1987; Beck 1992; see also Putnam 1993).

How can regimes be structured so as to encourage the former dynamic and inhibit the latter? This question becomes especially urgent in view of the theme of secular improvement that so forcefully emerges from this volume's survey of five environmental accords. For treaties change. The London Convention, as we see in chapter 5, evolved from a primary focus on ocean dumping toward integrated waste management, based in part on land. CITES arguably is moving from a preoccupation with endangered species toward a broader concern for biodiversity. These changes, moreover, are effective only when they secure the assent, and thus alter the behavior, of the enormously disparate parties whose behavior they must control, from a few large, mainly Northern, corporations under the Montreal Protocol to countless individual poachers, smugglers, and consumers around the world under CITES. Mere policing cannot guarantee the success of command-and-control regimes whose mandates do not correspond to the life circumstances of affected groups. Whole nations may default on treaty obligations when overwhelmed by economic and political crisis (see chapter 13 in this volume).

Our observations of science and expertise in the policy domain suggest that transparency is one part of the answer to questions of regime design, but it must be a transparency that reaches deeper than the offices and files of a regime's central secretariat. Access is needed, and continually, to the cognitive domains where issues are framed, agendas set, administrative approaches designed, and solutions or standards formulated. Treaties, like more material technologies, are forms of life that function best when mapped onto political and cultural structures that can effortlessly accommodate them, both cognitively and administratively. A treaty that relies heavily on technical standards, for example, will work best in settings where the capacity to set standards and implement them is already in place: where people's imagination already encompasses the idea of standards and what it means to enforce them. The ozone accord initially achieved fairly high levels of compliance in India (see chapter 12 in this volume) because, fortuitously in this case, India's centrally planned economy provided the infrastructure to support just such center-driven regulatory initiatives. Other examples of good and bad fits between prescriptions and institutions are scattered throughout the chapters that follow.

The contingency of knowledge about the environment leads us to recognize that implementation and compliance are not mere matters of imposing known truths on recalcitrant or dissident actors. Rather, failures of compliance can be thought of as windows onto mismatches in perception: between the rulers and the ruled, between the metropole and the periphery, between experts and laypeople. Mismatches can arise even within a shared cognitive framing of the problem addressed by the environmental regime (for example, divergent risk assessments of the same hazard in similarly situated countries), or they may reflect more fundamental moral and ideological separations (for example, about what constitutes a sustainable ecology). This perspective suggests caution in the mechanisms designed to achieve compliance, as well as humility in the assessment of breakdowns. The arguments in this chapter point on the whole toward a preference for compliance systems that leave room for negotiation among competing views of the world—a negotiated and networked compliance, promoting a horizontal politics (Putnam 1993) of the global environment, over a compliance that is too firmly imposed and monitored from above.

Contingency, however, is only half of the story told in this chapter. Equally important is the conclusion that, in spite of all indeterminacy and uncertainty, knowledge and social order are not perceived by human societies to be fluid at all points. Both can be made to hold still through institutions, material technologies, and shared norms or practices. The stabilizations brought about by international technical standards and transnational epistemic communities are particularly significant for environmental regimes. But the preceding analysis can be seen in effect as a mirror that reverses the conventional images of both standards and epistemic communities. It shows them more as indicators than as causes of evolution toward shared international norms, as dependent rather than independent variables in accounting for change.

It follows, then, that for analysts and designers of international regimes, the desired end point is not to devise decontextualized technical standards nor to create the legal machinery to ratify the views of some dominant epistemic community. It is the far more subtle and challenging task of creating the background conditions under which participants will see the need to standardize their own actions in order to protect the environment, and thus will come to build global communities of shared mission and belief.

Notes

I would like to thank Saul Halfon for helpful research and review of background literature at the early stages of this project, as well as the editors and chapter authors of this volume for their insights, encouragement, and critical advice.

5

The Five International Treaties: A Living History

Edith Brown Weiss

Since 1972, a new field of international law has emerged: international environmental law. The number of international legal instruments, both binding and non-binding or incompletely binding, has soared (Brown Weiss et al. 1992). Although it is tempting to view these instruments as static, they have in fact changed over time, just as political, economic, social, and technological conditions have changed over time. Identifying the patterns of change is essential to assessing national compliance and to understanding the factors that affect it (see Bilder 1981).

This study covers five global international treaties: the World Heritage Convention, the Convention on International Trade in Endangered Species, the London Convention of 1972 (formerly London Ocean Dumping Convention), the 1983 International Tropical Timber Agreement, and the Montreal Protocol on Substances That Deplete the Ozone Layer. The first three were directly linked to preparations for the 1972 United Nations Conference on the Human Environment in Stockholm, Sweden.

The treaties cover a range of environmental and natural resource issues: protection of the globally shared high-level ozone layer, protection against dumping of wastes in marine regions, and protection of natural and cultural resources located within national boundaries, such as World Heritage Sites, endangered species, and tropical timber. One of the Conventions, the International Tropical Timber Agreement (ITTA), is a commodity treaty that contains several increasingly important environmental provisions. The ITTA successor treaty, which went into effect on January 1, 1997, is far more concerned with ensuring sustainable forestry management than was its predecessor.

All the countries under study belong to nearly all of the treaties, as shown in table 1.3 in chapter 1. Two of the agreements continue to attract parties every year although they were concluded in the early 1970s: the World Heritage Convention and the Convention on International Trade in Endangered Species. In two of the agreements, the number of parties has increased marginally since the initial period for joining the convention: the London Convention of 1972 and the International Tropical Timber Agreement of 1983. Only the Montreal Protocol attracted early

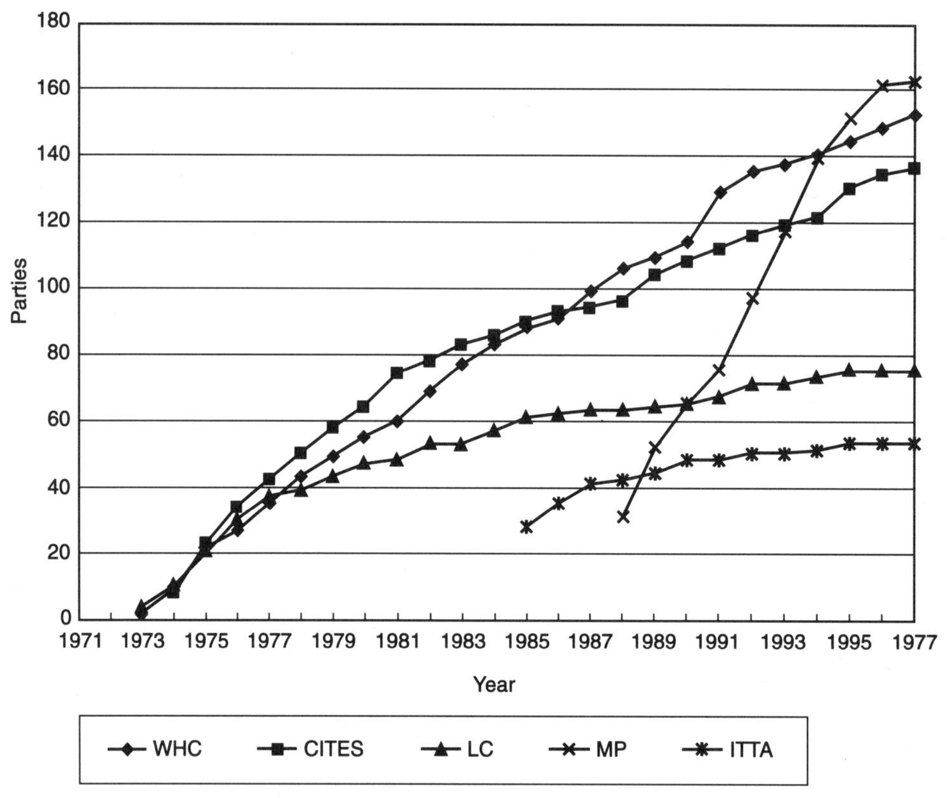

Figure 5.1
Total number of parties to treaties

widespread adherence. Figure 5.1 provides a comparative analysis of the rate of treaty ratification or accession.

The structures of the treaties vary. All but one provide lists of resources that must be conserved or pollutants that must be controlled. Table 5.1 compares the treaties on eighteen points.

The obligations in the five treaties vary from very specific commitments to phase out certain chemicals, prohibit trade in specified endangered species, and ban dumping of identified polluting materials in the oceans, to general commitments to conserve heritage sites and encourage sustainable management of tropical forests. Table 5.2 provides a comparative look at the most important substantive and procedural obligations in the treaties. These obligations formed the basis of the compliance work in the country chapters.

All of the treaties have a "living" history, in the sense that they have evolved over time. Parties have sometimes added new obligations, assumed new roles or

Table 5.1
Characteristics of the five treaties

	WHC	CITES	ITTA	LC	MP
Signed	1972	1973	1983	1972	1987
Entered into force	1977	1975	1985	1975	1989
Parties, July 1997	149	137	53	72	162
Secretariat affiliation	UNESCO	UNEP	none[a]	IMO	UNEP
1994 budget	\$2.9m	\$3.7m	\$3.9m	\$0.8m	\$3.1m
1994 staff	22	25	21	4	11[b]
Meeting of parties	annual	2 yrs.	6 mos.	annual	annual
Division of parties	no	no	producers consumers	no	developing countries
Provisions re major parties	no	no	yes	no	yes
NGOs formal partners	yes	yes	no	no	no
Dispute resolution	no	yes	no	no[c]	yes
Third-party trade control	no	yes	no	no	yes
Formal scientific committees	yes[d]	no	no	yes	yes
Special fund	yes	no	no[e]	no	yes
Periodic reports	yes[f]	yes	yes	yes	yes
On-site monitoring	yes	yes	yes	no	no[g]
Training	yes	yes	yes	no	yes (MP Fund)
Newsletter	yes	yes (USA)	yes	no	yes (*Ozone Action*)

[a] Agreement negotiated under the auspices of UNCTAD.
[b] Does not includes Montreal Protocol Fund staff of eighteen.
[c] The 1996 protocol provides for dispute resolution.
[d] Three designated outside organizations.
[e] Successor agreement provides for special fund.
[f] Agreement provides for periodic reports, but these are not completed in practice.
[g] On-site monitoring under Montreal Protocol Fund for countries receiving assistance.

Table 5.2
Treaty obligations of parties

	WHC	CITES	ITTA	LC	MP
Substantive					
General conservation obligation within own state	Ö		Ö	Ö	
General conservation obligation to protect resources beyond own state	Ö			Ö	
All states to protect/control items on international list	Ö	Ö		Ö	Ö
Restrictions on specific species or pollutants		Ö		Ö	Ö
No trade with nonparties, unless in compliance with treaty		Ö			Ö
Procedural					
National export/import permits		Ö		Ö	Ö[c]
National management authorities	Ö	Ö		Ö	Ö[a]
Penalties for national violations		Ö		Ö	Ö
National annual reports	Ö[b]	Ö	Ö	Ö	Ö
Required contribution to fund	Ö				Ö

[a] Article V developing countries receiving assistance from Montreal Protocol Fund must designate a National Ozone Officer.
[b] Agreement provides for reports, but these are not completed in practice.
[c] 1997 Amendment provides for national licensing systems for imports/exports of controlled substances.

assigned new functions to secretariats (or implicitly accepted their assumption of them), developed new implementation and monitoring activities, permitted nongovernmental organizations to become more heavily involved in implementing the treaties, and dispersed more information worldwide about implementation and compliance measures. Nearly all changes have been done without formal treaty amendment, often through guidelines, annexes, or decisions of the parties.

For purposes of analysis, the next two parts group the treaties into those primarily concerned with natural resources and those concerned with pollution. The final part sets forth some comparative observations about the development of the treaties from the period of negotiation through December 1996. The data reveal important common changes among the regimes in roughly the same time periods. These are then linked with the country analyses of compliance.

The Natural Resource Treaties

The earliest international environmental accords addressed conservation of commercially valuable fauna and of wildlife areas. These include the 1902 Convention

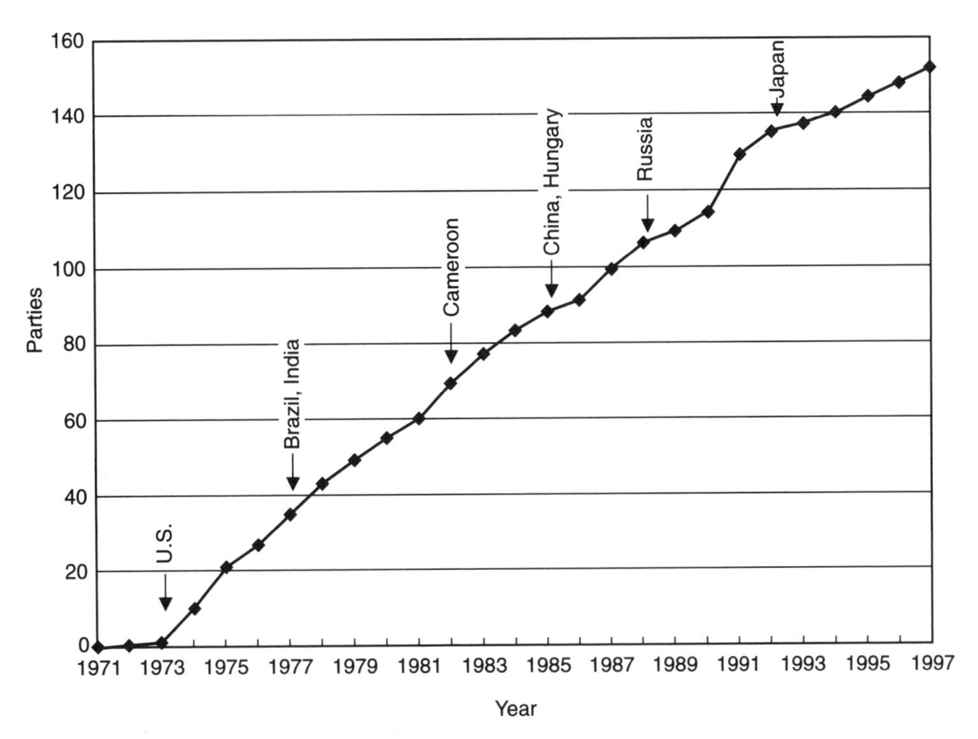

Figure 5.2
Total parties to World Heritage Convention, and ratification years for countries in this study

for the Protection of Birds Useful to Agriculture, the 1911 Treaty for the Preservation and Protection of Fur Seals, the 1940 Washington Convention on Nature Protection and Wildlife Preservation in the Western Hemisphere, and the 1946 Washington International Convention for the Regulation of Whaling. The three natural resource treaties in this study address natural resources located within states' sovereign territories and are global in scope: natural and cultural heritage sites, international trade in endangered species, and tropical timber. The treaties are addressed in the chronological order in which they were concluded.

The World Heritage Convention

The 1972 Convention for the Protection of the World Cultural and Natural Heritage provides that certain natural and cultural sites can be designated as World Heritage Sites, and conserved for present and future generations. As of June 1997, 149 countries were parties to the Convention; countries continue to ratify the Convention each year. Figure 5.2 shows the progress of ratification over time.

The World Heritage Convention was drafted at the same time as countries prepared for the 1972 Stockholm Conference. It derives from two separate initiatives: one to conserve cultural properties, and the other to preserve natural properties. After the destruction of priceless buildings and other structures in World War II, countries were determined to conserve important cultural properties for the future. In 1954 countries concluded the Hague Convention on the Protection of Cultural Property in the Event of Armed Conflict, which was followed by campaigns under the auspices of the United Nations Educational, Scientific and Cultural Organization (UNESCO) to save specific cultural sites, and in 1957 by the establishment of an international nongovernmental organization dedicated to conserving cultural properties—the International Council on Monuments and Sites (ICOMOS). In the 1960s UNESCO authorized the negotiation of a new international treaty to protect cultural properties; a draft text was ready by spring 1971 (Connally 1985).

At the same time the International Union for the Conservation of Nature and Natural Resources (IUCN) prepared a draft convention to protect national parks, some historic structures, certain natural sites, and important wildlife areas. The sites were to become part of a World Heritage Trust. The United States—Russell Train in particular—was instrumental in encouraging the creation of such a trust and a new convention. After considerable diplomatic exchange, the two drafts were combined in 1972 into a single convention that was opened for signature in November of that year and went into effect in 1977.

Countries are attracted to the Convention for different reasons. All countries attach some prestige to having sites on the World Heritage List. Developed countries see the treaty primarily as a way to preserve sites. Industrializing countries view it as a useful leverage for economic development through tourism and as a means to bring in foreign exchange. Moreover, some countries receive modest sums from the World Heritage Fund to help preserve the sites (see Nelson and Alder 1992).

Treaty Commitments and Implementing Structure. The centerpiece of the Convention is the World Heritage List. Individual countries nominate national natural and cultural sites to the List. The parties to the Convention decide whether to inscribe the sites nominated. As provided in the Convention, nongovernmental organizations prepare the background investigations and reports as to whether the proposed sites merit inclusion on the World Heritage List.

Under the Convention, states are obligated to protect their sites, and to refrain from actions damaging the sites of other countries. If sites on the World Heritage List become endangered, they may be put on the List of World Heritage Sites in Danger.

The Convention provides for three implementing institutions: the General Assembly, the Committee, and the Bureau. The powers of each are in inverse proportion to the number of members. The General Assembly, which includes all parties and meets biennially, has very limited powers. Its main task is to elect members of the World Heritage Committee, which consists of twenty-one member states serving six-year terms. The Committee, which meets annually (usually in December), approves the inclusion of sites on the World Heritage List; changes in the Operational Guidelines, including changes in the listing criteria; and programmatic developments, such as monitoring and conservation of sites. It has always operated by consensus. The Bureau, composed of the chairman of the Committee, five vice-chairs, and the rapporteur, is elected by members of the Committee. Members reflect the geographical distribution of states party to the Convention, which is the standard practice of UNESCO, the Convention's host organization. The Bureau, which meets twice a year, prepares the agenda for the upcoming Committee meetings, deals with emergency assistance requests and other problems, and serves as an important forum for considering new initiatives, such as the Strategic Plan in 1992.

Treaty Implementation. Since the Convention came into effect, states have expanded its scope and developed new implementing measures, all of which have been accomplished not through amendment but through changes to the Operational Guidelines for the Implementation of the World Heritage Convention. At the Bureau meeting in July 1995, parties recommended additional revisions to recognize the importance of local populations to inscribing sites and protecting them, to include technological heritage in the inscription criteria for cultural properties, to develop an annex to the Guidelines that would define particular terms (such as "canals" and "cultural routes"), and to clarify the role of consultative bodies in evaluating natural and cultural properties.

Until 1992, parties were largely concerned with whether to put nominated sites on the World Heritage List. Since 1992, the focus has shifted to monitoring the status of sites on the World Heritage List, to raising public awareness of the Convention, and to strengthening field management of sites. These trends are likely to continue.

As of December 1997, there were 552 properties on the World Heritage List in more than 100 countries. The list included 418 cultural sites, 114 natural sites, and 20 mixed cultural and natural sites. Between 1993 and December 1996, the World Heritage Committee added more than two dozen sites each year, with forty-six added in 1997. Figure 5.3 shows the types of sites listed in the countries in this study.

The number of sites added per year has generally increased since 1989, with the growth in cultural sites significantly exceeding that of natural sites. (See figures 5.4 and 5.5.)

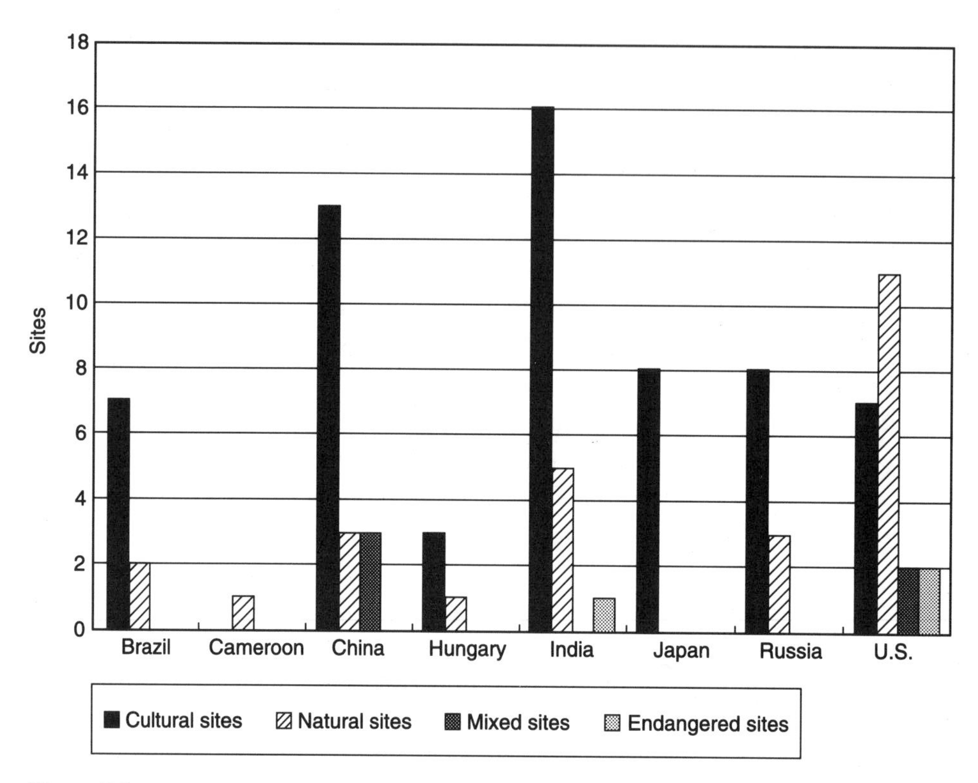

Figure 5.3
World Heritage Sites, by type listed, in countries in this study, as of December 1997

In 1993 the parties added cultural landscapes to the categories of properties that could be listed as World Heritage Sites. They also adopted new criteria for the listing of natural sites.

The problem of listing has changed over time. Initially there were many unique sites to put on the World Heritage List. Later, there were concerns about duplicating existing sites on the list, even if they are in different countries. More important, there was disagreement regarding the philosophy of placing sites on the list. Was it more effective to require conformity with strict conservation conditions before listing a site, on the theory that the carrot of approval as a World Heritage Site was the most effective instrument for ensuring that steps were taken to conserve the site? Or was it better to be more flexible in listing sites, on the basis that inclusion on the World Heritage List would encourage better conservation of the site? Not surprisingly, the nongovernmental organizations concerned with natural sites favored the former approach, whereas the Secretariat and parties adopted the latter, as shown in the greater number of sites added to the List each year (see figure 5.4). Moreover, a certain politicization of the process developed, in that countries joining the Con-

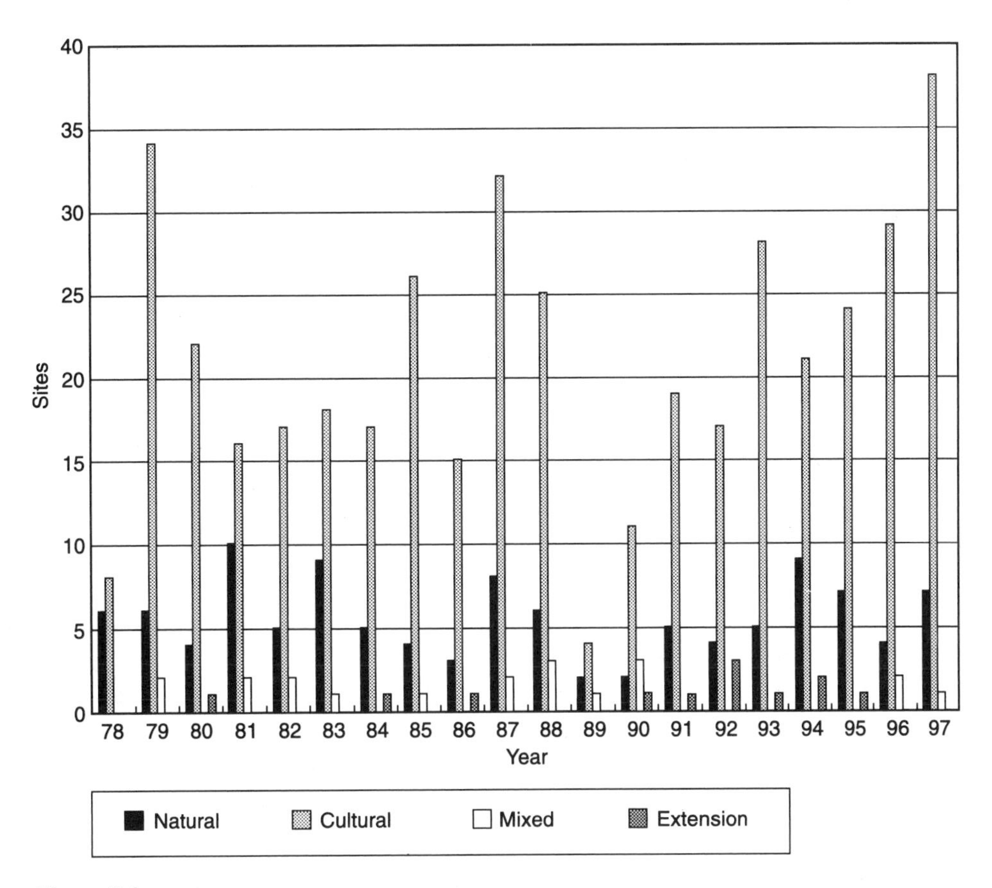

Figure 5.4
Annual additions to the World Heritage List, by type of site, 1978–1997

vention were generally deemed to be entitled to have one site on the list, even if by other standards the site might not fully qualify as a World Heritage Site.

There were new issues regarding whether sites should be reviewed periodically to determine eligibility to remain on the List. The IUCN adviser's proposal for a twenty-year sunset clause, in which sites would be removed from the List unless it were affirmatively shown that they continued to qualify, was not favorably received. Some argued it would discourage countries from putting sites on the List. Just as important, member states may not have wanted to be forced to judge regularly whether other countries' sites still qualified for the World Heritage List. Such judgments could have important political ramifications for member states' relations with each other.

Under the Convention, a threatened site can be put on the Danger List, which calls attention to its plight and makes the country eligible to receive support from the World Heritage Fund. The number of sites added to the Danger List each year has remained about constant, except for 1992, when many were added. As of December

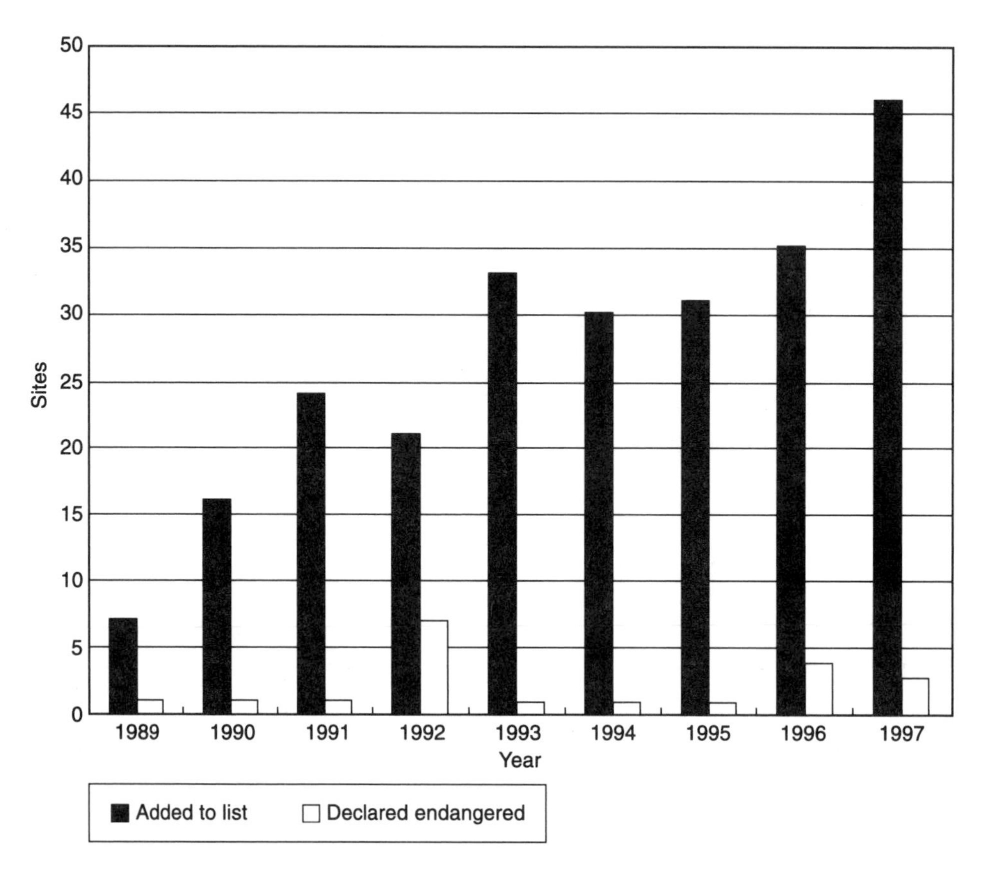

Figure 5.5
Annual additions to the World Heritage List and sites declared endangered, 1989–1997

1997, there were twenty World Heritage Sites on the Danger List, eight of them ecological sites (see figure 5.5). Although countries could always ask that a World Heritage Site within their borders be placed on the Danger List, only since 1991 have member states taken it upon themselves to put a site on the List against the wishes of the host country. In 1991, members unilaterally inscribed Dubrovnik in Croatia on the Endangered List, contrary to the wishes of the host country, and in 1992 unilaterally put India's Manas National Park there. In 1993, the World Heritage Committee placed the Everglades National Park in the United States on the Danger List, even though the United States had not requested this and abstained from the voting. Although this assertion of authority by the parties was initially controversial, it seems to have become accepted. The text of the Convention is arguably ambiguous on the point.

The outbreak of hostilities in the former Yugoslavia brought new attention to the problem of protecting World Heritage Sites during civil war or strife. In 1991 and 1992, the Secretariat and Bureau members devoted considerable time to this problem. Consultant visits to Dubrovnik to try to prevent damage to the World Heritage Site were unsuccessful. Similar problems arose regarding the protection of a natural site and gorillas during the Rwandan conflict.

Since 1990, the issue of whether and how to take a site off the List has arisen. No site has ever been taken off the List, although Germany, Thailand, and the United States in December 1993 indicated that this step should be considered for Manas National Park. Mount Nimba Nature Reserve in Guinea Côte d'Ivoire, which was put on the Danger List in 1992, poses the issue dramatically, because it has become a promising source of minerals since its inscription on the World Heritage List. At its 1994 meeting, the World Heritage Committee warned that India's Keolodo National Park (famed for its cranes) may need to be delisted if no cranes arrive.

The World Heritage Convention does not address the issue of delisting, but the Operational Guidelines provide detailed procedures that are intended to ensure that all possible measures are taken to prevent deletion but to authorize the World Heritage Committee to delist if these fail. Since there has been no case of delisting, there may still be controversy over whether delisting can take place over the host state's opposition and whether a site should be delisted at a host country's request.

In June 1992, the Bureau first considered a strategic plan for implementing the Convention. The Strategic Guidelines adopted in December 1992 emphasized monitoring and compliance, and called for promoting greater public awareness of the Convention. Since then, states have devoted more attention to monitoring World Heritage Sites, as detailed later in this chapter.

At the 1994 World Heritage Committee meeting, parties formally decided to make monitoring a more central component of the Convention. In February 1995, a new section was added to the Operational Guidelines on Monitoring (articles 69–75). It requests states to put on-site monitoring arrangements in place and "invites" member countries to provide monitoring reports to the World Heritage Committee (through the World Heritage Centre) on the conditions at their sites every five years. In 1997, it is too early to know whether countries will oblige. For specific sites that are threatened, the Guidelines call for what is termed "reactive monitoring": special reports and inquiries about the site. The World Heritage Committee's contribution to UNESCO's Medium-Term Plan for 1996–2001 includes monitoring of sites as a key element. The Secretariat has presented plans for regional offices of the World Heritage Centre that would assist in monitoring.

Starting in 1987, there was a marked shift in the activities initiated by the Secretariat, pursuant to the Convention, to emphasize training and technical assistance and, since 1993, emergency assistance. Whereas in 1987 there was only one training project, the number had risen dramatically by 1991 to twenty-one projects and, in

1992, to thirty-two projects. Technical cooperation projects numbered eleven in 1987; in 1991, sixteen; and in 1992, eighteen. Between 1987 and 1993, there was a limited number of emergency assistance projects. At the Committee meeting in December 1993, countries established an Emergency Assistance Fund, more than 60 percent of which had been spent by the time of the next Committee meeting in December 1994. Thus, over time there has been a sharp increase in projects devoted to building local capacity to comply with the treaty.

The World Heritage Centre also has launched a program with the Education Sector of UNESCO that is directed to educating young people about world heritage conservation through the development of "world heritage teaching and learning kits" for secondary schools and through international world heritage youth forums. The first was held at Bergen, Norway, in 1995.

Since the World Heritage Centre was formed in 1992, there have been new, innovative efforts to tap the private sector for funds to support activities of the World Heritage Convention. The Centre sells CD-ROMs, books, and films, and has been exploring other fund-raising activities. Although there was agreement to pursue possibilities of private-sector support at the World Heritage Committee meeting at Santa Fe, New Mexico, in 1992, the strategy has been controversial, as shown by parties' comments at the Committee meeting two years later.

Since 1990, the World Heritage Centre has made new efforts to make the public aware of the World Heritage Convention. By the summer of 1992, it had developed a CD of world heritage sites. In 1993, the Centre inaugurated a quarterly newsletter that provides information on the latest developments. It also produces films, which are available for sale.

The Centre is linked to the Internet and to the World Wide Web through an Internet "gopher" initiated by UNESCO. Through this, the World Heritage List, Operational Guidelines, and reports of experts' meetings are available on Internet. The reports from the 1993–1997 World Heritage Committee meetings have been available on-line, as has daily coverage of the 1992–1997 intergovernmental meetings. The Centre is considering a proposal for an electronic World Heritage Information Network that would integrate information supplied by parties to the Convention.

In the 1990s the World Heritage Centre is moving to establish regional centers to help implement the Convention. These centers will be established as national centers with regional jurisdiction. They are intended both to build national support for the Convention and to coordinate activities in regions, including assistance programs. The first one, which is being established in Oslo, Norway, would extend beyond the Scandinavian states to include the Baltic states, central and eastern Europe, and West Africa. There may soon be a regional training center for conservation of natural Sites in West Africa.

The new program for monitoring the condition of World Heritage Sites will also be regionally based. There is a draft work plan for implementing regional monitoring programs and reviewing syntheses of monitoring reports compiled regionally.

These shifts in activities are reflected in budget allocations. In 1980, the World Heritage budget allocated $170,000 to emergency assistance, $36,900 to promotional activity, $42,000 to advisory services, $149,234 to preparatory assistance for Meetings of the Parties, $165,400 to technical cooperation, and $204,700 to training. Whereas the amounts for preparatory assistance have declined over time (except for 1992 and 1995), the amounts for technical cooperation and for training increased dramatically in 1987–1990, then declined until 1993, and rose again in 1995 to $750,000 and $452,000 respectively. The shifts in amounts for technical cooperation may reflect declines and increases in the number of country applications, rather than the availability of funds, because amounts available for technical cooperation have not been fully used.

Whereas the budget for emergency assistance stayed more or less constant until the creation of the new Emergency Fund in 1993, that for promotional activity showed steady increases with a peak in 1992, the twentieth anniversary of the Convention. Advisory services have risen consistently since 1985, and reached $500,000 in 1995. Perhaps most significant, in 1993 the budget for the first time contained an explicit allocation of $189,000 for monitoring activities, which reflected the priorities in the 1992 Strategic Plan. At the December 1994 Committee meeting, the 1995 budget provided for $308,000 for monitoring; at the same time, the Centre's director indicated that countries should assume increased responsibility for monitoring sites. There were also increased funds for building local capacity to care for the sites.

The Secretariat. The Secretariat for the Convention has been housed in UNESCO. Until May 1992, there were separate secretariats for cultural properties and for natural sites, which were located in different sections of UNESCO and coordinated by the director of the Cultural Heritage Division. In May 1992, the Natural Sites Secretariat, in the Division of Ecological Sciences, and the Cultural Sites Secretariat, in the International Standards Section, were combined to create the World Heritage Centre within UNESCO. This meant that for the first time the Secretariat budget, including staff, was a separate item within the general UNESCO budget.

The Centre had a staff of twenty-two as of May 1995, and relied in significant part on secondments from national governments and short-term contracts. The World Heritage Fund paid for three; the general UNESCO budget for eight; and the others were on secondment or funded from other sources. Before 1992, all staff concerned with natural and cultural heritage were UNESCO staff or consultants, because the Centre did not exist. Six professional staff and three administrative staff were devoted to cultural sites before 1992.

Secretariat allocation of time also has changed since the Convention went into effect. Whereas previously the Secretariat spent much of its time on listing sites, meetings, and building capacity, as of 1993, according to Secretariat officials, staff divides its time somewhat evenly among listing sites, monitoring sites, building capacity, meetings, and, to a lesser extent, research.

The Role of Nongovernmental Organizations. The World Heritage Convention is an early and outstanding example of an international treaty that explicitly provides for nongovernmental organizations—the International Council of Monuments and Sites (ICOMOS) and the International Union for the Conservation of Nature and Natural Resources (IUCN)—and one intergovernmental organization, the International Centre for the Study of the Preservation and Restoration of Cultural Property (the Rome Centre, or ICCROM), to assist in treaty implementation (articles 13, 14). The Operational Guidelines (para. 57) expressly request these organizations to assist in monitoring.

The two nongovernmental organizations have played important roles. They evaluate requests to put cultural properties and natural sites, respectively, on the World Heritage List, review requests from countries for financial and technical assistance, and, more recently, monitor selected sites. While the IUCN has monitored selected natural sites since 1984, monitoring of cultural sites is recent. The IUCN uses its facility at the World Conservation Monitoring Centre in Cambridge, England, and an informal network of hundreds of people to provide information on the sites. The World Conservation Monitoring Centre maintains databases that include information on sites on the World Heritage List. The reports from IUCN help the members of the World Heritage Committee determine whether to include a site on the Danger List or to initiate other follow-up actions.

The three designated organizations have routinely participated in the meetings of the Bureau and of all the parties. Since they have special expertise in assessing and managing sites, they are welcomed. However, few other such organizations have been closely involved with the Convention.

Financial Arrangements. Until 1992, UNESCO provided staff to serve as the Secretariat for the Convention, and funds to administer it. In 1992, the World Heritage Centre was formed, and the administration of the Convention was no longer enveloped in the general budgets for natural sciences and cultural resources at UNESCO. It became a separate line item in the UNESCO budget. The parties must support the Centre's separate budget. Article 16 of the Convention obligates parties to provide 1 percent of their contribution to the regular budget of UNESCO to support the

World Heritage Convention. In addition, by 1995 UNESCO had again agreed to pick up the costs of seven central staff members of the Secretariat. From the budget figures available, the Convention suffered during the period 1983–1985 and generally has enjoyed increased resources since then.

The Convention provides for a World Heritage Fund to assist developing countries in maintaining World Heritage Sites; member states have always been responsible for contributing to this fund. Countries with a site on the Danger List are eligible for special funds to protect the site. The Secretariat also has tried to enlist the support of the Global Environmental Facility to fund natural heritage conservation projects, with limited success to date.

In 1993, parties created a new Emergency Fund to provide parties with more emergency assistance to conserve the sites. At about the same time, the director of the World Heritage Centre enlarged upon a program to enlist the private sector in supporting activities to protect World Heritage Sites.

Dispute Settlement. There is no provision in the Convention for dispute settlement. The major dispute over inscription of a property on the World Heritage List was in 1981, in the case of Jerusalem. The other disputes that have arisen among parties have involved the appropriateness of listing a site on the Danger List without the consent of the host country, as in Dubrovnik, Manas, and the Everglades, or the required procedures for delisting of a site. The issue of delisting has been raised twice—for Mount Nimba in Guinea and for Manas National Park in Assam, India—and continues to be controversial. Resolution of these issues has been handled by ongoing political dialogue among the parties.

Disputes over implementation and application of the Convention remain few but appear to have become more frequent since 1988, although none has been characterized as a formal dispute that needs to be resolved through settlement procedures. This is perhaps because there has been more attention since 1988 to conserving the sites, and hence to interpretive questions about putting properties on the Danger List, removing them from the Danger List, and ensuring that they are protected by host and neighboring countries. In no instance, however, has any party tried to bring another party to account by invoking formal dispute resolution procedures.

At the national level, there have been a few notable disputes about compliance with the Convention. In Australia, Tasmania's attempt to build a dam that would destroy a site on the World Heritage List pitted the federal government against the state government. The Australian Supreme Court, in *Australia* v. *Tasmania* (57 *Australia Law Reports* 450 (1983)), ultimately resolved the dispute in favor of the federal government and compliance with the obligations under the Convention.

A second dispute involved Canada and the United States. In 1992, Glacier Bay National Park, which traverses the border, was designated as a World Heritage Site.

A major mining operation planned for the junction of the Tatshenshini and Alesk rivers in British Columbia would have threatened the site, and was terminated due to the combined pressure of the United States government and the Canadian federal government. In 1997, the Convention was being used by those trying to protect Yellowstone National Park, a World Heritage Site, from degradation from proposed nearby mining operations and as a lure to protect the Carlsbad Caverns from similar threatening activity.

Monitoring and Compliance. Although the Convention calls for regular reporting by parties, the only effort to do so in the 1980s was not judged successful. As of December 1997, parties were considering initiating periodic reports on a four to six year basis. In the previous effort, many parties did not report, and those reports that were received varied enormously in detail and were not necessarily accurate. Countries want to report that the sites are fine, not that they are failing to conserve the sites as they should. Monitoring, to the extent that it was done, was done on-site.

Whereas there has been fairly systematic monitoring of natural sites, there has been comparatively little systematic monitoring of cultural sites. In 1984, the IUCN initiated regular monitoring of selected natural sites, using its facility at the World Conservation Monitoring Centre in Cambridge, England. This was a logical step, since the IUCN was responsible for preparing the assessments of whether sites should be put on the World Heritage List. The IUCN was able to draw upon a network of 6,000–7,000 volunteers that had been in place for a long time for other purposes. The 1993 workshop organized by IUCN in connection with the fourth World Congress on National Parks and Protected Areas provided expert reviews of more than sixteen natural heritage sites (IUCN/UNESCO 1992).

By contrast, there was no follow-up monitoring of cultural sites that were added to the World Heritage List, whose assessments for nomination were done by ICOMOS, often with the assistance of ICCROM (the Rome Centre). In 1991 the UNESCO–United Nations Development Program regional office in Lima, Peru, initiated an experimental program for the monitoring of cultural sites situated in Latin America and the Caribbean. Monitoring of cultural sites is difficult because in many cases there is little local information about the sites. The regional experience shows that monitoring should be a cyclical, continuous process, and that monitoring procedures should be developed in partnership with the host state.

Since 1992, with the initiation of the Strategic Plan, monitoring and compliance have received much more attention under the World Heritage Convention. In 1992, the World Heritage Centre had already started taking a more active role in monitoring and in coordinating monitoring missions (e.g., to Mount Nimba and Plitvice Lakes National Park in Croatia), and it continues to do so. The new concern with monitoring was reflected in the World Heritage Committee meetings, which in 1993 considered monitoring reports of ICOMOS and IUCN for sixty sites, and in 1997, forty-one reports, with the Bureau having examined thirty more reports. It was also reflected in the budget allocation for the Secretariat.

Monitoring projects in partnership with countries appeared for the first time in 1993 in the World Heritage budget, when there were twelve such projects; in 1994, there were forty-eight monitoring projects. In 1994, provisions were added to the Operational Guidelines inviting parties to submit reports on the condition of their World Heritage Sites every five years.

Nongovernmental organizations and individuals are important to the compliance process. They frequently report alleged threats to or impairments of sites to the Secretariat. Nongovernmental organizations in the United States, for example, have sent an extensive document to the World Heritage Committee outlining the threat from mining activity to Yellowstone National Park, a site on the World Heritage List. These communications have come mainly from the developed world. There is little informal communication of noncompliance from sources in developing countries.

Monitoring procedures continue to rely primarily on-site monitoring and on informal means of assessment, rather than on regular country reports.

Concluding Observations. Although the World Heritage Convention contains quite general obligations, compliance with the Convention is in many ways quite respectable. Perhaps this is in part because countries voluntarily nominate sites to be put on the World Heritage List and assume obligations to conserve them. The Secretariat carries forth this tenor, because it prefers to speak in terms of inducements to comply, rather than of violations of the Convention or enforcement measures. It looks to national pride and the prestige of being on the World Heritage List, and to media attention, as important means for securing protection of the sites.

After a period of almost a decade and a half in which parties implemented the Convention in more or less the same way, the parties in the 1990s seem to be expanding the operations of the Convention and looking to ways to make it more effective. It is no longer focused only on listing the world's jewels, but also on conserving them and managing threats from tourism, among other items. The years since 1990 are the same years in which countries mounted serious preparations for the Rio Conference on Environment and Development (the twentieth anniversary of the Stockholm Conference on the Human Environment), and countries began to implement Agenda 21, which was concluded at the Rio Conference. The international momentum associated with these developments seems to have affected the consciousness of parties to the World Heritage Convention.

Convention on International Trade in Endangered Species

International trade in wildlife has been going on for centuries. Modern efforts to control the trade emerged in the late nineteenth century in Europe and the United

States. International treaties for the next several decades focused largely on controlling birds, fur seals, and species prized by game hunters. The 1940 Washington Convention on Nature Protection and Wildlife Preservation in the Western Hemisphere required parties to control trade in the endangered species listed in an annex, but the Convention has never been effectively implemented.

In the late 1940s, trade in endangered fauna and flora expanded rapidly. By 1960, the IUCN began to lobby its member governments to restrict the import of animals and animal products (Wijnstekers 1992). In 1963, the IUCN General Assembly called for "an international convention on regulation of export, transit and import of rare or threatened wildlife species or their skins and trophies." Although the first draft of an agreement was available within a year, the producer and consumer countries of the species were unable to reach any consensus until the Stockholm Conference in 1972, when they adopted a recommendation on the issue as part of the Stockholm Programme of Action. In 1973, at a plenipotentiary conference hosted by the United States, countries concluded the Convention on International Trade in Endangered Species (CITES). The treaty entered into force on July 1, 1975. As of June 1997, 136 states were party to the Convention, with more than 600 species on the appendix I endangered list and more than 2,300 species of animals and 24,000 species of plants on the appendix II List of threatened species. Figure 5.6 shows the progress of ratifications over time.

The Convention must be seen in the broader context of the international treaties that are directed to conserving biological diversity, including the 1992 Biological Diversity Convention, regional conservation accords, and the nonbinding statement of forest principles concluded at the 1992 United Nations Conference on Environment and Development in Rio. As such, it targets only the profitable trade in identified species across national borders.

Treaty Commitments and Implementing Institutions. CITES was designed to protect species of fauna and flora from extinction by controlling commercial trade across national borders. Parties must provide varying degrees of protection, depending upon the status of the particular species. Regulated species are listed on one of three appendices: those threatened with extinction (appendix I); those that could face extinction if their trade is not controlled (appendix II); and those facing overexploitation in a particular country (appendix III). Member countries decide multilaterally to list a species in the first two appendices; a country may unilaterally put one of its species on appendix III (Favre 1989; Fitzgerald 1989; Lyster 1985).

The Convention operates through a permit system, which requires export and import permits for species listed in appendix I, and export permits for species in the other two appendices. Every country is to designate a Management Authority and a Scientific Authority to supervise the review of permit applications.

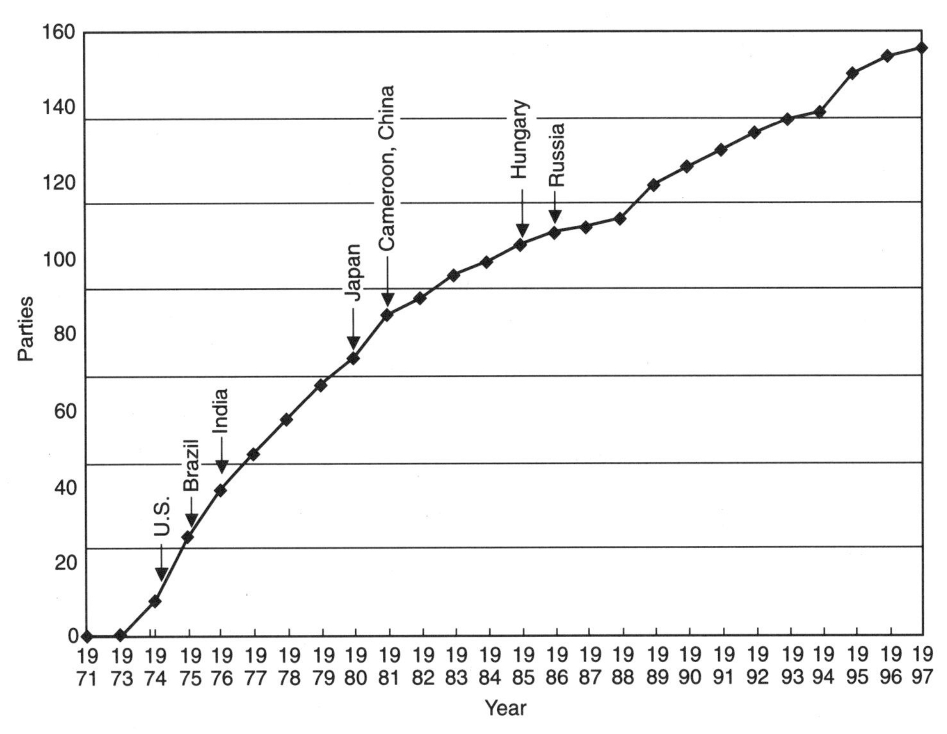

Figure 5.6
Total parties to CITES and ratification years for countries in this study

Any trader in appendix I species must present an export permit to the customs agency of the exporting country and both export and import certificates to the importing country's customs officials. In theory this provides a double check against illegal trade. In order to grant an export permit, a country's Scientific Authority must indicate that export will not adversely affect the survival of the species. A state may grant an import permit only if the Management Authority determines that the specimen will not be used primarily for commercial purposes, and the Scientific Authority certifies that the purpose of importing the species is not detrimental to its survival, and that the species will be properly cared for by its recipient. The provisions governing trade in species listed in appendix I are intended to shut down international trade in those species.

Species listed in appendices II and III receive less protection under the CITES permit system. An import permit is not required for trade in a species listed in appendix II or III, and the specimen may be used for commercial purposes. But the same conditions for granting an export permit for appendix I species apply to appendix II species, and importation of them requires an export permit or reexport certificate *prior* to import. If a country has placed a species on appendix III, an export permit is still required, but since the listing means only that the species is endangered in the listing country, no general assessment that the species is not endangered is needed for the export permit.

To implement the treaty, countries must designate ports of entry and exit, and create a detailed record of trade in specimens of listed species.

CITES provides six exceptions from the requirements of the three appendices, even if species are listed on them: exceptions for specimens owned as household or personal effects, bred or propagated in captivity for commercial purposes, acquired before regulation of them took effect, lent noncommercially between registered scientific institutions, transshipped while in customs' control, or transported as part of a traveling exhibition. Although these exemptions are intended to relieve the burden of CITES, the effects of the household use, captive breeding, and prior acquisition exceptions have been particularly prone to abuse and have undermined the objectives of CITES (Liwo 1991).

In order to encourage countries to participate in the CITES Convention, countries are allowed to opt out of any listing of species on appendix I or II—that is, make a reservation to the listing—within ninety days before the listing becomes effective. In this case, the reserving country is not treated as a party for trade in the particular species.

Like the Montreal Protocol, CITES controls trade in protected species with states not party to the Convention, except under conditions that approximate those demanded by the treaty. The state that is not a party to CITES must present documentation that "substantially conforms" to the required import and/or export permits. Because parties that enter reservations to the listing of certain species in appendices are considered nonparties with regard to the reserved species, the CITES prohibition on trade with nonparties applies to them for those species.

To implement the Convention, member countries have formed seven committees, six of which are still in existence: (1) Steering/Standing Committee; (2) Animals Committee; (3) Plants Committee; (4) Identification Committee; (5) Nomenclature Committee; (6) Ten Year Review Committee. The Technical Committee, established in 1979, was phased out in 1987. The Panel of Experts on the African Elephant and the Working Group on the Transport of Live Specimens also have been established to pursue specific issues of concern to the parties. The Secretariat services all the committees.

The Standing Committee, which meets every few months, is a relatively powerful body that carries out the functions of the Convention between the meetings of the Conference of the Parties. It functions much as the Bureau does for the World Heritage Convention. The Standing Committee reviews major infractions by countries, takes decisions relating to infractions, and develops action items for the Conference of the Parties.

Treaty Implementation. The main activity of the parties has been to decide whether particular species will be included on the lists appended to the Convention, trade in which will be controlled (Liwo 1991).

From 1975 to 1990, countries filed 289 reservations to changes in the listing of species in CITES's appendices. The highest-profile reservations were those taken by five African countries (Botswana, Burundi, Malawi, Mozambique, and Zimbabwe) to the October 1989 decision of the CITES parties to move the African elephant from appendix II to appendix I, effectively banning trade in ivory and other elephant products. As a result, these countries are nonparties to CITES for purposes of the African elephant. Japan also has been prone to opt out of the listing of particular species. By 1985, it had taken fourteen reservations to species listed in appendix I; in particular it has entered reservations to a number of species of endangered turtles.

Recently parties have been concerned with the criteria for listing species in the appendices and for delisting them. At the ninth meeting of the Conference of Parties in 1994, parties adopted new biological criteria for appendix I and new criteria for listing on appendix II that address sustainability of the species and closely related specimens or species.

The parties have amended the Convention once to provide a legal basis to assess parties for funding to support administrative operations pertaining to the Convention. The amendment was proposed in 1979 but did not enter into force until 1987. The parties approved a second amendment in 1987 that would allow regional economic integration organizations to accede to the treaty (i.e., the European Community), but it has never entered into effect because the required two-thirds of the parties has yet to ratify the amendment. This is partly due to concerns about diluting the pressure on individual countries to comply with the Convention.

The Secretariat. When the Convention was concluded, the IUCN, acting as an independent secretariat within the United Nations Environment Programme (UNEP), served as the host organization for the Secretariat. This seemed appropriate because the IUCN spurred the initial drafting of the Convention. However, in 1984, the Secretariat became part of UNEP, although it still retains its own bank account.

The Secretariat coordinates and prepares for the meetings of the parties and subcommittees; assists parties in implementing the Convention by responding to questions, and, more recently, by providing training seminars and assistance; administers special projects such as a survey of national implementing legislation; and helps with enforcement. The World Conservation Monitoring Unit (WCMU) formally tracks exports and imports of controlled species through its computerized facilities in Cambridge, England, and provides the data to the CITES Secretariat and interested parties.

Compared with ten years earlier, in 1995–1996 the Secretariat spent more time training officials in developing countries, responding to parties' requests regarding implementation, reviewing the validity of national permits (upon the country's request), and monitoring and enforcing the Convention. Approximately 30–35 percent of Secretariat time is spent on preparing for meetings of parties or carrying on activities flowing from the meetings, 35 percent goes to implementation measures, and 10–20 percent to training activities. Since being hired in 1988, the training officer had developed the training program to the point where he spent 50 percent of his time on training in the mid-to-late 1990s. The practice of submitting permits to the Secretariat for review was new in the mid 1990s, and may reflect a desire to have the Secretariat "take the hit" for denying the validity of a permit.

During the initial phase of CITES, the Secretariat had trouble keeping up with the increasing demands upon it. While parties to the Convention increased almost eightfold from 1975 to 1982, budget and staffing levels remained almost unchanged. Since the mid-1980s, the resources of the Secretariat have increased at a rate faster than that of membership. The Secretariat's budget increased from $438,000 in 1982 to $2.4 million in 1992. There was a sharp increase in 1985 to $806,000, followed by small declines, and then a sharp jump in 1988 to $2.2 million and in 1993 to $3.7 million. However, these budget figures may be misleading as an estimate of Secretariat resources, because some of the major funders of CITES, including the United States, were in arrears until recently.

The Secretariat staff grew from five full-time equivalent employees in 1982 to twenty-one in 1992. (The number of staff members on secondment—up to two people—did not increase.) By comparison, membership in the Convention grew by only about 50 percent during the same period. The Secretariat added an enforcement officer in the early 1990s, but the position has been vacant since October 1994.

Role of Nongovernmental Organizations. Although nongovernmental organizations (NGOs) are increasingly powerful in many environmental areas, their influence in CITES is unusual. Like the World Heritage Convention, the text of CITES explicitly provides for the participation of NGOs. They can attend the meetings of the parties, where they are not mere observers but powerful and numerous participants. They can express their opinions directly at the Conference of the Parties and receive copies of all reports compiled by the Secretariat; they lack only the right to vote. Over time the number of NGOs participating in Conference meetings has increased dramatically and at a faster rate than the increase in countries party to the treaty. For example, in 1976, there were thirty-two parties to the Convention (with 75 percent participating in the Conference) and twenty-seven NGO observers at the Conference of Parties in Bern, Switzerland. By 1985, NGOs exceeded parties, 128 to

87. By 1992, NGOs and parties numbered 163 and 113, respectively; 92 percent of the parties participated in the Conference. In 1994, 96 percent of the parties (119 countries) attended the Conference of the Parties, with 221 NGOs and several non-party governments attending as observers.

The strategy of NGOs has changed somewhat, from one relying primarily on participation in the meetings at the Conference of Parties to one of lobbying governments in the hallways and other forums. The NGOs push their own proposals and position papers, and at the November 1994 Conference in Fort Lauderdale, they published a daily bulletin on Conference developments that was made available electronically.

NGOs have a pivotal role in monitoring compliance with the Convention. The World Conservation Monitoring Unit has been responsible for the computerized tracking of exports and imports, and for maintaining the CITES trade database. TRAFFIC, a division of the World Wildlife Fund, also has been very active and effective in monitoring trade in illegal wildlife. In addition to their role in monitoring, NGOs assist with scientific and technical studies, and are a source of valuable expertise.

Whereas many countries welcome the participation and support of the NGOs, others are not so grateful. The increasing power of NGOs became especially controversial during the 1989 battle to move the African elephant from appendix I to appendix II, and again during the 1992 struggle to move it back again. The African countries that resisted the uplisting complained bitterly about the clout of the NGOs pushing for increased trade protection. In June 1997 the elephant was again transferred to Appendix II.

Financing. In contrast to the World Heritage Convention and the Montreal Protocol, CITES does not have a separate fund to support technical assistance and projects for member countries. Such assistance must come from the general UNEP budget for the Secretariat, and thus from governments.

Dispute Settlement. At least three methods for resolving disputes are prescribed by the Convention. Parties are first encouraged to resolve disputes through negotiation. If negotiation is unsuccessful, the parties may then submit the issue to binding arbitration at the Permanent Court of Arbitration at The Hague. The Court may resolve the dispute only if both parties consent to its jurisdiction. No dispute regarding CITES has ever been brought to The Hague or submitted to binding arbitration. Third, the dispute may be resolved at the biennial Conference of Parties under the CITES resolution procedure. The parties may adopt, by a two-thirds majority, resolutions expressing opinions and recommending action, but this has been seldom done.

Monitoring and Compliance. Parties to CITES must take "appropriate measures" to enforce the Convention. Such measures include penalizing illegal trade in or possession of protected species. Member countries are instructed to seize illegally obtained species and to return them to the exporting state. To encourage compliance with the Convention, countries adopted a resolution providing that trade can be suspended for certain species and certain countries until they take the recommended remedial measures. Countries agreed once to impose constraints on Italy if she did not comply, and considered such steps against China and Taiwan for trade in rhino horns and tiger parts. The United States, invoking national legislation, has unilaterally imposed sanctions against Taiwan for her trade in rhino horns, even though Taiwan cannot be a party to the Convention because she has not been recognized as a separate state in international law.

The primary method of monitoring compliance with CITES is through annual country reports, which provide export and import permit data. Countries must also file biennial reports on legislative, regulative, and administrative measures undertaken to improve implementation and enforcement. The Secretariat has repeatedly expressed frustration with the quality and quantity of reports received. Compliance with the reporting requirement improved from 50 percent in 1979 to almost 80 percent in 1987, but declined to under 30 percent in 1990. In 1991, the parties adopted a resolution that failure to file the required annual report on time is an infraction of the CITES treaty. A list of countries committing infractions was compiled and circulated to the parties. In April 1993, for the first time, the Secretariat sent a stiff letter to countries indicating that if they did not file reports, that failure would be considered a "major implementation problem" under the Convention. The letter further offered the assistance of the Secretariat in preparing and filing the reports.

As a result of these measures, the number of timely reports and the quality of reports increased. In 1991, only thirty-five reports were received on time; in 1992, fifty-two reports were received on time, and the deadline was formally extended for seven others upon request. As of July 1997, more than 70 percent of the reports due for 1991–1995 had been received. Figure 5.7 shows the reports due and received over time. Table 5.3 shows the extent to which the countries in this study have reported.

Even if reports are filed, however, they are not always complete. One major consuming country does not include any appendix III species, some do not include plant imports, and others do not include all the information required on the permits, particularly on the import forms. Theoretically the exports to a country of a given species should correlate with the figures reported from the importing country. But there is often little correlation. Where the reported imports are larger than the exports for species in trade between two countries, it usually indicates that there is a significant enforcement problem.

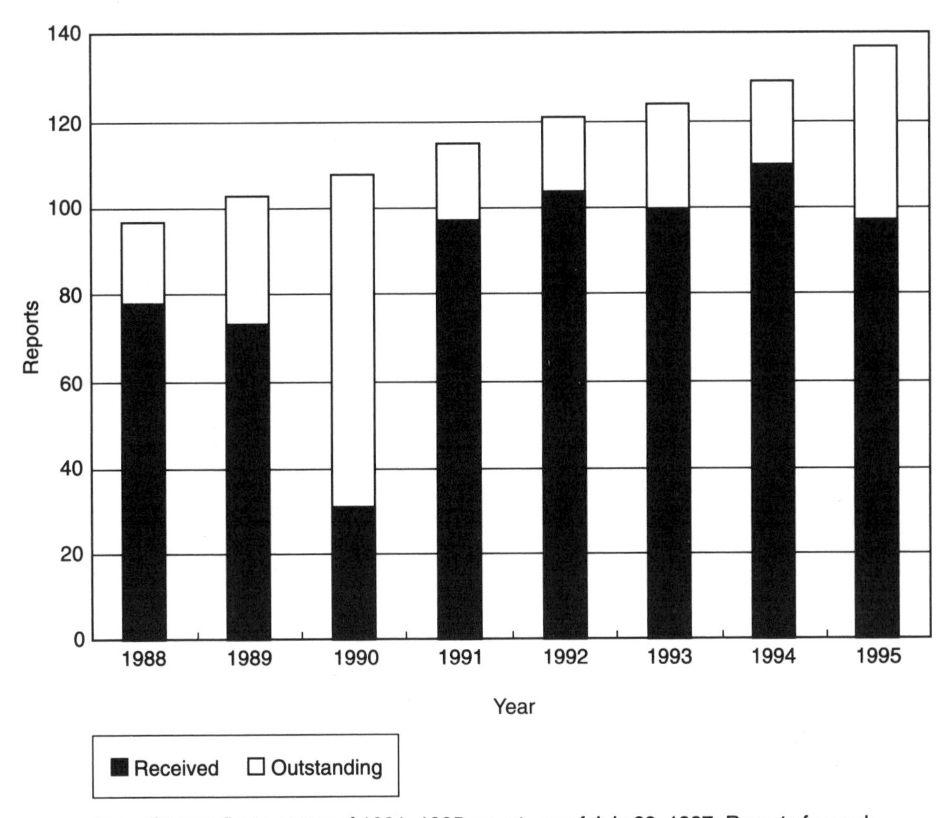

Note: Chart reflects status of 1991–1995 reports as of July 23, 1997. Reports for each year are due on October 31 of the following year; therefore 1996 reports are due on October 31, 1997. Nineteen hundred ninety-seven status of delinquent 1988–1990 reports is unknown; chart reflects status as of 1995 for those years' reports.

Figure 5.7
Annual reports required of CITES parties: Received or outstanding

A major enforcement problem is that the computerized correlations are not yet available in a timely manner. The WCMU, which tracked the reports on import and export permits, could provide only comparative tabulations for 1990 to the CITES Secretariat in 1993, which made it difficult to investigate countries with problems. As information technology advances and as countries begin to report their data on-line or in common permit formats, the timeliness should improve.

An important compliance tool is the Infractions Report that the CITES Secretariat publishes in connection with each Conference of the Parties. This report details countries that have violated the substantive controls on trade in listed species. The report is made public. Interviews revealed that conscientious scientific and management authorities within exporting countries have been able to use this report as an incentive to keep higher officials honest in controlling exports of endangered species.

Table 5.3
Annual reports of CITES parties: Transmission/receipt date

Country	Entry into force	1991	1992	1993	1994	1995	1996
Brazil	11/4/75	11/30/92	11/30/93[d]	1/23/95	11/27/95	12/12/96	N/R
Cameroon	9/3/81	9/12/94	7/26/94	N/R	N/R	N/R	N/R
China	4/8/81	9/22/92	9/25/93	9/28/94	10/19/95	10/30/96	N/R
Hungary	8/27/85	9/13/94	9/7/94	9/7/94	10/23/96	6/3/97	N/R
India	10/18/76	12/9/92	10/12/93	10/26/94	10/20/95	11/29/96	N/R
Japan	11/4/80	3/8/94	7/21/94[i]	3/7/95[i]	12/8/95[i,d]	1/16/97	N/R
			3/7/95[e]	5/1/95[d,e]	12/20/95[e,f]	1/8/97[f]	
Russian Fed.	1/13/92	7/27/93	11/27/93	12/26/94	1/29/96	1/17/97	N/R
United States	7/1/75	10/5/92	3/9/94[d]	10/25/94[a]	3/18/97[a]	4/2/97[a]	N/R
				5/20/96[p]	3/10/96[p]	12/11/96[p]	

a = animals.
d = extension of deadline granted.
f = fisheries and sea products.
e = exports.
i = imports.
p = plants.
Note: Annual reports from all CITES parties are due October 31 of the following year; thus, the annual report for 1991 is due on October 31, 1992. N/R indicates no report was received for that year as of July 1997.

Table 5.4
CITES training seminars

Year	Number of seminars	Number of people	Number of people/day
1989	2	39	117
1990	4	300	1,040
1991	6	350	1,073
1992	10	310	1,975
1993	6	285	1,043
1994	5	388	1,350
Total	33	1,672	6,598

As with other international treaties, there has been increasing attention to problems of compliance. Since the late 1980s concern with enforcement of the CITES has increased among parties and among Secretariat staff. The Secretariat has been requested to investigate an increasing number of alleged violations, with more than 300 claims brought to its attention during 1991 and 1992. At the November 1994 Conference of the Parties, enforcement was of central concern. Member countries advocated greater regional cooperation but debated whether to rely primarily on regional efforts or to develop an international network of government officials and Secretariat staff who are linked with Interpol in a new supranational enforcement authority. Since 1990 those responsible for trading in illegal drugs have become major traffickers in illegal wildlife, which makes the Convention even more difficult to enforce.

Regional enforcement efforts have been made. In September 1994, countries in southern Africa adopted the Lusaka Agreement on Co-operative Enforcement Operations Directed at Illegal Trade in Wild Fauna and Flora, which provides a forum for a pilot experiment in a regional enforcement network involving local police officers. The treaty went into effect on December 10, 1996.

One of the major changes in CITES efforts to secure compliance is a new and growing focus on in-country training and education. Although training programs began in 1989, they were not included in the Secretariat budget until 1993, when the Conference approved SF60,000 for 1993 and the following two years. In 1996, the budget was SF213,000, and in 1997, at least SF185,000. Table 5.4 shows the growth in training programs from 1989 to 1994. The figures for 1995 are significantly higher than those in 1994.

The CITES training strategy is to "train the trainers" in-country, by holding seminars and providing training materials. Countries are asked to contribute to the training program, so that they become engaged in its success. However, as in training programs in other areas, the effects may be short-lived because the targeted officials move on. More than 30 percent of the people trained reportedly have moved out of wildlife positions within a year after training.

CITES is inherently difficult to enforce. Compliance involves thousands of permits that must be accurately inspected, often at multiple customs points on the borders of the 136 member countries of CITES. Many species look alike, at least to the untrained eye. Until there are common permit forms, on-line access to national and international data to verify the validity of information, effective training of customs inspectors, and myriad other incentives, compliance will remain lower than desired. Advances in information technology may help (Birnie 1996).

Concluding Observations. It is difficult to assess the merit of the CITES Convention. There are data to show that the ban on trade in particular species, such as the elephant, has resulted in a significant drop in international trade (Glennon 1990). It is unclear how much trade in other species, particularly of plants, has been affected. But even if the Convention is highly effective in controlling international trade, it does not necessarily lead to the conservation of the endangered species because under the Convention, countries are free to decimate the species within their borders so long as no exports are involved. Thus, monkeys and other animals can be eaten for food even if the species becomes extinct, and other species can be wiped out through leveling of forests and destruction of soil productivity—all without leading to a violation of the Convention.

The International Tropical Timber Agreement

Trade in tropical timber began to boom after World War II, when Japan and Europe desperately needed cheap timber for reconstruction. Japan looked to Southeast Asia for its supply, and Europe found its requirements in Africa and South America. The developing countries on these continents were willing to trade their natural resources for much-needed foreign currency.

In the 1960s, several United Nations agencies voiced concerns about the inequity of trade between industrialized and developing countries and the decimation of tropical forests. In 1966, the United Nations Commission on Trade and Development (UNCTAD) raised the issue of increasing the price of tropical timber to better remunerate developing countries for their resources. In 1968, the United Nations Development Program (UNDP) became involved, and the International Trade Center hired Terence Hpay to prepare a report analyzing the situation and to suggest what could be done. Hpay produced a draft agreement and recommended that both producer and consumer countries be involved in developing an approach to the problem. In 1977, the International Trade Center held negotiations to establish the International Tropical Timber Bureau. However, contrary to Hpay's advice, consumer countries were not invited to the negotiations. The negotiations were a failure almost from the start.

In the 1970s there was mounting international concern about the need for developing countries to expand their ability to process their own raw materials (UNCTAD 1982), and about tropical deforestation. The latter meant the loss of an increasingly valuable commodity, for world trade in tropical timber amounted to an average of more than $7 billion per year during the period from 1978 to 1980 (Wassermann 1984).

In response to these concerns, several agencies and NGOs began to advocate trade controls and the raising of timber prices. UNCTAD requested that its secretary-general convene preparatory meetings for international negotiations on individual products; six preparatory meetings on tropical timber were held between May 1977 and June 1982. Developing countries hoped for higher prices; European and North American governments were motivated in good part by environmental pressure. Japan was the most active country, since she was especially concerned about arrangements that might affect her timber trade (see Dauvergne 1997). At the sixth Preparatory Meeting on Tropical Timber in 1982, the conference had before it draft texts submitted by Japan, the governments of the tropical timber-producing countries, the Nordic countries, and the United States. The 1983 United Nations Conference on Tropical Timber led to the International Tropical Timber Agreement (ITTA), concluded under the auspices of UNCTAD, and open to both producer and consumer countries.

The treaty benefited both producer and consumer countries. For the producer countries, it was expected to offer opportunities to expand trade and to receive technological transfers and financial assistance from consuming countries. Consumer countries, in turn, were to enjoy expanded and stable supplies of tropical timber as well as increased esteem in the eyes of domestic constituents and international organizations.

The ITTA was signed in November 1983 and went into effect in April 1985. No reservations to the treaty were permitted. However, because of a dispute over the location of the Secretariat, which was finally given to Yokohama, Japan, the treaty did not begin operations until 1987. As of June 1997, it had fifty-three member countries: twenty-six producer countries and twenty-seven consumer countries (including the European Union). Figure 5.8 shows the progress of ratification of ITTA over time.

The member countries account for 90 percent of the tropical forests and 90 percent of world trade in tropical timber. On January 26, 1994, countries concluded a successor treaty to the ITTA that entered into force on January 1, 1997. The new ITTA pays more attention to sustainable forestry and includes several new provisions similar to those found in the other treaties under study.

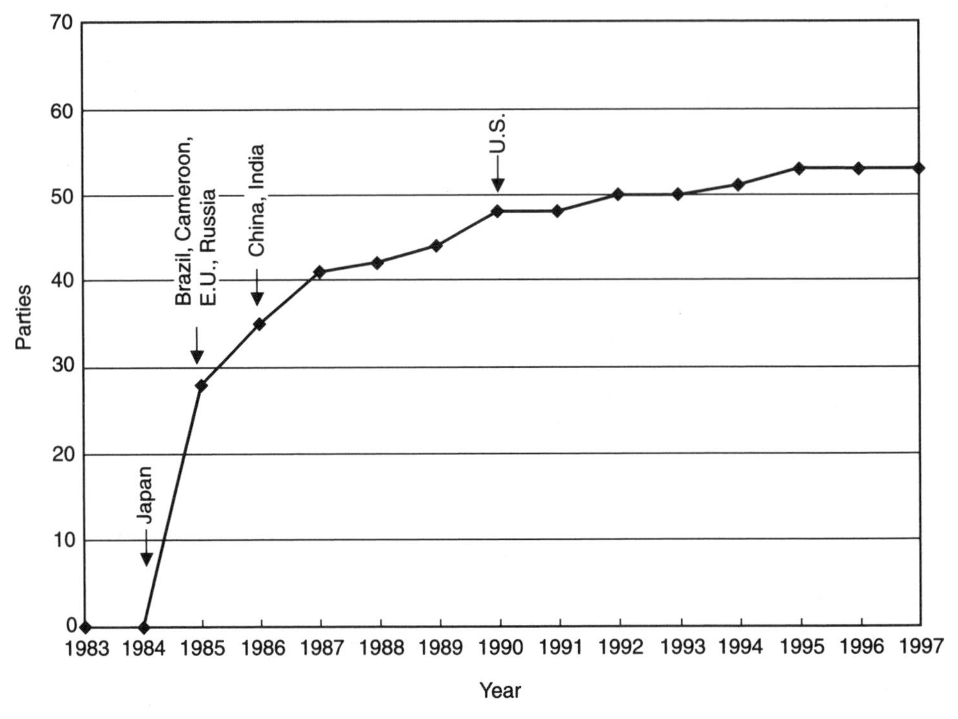

Figure 5.8
Total parties to International Tropical Timber Agreement and ratification years for countries in this study

The negotiations for the new treaty, which began in 1992, were very contentious, with talks between the producer and the consumer countries often nearly stalemated. There was no agreement on two key issues: whether countries would be required to engage in sustainable forestry by the year 2000 and whether this would apply to all forests. The vociferous debate that surrounded the development of forest principles for the United Nations Conference on Environment and Development in Rio suggested that it would be difficult to arrive at specific binding requirements for environmentally sound forest practices and to ensure that countries comply with them. Malaysia led the coalition of countries arguing that both temperate forests and tropical forests must be subject to any requirement of sustainable management. Some parties even appeared to be content to let the treaty lapse.

Environmental groups from Europe, the United States, and Japan were critical of the 1983 treaty and the International Tropical Timber Organization (ITTO). They argued that the ITTO lacked environmental safeguards and that a new treaty should include all timber, not just tropical timber. On the other hand, groups like the American Forest and Paper Association opposed the proposal to expand the scope of the ITTA and suggested that this would dilute the ITTO's mission.

In the new treaty, countries producing tropical timber agree to try to export wood only from sustainably managed forests by the year 2000. In return, consumer countries agreed to establish a new fund, the Bali Partnership Fund, to help producer countries meet the objective of sustainably managed forests. In addition, the consuming countries issued a separate formal statement pledging to respect comparable forest conservation guidelines for their own forests and committing themselves to the "national objective of achieving sustainable management of their forests by the year 2000" (UNCTAD 1994). The statement is not a formal part of the new treaty.

For many producer countries, forests represent a central resource. These countries have traditionally regarded their right of sovereignty over forest resources, in terms of international law, as sacrosanct. Yet in the end an international commodity agreement designed to facilitate trade, stimulated greater environmental awareness and provided a forum for negotiating a new, more environmentally friendly treaty.

Treaty Commitments and Implementing Institutions. The 1983 ITTA is designed to facilitate international trade in tropical timber; it is based on the 1982 International Agreement on Jute and Jute Products. Tropical timber is defined as nonconiferous tropical wood for industrial uses that grows or is produced in countries situated between the Tropic of Cancer and the Tropic of Capricorn. The term covers logs, saw wood, veneer sheets, and plywood. The treaty promotes research and development to improve forest management and wood utilization, to improve market intelligence, to encourage processing of tropical timber in producer countries, and to promote reforestation and forest policies to conserve tropical forests and their genetic resources. The treaty's preamble and article 1, "Objects," have served as the lever to advance conservation activities. Article 1 "encourages the development of national policies aimed at sustainable utilization and conservation of tropical forests and their genetic resources, and at maintaining the ecological balance in the regions concerned."

The ITTA tries to balance power between the producer and the consumer countries. Almost all producers are developing countries located in the tropical belts of Latin America, Africa, and Asia. The consuming countries are for the most part developed countries from North America, Europe, and the Far East. Since the treaty was concluded, five of the producer countries have become net importers of tropical timber (Friends of the Earth 1992).

The ITTA imposes modest obligations. Parties are to provide detailed information about their tropical timber trade to the Secretariat, contribute to the administrative account (and, for developed consumer countries, to the special projects account), and exert their best efforts to promote the objectives of the treaty and comply with decisions made under the treaty. The obligation to provide detailed trade information resembles the obligation under CITES, but unlike CITES, there are no sanctions for noncompliance. Of the five treaties under study, this is the only one that does not set forth binding substantive obligations related to conservation in the text.

To implement the treaty, parties established the ITTO, which functions through the International Tropical Timber Council and three permanent committees: the Committee on Economic Information and Market Intelligence; the Committee on Reforestation and Forest Management; and the Committee on Forest Industry. Both the Council and the committees meet twice a year, the Council for two weeks each time. The Council meetings are primarily concerned with reviewing and approving projects proposed for producer countries. The committees review suggestions for pre-projects, and projects, and make recommendations for project funding to the Council.

The voting system for the Council tries to balance power between producer and consumer countries and among the various factions of each group. Japan, a prodigious consumer of tropical hardwoods, possesses a significant share of the votes. Despite the elaborate voting system, or perhaps because of it, matters are settled by consensus. No decision has ever been put to a vote.

Implementation. The ITTO's primary activities have been the development of reports on the international trade in tropical timber, formulation of guidelines and criteria on various issues (including the sustainable management of forests), and the approval and implementation of forestry projects in producer countries. Attention to sustainable forestry began in the late 1980s and has continued to grow since then.

In 1989, at Malaysia's invitation, the Council decided to send a mission to Sarawak, Malaysia, to investigate whether forests were being sustainably managed and to recommend ways to improve sustainability. The field report noted many instances in which the forests in Sarawak were not managed on a sustainable basis, but the published report, after review and refinement, was allegedly less critical and forthright. The federal government of Malaysia, in requesting the mission, may have seen an opportunity to bring international pressure to bear on a recalcitrant province. Because there was considerable unease among some member governments about the propriety of on-site monitoring missions by the ITTO, no similar mission was undertaken in the ensuing five years. In 1995, however, there were plans to initiate another on-site visit in Bolivia, and several others were under discussion.

In 1990, when the Council reviewed the Sarawak report, it adopted Guidelines for the Sustainable Management of Natural Tropical Forests. In 1991 it agreed to a voluntary "Target 2000" for sustainably managed tropical forests. Member countries decided to confer annually on progress toward Target 2000, identify general criteria of sustainable management, and study what resources were needed to implement a Target 2000 goal. In May 1992, at its meeting in Yaoundé, Cameroon, the Council formally adopted guidelines for sustainable forest management (ITTO 1993a, 1993b) and received country reports on progress toward meeting Target 2000.

Most of the work of the ITTO is done through projects (pre-projects and regular projects) at the national level. Both the numbers of projects and the funding have varied sharply over time. The number of pre-projects and regular projects was high in 1987 (twelve and sixteen, respectively), then the numbers dropped sharply in the next two years. In 1990, the number of projects and pre-projects jumped to twenty-nine, and to forty-nine in 1991, before dropping to twenty-five in 1992. The funding for projects remained well below $5 million until 1990; funding dramatically increased to more than $17 million annually for the next two years, then declined to $12 million in 1992.

To strengthen the project proposals presented, in 1989 the Council authorized the establishment of Expert Panels for Technical Appraisal of Project Proposals, which may include representatives from both the permanent committees and NGOs, such as the World Wildlife Fund. In 1992 the Council decided to publish manuals to assist in formulating projects and in project monitoring, review, and evaluation.

One of the major ITTO tasks is to compile statistical data on the trade in tropical woods. The ITTO publishes an *Annual Review and Assessment of the World's Tropical Timber Situation*, which is prepared by the Secretariat and reviewed by a joint session of the three permanent committees. Each year the Secretariat sends a questionnaire to member countries, which they must return by the end of the calendar year. The percentage of countries returning reports ranges from 40 percent in 1990 and 1992, to 50 percent in 1991. In 1991, the ITTO introduced a much more detailed questionnaire. The responses to the 1996 Trade Enquiry showed improvement over previous years. Reporting is apparently hindered by a limited statistical capability in some member countries.

These reports compile statistical data; they do not address sustainable management practices. Nevertheless, they can provide a useful input for assessing how rapidly forests are being depleted, admittedly only in terms of depletion that results in declining trade across borders. One problem with the reports is that the data provided by countries are not necessarily reliable, partly because of limited statistical capability. The Secretariat relies on other available sources of information to fill gaps in reporting by countries, particularly the United Nations Food and Agriculture Organization's Forestry Department and Timber Section, the Economic Commission for Europe, several Japanese lumber and plywood associations and the International Trade Center (ITTO 1996). NGOs have begun to play an increasingly important role in monitoring the trade in tropical timber.

A new issue emerged on the agenda of the ITTO: ecolabeling of tropical timber. When the issue was initially raised in 1992, in response to Austria's attempt to impose ecolabels on tropical timber specifically from Indonesia and Malaysia, the Council called on members to refrain from unilateral action. This issue was shelved during the negotiations for the new ITTA, but it came back to the Council in the form of a discussion about certification of timber and timber operations as an incentive to promote sustainable forests. Producers demanded that temperate timber be included in any certification scheme, which was resisted by consumer countries. The ITTO conducted a 1994 study on certification and decided to continue to review the issue at the November 1997 Council meeting (ITTO 1996).

The ITTO has embarked on a public information program about tropical forests and indirectly about the Convention. In 1990 it launched a newsletter, a thirty-page publication that appears four times a year. It includes information on the trade in tropical timber and sustainable forestry management, a calendar of ITTO meetings, and a worldwide calendar of meetings on forestry and environmental conservation. The newsletter was initially produced under contract, but publication in the later 1990s has been done in-house.

The Secretariat publishes comprehensive reports of its meetings, which are available upon request for the cost of postage. As of July 1997, ITTO proceedings were not yet available in electronic form.

The Secretariat. The Secretariat, which is housed in Yokohama in a building donated by Japan, is freestanding and not attached to any international organization. As of 1996, it had a staff of approximately twenty persons and a budget of roughly $3 million. The budget increased from $2,178,000 in 1988 to $3,609,000 in 1993, at a relatively steady rate. Both the staff and the funding levels are about the same as for the CITES treaty.

The Secretariat grew from nine staff members (six professional) in 1986 to sixteen in 1988, to nineteen in 1992. From 1991 on, it has received a surcharge of up to 5 percent from every ITTA project to cover costs, which explains the additional staff by 1992. In every year several positions have been vacant. There have been two significant additions to the Secretariat. In 1990, Norway seconded a forestry ecologist, and in 1992, the Secretariat added a conservation officer. A staff person was added in 1995 to produce the newsletter.

The Secretariat staff has spent most of its time on administrative functions such as preparing for meetings of the Council or servicing the three permanent committees. They are responsible for gathering information on the tropical timber trade from governments, analyzing it together with other market intelligence, and issuing reports to governments on the trade in tropical timber. Some environmental NGOs criticized the Secretariat as inefficient, and in the early 1990s, member governments reportedly were reluctant to provide more funds until the administrative procedures improved.

While it is always difficult to determine how the allocation of time by the Secretariat to specific tasks has changed over time, a review of the activities of the personnel in the top four Secretariat positions from 1987 to 1992 reveals at least rough allocations of time among three activities: preparations for the Council meetings, ITTO diplomacy (attending meetings, promoting the work of the organization), and substantive research and project development. Most of the time has been spent on the first two, with increasing time devoted to preparing for Council meetings.

Nongovernmental Organizations. As with CITES, NGOs are playing an increasingly important role in monitoring the actions of governments and in pressing for sustainable management of forests. NGOs attend the meetings of the Council and attended the negotiating sessions for the successor treaty. TRAFFIC has begun to monitor the trade in tropical timber and has completed a very thorough review of the illegal trade in tropical timber. In general, the environmental NGOs are most active in the consuming countries, although new, active organizations have emerged in some of the producer countries as well.

One of the most notable conglomerate NGOs is the Forest Stewardship Council, which was established in Canada in 1993 by a diverse group of forestry trade, environmental, and certification organizations. Headquartered in Mexico, the Council has more than 130 members—both organizations and individuals—from 25 countries. It includes NGOs, timber industry groups, retailers, and indigenous peoples, among others. The Council has developed accreditation principles and protocols for sustainable forestry that apply to all tropical, temperate, and boreal forests. For now the international Secretariat will accredit timber producers, but the Council hopes to establish regional or national offices for this purpose.

As with CITES, NGOs are serving as watchdogs on the activities of governments and private parties engaged in the timber trade. But they have gone further by developing transnational environmental standards for forest practices and products when governments have been reluctant to act.

Financing. Member governments provide annual payments to finance the ITTO Secretariat. Producer and consumer countries each must contribute 50 percent of the costs, with each member's contribution in direct proportion to votes held. For example, in fiscal year 1992, Japan contributed more than $10 million to the ITTO, accounting for 33 percent of the body's budget. Most projects funded by the Council are paid for through voluntary contributions by governments, with some modest contributions by the Japanese timber industry and conservation NGOs (Friends of the Earth 1992).

Dispute Resolution. The ITTA makes no provision for the settlement of disputes. Decisions are made by consensus. If disputes arise, they will likely be handled by negotiation, mediation, or conciliation, rather than by the traditional confrontational resolution mechanisms.

Monitoring and Compliance. There is widespread criticism of the ITTA on the grounds that countries do not comply with it—whether the "it" be the guidelines or the criteria parties have adopted, reporting requirements, or even required payments to the ITTO. Less than half of the countries have filed reports on timber trade with the Secretariat, and these are not viewed as reliable. Although member states have adopted guidelines on sustainable management of forests and criteria for determining whether forests are sustainably managed, there has been little effort, other than through ITTO's forestry projects, to ensure that they are implemented or complied with. Producer countries have not adopted legislation according to the guidelines, although several have expressed interest in doing so.

Trade in tropical timber has declined over the course of the ITTA; log exports in 1995 were almost 10 percent less than in 1994 and continued the steady decline of the previous decade (ITTO 1996). Thus, a primary purpose of the treaty—namely, to facilitate trade in tropical timber—has not been fulfilled. This need not indicate that the treaty has failed. The decline in tropical timber trade may be due to declining consumer demand and the substitution of timber from other forests for many uses. In part, it may be because the tropical timber-producing forests are being depleted.

The goal of sustainable management of forests also has not been achieved. The ITTO reported that less than 1 percent of forests were under sustainable management.

Concluding Observations. The effort to use a commodity treaty directed to trade in tropical timber to force countries to engage in sustainable forest management seems to have had only marginal effect in directly changing forest management practices. Although the parties did develop and approve sustainable forestry guidelines, the one effort in Sarawak to monitor on-site practice remained controversial. The vociferous debate that surrounded the development of forest principles for the United Nations Conference on Environment and Development in Rio suggested that it would be difficult to arrive at specific binding requirements for environmentally sound forest practices and to ensure that countries comply with such obligations. For many producer countries, forests represent a pivotal resource over which their sovereign rights in international law are sacrosanct.

Yet, in the end the commodity treaty served as a stimulus to and as a forum for bringing pressure to negotiate a new, more environmentally friendly treaty, a key part of which is the agreement that export of tropical timber will be from sustainably managed forests and that funds will be available to support this. Moreover, the post–Rio United Nations Intergovernmental Panel on Forests, which proposed either a new international accord on forests or a commission on forests to the 1997 special session of the United Nations General Assembly to review Agenda 21, reflects the international community's continued concern with sustainable management of forests. The ultimate relationship between the United Nations forest initiative and the ITTA will need to be refined.

The Pollution Control Treaties: A Living History

The study looks at two international treaties that are directed to controlling pollution. Traditionally, states have been concerned with transboundary air and water pollution. The two treaties in this book address pollution that occurs in areas largely outside national borders. The London Convention of 1972 addresses the dumping of pollutants into the marine environment, and the Montreal Protocol of 1987, the depletion of the high-level ozone layer by emissions of certain chemicals. The two treaties are presented in chronological order.

London Convention of 1972

Between 1950 and 1977, marine pollution emerged as a serious global problem. Accidental oil spills, operational oil discharges from ships, and greater dumping of wastes into the oceans led to a growing international consciousness that marine pollution was a serious international problem. At the same time there was a growing realization that the oceans did not have an unlimited capacity to absorb wastes without harm to the marine environment. Because of this awareness, the international community was willing to regulate marine pollution long before states were prepared to negotiate accords to control air pollution or even, with a notable exception, freshwater pollution. No country seriously addressed the issue of land disposal of wastes.

By the time of the 1972 Stockholm Conference on the Human Environment, multilateral treaties existed to prevent marine oil pollution (1954); to govern use of the territorial sea and contiguous zones, the continental shelf, the high seas, and living resources of the high seas (the Geneva Conventions of 1958); to intervene on the high seas in case of oil pollution casualties (1969); to impose civil liability for oil pollution damage (1969); and to establish an international compensation fund for oil pollution damage (1971).

The 1972 Stockholm Conference served as an important catalyst for reaching additional accords to control marine pollution from the dumping of wastes and operational discharges from ships, and to protect regional seas. Within the next five years, countries negotiated the Convention on the Prevention of Marine Pollution by

Dumping of Wastes and Other Matters (the London Convention, 1972), the International Convention for the Prevention of Pollution by Ships (MARPOL, 1973), and the first of the United Nations Environment Programme's regional seas conventions, the Convention for the Protection of the Mediterranean Sea Against Pollution (1976). The long history of successful negotiations to protect the marine environment may have made it easier to conclude these treaties in the 1970s and may positively affect compliance with them.

The London Convention of 1972 (formerly known as the London Dumping Convention), which is the focus of this study, was a direct output of the 1972 Stockholm Conference preparations (IMO 1990). At the first preparatory meeting in June 1971, countries established the Intergovernmental Working Group on Marine Pollution, which recommended that a treaty be concluded to control ocean dumping. The United States submitted a draft text for a Convention for the Regulation of Transportation for Ocean Dumping, which served as the basic text from which states negotiated. Countries considered a draft text of the treaty at the Stockholm Conference, but in the end they decided only to recommend that a convention be finalized as soon as possible, leaving it to an intergovernmental meeting five months later to adopt the Convention. The London Convention was opened for signature in December 1972 and became effective in August 1975. As of June 1997, seventy-five countries were party. Figure 5.9 shows the progress of ratifications over time.

In the 1990s the London Convention was in transition, as reflected by the decision to delete any reference to dumping in the popular title of the Convention. Countries shifted the focus from controlling the marine disposal of a variety of wastes to one that encouraged "integrated land-based" solutions for most wastes, and marine disposal of those wastes where there are not suitable land-based options (i.e., dredged materials and sewage sludge). Countries committed to phasing out incineration of wastes at sea and agreed to phase out ocean disposal of industrial wastes worldwide by the end of December 1995. At a special meeting in the fall of 1996, parties adopted a protocol that supersedes the Convention for all member states that become parties to the protocol: the 1996 Protocol to the Convention on the Prevention of Marine Pollution by Dumping of Wastes and Other Matter.

The Protocol is an impressive updating of the 1972 Convention to keep pace with changing conditions regarding dumping and the oceans. It adopts the "precautionary" approach and permits dumping of wastes in the ocean only if the wastes fall into the categories listed in an annex and a permit has been obtained. It prohibits the incineration of wastes at sea, bans the export of wastes to other countries for dumping or incineration at sea, provides for establishing compliance procedures, obligates countries to promote technical cooperation and assistance to member countries so as to control dumping and eliminate pollution caused by dumping, establishes formal dispute resolution procedures, and in an annex pro-

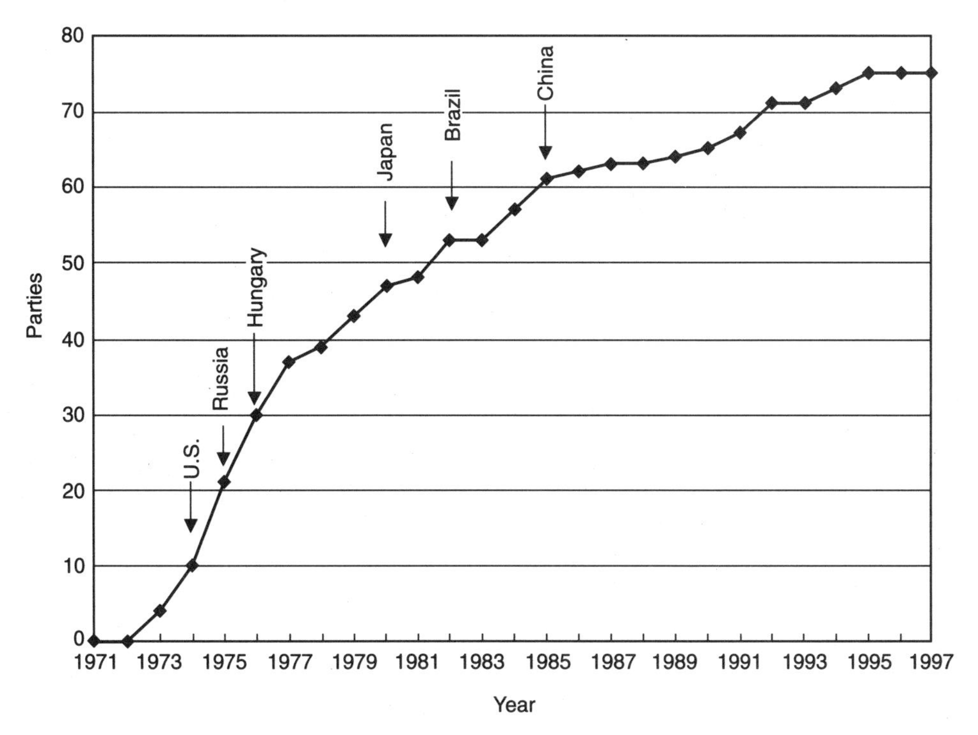

Figure 5.9
Total parties to London Convention and ratification years for countries in this study

vides for waste audits that include alternative forms of waste disposal. The Protocol opened for signature on April 1, 1997, and must be ratified by twenty-six states, including fifteen contracting parties to the present Convention, to go into effect.

Treaty Commitments and Implementing Institutions. The 1972 Convention regulates dumping according to a three-tier permit system that imposes increasing strictness in proportion to the potential environmental harm from the wastes. In the 1972 Convention, wastes falling in the most restrictive category, the Black List set out in annex I, may not be disposed of at sea even with a special permit. These wastes include highly toxic chemicals, persistent plastics and other synthetic materials, crude oil and its wastes, high-level radioactive matter, and materials produced for chemical and biological warfare.

Wastes that are not quite as harmful are included in annex II, the Grey List. This annex includes wastes containing significant amounts of heavy metals and toxic chemicals, acids and alkalies containing heavy metals, containers and other bulky items that may impede navigation, and low-level radioactive wastes.

All other matter falls into the least restrictive tier, annex III, for which only a general dumping permit is required. Annex III lists factors the permitting authorities should take into account when evaluating whether to grant a general permit: characteristics of the matter to be dumped and of the dumping site and method, and some general considerations and conditions.

Like the Convention on International Trade in Endangered Species, the London Convention provides a number of exceptions for wastes that are not included. These include disposal of wastes incidental to normal operations of vessels, dumping necessary to protect human life or property, and dumping related to offshore processing of seabed minerals. There is also an exception for items on the Black List if they are "rapidly rendered harmless by physical, chemical or biological processes in the sea." As with CITES, these are exceptions potentially subject to abuse.

The institutional framework for implementing the Convention consists of an annual Consultative Meeting of the parties, a permanent scientific committee (since 1983), ad hoc scientific advisory bodies that meet on a regular basis, and a Secretariat. The most important scientific groups have been the ad hoc Scientific Group on Dumping, the ad hoc Working Group on Dredged Material Disposal, the ad hoc Working Group on Incineration at Sea, the ad hoc Group of Legal Experts on Dumping, and the Panel on Sea Disposal of Radioactive Wastes.

Treaty Implementation. The London Convention was negotiated to control dumping of specific pollutants into the marine environment. From the beginning until 1987/1988, member countries implemented the Convention in more or less the same way. In 1988 parties started to consider new directions for the Convention. In 1990, in parallel with preparations for the United Nations Conference on Environment and Development, parties discussed new issues that would be before the Rio Conference. By 1992, they were eager to tie the Convention into Agenda 21 and the emerging global focus on land-based sources of marine pollution.

In 1991, parties introduced the precautionary approach and adopted language defining the approach in the context of the Convention. They also established a Twentieth Anniversary Fund, to which China, Denmark, the Netherlands, Portugal, South Africa, and the United Kingdom contributed $28,000 (in total) for purchasing computers for the Secretariat and initiating a global survey of countries on waste disposal practices. In 1993, the parties initiated the World Waste Survey to provide an integrated assessment of waste disposal patterns. They established a group to consider amending the Convention in 1992, and in 1993 agreed to hold a conference in 1996 to consider a formal amendment to the Convention.

As part of rejuvenating the Convention in the 1990s, there was increasing attention to building capacity in the developing countries and to getting technical assistance to countries through the Global Environmental Facility. The 1996 Protocol formalized the commitment to technical assistance and support, but it did not establish a separate fund to provide such assistance.

Throughout the life of the Convention, parties have "amended" or modified the annexes to add or delete substances or to make other adjustments to the agreement's operation. At least four times they have significantly changed the treaty this way. In 1978 an adjustment to annexes I and II allowed the incineration at sea, under certain strict conditions, of substances otherwise on the Black List. In 1980, countries broadened the language of the annex I prohibition against dumping crude oil to specify that all petroleum products and distillate residues were included, and they added a new provision to annex II bringing characteristically nontoxic substances that became harmful when dumped in too large quantities under annex II controls. In 1989, annex III was modified to urge parties, in deciding whether to issue a general dumping permit, to consider whether an adequate scientific basis exists to assess the impact of the dumped material.

In 1993, the parties adopted several far more encompassing resolutions that effectively shift the emphasis of the Convention to integrated waste management by discouraging marine disposal and encouraging sound disposal of wastes on land. They decided to phase out sea disposal of industrial wastes, to prohibit incineration at sea, and to suspend the disposal of radioactive wastes and other radioactive materials. This modification or "amendment" entered into force on February 20, 1994, 100 days after it was adopted by the parties (except countries that filed a declaration indicating they did not accept the change). Parties also agreed to reassess the controversial ban on the dumping of radioactive waste in twenty-five years.

Before 1996, parties proposed only one formal amendment to the Convention: on dispute resolution, in 1978. Formal amendments require ratification by two-thirds of the parties. As of early 1994, only eighteen countries had ratified the amendment, far short of the number needed. No countries ratified the proposed amendment between 1980 and 1989, and it languished. Since then, several countries have ratified it. The 1996 Protocol makes formal provision for the settlement of disputes, and hence preempts the proposed amendment.

For many years participation by member countries in Meetings of the Parties has been low. Only 60–65 percent of the parties attended the Consultative Meetings, and only about 25 percent participated in the meetings of the more specialized groups. As recently as 1990, only forty-three of the then sixty-seven members sent delegations to the Consultative Meeting. Of these, only nineteen were developing countries, perhaps in part because there is no financial assistance to help defray the costs of participation in the meetings. In 1993, the rate of member participation increased to fifty states, but has not risen since then. The number of observers from nonmember states remained roughly constant, about six, until the 1996 Special Meeting, when sixteen states attended. NGOs' participation has always been limited (nine or ten from 1990 to 1993, with three of them environmental NGOs); five were present in 1996, with Greenpeace the only environmental group. In some cases, nonmember observer states eventually become parties, such as China, whereas others, such as India, continue as observers.

Now that the Law of the Sea (LOS) Convention has gone into effect, parties have been concerned about the interaction of the two treaties, since both address marine pollution. More than forty of the LOS parties do not belong to the London Convention. At the Meeting of the Parties in October 1994, just before the LOS entered into force, countries observed that article 210 of the LOS obligated states to take measures to "prevent, reduce and control pollution by dumping [that] must be no less effective than the global rules and standards." This was interpreted to mean that all parties to the LOS must respect the standards in the London Convention. Countries party to the LOS, but not yet party to the London Convention, are encouraged to join the latter. Representatives from countries party to the London Convention have been concerned that their treaty not be marginalized by the LOS. The 1996 Protocol addresses the issue of dispute settlement under both the Protocol and the LOS Convention.

The Secretariat. The International Maritime Organization (IMO), before 1981 the Intergovernmental Maritime Consultative Organization, is responsible for the Convention's Secretariat. Secretariat staff prepare for the consultative meetings of the scientific advisory bodies, respond to questions from member states, disseminate notices received from parties, develop procedures for safe dumping in emergency situations, sponsor international symposia for exchanging relevant scientific and technical information, and in general service governments in helping to implement and monitor the Convention. The Secretariat presented a survey questionnaire to the parties at the 1993 Consultative Meeting that asked them for guidance in defining the Secretariat's role. The inquiry focused on the role of the Secretariat in enforcement, in providing assistance to parties, possibly in managing a fund to help parties, and in other implementing and compliance activities.

In 1992, the Secretariat had a budget of about $760,000 and the equivalent of five full-time positions. In 1995, the figures were comparable. But only one professional person was paid directly by the IMO; the rest were on secondment.

Nongovernmental Organizations. NGOs have participated in the Convention as observers since the beginning. These organizations include port and harbor associations (dredgers), chemical manufacturers, an association of oil companies, and environmental organizations, most notably Greenpeace. In response to problems about which organizations should be selected to participate, parties adopted a special annex setting forth procedures governing invitation, documentation, and participation.

Nine to ten NGOs, including both business and environmental groups, attended the Consultative Meetings each year from 1990 to 1993; the data are comparable for earlier and later years. They also have attended meetings of the scientific working groups and in some cases have become involved in the process. Greenpeace, for example, has participated in the scientific working groups by fielding scientists to review scientific justifications for dumping, providing scientific advice to certain states to help them articulate their positions, lobbying delegations, and even drafting resolutions that states can submit. It has joined in deliberations concerning proposed modifications of annexes to the Convention. When member states held the first meeting of the Long-Term Strategy Steering Committee in 1990, NGOs had access to the deliberations.

NGOs have also served as watchdogs in enforcing compliance with the Convention. In 1988, when the Portuguese authorities licensed the scuttling of a fire-damaged vessel containing many listed substances and did not report it to the London Convention Secretariat, Greenpeace brought the issue to the attention of member countries for discussion. The vessel was salvaged (Stairs and Taylor 1992).

Financing. The International Maritime Organization, the host organization for the London Convention of 1972, provides the budget for implementing the treaty. This contrasts with other conventions, in which funding comes directly from the parties to the convention. Because funding comes directly from the IMO, there are problems in securing appropriate amounts. Ship owners and states pay to the IMO budget on the basis of tonnage. Not surprisingly, ship owners are not sympathetic to the Convention. Countries of flags of convenience, such as Liberia and Panama, that make significant payments to the IMO are not eager to see significant allocations to the London Convention. Liberia is not a member of the London Convention.

The Convention has never had a separate fund attached to it for technical assistance or special projects. However, the 1996 Protocol has a new provision on technical cooperation and assistance (article 13) that provides for coordinating assistance and making it available where feasible. No special fund is created. In 1991, parties established a special one-time fund, the Twentieth Anniversary Fund, to computerize the Office of the London Convention at the IMO. The Office receives limited voluntary support from parties for specific projects to implement the Convention.

Dispute Resolution. The Convention does not contain any provision for dispute resolution. Article XI directs the parties to develop dispute resolution procedures, which they did in an unratified 1978 amendment that never took effect. The proposed amendment directed parties to settle disputes by negotiation; and if that failed, upon agreement, to submit the dispute to the International Court of Justice or, at the request of one party, to an Arbitral Tribunal established under the amendment. The

1996 Protocol contains a separate article on settlement of disputes (art. 16) that calls for negotiation, mediation, or conciliation in the first instance. If that is not possible within twelve months, parties may use either the arbitral procedure set forth in the Protocol's annex or the dispute resolutions procedures in the LOS Convention.

Monitoring and Compliance. In assessing compliance with the substantive provisions of the Convention, industrialized countries must be treated separately from those that are industrializing. Most of the industrialized countries have agreed to phase out dumping at sea and are in the process of doing so. Many of the member states either lack the capacity to do so, or are largely unaware of the Convention or the obligations they have assumed under it. This problem is exaggerated because often the contact point within countries is a marine transport department, such as the Coast Guard in the United States. The IMO normally communicates with countries through the naval attachés in London, who may or may not forward messages relating to the London Convention and environmental concerns to the appropriate officials within their countries.

Procedurally, countries are required to provide annual reports to the Secretariat on how they are implementing the Convention. These reports are to indicate, among other things, the nature and quantities of all matter permitted to be dumped, and the location, time, and method of dumping; the permits issued for incinerating wastes at sea; and the wastes dumped without permit because of weather conditions or because of a danger to human life.

From 1991 to 1995, less than 50 percent of the member states fulfilled the annual reporting requirement (as of July 1997). Industrialized countries had the best record in filing annual reports, possibly because they have more resources available for reporting, or perhaps because many developing countries, which do little dumping at sea, simply have nothing to report and have failed to file a report indicating that. As of September 1997, eighteen countries have never filed reports, and eleven have reported irregularly in recent years (IMO 1997). The Secretariat has changed its procedures to provide guidelines on reporting and to regularly remind countries of their obligation to report. States do not publicly chide each other during Convention meetings about the failure to report. Figure 5.10 shows the reports due and received over time.

Table 5.5 shows the record of reporting for the countries in this study that are parties to the Convention. Whereas Brazil, China, Japan, and the United States had filed their required reports through 1995, the Russian Federation and Hungary had not. Hungary filed its first reports for the years 1993 and 1994 on June 30, 1997. Russia has filed a report only for 1993.

It is not clear what failure to report or to report fully means for parties to the London Convention. It could mean the country has not dumped any of the con-

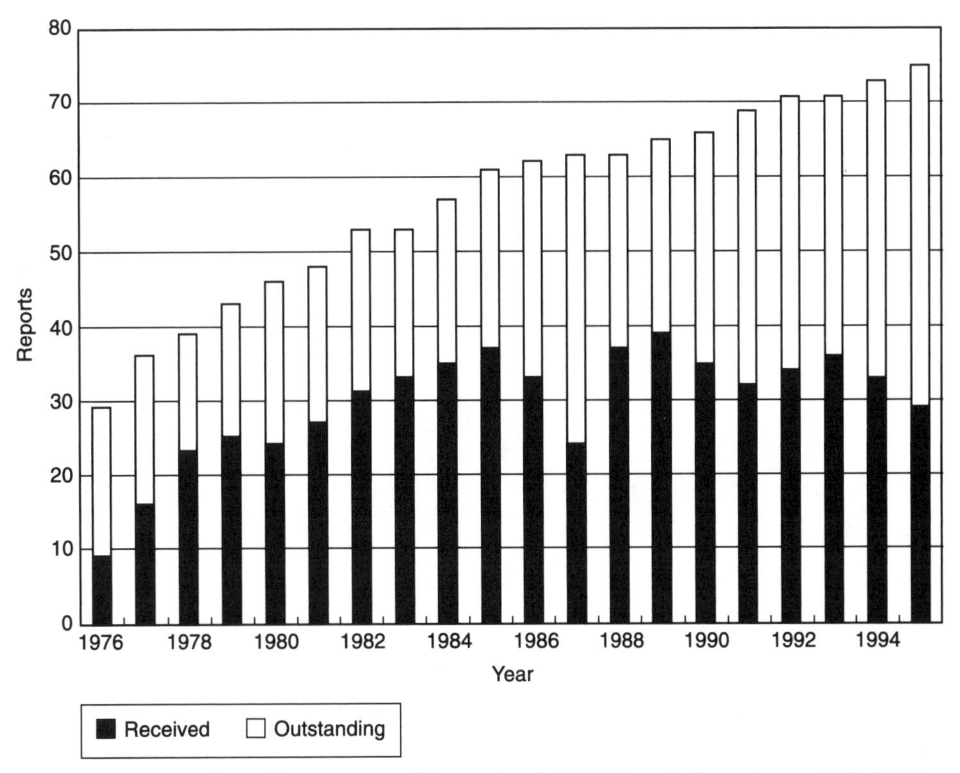

Figure 5.10
Annual reports required of London Convention parties: Received or outstanding

trolled substances, perhaps because it is landlocked; that it has dumped some substances properly but has not reported; or that it has dumped illegally. Only by knowing the economic and political characteristics of the countries involved, or by carefully monitoring the oceans or activities within these countries, can the significance be known.

Significantly, the majority of countries that are the most important for the dumping problem report as required under the Convention. (The Russian Federation has not.) This means that we must reassess how we view noncompliance in filing reports. If the major contributors to a problem are complying with the reporting requirement, it is misleading to look at the overall percentage of parties complying to determine the effectiveness of a convention. In the future, it is likely that parties will be asked to report data in computer-readable form, as is happening with CITES reports.

The first and only published documented case of serious contravention of the spirit of the treaty is the Russian dumping of high- and low-level radioactive waste

Table 5.5
Annual reports of London Convention parties: Transmission/receipt date

Country	Entry into force	1991	1992	1993	1994	1995	1996
Brazil	8/25/82	1/24/96	3/29/96	3/29/96	3/29/96	3/29/96	7/2/97
Cameroon	Not a party						
China	12/14/85	reported, date unknown	9/7/93	8/23/94	6/19/97	6/19/97	N/R
Hungary	3/6/76	N/R	N/R	6/30/97	6/30/97	N/R	N/R
India	Not a party						
Japan	11/14/80	9/10/92	9/2/93	9/13/94	10/4/95	10/18/96	N/R
Russian Fed.	1/29/76	N/R	N/R	10/5/93	N/R	N/R	N/R
United States	8/30/75	dated Oct. 93	12/16/94	112/16/94	dated Dec. 95	dated Oct. 96	N/R

Notes: Annual reports from all London Convention parties were requested by August 1 of the following year. Thus, the annual report for 1991 was requested by August 1, 1992. Beginning in 1997, reports will be expected by November 1. N/R indicates no report was received for that year as of July 1997.

into the oceans. This dramatic case has energized countries into focusing more on monitoring and treaty compliance, as reflected in the 1996 Protocol.

Concluding Observations. The Convention has passed through its first phase—controlling dumping—to a phase focused on controlling marine disposal of dredge material and sewage sludge, and integrated waste management, including controls on land-based sources of marine pollution. Dumping of industrial wastes, wastes incinerated at sea, and low- and medium-level nuclear wastes has either been banned or is in the process of being phased out. Although some dumping continues, much of it is allegedly due to the lack of capacity in some countries to dispose of the wastes elsewhere.

The reports of the IMO and the experts convened to consider compliance for the United Nations Conference on Environment and Development in Rio noted that the Convention has achieved considerable success in combating marine pollution (Boyle et al. 1992). The permitting regime, in addition to restricting dumping, has encouraged predumping changes to meet the new requirements. These changes include altering industrial processes to reduce output of hazardous wastes, increasing the use of recycling methods, and adopting new waste treatment techniques. As a result of these changes, the dumping of industrial waste decreased from 17 million tons in 1979 to 6 million tons in 1987. Even dredged material and sewage sludge, which are still permissible under a general permit, have declined over roughly the same period, from 17 million tons in 1980 to 14 million tons in 1986, and have continued to decline since then (Boyle et al. 1992).

This relative success has occurred despite the fact that there is less than 50 percent compliance with reporting requirements and, as revealed by the discovery of Russian dumping of nuclear wastes, still ineffective global monitoring. It has occurred even though there is no technical assistance fund that makes grants to the parties or provides the Secretariat with an independent source of funds for technical assistance. Nor have NGOs exerted the extraordinary influence seen in CITES or the World Heritage Convention. The life history of the Convention raises the question of why countries have been willing to control marine dumping of wastes relatively effectively. Familiarity with regulating the marine environment, integral use of scientific advisory bodies, availability of other waste disposal options, and technological advances seem to have been important in facilitating implementation and compliance.

Montreal Protocol on Substances That Deplete the Ozone Layer

The Montreal Protocol is a path-breaking example of countries' willingness to address a problem in which the harmful consequences of our activities today will be felt primarily by our children and their descendants. In the early 1980s jurists wrote of the unlikelihood of reaching agreement to control the substances that deplete the ozone layer (Williams 1986). Yet by 1987 a treaty had been concluded. Almost ten years later it has evolved from one primarily concerned with identifying risks to the ozone layer and adding new substances to one focused primarily on managing the complete phaseout of most of the controlled substances. Moreover, it is one of the most universally adhered-to international environmental accords.

In 1974, Mario Molina and Sherwood Rowland, scientists at the University of California at Davis, published findings that chlorofluorocarbons (CFCs) can remain in the atmosphere and release large quantities of chlorine into the stratosphere when broken down by the sun (Molina and Rowland 1974). The chlorine in turn breaks down the high-level ozone layer protecting Earth from ultraviolet radiation. In March 1977, at United States initiative, government experts, intergovernmental organizations, and NGOs met in Washington, D.C., to draw up a World Plan of Action on the Ozone Layer (Benedick 1991). The Plan recommended intensive international research and monitoring of the ozone layer, and charged the United Nations Environment Programme (UNEP) with central responsibility for promoting research and gathering relevant data.

In May 1981 the Governing Council of UNEP established the Ad Hoc Working Group of Legal and Technical Experts to draft a global framework convention on stratospheric ozone protection (UNEP 1981a, 1981b). In March 1985, states adopted the fifth draft convention with technical annexes as the Vienna Convention for the Protection of the Ozone Layer, which entered into force on September 22, 1988. As of June 1997, 164 countries were party to the Convention.

The Vienna Convention is a framework treaty in which states agree to cooperate in conducting research on and scientific assessments of the ozone problem and in exchanging information, and to adopt "appropriate measures" to prevent human activities that adversely affect the ozone layer. The obligations are by nature general and contain no precise limits. The Convention contemplates that states will adopt protocols to implement it.

During the negotiations, states discussed the drafting of a protocol controlling chemicals that deplete the ozone layer, which would accompany the Vienna Convention. All of the UNEP regional seas accords provided a framework convention and at least one protocol, which countries were required to ratify at the time they joined the framework treaty. However, no consensus could be reached on a protocol to the Vienna Convention, so the Convention went forward on its own. The executive director of UNEP established a working group to begin the drafting of a protocol. Although this process has become common today, the separation of the framework treaty from any accompanying protocol was novel at the time. The Montreal Protocol on Substances That Deplete the Ozone Layer was concluded in September

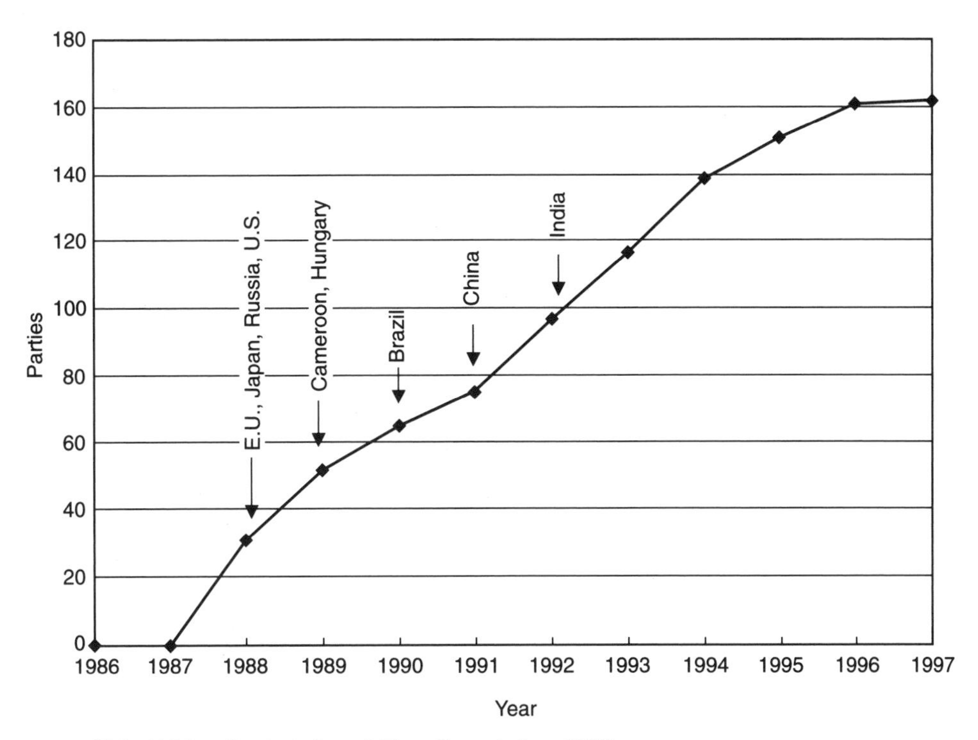

Figure 5.11
Total parties to Montreal Protocol and ratification years for countries in this study

1987 and became effective in January 1989. As of June 1997, 162 countries were party to it. Figure 5.11 shows the rate of ratification of the Protocol over time.

The Montreal Protocol reflects a convergence of interests of different actors: scientists who warned of growing threats to the ozone layer from certain chemicals, private industry that wanted a level international playing field as companies responded to national legislation controlling the chemicals, NGOs eager to promote environmental protection, and governments prepared to further what they perceived as national interests in reaching an agreement (Soroos 1997; Benedick 1991). It took less than four years to negotiate the Protocol and to put it into effect, a very short period at the time.

During the negotiations for the Montreal Protocol, there was heated controversy over whether targets should be based on consumption or production of the controlled chemicals. The CFC industry was very concentrated, with five major companies in the United States, five in Japan, and nine in the European Community. Du Pont, the largest United States producer, accounted for about 50 percent of United States CFC production and 25 percent of global production at that time; Imperial Chemical Industries was the largest European producer.

Consumption of CFCs, by contrast, was and is diffuse, because there are many different uses for the chemicals. In 1986 the European Community and the United States each consumed about 30 percent of CFCs, with Japan consuming another 13 percent.

Although both the United States and the European Community were leading producers of CFCs, only the European Community was a large net exporter. It is perhaps not surprising, then, that the European countries argued strenuously for a production-based control system, and the United States for a consumption-based system (Enders and Porges 1992). The resulting formula equates consumption to production minus exports plus imports of controlled ozone-depleting substances.

Treaty Commitments and Implementing Institutions. The Montreal Protocol controls the production and consumption of specific chemicals that deplete the ozone layer: CFCs, halons, fully halogenated CFCs (HCFCs), methyl bromide, and similar chemicals. None of the chemicals occurs naturally. The 1987 text of the Protocol required parties other than developing countries to freeze their consumption and production of CFCs at 1986 levels (the base year), to reduce them by 20 percent and then another 30 percent by 1999, and, further, to freeze the consumption of halons at 1986 levels.

Since 1987 there have been significant amendments to the Protocol to add controlled substances and to develop new institutional features of the regime and formal adjustments to ratchet up the targets and timetable for controlled substances already listed in the annexes.

Less than eighteen months after the Protocol came into effect, parties adopted the London Amendments of 1990, which were treated as a package that had to be accepted or rejected in their entirety. The Amendments provided, most importantly, for an Interim Multilateral Fund to provide assistance to qualifying developing countries, for noncompliance procedures, for the addition of new chemicals (carbon tetrachloride, methyl chloroform, and other fully halogenated, and for miscellaneous changes to the Protocol. The London Amendments entered into force on August 10, 1992, and, as of March 1997, had been ratified by 114 states. In 1990, states also adopted adjustments, which are not amendments, requiring parties to phase out the CFC chemicals listed in 1987 by the year 2000 and to phase out halons by then except for certain essential uses.

In November 1992, at Copenhagen, the parties amended the Protocol to make the Interim Multilateral Fund permanent, to create the Implementation Committee, and to control additional chemicals (HCFCS, hydrobromide fluorocarbons, and methyl bromide). The Copenhagen Amendment went into effect on June 14, 1994, ninety days after twenty states had ratified it. As of March 1997, sixty-five states were parties to the Copenhagen Amendment. In 1990, 1994, and 1995, parties adjusted the targets and timetables for controlling chemicals that were already listed in order to hasten their phaseout (UNEP 1996a). Further adjustments and amendments were adopted at the ninth Meeting of the Parties in September 1997. They included rational licensing systems for exports and imports of ozone-depleting substances, controls on trade by parties not in compliance with the Protocol, trade restrictions on used, recycled, and reclaimed substances, adjustments on CFCs reduction for Article V countries, accelerated phaseout of methyl bromide, controls over trade in methyl bromide with nonparties, and other items (UNEP 1997). Existing obligations, indicating targets and timetables as of December 1997, are presented in table 5.6.

Anticipating that our scientific knowledge about ozone-depleting chemicals will constantly change, the treaty negotiators provided parties with two ways to alter obligations to control substances: an adjustment process whereby parties may adjust the targets and timetables for existing chemicals without having to go through the formal amendment procedure, and a formal amendment process. Adjustments enter into force six months after parties receive formal notice of them; they bind all parties to the Protocol. For amendments, article 9(4) of the Vienna Convention on the Protection of the Ozone Layer governs and provides that the Protocol can be amended by a two-thirds vote and then submitted to parties for ratification. Amendments bind only the countries that ratify them.

As a result of the amendment process, different states are bound by different obligations. Countries becoming party to the Protocol after the amendments take effect automatically become party to the fully amended treaty. But for states that were already parties, the pattern of obligations is complicated. Some states are still party only to the original Protocol, others to the Protocol as amended in London, and others to the Protocol as amended both in London and in Copenhagen.

The Protocol provides flexibility to certain groups of countries in meeting obligations. To attract the former Soviet Union as a member, it provides in article 2(6) that any country that prior to September 1987 had facilities to produce the controlled chemicals under construction or contracted for may add this production to its 1986 base level for purposes of calculating its compliance with base year production. The Protocol also incorporates "industrial rationalization" among countries. Under this policy, a country can transfer part of its calculated level of production of controlled chemicals to another country. Thus, Canada, which produces at less than 20 percent of allowable levels, transferred its allowance to the United States. While industrial rationalization can take place between developed states or between developing states, it was not intended for transfers between developed and developing countries.

The Protocol makes special provision for developing countries in article V: a ten-year delay in required compliance with targets and timetables, a separate con-

Table 5.6
Montreal Protocol on Ozone-Depleting Substances

Conference of the Parties, London, 1990	
Substance	Adjustments/amendments
CFCs	Adj: After the first day of the seventh month following entry into force of MP, annual consumption or product levels will not exceed 1986 levels. Adj: After January 1, 1995, annual consumption or production will not exceed 50% of 1986 levels. Adj: After January 1, 1997, annual consumption and production levels will not exceed 15% of 1986 levels. Adj: After January 1, 2000, annual consumption and production levels will not exceed zero.
Halons	Adj: After January 1, 1992, annual consumption or production levels will not exceed 1986 levels. Adj: After January 1, 1995, annual consumption or production levels will not exceed 50% of 1986 levels. Adj: After January 1, 2000, annual consumption and production levels will not exceed zero.
Other fully halogenated CFCs	Amend: After January 1, 1993, annual consumption or production will not exceed 80% of 1989 levels. Amend: After January 1, 1997, annual consumption or production will not exceed 15% of 1989 levels. Amend: After January 1, 2000, annual consumption or production will not exceed zero.
Carbon tetrachloride	Amend: After January 1, 1995, annual consumption or production will not exceed 15% of 1989 levels. Amend: After January 1, 2000, annual consumption or production will not exceed zero.
Trichlorethane (methyl chloroform)	Amend: After January 1, 1993, annual consumption or production will not exceed 1989 levels. Amend: After January 1, 1995, annual consumption or production will not exceed 70% of 1989 levels. Amend: After January 1, 2000, annual consumption or production will not exceed 30% of 1989 levels. Amend: After January 1, 2005, annual consumption or production will not exceed zero.

Table 5.6 (continued)

Conference of the Parties, Copenhagen, 1992	
Substance	Adjustments/Amendments
CFCs	Adj: After January 1, 1994, annual consumption or production will not exceed 25% of 1986 levels. Adj: After January 1, 1996, annual consumption or production will not exceed zero.
Halons	Adj: After January 1, 1994, annual consumption or production will not exceed 25% of 1989 levels. Adj: After January 1, 1996, annual consumption or production will not exceed zero.
Other fully halogenated CFCs	Adj: After January 1, 1994, annual consumption or production will not exceed 1989 levels. Adj: After January 1, 1996, annual consumption or production will not exceed zero.
Carbon tetrachloride	Adj: After January 1, 1996, annual consumption or production will not exceed zero.
Trichlorethane (methyl chloroform)	Adj: After January 1, 1996, annual consumption or production will not exceed zero.
Hydrochloro-fluorocarbons	Amend: After January 1, 1996, annual consumption or production will not exceed the sum of 3.1% of calculated level of consumption in 1989 of the controlled substances in group 1 of annex A and the calculated level of consumption in 1989 of the controlled substances in group 1 of annex C. Amend: After January 1, 2004, annual consumption will not exceed 65% of the sum above. Amend: After January 1, 2010, annual consumption will not exceed 35% of the sum above. Amend: After January 1, 2015, annual consumption will not exceed 10% of the sum above. Amend: After January 1, 2020, annual consumption will not exceed 0.5% of the sum above. Amend: After January 1, 2030, annual consumption will not exceed zero.
Hydrobromo-fluorocarbons	Amend: After January 1, 1996, annual consumption will not exceed zero.
Methyl bromide	Amend: After January 1, 1995, annual consumption will not exceed 1991 levels.

Table 5.6 (continued)

Seventh Meeting of the Parties, Vienna, 1995	
Substance	Adjustments to non-article V countries and to article V countries[a]
CFCs[b]	Adj: After July 1, 1989, annual consumption and production of annex A CFCs will be frozen.
	Adj: After January 1, 1993, annual consumption and production of annex B CFCs will be reduced by 20% from 1989 levels.
	Adj: After January 1, 1994, annual consumption or production of annex A CFCs will be reduced by 75% from 1986 levels; annual consumption or production of annex B CFCs will be reduced by 75% from 1989 levels.
	Adj: After January 1, 1996, annual consumption and production levels of annex B and annex A CFCs will not exceed zero.
	Adj: After July 1, 1999, annual consumption and production levels of annex A CFCs will be frozen at 1995–1997 average levels (base).
	Adj: After July 1, 2003, annual consumption and production levels of annex B CFCs will be reduced by 20% from 1998–2000 average levels (base).
	Adj: After July 1, 2005, annual consumption and production levels of annex A CFCs will be reduced by 50% from base levels.
	Adj: After January 1, 2007, annual consumption and production levels of annex A and annex B CFCs will be reduced by 85% from base levels.
	Adj: After January 1, 2010, annual consumption and production levels of annex A and annex B CFCs will not exceed zero.
Halons[c]	Adj: After January 1, 1992, annual consumption or production will be frozen.
	Adj: After January 1, 1994, annual consumption or production will not exceed zero.
	Adj: After January 1, 2002, annual consumption and production will be frozen at 1995–1997 average levels.
	Adj: After January 1, 2005, annual consumption and production will be reduced by 50% from 1995–1997 average levels.
	Adj: After January 1, 2010, annual consumption and production will not exceed zero, as per the London Amendments.
Carbon tetrachloride	Adj: After January 1, 1995, annual consumption and production will be reduced by 85% from 1989 levels.
	Adj: After January 1, 1996, annual consumption and production will not exceed zero.
	Adj: After January 1, 2005, annual consumption and production will be reduced by 85% from 1998–2000 levels.
	Adj: After January 1, 2010, annual consumption and production will not exceed zero, as per the London Amendments.

Table 5.6 (continued)

Substance	Adjustments to non-article V countries and to article V countries[a]
Trichlorethane (methyl chloroform)	Adj: After January 1, 1993, annual consumption and production will be frozen.
	Adj: After January 1, 1994, annual consumption and production will be reduced by 50% of 1993 levels.
	Adj: After January 1, 1996, annual consumption or production will not exceed zero.
	Adj: After January 1, 2003, annual consumption and production will be frozen at 1998–2000 average levels (base).
	Adj: After January 1, 2005, annual consumption and production will be reduced by 30% from base levels.
	Adj: After January 1, 2010, annual consumption and production will be reduced by 70% from base levels.
	Adj: After January 1, 2015, annual consumption and production will not exceed zero.
Hydrochlorofluorocarbons[d]	Adj: After January 1, 1996, annual consumption or production will be frozen at 1989 levels plus 2.8% of 1989 consumption of CFCs (base).
	Adj: After January 1, 2004, annual consumption or production will be reduced by 35% from base levels.
	Adj: After January 1, 2010, annual consumption or production will be reduced by 65% from base levels.
	Adj: After January 1, 2015, annual consumption or production will be reduced by 90% from base levels.
	Adj: After January 1, 2016, annual consumption and production will be frozen at 2015 levels.
	Adj: After January 1, 2020, annual consumption or production will be phased out, allowing for a service tail of up to 0.5% until 2030 for existing refrigeration and air-conditioning equipment.
	Adj: After January 1, 2040, annual consumption and production will not exceed zero.
Hydrobromofluorocarbons[e]	Adj: After January 1, 1996, annual consumption and production levels will not exceed zero.
Methyl bromide	Adj: After January 1, 1995, annual consumption and production will be frozen at 1991 levels.
	Adj: After January 1, 2001, annual consumption and production will be reduced by 25% from 1991 levels.
	Adj: After January 1, 2005, annual consumption and production will be reduced by 50% from 1991 levels.
	Adj: After January 1, 2010, annual consumption and production will not exceed zero.

Table 5.6 (continued)

Montreal 1997 9th Meeting of the Parties	
Substance	Adjustments to Non-article V countries and to article V countries
CFCs[b]	**Adj: The lower of the average of annual calculated level of production for the period 1995 to 1997 or a calculated level of production of 0.3 kg/capita will be used for determining compliance levels of annex A CFCs.**
	Adj: The lower of the average of annual calculated level of production for the period 1998 to 2000 or a calculated level of production of 0.2 kg/capita will be used for determining compliance levels of annex B CFCs.
Methyl bromide	Adj: After January 1, 1999, annual consumption or production will not exceed 75% of 1991 levels.
	Adj: After January 1, 1999, annual consumption or production will not exceed 85% of 1991 levels.
	Adj: After January 1, 2001, annual consumption or production will not exceed 50% of 1991 levels.
	Adj: After January 1, 2001, annual consumption or production will not exceed 60% of 1991 levels.
	Adj: After January 1, 2003, annual consumption or production will not exceed 30% of 1991 levels.
	Adj: After January 1, 2003, annual consumption or production will not exceed 40% of 1991 levels.
	Adj: After January 1, 2005, annual consumption or production will not exceed zero.
	Adj: After January 1, 2005, annual consumption and production will not exceed 15% of 1991 levels.[g] Annual consumption or production will also not exceed 80% of the average of its annual calculated levels of consumption and production for the period of 1995 to 1998 inclusive.
	Adj: After January 1, 2015, annual consumption or production will not exceed zero.

[a] Adjustments in bold apply to article V countries.
[b] Annex A CFCs are CFCs 11, 12, 113, 114, and 115. Annex B CFS are CFCs 13, 111, 112, 211, 212, 213, 214, 215, 216, and 217.
[c] Halons 1211, 1301, and 2402.
[d] 34 hydrochlorofluorocarbons.
[e] 34 hydrobromofluorocarbons.
[f] With exceptions for essential uses. Consult UNEP, *Handbook on Essential Use Nominations Prepared by the Technology and Economic Assessment Panel* (UNEP, 1994) for more information.
[g] Will apply save to the extent that the parties decide to permit the level of production or consumption that is necessary to satisfy uses agreed by them to be critical uses.

sumption limit of 0.3 kilogram per capita, access to the Interim Multilateral Fund (1990 amendment) to assist with costs of compliance, and the promotion of bilateral technical assistance programs. The classification of a developing country as deserving special treatment (i.e., operating under article V(1) of the Protocol) is not permanent, but is subject to the recommendations of an open-ended working group of parties. For example, in 1993, Cyprus, Kuwait, Korea, Saudi Arabia, Singapore, and the United Arab Emirates were declassified as article V countries (UNEP 1993b). Other countries have been reclassified to become article V countries, most recently Georgia, in November 1996. As of March 1996, eighty-one countries were categorized as operating under article V(1), and thirty countries were temporarily categorized as article V(1) pending receipt of complete data.

To induce developing countries to join the Protocol, in 1990 the parties established the Interim Multilateral Fund to provide assistance to developing countries (those qualifying under article V) to meet the additional costs of complying with the Protocol. This step was essential to convincing countries such as China and India to join the Protocol. The Fund's Secretariat is based in Montreal, Canada, rather than in Nairobi, with the Protocol's Secretariat.

The Protocol prohibits exports and imports of controlled substances involving countries that are not party to the treaty. In a second stage the trade ban with nonparties would extend to products containing the controlled substances and, upon agreement of the parties within five years, to goods produced with the controlled chemicals. However, parties have accepted the recommendation of the Technology and Economic Assessment Panels and decided not to implement the ban based on process (UNEP 1992a).

Critics have questioned whether the provisions restricting trade with nonparties are consistent with the General Agreement on Tariffs and Trade (GATT), later incorporated into the World Trade Organization. GATT prohibits import quotas, requires national treatment for both imported and domestic "like" products, and requires most-favored-nation treatment among countries (i.e., the most favorable treatment offered to one exporting country must be granted to all). Critics argued that the trade provisions would cause states to discriminate in their trade on the basis of whether a state was a party to the Protocol and did not come within the article XX(b) and (g) exceptions (Brack 1996).

The protocol's Ad Hoc Working Group of Legal Technical Experts (UNEP 1987) contended that the protocol's trade ban provisions fell under certain GATT article 20 exceptions that allowed for measures "necessary to protect human, animal or plant life or health" and/or "relating to the conservation of exhaustible natural resources," although questions have been raised about the interpretation of both exceptions. The Working Group based its opinion on precedents pertaining to the Convention on International Trade in Endangered Species of Wild Flora and Fauna

(CITES) and the London Convention of 1972 resolution 29(10), on export of waste for disposal at sea. The Working Group argued that the GATT prohibitions against discrimination "would not arise at all, if the trade restrictions regarding non-parties did not apply to non-parties that were able to demonstrate full compliance with the control measures provided for in the protocol." Article 4(8) of the Protocol permits such trade with "complying nonparties," although this requires a decision of the parties that the nonparty is in fact complying.

At the fourth Meeting of the Parties in Copenhagen, the parties adopted a procedure for identifying such nonparties exempt from the usual trade controls (UNEP 1992e). At the fifth Meeting of the Parties, this procedure was reviewed for the first time, and the Secretariat received complete data from twelve nonparty states, partial data from an additional four, and data from eleven states that were parties to the Montreal Protocol but not to the London Amendments.

Since all major producers and consumers of the controlled substances are parties to the treaty, it is very unlikely that the restrictions on trade with nonparties will be important for trade in chemicals listed in the 1987 annexes. However, to the extent that countries are not parties to the two amendments that add new chemicals to the list of controlled substances, the restrictions could have a significant effect if nonparties cannot demonstrate to the satisfaction of the parties that they are in compliance with the provisions of the amendments. The issue of controlling trade in methyl bromide—a chemical added by amendment in 1992—with nonparties was addressed in the 1997 Conference of the Parties.

Treaty Implementation. States have been unusually aggressive and effective in implementing the Montreal Protocol. By the time it came into effect in 1989, countries were already contemplating its modification. Since then, parties have amended the Protocol three times: 1990, 1992 and 1997. They have made numerous adjustments to the timetables for phasing out consumption of listed chemicals, as well as very significant institutional changes: the addition of the Montreal Protocol Fund, establishment of the Implementation Committee, the development of noncompliance procedures, and the expansion of the Technology and Economic Assessment Panels. They have addressed new issues as they have arisen, such as recycling and smuggling of CFCs.

At the first Meeting of the Parties, states established four Technical and Economic Assessment Panels: Scientific Assessment, Technology Review, Environmental, and Economic. The Technology and Economic Assessment Panels involve several hundred scientists worldwide. They have played a crucial role in causing parties to ratchet up targets and timetables, add chemicals to the list of controlled substances, and address problems such as recycling. They also have provided credible risk assessments and evaluations of control options.

As the Protocol has matured, attention has shifted from primarily listing the culprit chemicals and advancing phaseout timetables for developed countries to making the regime effective in controlling chemicals that deplete the ozone layer. This means greater attention to monitoring and reporting, country programs under the Montreal Protocol Fund, and new issues such as recycling and smuggling of ozone-depleting substances.

One of the most significant innovations under the Protocol is the regime established to address issues of noncompliance, which includes the Implementation Committee of ten states and specific noncompliance procedures (Koskenniemi 1992; W. Lang 1995; Szell 1995; Victor 1995). The Committee meets biannually. It was established in 1990 (with only five members) and finalized in 1992. The Committee has become increasingly important to securing parties' compliance, as detailed in the discussion of monitoring and compliance.

Noncompliance procedures can be activated by one party against another, by the Secretariat, or by a party in respect of itself. Proposals to let NGOs activate the Non-Compliance Procedure were greeted warily by developing countries. In the first most significant cases of substantive violations, the parties themselves initiated the procedures because other parties and the Secretariat were reluctant to take responsibility for doing so, and preferred to exert pressure on the countries potentially in violation to do so.

The Implementation Committee reviews reports submitted by parties and addresses possible violations of targets and timetable obligations. It can make on-site visits to countries believed to be in noncompliance. In response to noncompliance, the Committee can turn either to incentives (assistance to enable countries to comply) or sticks (warnings or suspension of rights and privileges under the Protocol).

As with other environmental treaties, there have been efforts to disseminate information about the treaty and to engage industry and interested publics. In 1992, UNEP, in collaboration with industry, launched a quarterly newsletter, *OzonAction*, which provides information on relevant developments in the public and the private sectors, especially on new technology advances and technology cooperation with developing countries. It is available in six languages. The UNEP Industry Environment (IE) OzonAction Programme, based in Paris also holds symposia that bring together government officials, industry officials, nongovernmental experts, and scholars who are directly concerned with controlling ozone-depleting substances (UNEPIE 1996a).

The Secretariat. There are two separate secretariats associated with the Montreal Protocol: one for the Protocol, and the other for the Montreal Protocol Fund. The main secretariat is located at UNEP headquarters in Nairobi, Kenya. The Montreal Protocol Fund's Secretariat is in Montreal, Canada, but is associated with the Global

Environmental Facility at the World Bank in Washington, D.C. As of 1995, there were eleven people in the Montreal Protocol Secretariat at UNEP, with a budget of $3.4 million. In 1990, by comparison, there were five staff members and a very large consultant budget that no longer exists. The Montreal Protocol Fund Secretariat has remained constant at eighteen people.

The Protocol Secretariat prepares materials for meetings, compiles and monitors the reports from parties, answers inquiries, and in general tries to service the parties. It is actively engaged in encouraging companies, countries, and others to monitor and facilitate compliance with the Protocol, and this pattern of engagement has appeared to increase significantly over the lifetime of the Protocol. Whereas in 1990, the Secretariat divided its time about equally between reporting by the parties and preparing background documentation, in 1994 it spent about 30 percent of its time on reporting of the parties, 25 percent responding to questions, 20 percent providing background documentation for meetings, and 25 percent coordinating activities with the Montreal Protocol Fund and Global Environmental Facility. The Secretariat appears to play an increasingly important role in facilitating compliance with the Protocol.

Nongovernmental Organizations. From the beginning, industry associations of producers of the controlled substances have been important actors in the negotiation and implementation of the Convention. In the United States, the CFC Alliance ultimately became one of the foremost proponents of an international treaty. A private-sector network of producers in the United States has provided information to countries around the world about substitutes for CFCs.

The private sector also actively monitors compliance with the obligations to phase out production and consumption of controlled chemicals. Producers want to maintain a level playing field, and hence have a strong economic incentive to be sure that all companies are following the rules. Although this monitoring function operates on an informal basis and is rarely acknowledged, several officials interviewed in both the public and the private sector have confirmed its importance.

Environmental NGOs have also been active participants. They pressed hard for a treaty to limit CFC production and launched an effective public drive in their respective states to get the Protocol ratified. They have been active at subsequent Meetings of the Parties, where they have pressed for stronger measures. Within the United States, environmental NGOs have been critical to ensuring that the government has developed regulations to phase out controlled substances.

In contrast to CITES, environmental NGOs have not been given primary responsibility for tracking compliance with the Protocol. Most such organizations do not possess the scientific sophistication and bureaucratic size needed to monitor compliance with the Montreal Protocol. Moreover, production data are classified as a trade secret in many states, which hinders monitoring efforts.

In the Montreal Protocol Fund, NGOs generally have had even less of a role. They have been excluded from Executive Committee meetings. Officials brief them following the meetings and provide special opportunities to address the Committee.

Financing. To assist countries in complying with the Protocol, parties agreed to establish the Montreal Protocol Fund in 1990. Only countries that qualify as article V developing countries are eligible for assistance from the Fund.

In order to receive assistance, countries must submit for approval a country program that covers production and consumption of the regulated substances, and sets forth an institutional structure that will enable the country to comply. Each country designates a person to serve as the point person for the country's program implementation.

The Executive Committee of fourteen states party to the Protocol (equally divided between developed and developing countries) works with the Secretariat to manage the Fund. The Committee approves project funding, the administering guidelines, and the work programs of the intergovernmental organizations that implement projects for the Fund. Voting is consensual, but if no consensus can be reached, a two-thirds vote of both the developed and the developing countries is needed to approve projects larger than $500,000.

The Executive Committee allocates funds to the World Bank, the United Nations Development Program, the United Nations Environment Programme, and, since 1993, the United Nations Industrial Development Organization to execute projects in countries. Figure 5.12 indicates the distribution of funds among the four implementing organizations.

Contributions to the Fund are based on the formula for United Nations assessments. Thus, the United States pays 25 percent; the European Union countries, 34.57 percent; and Japan, 13.39 percent. These are also the countries that have contributed most to creating the ozone depletion problem. Individual states may decrease their assessed contributions by entering a bilateral or regional cooperation agreement with an article V(1) country that "strictly relates to compliance with the provisions of the protocol, provides additional resources, and meets agreed incremental costs."

The Fund is linked to the Global Environment Facility (GEF), a $1.2 billion, three-year fund created in 1990 to provide grants and concessional loans to developing countries so that they can address environmental issues, including ozone depletion. The GEF was restructured in 1994 to include developing countries in the governing body and has been made permanent.

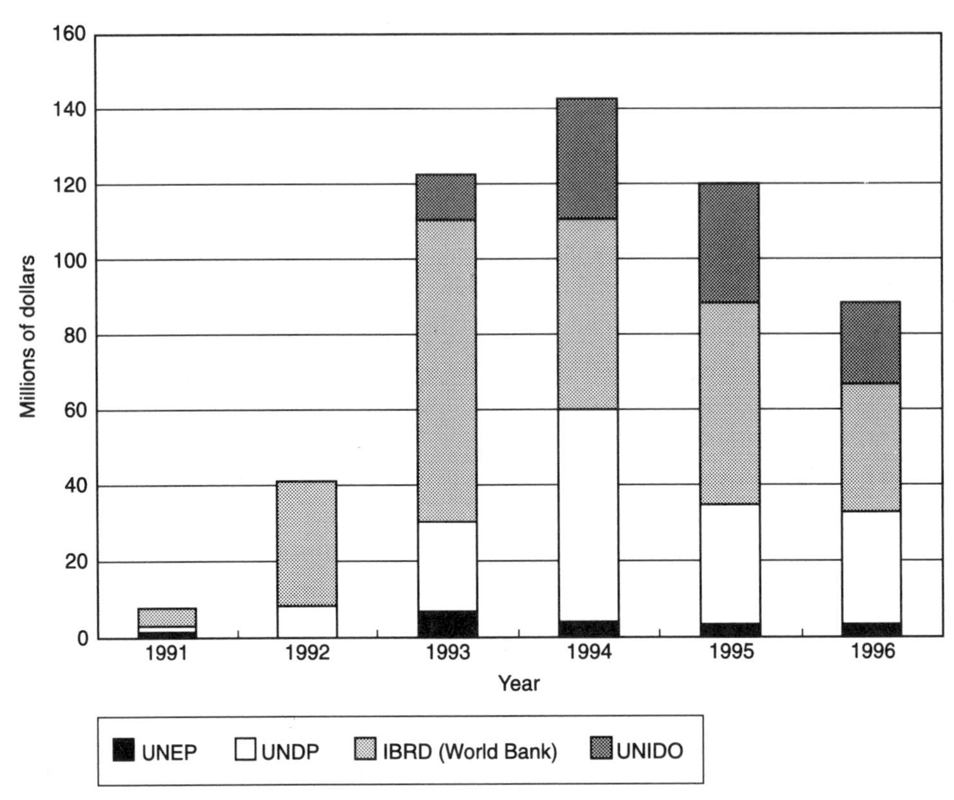

Figure 5.12
Montreal Protocol Fund by disbursing agency, 1991–1996

As indicated, the Executive Committee must approve all country programs, which provides an important opportunity for monitoring and review. Surprisingly, however, the Committee has not required, as a condition of approving country programs, that countries file annual reports on consumption of controlled chemicals, as required by the Protocol. In 1994 the Meeting of the Parties decided that if countries have not filed baseline data reports within a year after approval of their country report, they will no longer be considered article V developing countries entitled to delayed compliance deadlines and special assistance.

Implementation of the Montreal Protocol Fund has gone from an initial period in which funds were available but there were few projects, to the situation in 1994 and thereafter, in which many projects have been proposed but funding is scarce. There are many reasons for this. Initially there were problems in developing projects that could be funded. World Bank procedures were cumbersome and lengthy for small projects, and many countries did not know what projects to propose or how to

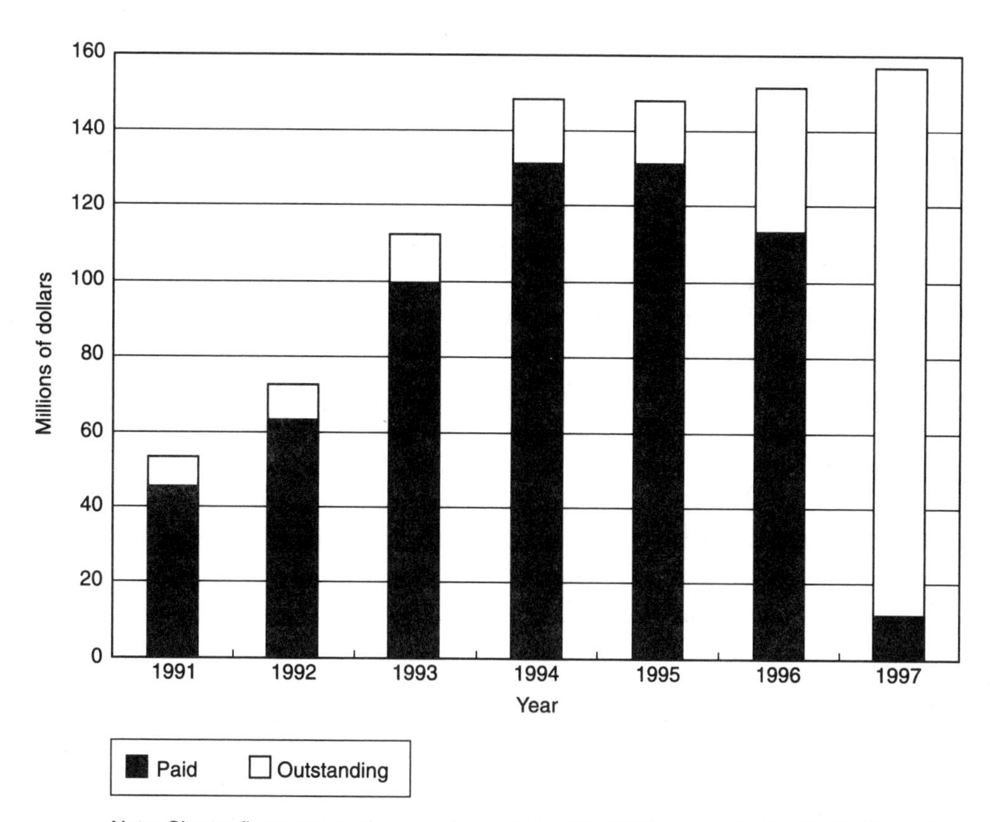

Note: Chart reflects status of payments as of May 30, 1997, as reported by the Multilateral Fund for Implementation of the Montreal Protocol, July 15, 1997. Nineteen hundred ninety-six total includes $8,098,267 in disputed contributions.

Figure 5.13
Pledges and payments to Montreal Protocol Fund, 1991–1997

propose them. By contrast, the United States Congress and national parliaments wanted to know that the funds had been spent before they allocated more, thus creating pressure to fund as many projects as possible as quickly as possible. Countries were thus encouraged to come up quickly with projects. By 1995, although proposals reached $200 million, available funds amounted to less than $20 million, which put great pressure on countries asking for funds. Figure 5.13 indicates the level of funding by year and the extent to which donor countries have outstanding pledges to the fund.

At first, countries were slow to pay the assessed contributions. By the time of the third Meeting of the Parties, only $12.7 million had been received out of the $80 million committed by the parties. It has been argued that, given "the legally binding nature of the duty to contribute," the failure by a developed country to pay its contribution could be deemed an instance of noncompliance. However, nearly all of the assessed contributions were eventually received. In the second phase of the Fund's operations, countries also were slow to provide their assessed contributions, and substantial agreed contributions are outstanding. As of November 22, 1996, of the $152 million of agreed contributions to the Fund, $78 million were outstanding (although countries had until the end of January 1997 to pay) (UNEP 1996b).

The Secretariat of the Montreal Protocol Fund, together with the UNEP IE OZONAction Programme has helped develop and review country programs and proposed projects. The review process appears to be a tough one, relatively free from political machinations.

In 1992, the Executive Committee of the Fund approved funding for the preparation of thirty-seven country programs at a total cost of $1.7 million; four country programs (in Chile, Ecuador, Malaysia, and Mexico); sixty-seven projects in thirteen countries at a cost of $35 million; and disbursements to implementing agencies and the Fund Secretariat (UNEP 1992d). By 1996, most countries had developed country programs or were in the process of doing so.

The country ozone action officers designated as part of the country programs meet regularly on a regional basis to discuss implementation problems and to exchange views. This has led to a new transnational network of officials who are known to each other and within their own countries are engaged in pushing for implementation of the Protocol.

Dispute Settlement. The Vienna Convention for the Protection of the Ozone Layer contains an elaborate article on settlement of disputes involving its interpretation or application. These procedures also apply to disputes regarding the Montreal Protocol. The disputing parties are to resolve their differences through negotiation, and if that does not succeed, they are to involve a third party. States also can agree, at the time they ratify the treaty, to accept compulsory dispute settlement procedures of arbitration or referral to the International Court of Justice; if they have not done so, then the dispute goes to a conciliation commission, set up for the dispute, that is charged with giving a final recommended award. These provisions have never been invoked. Rather, countries have invoked the Protocol procedures for noncompliance, which provide a nonconfrontational means of resolving disputes.

Monitoring and Compliance. Monitoring of compliance is done by governments, the major industries in the private sector, and, to a much lesser extent, NGOs. The handful of large companies that produce ozone-depleting substances have an important financial stake in ensuring that competitors abide by the treaty, as well as the resources to monitor compliance, albeit quietly.

Since the Protocol has come into effect, there have been four important noncompliance issues: failure to report or to report fully on a timely basis; failure to meet targets and timetables for controlled chemicals (in Russia and several central and east European countries); smuggling of CFCs into Western countries; and anticipated compliance problems by several article V developing countries in meeting targets and timetables when the ten-year delay period expires.

The primary tool that parties have under the Protocol to monitor compliance with the obligations to limit or phase out consumption of controlled chemicals by given dates is annual country reports. These are sent to the Secretariat for compilation and review by parties. The Implementation Committee reviews cases of noncompliance with the reporting procedures and with the substantive obligations. Early on, it was apparent that many countries either were not providing the required annual reports or were failing to provide complete reports on time. This made the reports less relevant and useful. In 1991, of the forty-eight parties required to report 1989 data, only twenty had fully complied as of March (UNEP 1991a). In November 1996, 104 of the 141 countries that should have reported data for 1994 had done so; only 61 parties had reported for 1995 (UNEP 1996b). This is a substantial percentage improvement over 1990. The countries reporting include the major producers and consumers of the controlled chemicals. However, there have been indications over time that at least some of the data reported have been unreliable. Figure 5.14 shows the overall reporting record.

Table 5.7 shows the reporting records of the countries in this study who are parties to the Montreal Protocol. As of July 1997, all had filed the required reports through 1995. In 1994, parties required countries to report on facilities for recycling controlled chemicals and on exports of recycled chemicals.

The Implementation Committee took as its first compliance issue the failure of countries to report or to report completely and on time. The Committee chose to view such cases of noncompliance by article V countries as ones in which the countries lacked the capacity to comply. The resolution was to provide "incentives" to countries to report by helping them to develop the capacity to report and to ask the Fund Secretariat and implementing bodies to indicate how their assistance was helping to develop this capacity.

However, the Implementation Committee also initially expressed concern at the failure of some of the members of the European Community (Belgium, Greece, Italy, and Portugal) to report production data, and of the Commission of the European Communities to report consumption data. The Committee invited a European Commission representative to its next meeting "for an exchange of views on this issue." The Committee also expressed concern regarding the nonreporting of central and east European members and "the unreliability of data" reported by some countries. The Russian reports, for example, have been alleged to have "major inconsistencies."

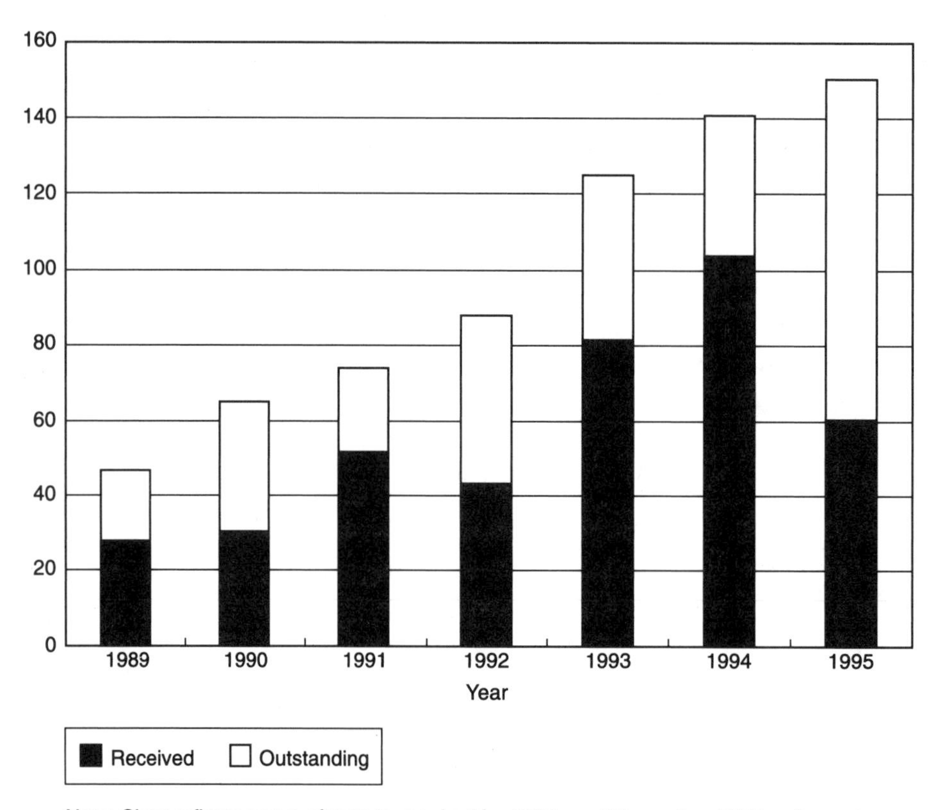

Figure 5.14
Annual reports required of Montreal Protocol Parties: Received or outstanding

Noncompliance with Targets and Timetables. In 1995, for the first time, the Implementation Committee addressed projected failures to meet targets and timetables. Five countries—Belarus, Bulgaria, Poland, Russia, and Ukraine—could not comply with the January 1, 1996, deadline for phasing out consumption of fifteen CFCs, carbon tetrachloride, and methyl chloroform. Since neither other parties nor the Secretariat was eager to take responsibility for accusing parties of noncompliance, the states themselves were persuaded to initiate the noncompliance procedures before the Implementation Committee. Treatment of the countries differed. For example, Belarus would receive funding from the Global Environmental Facility so that it could meet the deadline. The committee was to work with Poland to meet the deadline.

Russian compliance remains unclear as of 1997. In 1996 the Russian Federation indicated that the treaty was negotiated by a predecessor government and that she

Table 5.7
Annual reports of Montreal Protocol parties

Country	Entry into force	1991	1992	1993	1994	1995	1996
Brazil	6/17/90	R	R	R	R	R	N/R
Cameroon	11/28/89	R	R	R	R	R	R
China	9/12/91	R	R	R	R	R	R
Hungary	7/19/89	R	R	R	R	R	R
India	9/24/92	—	R	R	R	R	R
Japan	1/1/89	R	R	R	R	R	R
Russian Fed.	1/1/89	R	R	R	R	R	R
United States	1/1/89	R	R	R	R	R	R

Notes: N/R indicates no report filed as of July 28, 1997. R indicates that a report was filed for that year, according to the Secretariat, at least prior to 7/28/97. These reports include on-time, late, and subsequently revised reports; some countries met reporting requirements for two or more years with a single report.

would need very substantial financial assistance to comply. The Implementation Committee and the Russian Federation set forth a phaseout schedule that would require Russia to meet her commitments by 2004. The Committee indicated that it would monitor production and consumption, and would require that specific information be provided; that negotiations would continue later. The Russian Federation must still contribute her obligated amount to the Montreal Protocol Fund.

From the temporal perspective, within a period of less than ten years, the Implementation Committee has been established and has evolved from a rather cautious institution addressing issues of reporting to one that is actively engaged in trying to promote and ensure compliance with targets and timetables by one of the major parties.

Noncompliance in Ten Years for Developing Countries. The Montreal Protocol is commonly regarded as an effective international treaty, because consumption levels of the controlled substances have declined sharply, as required by the Protocol. As of March 1993, reporting by developed countries had already indicated a decline of 51.52 percent in production and 62.05 percent in consumption of controlled substances. The reports of countries operating under article V(1) showed a decline of 3.17 percent in consumption of controlled substances. By 1996, reports indicated that the countries that are the major producers of the controlled chemicals are meeting their target timetables for phasing out production and consumption.

However, it is unclear whether in ten years, when the developing countries are required to phase out controlled substances, compliance with the Protocol will be as good as it is today. In important countries such as China and India, with large internal markets, there are numerous producers of controlled substances. The number of producers has been growing. The danger is that developing countries with large markets for the controlled substances will find it hard to phase them out when the time comes. The demand on the Montreal Protocol Fund could become enormous, as firms ask for help in switching to alternative chemicals or, as is likely in many cases, for compensation for closing production facilities.

In an effort to address this problem, parties decided that no plants built after August 1995 would be eligible for Montreal Protocol funds to phase them out, and that after December 1995, no article V countries could build new plants. Despite these decisions, there are still reasons for concern about whether countries in fact will phase out production at the end of ten years, whether newly listed chemicals such as methyl bromide can be effectively curtailed, and whether the absence of participation by Taiwan may effectively undercut the treaty in the long term. Moreover, there are controversial issues about how to value production facilities that are closed down for purposes of compensation under the Montreal Protocol Fund: whether it should be lost profit, book value, replacement cost, or some other measure. This is particularly critical for the many small firms that are operating at only a small percentage of capacity and could justifiably be closed as inefficient apart from the availability of funds from the Montreal Protocol.

Smuggling Regulated Chemicals. Since 1992, a new problem has arisen: smuggling of original CFCs into the United States and other industrialized countries. Original CFCs are disguised as recycled CFCs to avoid United States taxes levied on virgin production. Reportedly the Russian Federation is a major source of the smuggled goods; production facilities in China and India reportedly also contribute to the flows. This is in part due to a lack of central control over border areas and to a lack of local capacity to control exports. Importing countries also face difficulties in controlling the smuggling because the resources to do so must compete with demands to control smuggled drugs, arms, and other contraband.

Smuggling of controlled substances under the Montreal Protocol has begun to resemble the problems in trying to control the trade in endangered species. Although as of 1997, the smuggling problem affects only a small percentage of the CFC trade, it could grow unless the markets for the chemicals are phased out. To address this problem, countries decided in November 1996 to require national systems of validation and approval of the imports of any used, recycled, or reclaimed ozone-depleting substances before importing them. (UNEP 1996b). They adopted at the 1997 Meeting of the Parties an amendment requiring rational export and import licensing of such materials (UNEP 1997). Parties must ratify the amendment.

Concluding Observations. The Montreal Protocol has proved thus far to be an unusually effective treaty with which most countries are increasingly complying over time. The leadership of the countries that have been the major producers of the controlled substances has been crucial in this process, as has the support of the major industrial producers and the input from an expert technology and economic assessment network reporting to the parties. The concern for the future is whether the parties to the treaty and the private sector will be able to maintain the market for the alternatives to the controlled substances and choke off the markets for the controlled substances.

The Comparative Perspective Across Time

International treaties resemble living organisms, in that they evolve and develop their own histories. The snapshot of an international treaty fails to capture the important changes that occur over time. Sometimes change is continuous; at other times, treaties may remain more or less the same for years, then begin a period of change.

The five treaties studied indicate that the pace of change is accelerating, that treaties may become obsolete in the face of new challenges and need to be replaced, and that most of the changes that occur in treaties are accomplished through procedures and processes other than formal amendments requiring ratification by states. In two of the five treaties, the International Tropical Timber Agreement and the London Convention of 1972, new instruments that replace the existing treaties were concluded during the course of the study.

The five treaties have evolved in response to problems that parties face in achieving the treaties' objectives. The treaty commitments have expanded, the institutional arrangements have grown and diversified, and the concern with monitoring and compliance among the parties has very substantially increased.

The latter has taken the form of more incentives or inducements to comply (funds, technical assistance, and training programs), of new efforts to promote public awareness and involve the private sector, and of greater emphasis on dealing directly with parties who do not comply.

Notably, in the two treaties addressing pollution—the London Convention and the Montreal Protocol—parties have been quite successful in controlling the pollutants they initially targeted. The treaties have been in transition from ones concerned largely with identifying and limiting specific pollutants to ones focused on ensuring continued success in controlling the pollutants already identified, as well as addressing new pollution issues that may arise. In the three natural resource treaties, success has been less clear. To be sure, more sites have been added to the World Heritage

List, and particular endangered species or World Heritage Sites have improved to the point that they are no longer endangered.

The pace of change for all five treaties has increased since the late 1980s. Two of the treaties, the World Heritage Convention and the London Convention of 1972, were more or less frozen in time until the end of the 1980s, when parties and the Secretariat became much more active in developing new implementing initiatives. The 1983 International Tropical Timber Agreement has gradually become more concerned with sustainable forestry. The Montreal Protocol has seen constant change. Parties have adjusted their commitments at each Meeting of the Parties, added new institutional features, and subjected new chemicals to the protocol's controls. Only in the CITES Convention has the pace of change been slower, although even here, change has accelerated since the early 1990s.

Although all the conventions could be amended, most of the changes to them have occurred by procedures other than formal amendment. Once in effect, the conventions offer a forum for parties essentially to legislate changes in both substantive and procedural obligations. Generally this is done through the formulation of guidelines, decisions, and technically nonbinding measures rather than through formal amendment (except for the amendments to the Montreal Protocol). A treaty can be significantly transformed through this informal legislative process.

The following material compares the five treaties for each of the categories highlighted in the preceding two sections. Inevitably there are certain exceptions to general trends, and these are noted.

Expanded Treaty Commitments

In all five treaties, parties expanded their obligations after the negotiations had been concluded and the treaty had come into effect.

In the World Heritage Convention, parties created a new category of sites—cultural landscapes—and took on new commitments regarding monitoring. In CITES, parties adopted a resolution in 1994 calling upon Asian member parties to control internal demand for tiger parts and rhino horns, which has taken commitments beyond measures strictly focused on trade across borders. Countries in southern Africa have assumed new enforcement obligations for CITES under the Lusaka Agreement on Enforcement, although technically this accord is independent of CITES. The ITTA has changed from an international commodity treaty, with a few general provisions related to sustainable forestry, into one that commits to sustainable forestry by the year 2000. Under the London Convention of 1972, parties agreed to phase out sea disposal of industrial wastes, to suspend marine disposal of radioactive wastes, to ban incineration of wastes at sea, and, most recently, to tackle other issues. In the Montreal Protocol, parties have enlarged or strengthened their substantive commitments at each Meeting of the Parties and have agreed to establish the Montreal Protocol Fund to help developing countries in complying with the Protocol.

Some of the changes in the treaties reflect a newly emerging focus on environmental issues: integrated environmental management. Frequently environmental problems are seen as unidimensional: protect against lake pollution by controlling direct emissions of pollutants into lakes; conserve biological diversity by controlling trade in endangered species; guard against marine pollution by preventing ocean dumping of specific pollutants; and safeguard particular sites by listing them as World Heritage Sites. But increasingly, the international community is recognizing that environmental problems are multidimensional and that effective international legal instruments must be implemented so as to address the suite of issues associated with the problem. Sometimes this requires attention to multimedia pollution or ecosystem conservation; at other times it means attention to monitoring, financial incentives, and a suite of other measures in addition to the specific regulatory controls. Thus, protecting against lake pollution requires attention to the ecosystem and airborne transport of pollutants. Conserving biological diversity requires attention to internal demand for species and to habitat preservation. Safeguarding sites requires efforts to persuade local communities that it is in their interest to help preserve the sites and to engage them in conservation activities.

In the five treaties under study, parties have given increasing attention over time to considering the problems within a more integrated environmental context, and to broadening the range of issues that are considered essential to addressing a particular problem.

Implementation and Institutional Changes

As countries have implemented the five treaties, they have developed new institutional features and refined and expanded others. The major developments include regional networks for implementing the treaties, implementation committees and noncompliance procedures, active secretariats (including the addition of an enforcement officer), increased use of scientific and technical bodies, and enhanced roles for NGOs and the private sector. Table 5.8 provides a comparative look at institutional measures in the five treaties that encourage compliance. There has been no use of formal dispute settlement mechanisms in the five treaties studied.

Regional Implementation Centers. As the executive director of UNEP noted in 1995, countries are much stronger when they speak as a region than when they speak individually. Regional centers or other regional measures help to engage local communities and to ensure that various requirements and functions are properly tailored to local needs and properly monitored in light of local conditions.

Table 5.8
Institutional measures in the five accords that encourage compliance

	WHC	CITES	ITTA	LC	MP
Implementation committee					yes
Noncompliance procedure				1996 Protocol only[a]	yes
Enforcement staff officers		yes			
Formal NGO role in implementation	yes	yes			
Formal industry role					yes (OzonAction)
Scientific assessment body	yes[b]			yes	yes
Regional Center/Network	yes	yes (Lusaka)	yes	yes (UNEP regional seas accords)	yes (art. V, MP Fund/ OzonAction)

[a] Under the 1996 Protocol, such procedures must be established by a Meeting of the Parties within two years of entry into force.
[b] Three designated outside organizations.

In four of the five treaties under study—the World Heritage Convention, CITES, ITTA, and the Montreal Protocol (through the Fund)—parties and the secretariats have turned to regional centers or networks to secure better implementation. In the fifth, the London Convention of 1972, some countries have long been joined in regional arrangements through the UNEP regional seas accords.

Both the World Heritage Convention and the ITTA are establishing regional centers. The ITTA has established three thus far, most recently in Cameroon. Their purpose is to monitor and review projects funded under the ITTA. CITES has developed a regional focus in southern Africa through the Lusaka Agreement. Although there is no regional center under the Montreal Protocol, the regional meetings of nationally designated officials responsible for implementing the Protocol in countries receiving assistance from the Montreal Protocol Fund have provided an important resource for exchanging information and addressing common problems in complying with the Protocol.

The three global conventions that have regional centers—ITTA, CITES, and the World Heritage Convention—address resources, located within national borders, that would be matters of national policy were it not for the international treaties. Compliance by neighboring countries is especially important in those treaties that involve trade across national borders, as CITES does.

The London Convention of 1972 expressly obligates states with common interests in a geographical area to "endeavor ... to enter into regional agreements consistent with [the] convention for the prevention of pollution, especially by dumping" (article VIII). Although the Convention has not been implemented on a regional basis, the UNEP regional seas accords have provided the comparable regional focus. The coexistence of the global convention with the separate regional conventions has raised the possibility of differing standards between the treaties. For example, in responding to the movement to ban dumping from offshore oil platforms (e.g., the United Kingdom's in the North Sea), countries legally could adopt a stricter standard under the 1972 Oslo Convention for the Prevention of Marine Pollution by Dumping from Ships and Aircraft than under the global London Convention. However, the stricter regional standards should foster compliance with the global convention.

Noncompliance Procedures. Procedures for addressing noncompliance issues have been important. Under the Montreal Protocol, in 1990 parties established the Implementation Committee as a standing body and developed formal procedures to address noncompliance. The model reflects procedures in international human-rights law. The Implementation Committee has become important in building compliance with the protocol (W. Lang 1995). In the initial years the Committee focused largely on procedural obligations (i.e., countries' failure to report data) or on incomplete reporting, but since 1995 it has addressed significant substantive noncompliance issues, such as those related to targets and timetables. The model of an Implementation Committee and noncompliance procedures developed under the Montreal Protocol has been incorporated into the 1996 Protocol to the London Convention of 1972.

Increasing Use of Scientific and Technical Bodies. The treaties studied rely on scientific and technical advice, although to varying degrees. The Montreal Protocol provides for regular technical assessments, which have been carried out through the Technology and Economic Assessment Panels and their subcommittees. These assessments have expanded over time. The Panels have been very important in assessing the feasibility of proposed treaty amendments and in linking the private sector and governments in evaluating needed technology for article V developing countries. The London Convention has a Scientific Committee that meets regularly, the World Heritage Convention uses the IUCN for technical evaluations of natural sites, and CITES considers IUCN and other scientific input to criteria for listing species. The ITTA employed technical advice in developing guidelines for sustainable forestry management. The experience suggests that it is important for parties to be able to draw on regular scientific advice so that they can respond flexibly to changes in scientific understanding and can assess the technical adequacy of their response.

More Responsive Secretariats. For all five treaties, the secretariats play important, though often understated, roles in implementing the treaties. Over time, they have generally acquired new powers and become more powerful actors, although the staffs remain small and generally have not increased significantly. In all the treaties, the Secretariats serve as the hub in an extensive informational network linking parties, NGOs, and, for most treaties, the private sector. All secretariats are responsible for setting the agendas of meetings, in consultation with member countries.

The traditional image of secretariats is that they have little power because they operate at the direction of the parties to the treaty. This is technically correct, and indeed secretariat officials are often careful to convey this. However, in practice secretariats can be powerful actors, because their staffs may be the only people with comprehensive knowledge of whether states and other actors are complying with the treaty, and of the problems that have arisen at the national level. Secretariat officials communicate not only with governments but also with private-sector officials, NGOs, and interested individuals who seek information or advice. The power of the secretariats seems to increase over time, as they gain the confidence of member states and of private actors, including both industry and NGOs.

The interaction between governments and secretariats is complicated and dynamic. Secretariats may generate ideas that are introduced by a member government. Governments may induce secretariat staff to float ideas that may be more acceptable to other parties if offered by staff. Secretariats work directly with national implementing agencies; and national agencies seek the assistance of secretariat staff. Again, this process of interaction seems to become more routine and deep-rooted over time.

The budgets for the treaty secretariats are minuscule in comparison with the budgets of formal international organizations. They either have stayed constant or have enjoyed modest increases, although the project monies available in the World Heritage Fund, the ITTO, and the Montreal Protocol Fund show a more varied picture. Figures 5.15 and 5.16 provide budgetary data for 1989 and 1994—years before and after the 1992 Rio Conference on Environment and Development.

Despite the modest budgets and small secretariat staffs, the treaty regimes operate in many ways as mini-institutions. One of the most disturbing new trends, however, is the increasing use of short-term contracts (one to six months) for staff positions in at least two of the conventions: the CITES and the World Heritage Convention. This is also the case in other environmental conventions not under study here. This development reflects budget uncertainties and funding constraints in the United Nations; over the long term it will likely harm the quality of the secretariat staffs and their productivity. The problem needs urgently to be addressed.

Depending upon the treaty, formal enforcement officers may be included in the secretariat or links with Interpol may be made. In the early 1990s, CITES had an

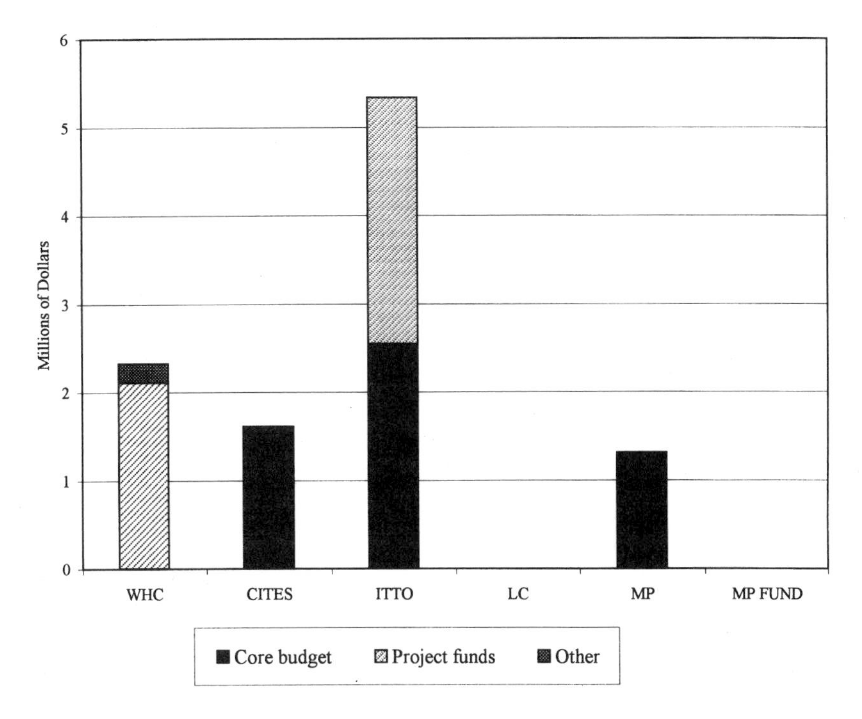

Notes: Other reflects $210,700 of the World Heritage Fund budget, provided as temporary assistance to the Secretariat. WHC secretariat personnel are members of UNESCO staff. London Convention secretariat personnel are members of IMO staff; budgetary figures are not available. Montreal Protocol Fund was established in 1990.

Figure 5.15
1989 Budget by treaty

enforcement officer formally attached to its Secretariat who could conduct on-site investigations. Under the Lusaka Agreement, CITES enforcement officers in member countries could link directly with Interpol. In several of the treaties studied, consultants reporting to the parties and secretariats could serve indirectly to assist compliance by detecting violations and corrupt behavior.

Increased Role for NGOs and Industry. In general there has been greater NGO participation in the Meetings of Parties, with the exception of the Montreal Protocol Fund, where NGOs are excluded. NGOs also have been increasingly active in monitoring compliance and, in some cases, in assisting secretariats. This is consistent with the increasing role of NGOs in global environmental issues (Cameron 1996).

In the World Heritage Convention, the NGOs given formal status under the convention have been active in developing the reports on sites that are nominated for listing on the World Heritage List and in monitoring the conservation of the sites.

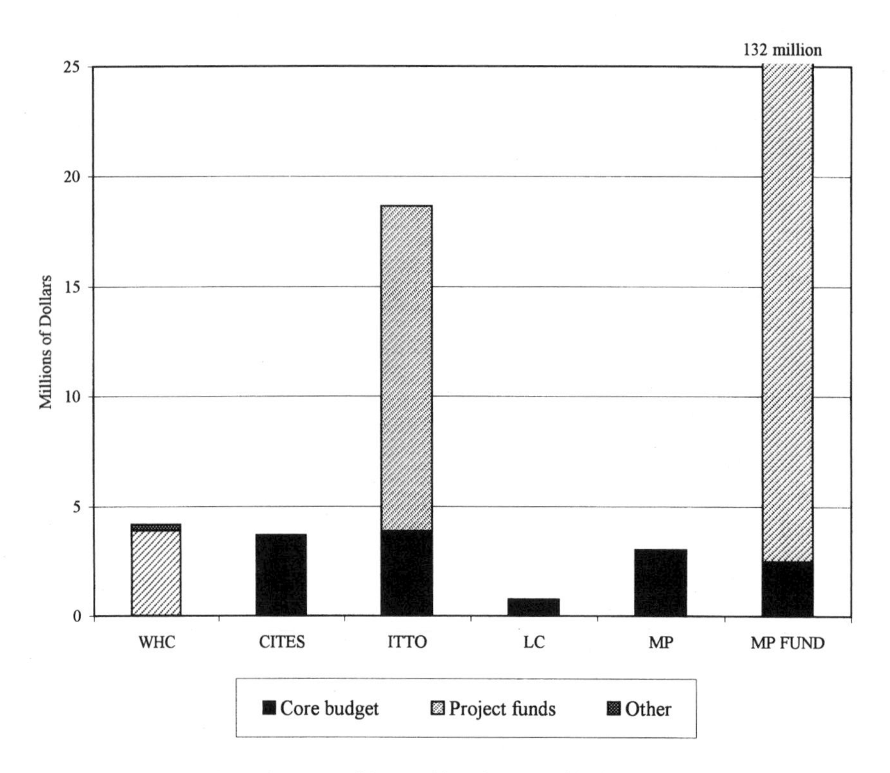

Note: Other reflects $280,000 of the World Heritage Fund budget, provided as temporary assistance to the secretariat. WHC secretariat personnel are members of UNESCO staff. London Convention secretariat personnel are members of IMO staff or on secondment.

Figure 5.16
1994 Budget by treaty

They have become more involved in monitoring site conservation since the early 1990s. In addition, TRAFFIC, in Cambridge, England, monitors World Heritage Sites. Generally, NGOs are attending more committee meetings and bringing more matters to the attention of the Secretariat. In many cases, international and national NGOs alert the World Heritage Committee to violations of the Convention.

In CITES the number of NGOs that attend the Conference of the Parties continues to grow. Before 1985, the number of governments exceeded the number of NGOs. But since then the number of NGOs attending as observers has exceeded the number of governments. NGOs participate in the formulation of national policies prior to the Conference of the Parties. Their use of the media helps to define the agenda of the Conference of the Parties. TRAFFIC has been a major source of information about trade in endangered species, as well as for baseline data in countries. The World Conservation Monitoring Unit in Cambridge, England, has collated CITES export and import permit data and provided them to the Secretariat.

In the ITTA, NGOs have been crucial in defining future work under the treaty. They are largely responsible for shifting the focus from a commodity treaty to a treaty that in part focuses on sustainable forestry. They had an instrumental role in setting the agenda for the negotiation of the new treaty. NGOs also participate in funding numerous conferences and training projects, and the commercial sector has been increasingly involved in funding projects. The new Forest Stewardship Council, which includes both forest trade organizations and some environmental organizations, is helping to push the agenda for the ITTA, in the sense that it is developing its own certification scheme for timber that has been harvested in an environmentally sustainable way. The environmental NGOs have been increasingly active at the national and international levels in pushing compliance with sustainable forestry guidelines.

In the London Convention of 1972, NGOs have been instrumental in setting the agenda. They have circulated drafts among parties and prepared position papers that have influenced negotiations. They have engaged in extensive lobbying efforts at the national and international levels. Greenpeace, an international NGO, played an essential role in gathering information about Russia's dumping of nuclear materials into the ocean, and has been very active on the question of oil platform disposal in the oceans. Although both environmental and industry organizations attend the Meetings of the Parties, their numbers are always limited. This is because these organizations need permission to attend, and the parties have kept the numbers that could attend low.

Although NGOs have been involved in the Montreal Protocol, their involvement is probably more modest than in any of the other treaties. They were very important in the negotiation of the Protocol, yet their relative influence declined subsequently as the action shifted to governments and the commercial sector. Nonetheless, they were important actors in developing the London and Copenhagen Amendments to the Protocol. As of mid-1997, only a limited number of NGOs were active in the ozone issue. In the 1970s, the NGOs were grassroots organizations. Since then they have become active partners with governments in pushing forward the Montreal Protocol agenda and with industry by participating in training and other technical assistance projects sponsored by the Montreal Protocol Fund.

Of the five treaties studied, only the Montreal Protocol provides direct links with industry, through *OzonAction*. This linkage has been highly effective in enlisting industry to assist with compliance, in providing an invaluable source of information on country profiles and industrial substitutes for regulated chemicals, and in bringing together at an operational level the community concerned with implementing the treaty.

No Formal Dispute Settlement. It is useful to consider those measures that have not developed over time. None of the treaties provide for formal dispute settlement, such as by arbitration, conciliation commissions, or the International Court of Justice, although parties to the Montreal Protocol have the option of referring a dispute to the formal procedures specified in the Vienna Convention on the Protection of the Ozone Layer. In the London Convention, parties proposed an amendment on dispute settlement that was never ratified. The 1996 Protocol provides, however, for formal dispute settlement. States objected to including provisions on dispute settlement in the World Heritage Convention.

Disputes that have arisen under the five conventions have been resolved by discussion among the parties, as in a Meeting of the Parties, or by measures initiated by the parties or secretariats. The Montreal Protocol has its own highly developed noncompliance procedures. In general, NGOs may bring complaints to the attention of the Secretariat or of parties. Parties informally bring complaints against other parties or otherwise arrange for the offending actions to be brought to parties' attention. States and NGOs bring pressure against other states to comply, and continue that pressure until the issue is resolved. In the World Heritage Convention, media pressure from press releases or articles in local newspapers has been important in securing compliance.

Given this practice, it is questionable whether the strong emphasis in environmental treaty negotiations on including formal binding dispute-resolution mechanisms is always useful. It is, of course, arguable that the existence of such agreed mechanisms encourages parties to settle their disputes in other ways, although the experience in the five treaties under study does not demonstrate that.

Monitoring and Compliance

Over time, there is a distinct trend across treaties to find new strategies for implementing and complying with the treaties. This seems to be an almost unconscious process that is accelerating. It appears to be linked to advances in information technology that have rapidly penetrated national borders with new scientific, economic, and cultural information, and have greatly facilitated access of states and other actors to relevant information.

In chapter 15, we distinguish three strategies of compliance: sunshine, incentives, and sanctions. The first includes techniques such as reporting, on-site monitoring, participation by nongovernmental actors, and related methods that give transparency to the process and mobilize public support. The second focuses on financial and technical assistance and training programs. The comparative review of the five treaties indicates strong reliance on methods associated with the sunshine approach, particularly reporting; increased use of incentives; and little employment of sanctions. This is consistent with methods used in other international environmental treaties (Shihata 1996; Boisson de Chazournes 1993).

More Attention by Parties to Compliance. In all the treaties, there has been increased concern with monitoring and compliance with the obligations in the Conventions. This new pattern dates to 1991, and thus corresponds to the initiative in the preparations for the Rio United Nations Conference on Environment and Development asking states to report on their enforcement of international environmental treaties.

In the World Heritage Convention, the 1992 Strategic Plan stressed the importance of monitoring. The amended Guidelines explicitly incorporate provisions on monitoring that focus on regional and local on-site monitoring, and invite countries to report every five years. In 1993, for the first time, part of the budget was allocated to monitoring and compliance. There has also been a marked shift in the allocation of time devoted to monitoring at the meetings of the World Heritage Committee, from almost none in the late 1980s to 30–50 percent of the time in 1996.

CITES gives much greater attention to monitoring and enforcement in 1997 than in 1990. As noted, the Secretariat added an enforcement officer (he departed in 1994), and parties have discussed seconding other enforcement personnel to the Secretariat. The World Conservation Monitoring Centre computerized information on export and import permits with the intent of providing faster tracking of trade in endangered species. In 1991, parties decided that the failure to provide an annual report of export and import permits would be a violation of CITES that could trigger action by the parties, including sanctions. The Secretariat's Report of Infractions to parties discourages noncompliance, and there has been increased emphasis on penalizing violators of CITES. In 1994, African countries concluded a regional enforcement agreement involving African member countries, the CITES Secretariat, and interpol.

Of the five treaties studied, parties to the ITTA probably have devoted the least explicit attention to compliance. Nonetheless, increasingly there are measures that encourage compliance: country reports on progress toward the goal of sustainable forest management by the year 2000; and on-site monitoring missions, notably that to Sarawak, Malaysia, to examine whether the province was following the sustainable forestry guidelines.

Parties to the Montreal Protocol probably have devoted the most attention to compliance of any of the five treaties studied. At the 1990 Meeting of Parties, countries provided for noncompliance procedures and set up the Implementation Committee. They established the Montreal Protocol Fund to build countries' capacity to comply. They also have focused on compliance with reporting requirements by deciding in 1994 that article V countries had to fulfill the reporting requirement in order to continue their status and hence be entitled to assistance from the Montreal Protocol Fund.

In the London Convention, there is also increased concern with monitoring and compliance. In 1990 parties called for prompt reporting, and since then there has been a new effort to track the filing of the annual country reports and the completeness of these reports. The 1996 Protocol directly addresses monitoring and compliance issues.

Reporting by Parties. All five treaties provide for regular reporting by parties, although the obligation has not been implemented in the World Heritage Convention. In the ITTA, the reports required are of exports and imports of tropical timber, and not of compliance with environmentally related provisions. In the remaining three treaties, compliance with reporting obligations varies. The percentage of compliance has changed over time, and not always in the same direction.

Figure 5.17 provides data on the percentage of parties complying with reporting requirements in the CITES, the London Convention of 1972, and the Montreal Protocol.

Although it is tempting to focus on the percentage of noncompliance with reporting requirements, that would be very misleading. The important question is whether the countries that are the major actors for a given convention are meeting their reporting requirements fully and in timely fashion. From this perspective, there is good compliance with reporting requirements, since nearly all of the most significant members of the conventions file relatively timely and accurate reports.

There are new efforts in the secretariats of at least four of the treaties to develop common formats for reporting data. This would ease the burden of reporting, provide easier comparison among countries, and make the reports more accessible to the public, if the parties agree.

Greater Transparency, Public Awareness, and Use of Information Technologies. In all of the conventions there is much greater transparency. This is connected to much greater public awareness of the conventions, much greater participation by NGOs in the treaty regimes, and the development of advanced information technology.

The CITES Secretariat has routinely made available all the documents from the Conference of the Parties, a practice that has become routine for most conference documents in all of the treaties.

In three of the conventions, the secretariat or an associated organ publishes a newsletter that provides detailed information on implementation: the *World Heritage Newsletter*, *OzonAction*, and the ITTA newsletter. These started during the 1990s, and have grown in size and distribution. They are published either quarterly or three times a year. In addition, in the United States the Federal Fish and Wildlife Service publishes a newsletter on CITES that in 1995 was circulated to over 700

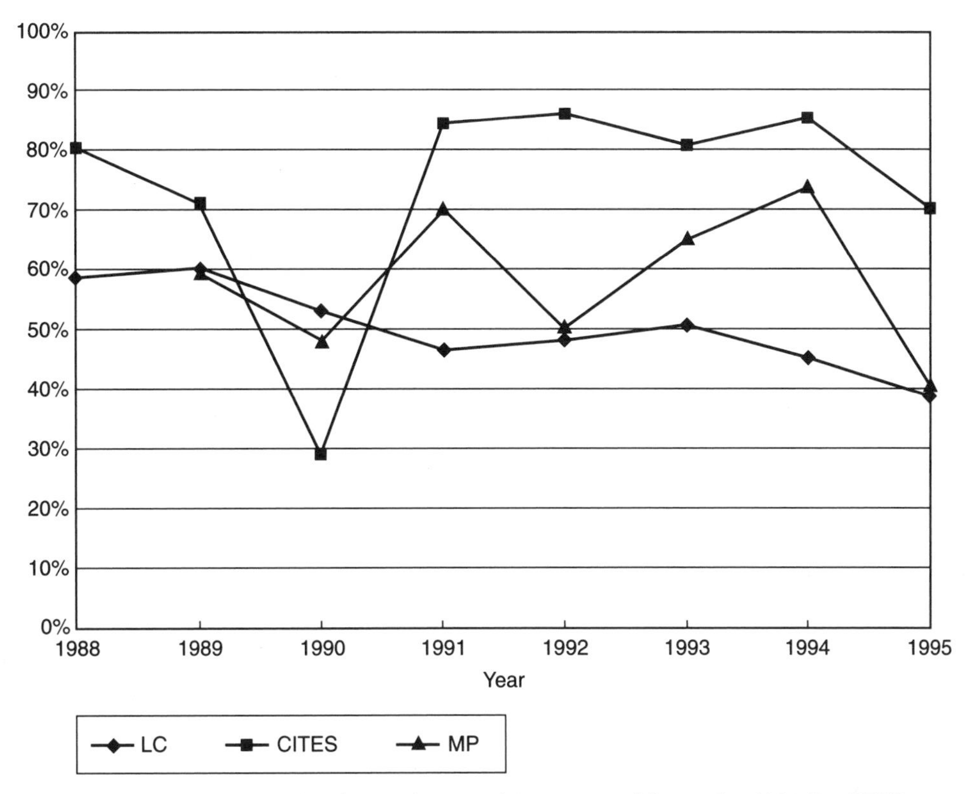

Note: Chart reflects London Convention reporting status as of September 1997. For CITES reports 1988–1990, chart reflects 1995 status; however, chart reflects status as of July 1997 for 1991–1995 CITES reports. Montreal Protocol reporting status is as of November 1996 for reports due through 1994, and as of December 1996 for reports due in 1995.

Figure 5.17
Percentage of countries reporting as required by three agreements

interested organizations (compared with forty organizations in 1990). The newsletter is delivered on the Internet; a World Wide Web page is being developed. Only the London Convention of 1972 does not yet publish a newsletter.

Similarly, new information technologies are being used to bring treaty operations to the attention of the public and presumably to enlist their support for implementation. The proceedings of the World Heritage Committee meetings are available on the Internet. TRAFFIC, in Cambridge, England, has made the proceedings and documents from the 1994 CITES Conference of the Parties available on-line on the Internet. The World Heritage Centre has produced a CD-ROM on World Heritage Sites. Documents are available on-line from the Montreal Protocol Secretariat. Parties to the Montreal Protocol Fund have discussed transforming the Fund process from an exchange of papers to electronic mail.

Incentives: Greater Emphasis on Training and Building Capacity. All five treaties have given greater attention over time to training and building capacity. This is tied to financial incentives and support.

Three treaties have technical assistance projects as either central or very significant elements: the ITTA, the Montreal Protocol, and the World Heritage Convention. The ITTA operates primarily through the development and approval of technical assistance projects. The number of projects has increased over time, many of them funded by Japan. Under the Montreal Protocol, qualified countries receive support from the Montreal Protocol Fund to develop programs for controlling CFC production and consumption, and to move from use of ozone-depleting substances to use of ozone-friendly substitutes, or to close their production facilities. The Fund requires that each country have a program in place before a project is approved, which builds local capacity to comply with the Protocol. The World Heritage Convention, which from the beginning has had a modest World Heritage Fund, has been providing technical assistance to countries to protect sites. Since 1992, there has been increased emphasis on training programs and building capacity.

Until 1989, CITES paid very little attention to training. Since then, the Secretariat has had an officer dedicated half-time to training programs. The budget for training has grown dramatically, with a tripling expected in 1996 and the years immediately following.

The IMO Office for the London Convention of 1972 has done little training. Parties have not established any special fund for technical assistance, although the 1996 Protocol specifically addresses this issue and encourages countries to provide funds for training and technical assistance.

Enhanced Financial Incentives. As noted earlier, the development and use of special funds to provide technical assistance to countries or otherwise to facilitate the effective operation of the Conventions has grown.

Two of the five treaties, the World Heritage Convention and the Montreal Protocol, have special funds to develop local capacity to comply with the treaty. A third, the newly negotiated successor ITTA, provides for a special fund, the Bali Partnership Fund. The London Convention established a one-time special fund, the Twentieth Century Fund, to computerize the IMO office servicing the treaty. CITES does not have a special fund attached to it.

The budgets for the funds vary widely among the conventions. In the World Heritage Convention, which provided for a special fund when it was concluded in 1972, the budget has remained more or less constant over time—about $2.4 million for seventeen years and slightly higher since 1989. Parties are assessed 1 percent of their contribution to UNESCO, which explains this trend. However, for subsequent treaties, the budgets for the special funds are much larger and have increased over time—dramatically for the Montreal Protocol.

The Global Environmental Facility is a source of funding for measures that can bring countries into compliance with the treaties. Its projects that have been approved help countries comply with the London Convention, the Montreal Protocol, and at least once, indirectly, the World Heritage Convention. The GEF Council decided in its May 1995 meeting that GEF funding would be withheld from any countries that failed to meet their obligations under the relevant treaty for which funding was being provided. This includes substantive commitments, procedural obligations such as reporting, and payment of assessed contributions under the treaty (Sand 1995).

Sanctions as a Last Resort. Sanctions have not been important in implementing the five conventions. However, in the two treaties that deal with trade across national borders—CITES and the Montreal Protocol—the threat of sanctions has occasionally been important, particularly in CITES. Countries have endeavored to induce compliance by other countries through exposing their actions to "sunshine" by reporting and on-site monitoring, and through providing technical assistance and financial support that will enable them to build the capacity to comply with the convention.

The Enhanced Role of Markets. Markets both help and hinder compliance with the international treaties studied. In the Montreal Protocol, markets have been essential for ensuring the developing of substitute chemicals. On the other hand, the market for recycled chemicals undermines compliance with the Protocol. Virgin CFCs are illegally entering the United States as recycled CFCs; their lower price has created a widely dispersed market for them. In the World Heritage Convention, markets help draw tourists to World Heritage Sites, and thereby to bring in economic resources that facilitate compliance with the Convention. On the other hand, the tourist market also imposes costs on countries in managing the site for tourism, coping with the influx of tourists, and preventing or dealing with site degradation.

Dampening demand for endangered species helps compliance with CITES, but lucrative markets in species whose trade is prohibited under the CITES Convention undercuts compliance with the Convention. Under the ITTA, market demand for tropical woods for furniture, chopsticks, and other consumption items has facilitated the demise of tropical forests because countries often clear-cut areas for only one or two species that bring a high price in the export market.

One of the growing, disturbing developments is the apparent link between organized crime, drugs, and shipments of illegal goods under CITES, ITTA, the

London Convention, and the Montreal Protocol. In the case of CITES, Interpol was called in to assist in enforcement in the Lusaka Agreement.

In forestry, there is growing pressure to provide labels certifying that woods have been harvested in accordance with sustainable forestry practices. This strategy relies on informed consumers to make the market push countries to engage in sustainable forestry if they wish their timber to be traded internationally. Similar strategies could apply to trade in endangered species, if reliable methods could be found to effectively distinguish the harvesting of endangered members of a species from harvesting of nonendangered members, as in the case of elephants found in eastern and in southern Africa.

While the public in industrialized countries often assumes that NGOs always assist in getting the market to protect the environment, this is sometimes inaccurate, particularly in certain countries. NGOs may in fact assist the market in undermining compliance or themselves undermine compliance. Their stance often depends upon their source of financial support and their political allegiances.

Concluding Observations: The Patterns of Change

All five treaties were concluded less than twenty-five years ago. Yet, within this short time frame, two have already been replaced with new treaties. In the others, parties have effectuated significant changes in implementing the treaties. The patterns of change in the five treaties indicate that countries have created important new mini "international institutions" that are problem-oriented and that increasingly engage civil society over time in the compliance process. These institutions have responded rather well to changes in the international environment. They are able to provide a voice to governments, to nongovernmental actors, and to the private sector. Perhaps because these institutions remain small, they are able to adapt more readily than larger organizations to change. They may become key instruments in the future world order.

Notes

Many people from governments, secretariats, universities, nongovernmental organizations, and industry have been helpful in putting together this chapter. I am deeply grateful to everyone. Special thanks for sharing information, insights, and critical appraisals go to Jonathan Barzdo, Jacques Berney, John Caldwell, Laura Campbell, René Coenan, James Cook, Bernd von Droste, George Furness, B. Freezalah, John Gavitt, Paul Horwitz, Winfried Lang, Susan Lieberman, Sun Lin, Manfred Nauke, Lyndel Prott, Iwona Rummel-Buska, M. Sarma, Rajendra Shende, Vivian Sheridan, Alan Sielen, T. Subramanian, Paul C. Szasz, Maya Thomas, and James Thorsell. They are not responsible, however, for the views expressed here.

6

The United States: Taking Environmental Treaties Seriously

Michael J. Glennon and Alison L. Stewart

This chapter assesses United States participation in five international agreements to protect natural resources and prevent pollution. Four of the agreements were entered into by the United States as treaties: The London Ocean Dumping Convention, the Convention on International Trade in Endangered Species of Wild Fauna and Flora (CITES), the World Heritage Convention (WHC), and the Vienna Convention Montreal Protocol on Substances That Deplete the Ozone Layer. The United States entered into the fifth agreement, the International Tropical Timber Agreement (ITTA), as an executive agreement.

Under the U.S. Constitution, treaties are made by the President with the advice and consent of the Senate. Executive agreements are of two principal varieties: the "congressional–executive" agreement, which is entered into with congressional approval, and the "sole" executive agreement, which is entered into pursuant to the President's plenary constitutional power. Well over 90 percent of international agreements to which the United States is a party are congressional–executive agreements. The ITTA falls into this category.

In international law, all such agreements are equally binding; whether an agreement is obligatory does not depend upon the level of formality imposed by a party's domestic approval processes. In U.S. domestic law, however, some agreements are given force by the courts and some are not. The distinction does not depend upon whether they are treaties or executive agreements. The courts give force to agreements that are "self-executing"; that is, susceptible of judicial enforcement absent implementing legislation. "Non-self-executing" agreements, on the other hand, are enforced by the courts only if and when Congress enacts legislation to implement them. Except for the WHC, which may be self-executing (see discussion of World Heritage Convention), each of the treaties under consideration has been implemented by the enactment of legislation. Whether the ITTA, an executive agreement, is self-executing or non-self-executing is not clear.

The dates of signing and ratification of the agreements by the United States are listed in table 6.1.

Table 6.1
Treaty membership of the United States

Treaty	Year treaty came into effect	Signed by United States	Ratified by United States
WHC	1977	1973	1973
CITES	1975	1973	1975
ITTA	1985	1985	1990
LC	1975	1972	1975
MP	1989	1987	1989

Table 6.2
U.S. compliance with treaty reporting requirements, 1991–1996: Report transmission/receipt date

Treaty	Entry into force	1991	1992	1993	1994	1995	1996
CITES	7/1/75	10/5/92	3/9/94 (extension granted)	animals 10/25/94 plants 5/20/96	animals 3/18/97; plants 3/10/96	animals 4/2/97; plants 12/11/96	N/R
LC	8/30/75	report dated Oct. 93	12/16/94	12/16/94	report dated Dec. 95	report dated Oct. 96	N/R
MP	1/1/89	R	R	R	R	R	R

Notes: N/R indicates that no report was received by July 1997. R indicates that a report was filed, but the date of receipt is unavailable. Montreal Protocol reports received include those filed on time, late, or subsequently revised; in some cases two or more years' reporting requirements may have been met with the submission of a single report.

As will be discussed, the United States has proceeded to implement and comply with each agreement.

A few conclusions regarding U.S. compliance with the five agreements can be stated at the outset. Most important, the United States has met the primary obligations of each of the agreements. With differing levels of specificity, the United States has implemented each agreement through domestic legislation, put appropriate management structures and agencies in place, and, with certain delays, contributed funds and submitted reports where required. Table 6.2 shows the U.S. record with respect to reporting.

Furthermore, the United States has taken enforcement actions against individuals who have violated either the agreements or their implementing domestic legislation. However, although the degree of U.S. compliance can generally be characterized as good, varying degrees of financial support, personnel, and political attention have been dedicated to each agreement.

Ironically, preexisting domestic legislation complicates the effort to assess U.S. enforcement and compliance. In every case except the ITTA, such legislation regarding the same or similar subject matter was already in place before the United States committed itself to participate in the agreement. In at least one instance, the United States exceeded the performance obligations of the agreement and took steps in advance of the treaty's timetable. (See discussion of Montreal Protocol.) The question, therefore, becomes this: What factors influence the United States' decisions to comply with its international agreements in the first place and, sometimes, to exceed the scope of its commitments under those agreements?

By considering the strength of the U.S. role in the negotiation of each agreement, examining the methods for implementing the agreements, and exploring the variations in U.S. compliance with each agreement, this chapter will attempt to identify some of the most important elements influencing the strength of U.S. compliance with the five environmental agreements in particular and with environmental treaties in general.

U.S. Role in Treaty Negotiations

The United States actively participated in the negotiation of each of the agreements. In addition, credit for initiating four out of the five can be attributed largely to the United States.

The concept of the WHC can be partially credited to an American, Russell Train, chairman of the U.S. Council on Environmental Quality (CEQ) and cofounder of the U.S. World Wildlife Fund. Speaking at the second World Conference on National Parks in 1972 and at other international meetings, Train emphasized the need to extend national parks to an international level and to create "a world heritage trust." He stressed the existence of "sites of such outstanding universal importance that they belong to the heritage of the world and should be accorded a special status of protection" (Everhart 1983). Train's personal leadership caused the United States to propose the WHC at the Stockholm Conference on the Environment in 1972. The United States then argued in favor of Train's idea to combine a "natural" treaty, being drafted by the International Union for the Conservation of Nature and Natural Resources (IUCN), and a parallel "cultural" treaty, being drafted by the United Nations Educational, Scientific and Cultural Organization (UNESCO). Later in 1972, UNESCO adopted the WHC. In December 1973, during President Nixon's second term in office, the United States became the first party to the WHC.

Also during Nixon's second term, the United States initiated the negotiations that led to CITES. In 1972, a groundswell of public support arose for environmental preservation in general and for endangered species protection in particular. At that time, President Nixon declared the need for legislation that would make it a federal offense to trade in any endangered species and that would provide protection *before* a species faced critical danger of extinction (Kohm 1991). The U.S. delegation played a leadership role during the conferences that led to the conclusion of CITES (Campbell 1979). Led by Russell Train, the delegation was made up of government officials and nongovernmental organization (NGO) representatives, and had a strongly conservationist leaning. It was the only delegation composed of both government officials and individuals from private environmental organizations (Layne 1973).

Negotiations that led to the London Convention of 1972 were initiated by the United States during Nixon's second term. The impetus for these talks arose from a 1970 report by the CEQ concluding that national and international actions were necessary to control ocean dumping. Following the report, President Nixon instructed the U.S. State Department and the CEQ to develop and pursue international initiatives directed toward regulating ocean dumping worldwide. The conference that spawned the London Convention of 1972 began in October 1972, one week after the U.S. domestic law on ocean dumping, the Marine Protection, Research, and Sanctuaries Act (the MPRSA), went into effect.

Ten years later, the United States led domestically and internationally in developing a policy regarding stratospheric ozone depletion. Initially, the United States assumed a primary leadership role in the negotiations surrounding the Vienna Convention and the Montreal Protocol during the Reagan administration in the mid-1980s. U.S. delegation head Richard Benedick attributes part of the Protocol's success to U.S. leadership and policies (Benedick 1991). According to Benedick, the United States set an example for other nations by being the first nation to take domestic regulatory action against ozone-depleting chemicals. The United States then developed a global plan for ozone layer protection and campaigned for its international acceptance. According to Benedick, the United States' innovative policies on ozone, coupled with its size and importance as the largest producer and consumer of chlorofluorocarbons (CFCs) and halons, gave it considerable influence during the negotiations on international controls. Negotiations to modify the phaseout schedules in the Protocol for the original and additional ozone-depleting substances occurred primarily during the Bush administration.

In contrast, the United States was neither the creator nor the chief negotiator of the ITTA. Although the movement toward a deforestation agreement was encouraged in part by U.S. reports on the depletion of the world's tropical forests in 1980, the impetus for the treaty was international. (For a more complete discussion of the negotiations surrounding the ITTA, see chapter 5.) Although the United States submitted a draft text at the sixth Preparatory Meeting on Tropical Timber in 1982 and helped shape the nature and content of the final agreement, the ITTA was initiated by the United Nations Conference on Trade and Development (UNCTAD). The United States participated in drafting the ITTA from 1982 to 1983, and strove to ensure that the ITTA merely established an information-gathering body with information-sharing requirements rather than an agreement with price-support mechanisms. President Ronald Reagan signed the ITTA in April 1985.

Initial U.S. involvement in both the creation and the formation of four out of the five agreements virtually guaranteed preliminary observance of the agreements' obligations, and the United States acted quickly to supplement its existing legislation to implement the agreements. With several exceptions, there were few differences that required extensive reworking of domestic legislation.

Methods of U.S. Implementation and Extent of Compliance

World Heritage Convention

Implementation. Before signing the WHC in 1973, the United States already had an extensive domestic preservation network in place. For example, Congress enacted the National Historic Preservation Act (NHPA) in 1966. The Act responded to public concern that arose in the 1950s and 1960s, when hundreds of federal highways, dams, and urban renewal projects were completed with little regard for historic areas. The NHPA set forth criteria and guidelines for administering a historic preservation program for properties in the United States. It also established the Advisory Council on Historic Preservation to advise the President and Congress, and to develop policies to resolve conflicts between various programs affecting historic preservation. In the mid-1990s the interagency panel consisted of twelve agencies and met several times per year to advise the Secretary of the Interior on nominations for domestic national historic preservation.

The NHPA also established the National Register of Historic Places, which in 1986 contained more than 45,000 listings (Advisory Council on Historic Preservation 1986). Properties on the list are accorded special protection, tax treatment, grants, and other funding. The Secretary of the Interior delegated the responsibility for administering the National Historic Preservation Program to a bureau within the Department of the Interior, the National Park Service. Other laws addressing some of the same issues as the WHC include the Antiquities Act of 1906, the Archeological Resources Protection Act of 1979, and the National Park Service Organic Act of 1916, whose stated purpose is to "conserve the scenery and the natural and historic objects and the wildlife therein, and to provide for the enjoyment of the same in such manner and by such means as will leave them unimpaired for the enjoyment of future generations" (Rothman 1989).

Through the NHPA and other preexisting laws, therefore, many of the World Heritage Sites in the United States were already monitored, and in some cases funded, by the National Park Service before the WHC came into being. In addition, some monuments were under individual state administration. In those instances, state authorities assumed responsibility for the overall care and presentation of the sites.

To implement the WHC, 1980 amendments to the NHPA authorized the Secretary of the Interior to nominate "districts, sites, buildings, structures, and objects significant in American history, architecture, archeology, engineering, and culture" for WHC designation. Because the NHPA does not by its terms apply to natural sites, the Secretary of the Interior promulgated regulations to designate natural sites as national natural landmarks under the authority of the National Historic Sites Act of 1935. The National Park Service is responsible for monitoring these sites.

The 1980 amendments to the NHPA required the Secretary of the Interior, in cooperation with the Secretary of State, the Smithsonian Institution, and the Advisory Council on Historic Preservation, to ensure U.S. participation in the WHC. This group, called the Federal Interagency Panel for World Heritage, makes recommendations to the Department of the Interior on proposed U.S. nominations and related matters. According to Richard Cook, former head of the Office of International Affairs of the National Park Service, the U.S. law that implements the WHC is the strongest of any member nation to date (R. Cook 1993a).

The 1980 amendments set forth three requirements for a site to be nominated: (1) the site must be of "national significance" (i.e., the property must be designated as a National Historic Landmark or National Natural Landmark by the Secretary of the Interior under the provisions of the 1935 National Historic Sites Act, or be an area of national significance as established by Congress or by presidential proclamation under the Antiquities Act of 1906); (2) there must be written concurrence from the owner of the site; and (3) the property owner must promise to comply with the WHC. These requirements are intended to avoid later problems, such as disputes over government interference with private property. Thus, the treaty covers only properties that are already regulated or properties that the owners have already pledged to protect. This distinguishes the WHC from other agreements that seek to control behavior that was not regulated previously.

At this point, two World Heritage Sites in the United States are privately owned: the Taos Pueblo is owned by the Taos Indian tribe, and the Monticello portion of the Monticello/University of Virginia site is owned by the Thomas Jefferson Memorial Foundation. There are three state-owned sites: the University of Virginia portion of the Monticello/University of Virginia site, the La Fortaleza and San Juan Historic

Site, and the Cahokia Mounds State Historic Site. In each case, the state governor drafted a letter authorizing the site's inclusion on the World Heritage List.

The Department of the Interior's mandate grants it sufficient authority to ensure full compliance with the WHC's obligations. In practice, however, the Secretary of the Interior has delegated substantial powers to the National Park Service, which is primarily responsible for implementing the WHC. Richard Cook had the main responsibility for overseeing compliance with the WHC for many years. The superintendents of the parks have day-to-day responsibility for complying with the obligations of the treaty.

In turn, the National Park Service has delegated some of the responsibility for WHC compliance with regard to cultural sites to the U.S. National Committee of the International Council on Monuments and Sites (US/ICOMOS). According to its president, Terry Morton, since 1984 US/ICOMOS has carried out a number of tasks under a line item in the budget of the National Park Service that amounts to approximately $35,000 a year. Such tasks include (1) drafting nominations of cultural sites and monuments to the World Heritage List; (2) preparing exhibitions to publicize U.S. sites on the List; and (3) preparing other educational and publicity materials (Morton 1993). Two staff members and two volunteers carry out the above tasks.

The Department of State acts as a liaison between UNESCO, the WHC Secretariat, and the Department of the Interior. It provides overall political guidance so that the United States' positions under the WHC are consistent with U.S. foreign policy. The Department of State had been voluntarily contributing approximately $250,000 per year to the World Heritage Fund for efforts to enforce the WHC; in 1991 it raised its contribution to $450,000 and was contributing that amount as of 1993 (U.S. Department of State 1993; Morton 1993).

Compliance. There are now twenty sites in the United States on the World Heritage List (see table 6.3).

In some cases, World Heritage designation has improved maintenance and stimulated the development of programs for the sites.

The administration of Mesa Verde National Park in Colorado represents one example of U.S. compliance with the treaty. Mesa Verde was established as a national park in 1906 to preserve the ruins of prehistoric cliff dwelling's once inhabited by the Anasazi Indians. The site represents the first national park in the world devoted to human culture, and was placed on the World Heritage List in 1978. Mesa Verde is considered by some to be a "model" World Heritage Site.

According to Mesa Verde park staff, no federal funding accompanies World Heritage designation. Rather, the designation gives the park an edge in asking Congress for additional funds as needed. In addition, the park can request that

Table 6.3
World Heritage Sites in the United States and years listed

Year	Natural site	Cultural site
1978	Yellowstone National Park	Mesa Verde National Park
1979	Everglades National Park	Independence Hall
	Grand Canyon National Park	
	Tatshenishi-Alsek/Kluane National Park/ Wrangell-St. Elias National Park and Reserve, and Glacier Bay National Park*	
1980	Redwood National Park	
1981	Mammoth Cave National Park	Olympic National Park
1982		Cahokia Mounds State Historic Site
1983	Great Smoky Mountains National Park	San Juan National Historic Site and La Fortaleza
1984	Yosemite National Park	Statue of Liberty
1987	Hawaii Volcanoes National Park	Monticello/University of Virginia
		Chaco Culture National Historical Park
1992		Pueblo de Taos
1995	Carlsbad Caverns National Park	
	Waterton Glacier International Peace Park*	
Total	11	9

* Site shared by Canada and the United States.

UNESCO provide funding from the World Heritage Fund to support preservation and conservation activities. The present Mesa Verde park superintendent has considered making such a request.

Mesa Verde is formulating several innovative projects both to preserve the park and to improve visitors' experiences. For example, the staff is exploring the optimum carrying capacity at the site per day as part of the VERP (Visitor Experience and Resource Protection) program. It is also considering measures that would require ticketing for guided tours, limit access to certain areas of the park, or allow visitors to observe the site only from above. Finally, plans are being made to send out information about the park over the Internet so that individuals can learn about the park and its history without actually visiting it.

Along with its domestic efforts, such as those at Mesa Verde, the United States has supported international projects financed by the World Heritage Fund and the U.S. Agency for International Development for the restoration of World Heritage monuments. For example, US/ICOMOS assisted the government of Ghana in the preservation of its colonial period forts. The United States has provided experts for overseas work on a number of World Heritage Sites; for example, it sent a planner to Mt. Kilimanjaro and an ecologist to Lake Baikal.

There is a growing trend toward using the WHC as a tool for the protection of natural and cultural sites within the United States. Environmental groups have called for Yellowstone National Park to be placed on the List of World Heritage Sites in Danger in order to draw attention to the increasing deterioration of the park and to the proposed mining activity nearby that may threaten the park's integrity. Department of the Interior reports on the condition of the Everglades recently caused it to be added to the list of endangered sites. The Department has also submitted a report on the Olympic Peninsula, where oil spills off the coast may harm the ecosystem.

It is becoming apparent that the WHC can also be used as a diplomatic tool for preventing damage to U.S. sites by actions originating in other nations. An example is the interaction between the United States and Canada regarding the treatment of a World Heritage Site. Canada had plans for copper mining at a site called the Windy Craggy Mine, adjacent to the World Heritage Sites of Glacier Bay and the Canadian site of Kluane Park on the Yukon–Alaska border. However, because Canada is prohibited from damaging a U.S. site, Canada was pressured by the United States and other parties to the Convention to keep the land untouched (R. Cook 1993b). The conclusion of the dispute was the creation of a new Canadian nature preserve, the Tatshenshini–Alsek Wilderness Park, completing the 24-million-acre expanse between the two World Heritage Sites.

Despite the many actions taken by the United States to protect its natural and cultural sites, however, U.S. compliance with several of the obligations of the WHC is incomplete. The WHC requires that the public be kept broadly informed of the dangers threatening the sites and of activities carried on in pursuance of the WHC. However, it was almost twenty years before the National Park Service began to provide information about the WHC to site visitors. The level of public awareness regarding the Convention is still relatively low compared with treaties like CITES and the Montreal Protocol. The National Park Service's admission that it and the Department of the Interior had been deficient in publicizing the WHC resulted in a meeting in Santa Fe, New Mexico, in 1992, at which U.S. park staff personnel discussed ways to increase the use of the WHC in the planning and management of parks.

According to Richard Cook, officials within the Park Service would like to make the WHC more a part of the national historic preservation process. They want to incorporate the WHC into the rationale for each park's existence, then set specific guidelines for each park's operation based on the Convention. These officials would also like to increase public awareness of the WHC through incorporating it into the Department of the Interior's short films and other park information.

The legal effect of the WHC in the United States, though never challenged, is questionable. It is undisputed that the Convention is binding on the parties as international law. With respect to enforceability in U.S. domestic courts, however, a question still remains as to whether the Convention is binding (Magraw 1988). Under U.S. law, any treaty that is not "self-executing" must be implemented by U.S. domestic law in order to create legal obligations. If the Convention is not self-executing, it will not be effective as domestic law within the United States unless implementing legislation is passed. According to Daniel Magraw of the U.S. Environmental Protection Agency's (EPA) Office of International Activities, the U.S. position is that the World Heritage Convention is not self-executing and cannot be enforced in a U.S. court without legislation to activate its terms (Magraw 1994). At present, there are only two sections within the NHPA that address this very general treaty.

According to Magraw, because the sections of the NHPA addressing the WHC are very general, they don't give any directives that are enforceable in a U.S. court. Since they do not cover the whole entirety of the Convention, it is unclear whether the terms of the Convention would be enforced in a federal court, or whether it would be necessary to look to some other statute. To date, there have been no court cases interpreting or enforcing the Convention.

Furthermore, the regulations implementing the WHC do not yet contain the key requirement of the 1980 amendments to the NHPA that, prior to the approval of any federal undertaking outside the United States that may directly and adversely affect a World Heritage Site, the head of the agency with jurisdiction over the undertaking must consider the possible effect on that site (R. Cook 1993a). No attempt has been made thus far to apply that provision to any federal undertaking abroad.

Finally, although the United States was a key player in the creation of the treaty and still is its largest financial supporter (Hoffman 1993), the United States has opted under the treaty to make its contribution obligation purely voluntary. In the 1980s, when Jordan proposed that the Old City of Jerusalem and its walls be put on the endangered list, the United States ceased making contributions to the World Heritage Convention for three years and cast the only dissenting vote (Douglas 1983). The United States then lost its seat on the World Heritage Committee.

U.S. compliance with the WHC took a marked turn for the better in the late 1980s when Congress resumed appropriating funds to the Convention and increased its efforts to improve public awareness of the significance of the WHC. The United States now provides about 25 percent of all annual contributions. Despite the small size of the U.S. contributions to the WHC relative to its contributions to other agreements (the U.S. contribution to the WHC was $450,000 in 1991), the Director of the WHC under UNESCO asserts that the United States continues to contribute more than is required under the Convention (Prott 1993).

However, at least one State Department official has advocated increased financial or staff support for ensuring compliance with the WHC. He asserts that the amount of money the United States devotes to the treaty for both domestic and international purposes is inadequate (Williamson 1994).

CITES

Implementation. U.S. involvement in CITES can be traced to a long history of concern about species protection. Beginning in the early 1900s, Congress passed a series of laws relating to endangered species. In 1963, the IUCN called for an international convention on the "regulation of export, transit and import of rare or threatened wildlife species or their skins and trophies" (Lyster 1985). Such a treaty was also advocated within the U.S. Endangered Species Conservation Act of 1969.

In part to enforce the requirements of CITES, the United States enacted the Endangered Species Act (ESA) of 1973. At the official signing, President Nixon announced that the ESA put CITES into effect in the United States. With the ESA's enactment, the United States became the first country to implement CITES (Kosloff and Trexler 1987). In addition to implementing CITES, the ESA remedied weaknesses in the 1966 and 1969 Endangered Species Preservation and Conservation Acts.

Since its enactment in 1973, the ESA has come under fire for being too draconian in its requirements to protect threatened and endangered species. However, its provisions that implement CITES seem to have evaded the scrutiny that has attached to the other species-protecting measures in the Act.

Under the ESA, Congress recognized the U.S. pledge to conserve endangered wildlife pursuant to CITES. Section 8 of the ESA directed the Department of the Interior to carry out the U.S. CITES obligations and to encourage foreign governments to establish endangered species programs. To do so, the Secretary of the Interior was authorized to provide financial and technical assistance in support of foreign fish, wildlife, and plant conservation programs. In addition, section 8 authorized the Secretary to implement the related Convention on Nature Protection and Wildlife Preservation in the Western Hemisphere.

Section 9 prohibited the import or export of ESA-listed species or any products made from a listed species. Under section 9, it became illegal even to possess a listed species taken in violation of the Act. Section 9 also made CITES violations specifically punishable under the ESA. The key provisions penalizing trade of endangered species and regarding ESA management were refined by amendments made in 1978, 1982, and 1988.

The ESA fulfilled the two main obligations of CITES. First, it established a system for penalizing any person trading in, or possessing, specimens contrary to the provisions of CITES. Second, the ESA designated the Secretary of the Interior to act as both management and scientific authority under the Convention. Subsequently, the functions of the management and scientific authorities were delegated in part to the Fish and Wildlife Service (FWS), a bureau within the Department of the Interior.

In the 1990s the FWS administers the listing of most plants, both freshwater and land species. The National Marine Fisheries Service (NMFS) lists most marine species. The Animal and Plant Health Inspection Service (APHIS) in the Department of Agriculture enforces import/export restrictions on listed plants. Federal land management agencies, primarily the Forest Service (USFS) and Bureau of Land Management (BLM), share responsibility for the protection of endangered species and their remaining habitat (see US GAO 1992a).

In February 1977, the FWS issued regulations to implement CITES. They include CITES definitions, the format and criteria for permits and certificates, and the prohibitions and exceptions discussed in CITES, and duplications of CITES appendices I, II, and III. They also list the twelve designated ports of entry for wildlife and sixteen for plants in the United States.

The Federal Wildlife Permit Office within the FWS files an annual report detailing import, export, and reexport activity through U.S. ports. These reports provide statistics on violations of CITES as well as of the Lacey Act, the Marine Mammal Protection Act, the Migratory Bird Treaty, and the ESA. The FWS and APHIS levy both civil and criminal fines for CITES violations. These agencies are also permitted to seize illegal plant and animal products.

In addition to the ESA, the United States can be considered to implement CITES in a variety of other ways. Laws relating to customs procedures, conspiracy, smuggling, and mail fraud have all been used at times to enforce the provisions originating in CITES (Kosloff and Trexler 1991). Some observers consider the United States to have the best implementation infrastructure for CITES among the 102 member nations (Kosloff and Trexler 1991).

Compliance. U.S. federal courts have actively enforced the terms of CITES. For example, the Fifth Circuit applied the provisions of CITES in a 1991 case in which defendants failed to obtain valid export permits for animal products they acquired from Mexico (*Wallace* v. *U.S.*, 949 F.2d 759 (5th Cir. 1991)). In 1993, the Ninth Circuit affirmed the convictions of two defendants for smuggling bird eggs from Australia in violation of U.S. law and CITES (*U.S.* v. *Parker*, 991 F.2d 1493 (9th Cir. 1993)).

The United States has also taken unilateral action internationally to protect species listed under CITES. In 1991, the Department of Interior banned all trade with Thailand in wildlife and wildlife products after the CITES Secretariat asked all parties to take all measures to prohibit trade with Thailand. This action affected an estimated $18 million in annual trade in CITES-regulated wildlife between the United States and Thailand.

To date, U.S. actions to curb illegal trade in plants and animals have frequently taken place outside the CITES regime. For example, the Lacey Act prohibits the importation of wildlife into the United States in violation of the law of the source nation. In addition, under the Pelly Amendment to the Fisherman's Protective Act of 1954, when the Secretary of the Interior determines that nationals of a foreign country are engaging in trade that diminishes the effectiveness of any international program for endangered or threatened species, the Secretary must certify that fact to the President of the United States. Upon receipt of a Pelly certification, the President may order the Office of the U.S. Trade Representative (USTR) to prohibit imports of wildlife products originating in the offending country into the United States.

The President has complete discretion over the imposition of any Pelly Amendment sanctions, and Pelly Amendment certifications have therefore served as important diplomatic negotiating tools. These certifications have been issued over twenty times since 1974, without sanctions. For example, in 1991, the United States threatened to impose a comprehensive ban on all animal products from Japan under the Pelly Amendment because of Japan's trade in endangered hawksbill sea turtles. Japan yielded to the pressure and announced it would gradually end its trade in the endangered sea turtles (Abramson 1991).

It has been asserted that U.S. participation in the General Agreement on Tariffs and Trade (GATT) may jeopardize its ability to enforce its laws extrajurisdictionally because both the Lacey Act and the Pelly Act sanctions may be inconsistent with GATT (Pope 1993). Some observers, expecting the Clinton administration to engage in more activist unilateral actions against nations that exploit endangered species, have thus far been disappointed at its slow movement toward imposing Pelly Amendment sanctions (Schorr 1994). However, the Clinton administration did announce trade sanctions against Taiwan for its lack of progress in eliminating its illegal trade in tiger parts and rhinoceroses horns (Clinton 1993). That was the first time the United States imposed trade sanctions under the Pelly Amendment to protect endangered species (Raclin 1994).

U.S. law enforcement authorities have planned and executed successful large-scale CITES enforcement actions with some frequency. Following a two-year investigation, for example, federal agents broke a ginseng smuggling operation in 1992 that included falsification of state certification documents to obtain export permits. In a 1990 case, the FWS recruited a falconer to work undercover to expose illegal trafficking in birds. The investigation led to 68 convictions, mostly for misdemeanors, entailing 3 years of jail time, $500,000 in fines, 1,615 hours of community service, and 78.5 years of probation (Roberts 1990).

Although most observers assert that the United States has one of the most sophisticated programs in place for implementing CITES, the United States has been unable to prevent all trade in endangered species. Enforcement is difficult due to the large amount of international trade that comes into and goes out of the United States. As in the drug trade, many opportunities exist for smuggling items into the United States. In fact, according to one TRAFFIC U.S.A. representative, drug smugglers and wildlife smugglers often use the same techniques and are involved with both trades (Robbins 1995).

The FWS lacks the resources to conduct random searches at all designated and undesignated points of entry into the United States, and large shipments can be transferred clandestinely across the Canadian and Mexican borders. In fact, some FWS Division of Law Enforcement officials claim that funding and staffing are inadequate to carry out necessary investigations (Speart 1993). In 1993, only 72 FWS wildlife inspectors were stationed among all of the major U.S. ports, and only 5 inspectors were assigned to the entire 2,000-mile Mexican border area (Speart 1993). In addition, although high-profile prosecutions for CITES violations do occur from time to time, enforcement efforts generally focus on tourist, souvenir, or personal transactions rather than on commercial transactions.

Although one FWS inspector has described the customs officials stationed at the ports of entry as the "eyes and ears" of the FWS, such customs officials are trained merely to control exports and imports and normally are not experts in biology (as are many FWS inspectors). Consequently, the FWS generally ends up examining only the permits of documented shipments for validity. Containerized cargo shipments not accompanied by CITES permits usually are not inspected by the FWS.

Anecdotal reports indicate that substantial quantities of wildlife products enter the United States undetected. One official in the Department of Commerce estimated that between $100 million and $250 million in illegal wildlife enters the United States annually, and that the amount has increased since 1987. In 1987, the U.S. market was estimated to account for up to one-third of the declared value of the $5 billion market in illegal wildlife trade (Trexler and Kosloff 1991). An official within the FWS Division of Law Enforcement has stated that of 80,000 wildlife shipments that come into the United States each year, 95 percent are never inspected but are cleared on paperwork alone (Speart 1993).

According to a Department of Agriculture official at an official U.S. Customs inspection point on the California–Mexico border, it is difficult to estimate exactly how many articles are imported illegally into the United States by individuals traveling from Mexico. Customs officials generally spot-check the luggage or vehicles only of persons who "look like they are hiding something." However, according to that official, most people do not know that the cactus, aloe vera, sea turtle article, or mammal bone they are bringing home for their personal use is a CITES-listed species. Therefore, they might not be stopped for looking "suspicious." In 1994, the official estimated that approximately twenty plants or animal goods were confiscated each month at that customs checkpoint. A number of law enforcement officials have noted deficiencies in the effort to educate U.S. travelers about CITES.

With respect to spending for CITES alone, specific numbers are hard to come by. It is extremely difficult to determine what federal funds for the FWS Endangered Species programs are dedicated solely to CITES administration. The budget for enforcement of the ESA has increased over time, with the FWS ESA budget increasing from $18.8 million in 1988 to $42.3 million in 1992 (USGAO 1992a). The Division of Law Enforcement's budget in 1992 was $31.5 million (Speart 1993). According to Kosloff and Trexler, as of 1991 the United States expended $2 million per year on the FWS inspection program alone. However, one must compare these figures against the FWS's total budget, which in 1993 was close to $1 billion (Speart 1993). APHIS's CITES funding was approximately $1.5 million in 1993.

As an additional note, a 1994 UNEP-sponsored report lists the United States at 88 percent for "CITES Reporting Requirements Met" in the years since becoming a party to the Convention through 1991 (Report by the World Resources Institute 1994). However, according to one NGO representative, preparation of the annual reports is a colossal task that is not supported by adequate staff. Therefore, the reports are often late (Campbell 1993).

The International Tropical Timber Agreement

Implementation. The United States signed the ITTA in 1985 and participated fully as a provisional member until Congress approved membership in 1990. The ITTA represents an executive agreement, rather than a treaty, for the United States. Therefore, the U.S. Senate did not approve the agreement before its entry into force.

In the United States, a bureaucratic structure sufficient for enforcing the new agreement was in place prior to the ITTA's entry into force. Primary responsibility for U.S. participation in the International Tropical Timber Organization (ITTO) and compliance with the ITTA rests with the U.S. Department of State and the USTR. Technically, the USTR serves as the lead agency for the ITTA because the ITTA is a commodity agreement (Drucker 1994). However, according to the head of the U.S. delegation to the ITTO, in practice the Department of State has assumed the functional lead in working on ITTA issues (Drucker 1994). In addition, the U.S. Forest Service is responsible for providing technical assistance regarding forestry issues (Hicks 1994). The EPA participates as part of the U.S. delegation to meetings of the

ITTO and the International Tropical Timber Council (ITTC), and conducts special projects on sustainable forest management.

All of these federal agencies cooperate to develop U.S. policy on tropical timber trade in preparation for meetings of the ITTO and ITTC. In addition, representatives from the U.S. Department of Commerce, the International Trade Administration, and the Agency for International Development serve on the U.S. delegations to the meetings of the ITTO and the ITTC.

The ITTA's annual contribution was the only obligation that made if necessary for the United States to pass implementing legislation. Initially, the United States attempted to meet its obligations to the ITTO through the Department of State's contingencies account. However, this account has been described as "vastly oversubscribed" because it supports U.S. participation in many international organizations. The United States underfunded the ITTO by relying on the contingencies account and lost its voting rights in 1988. At one point, it was in arrears for more than $220,000 of its dues to the organization for the years prior to 1990 (U.S. International Trade Commission 1991). To remedy this situation, the United States resumed membership in the ITTO in 1990 by authorizing $231,000 to the ITTO for dues and arrearages. The United States allocated $119,000 for this purpose in 1993. In 1991, it made its initial pledge of almost $1 million to the Special Account (ITTO Annual Reports 1988–1992).

Compliance. The United States assumed obligations under the ITTA only as a major tropical timber consumer, and not as a producer, because it contains only temperate forests. As such, the United States is required to cooperate with other parties to achieve ITTA objectives. It also must abide by decisions of the ITTC and is required to pay an annual contribution for administration of the ITTA. The ITTA requires member countries to encourage development of national policies aimed at sustainable utilization and conservation of tropical forests. It also asks member countries to comply with information-sharing requirements.

Under the ITTA, member countries are requested to "furnish, within a reasonable time, statistics and information on tropical timber requested by the Council." Consensus does not exist as to whether these reporting obligations are mandatory or voluntary (Fletcher 1994). Generally, the ITTC seeks data on U.S. production, trade balance, domestic wood requirements, and by-products for veneer, plywood, softwood, hardwood, and tropical hardwood. It also seeks data on U.S. tropical timber imports and exports of products.

The United States has consistently filed complete reports to the ITTO on an annual basis and in a timely manner. According to one U.S. Department of Agriculture official, the ITTA's reporting requirements are not burdensome for the United States because the relevant information is readily available. The ITTO reports take approximately two weeks for one staff person to prepare. Responsibility for preparing the reports may shift between federal agencies when the officials responsible for preparing the reports change agencies (Hicks 1994). The Foreign Agricultural Service within the Department of Agriculture prepared the reports in 1991, 1992, and 1993.

The United States has taken several important steps to support the objectives of the ITTA. In 1986, Congress enacted legislation that called on the President to make sustainable tropical forest management a priority. In addition to various other goals, the legislation required the President to support the establishment of institutions that would help developing countries draw up forest policies and improve forest management. Also, the United States has been an active, if sometimes reluctant, participant in international policy discussions intended to conserve forests in temperate regions.

Article 30 of the ITTA requires the United States to accept the decisions of the ITTC as binding. Perhaps the most important ITTC decision to date involved adopting The ITTO Guidelines for Sustainable Management of Natural Tropical Forests (World Wildlife Fund 1992). These "Target 2000" guidelines direct that all international tropical timber trade should be based on sustainably managed forests by the year 2000.

Under the guidelines, parties to the ITTA must develop their own national policies for sustainable tropical forest management (Sand 1992). Parties are then responsible for implementing and enforcing concrete measures. Although the United States initially resisted expansion of the ITTA to cover temperate forest products because it has one of the world's largest industries in those products, it finally indicated its willingness to commit even more fully to the "Target 2000" goal by agreeing to the Formal Statement of the Consumer Members, which accompanies the successor agreement to the ITTA. Although this statement is not a formally binding part of the new agreement, agreement to it indicates the United States' willingness to apply the "Target 2000" goals to its own, nontropical forests. The United States filed its first report on its progress toward the "Target 2000" goal in November 1993.

London Convention of 1972

Implementation. Congress began to regulate ocean dumping in the early 1970s, partly in reaction to the CEQ's conclusion that dumping of radioactive waste presented a serious threat to the marine environment (Guarascio 1985). According to Bakalian (1984), the amount of waste dumped into the world's oceans increased by 335 percent from the early 1950s to the mid-1960s. In 1972, Congress enacted the Marine Protection, Research and Sanctuaries Act (MPRSA; also called the Ocean Dumping Act), which outlawed dumping of certain substances and allowed the dumping of other substances only with permits. The Ocean Dumping Act was signed into law by President Nixon one week before the London Conference began.

Three years later, in 1975, the London Convention took effect in the United States. In 1974, the MPRSA had been amended to include changes conforming domestic U.S. law to the London Convention.

A number of commentators have observed that the London Convention of 1972 is patterned after the U.S. law (Thorne-Miller 1993; Guarascio 1985). However, it is unclear exactly how much existing U.S. law actually influenced the drafting of the Convention. Although U.S. law already banned the dumping of radiological, chemical, and biological warfare agents and high-level radioactive wastes, and provided permitting rules for other categories of materials covered by the Convention, it contained no annexes analogous to those in the Convention. We do know that in June 1971, the United States prepared the first draft of a convention that closely paralleled the Ocean Dumping Act, and presented it at the first meeting of the International Working Group on Marine Pollution, which had been convened to prepare for the U.N. Conference on the Human Environment in Stockholm. However, at that time, the Working Group followed the approach in the regional Convention for the Prevention of Marine Pollution by Dumping from Ships and Aircraft (Oslo, February 15, 1972). In any event, the United States was a major contributor to the negotiations for the London Convention of 1972 and continues to participate at Consultative Meetings and meetings of subsidiary bodies.

One commentator observed that amendment of the MPRSA "has become routine"—primary amendments have been enacted in 1974, 1977, 1980, and 1988 (Moore 1992). The 1974 amendments conformed the MPRSA to the standards of the London Convention of 1972. The 1977 amendments directed the EPA to end dumping of sewage sludge by the end of 1981. The 1980 amendments to the MPRSA directed the EPA to end dumping of industrial waste by the end of 1981 and extended the MPRSA to cover disposal of dredged soil in Long Island Sound. The 1988 amendments unequivocally banned the dumping of all sewage sludge and hazardous or toxic wastes ("industrial wastes") at sea by 1992, and prohibited the dumping of medical wastes into the oceans.

The 1988 amendments to the MPRSA also banned dumping of low-level radioactive waste. In December 1993, the signatory nations agreed to amend the London Convention of 1972 to ban such dumping.

Compliance. According to Coenan (1993), the United States complies well in most areas of the Convention. Even the environmental organizations that are generally the most critical of U.S. government actions tend to conclude that U.S. compliance is above average.

The U.S. record on reporting under the treaty has been good. Until the early 1990s, the United States failed to comply only with its 1983 reporting requirement. According to Coenan (1993), the United States has not reported since that time because it no longer dumps materials that require reports. Also, an encouraging example of compliance with the goals of the treaty occurred when the United States reversed its policy on dumping of low-level radioactive waste to support a ban on all such dumping. The long-standing U.S. position had been not to support a permanent ban, but to keep the option open for the future. Formerly only the subject of a moratorium by the treaty parties, the ban is now officially a provision of the treaty.

However, there remain some areas in which U.S. compliance could be improved. Obvious enforcement problems exist with respect to the Convention, particularly outside U.S. waters and against non-U.S.-flag vessels. As with CITES, these problems stem from the practical difficulty of monitoring vast expanses of territory—or, in this case, ocean area. To address this difficulty, an interagency agreement with the U.S. Coast Guard for surveillance of illegal dumping has been suggested. Article 7(2) of the Convention calls upon parties "to take appropriate measures to prevent and punish conduct in contravention of the provisions of [the] Convention." To comply with this obligation, the United States must institute a better system to catch illegal dumping.

Diehl (1993a) points to an example of illegal dumping by a U.S. vessel that illustrates the inability of the United States to control all illegal dumping. In the first criminal case brought in response to dumping in international waters, the U.S. Department of Justice filed a claim against the agents of the ship *Khian Sea*. In 1986, the *Khian Sea* left Philadelphia, loaded with 15,000 tons of municipal incinerator ash, bound for the Bahamas, where the load was to be sold for landfill. On the way, the ship was turned away at gunpoint from two ports and threatened with attack by Greenpeace. Eventually, the ship disposed of its cargo at unknown points in the ocean before returning to the United States. The defendants were convicted of perjury and violation of the MPRSA in 1993.

Furthermore, courts have not consistently enforced the terms of the Convention. Despite the 1977 amendments to the MPRSA, which led to the abandonment of ocean dumping by more than 200 East Coast sewage treatment plants, the statute's definition of "sewage sludge" allowed some municipalities to continue dumping sewage sludge long after the deadline passed. Perhaps the most controversial decision rendered by the courts regarding ocean dumping was *City of New York* v. *EPA* (543 F.Supp. 1084 (S.D.N.Y 1981). In that case, New York City challenged the EPA's decision to deny the city continuance of its interim sludge-dumping permit. The court ruled in favor of the city, finding that the EPA's contention that the amendment required an absolute end to all ocean dumping was erroneous. Neither the EPA nor the court considered the London Convention of 1972 in *City of New*

York. The *City of New York* decision had a considerable effect on ocean dumping until the late 1980s because the EPA did not oppose the decision. In 1988, New York City and eight New Jersey sewerage agencies were still dumping sewage sludge into the ocean.

In fact, the last sewage sludge was actually dumped by a private interest in 1992, at a site off the East Coast. This was accomplished pursuant to a statutory extension to allow the dumping; then, upon the extension's expiration, under the imposition and payment of fines for noncompliance with the dumping ban. Actual ocean dumping by municipalities has ceased since the 1988 amendments, although wastes continue to enter the oceans through indirect sources such as runoff or land outfall pipes. These wastes account for up to 90 percent of all contaminants entering the oceans each year (Moore 1992).

If the purpose of the London Convention of 1972 is to contribute to a reduction in marine pollution, rather than simply to reduce ocean dumping, questions arise concerning whether the United States has achieved an overall reduction in the pollution that reaches and degrades the marine environment (Diehl 1993b). Marine pollution continues from sources other than marine vessels, despite the Convention's article 1 obligation that parties are individually and collectively to promote the effective control of "all sources of pollution of the marine environment" (see also article 12).

Montreal Protocol

Implementation. Authority for the regulation of chlorofluorocarbons (CFCs) has existed in the United States since 1978. When negotiations for an ozone protocol began at Geneva in 1986, the United States supported either a major reduction or a full ban on nonessential CFC uses and exports over a six-year period (Sand 1985). The United States could safely advocate this position because it had already banned nonessential uses in 1978 under its own Clean Air Act.

However, despite progress on the Montreal Protocol negotiations in 1986, an anti-regulatory backlash developed in the United States in early 1987 (Doniger 1988). By that time, the EPA Administrator had attained support for a total CFC phaseout from all relevant federal agencies. However, opponents within the Reagan administration came out strongly against the phaseout policy. The most spectacular example of the backlash occurred in May 1987, when Secretary of the Interior Donald Hodel advocated a personal protection strategy of wearing hats and sunglasses in lieu of an international policy to prevent ozone depletion.

Hodel's statement came during the critical period when the EPA was attempting to convince the Reagan administration to support the Montreal Protocol (Horowitz and Seidel 1993). The public outcry against Hodel's position was dramatic, and may have guaranteed that the official U.S. position would be to back stringent CFC controls (Shimberg 1991). Senators John Chaffee and Albert Gore introduced legislation to phase out ozone-depleting chemicals completely in the United States. Following Hodel's statement, the Senate responded by quickly passing Gore's and Chaffee's resolutions (Horowitz and Seidel 1993). At hearings on ozone depletion, members of the Senate strongly encouraged the U.S. delegation to negotiate for a CFC phaseout.

As the Montreal Protocol negotiations proceeded through the summer of 1987, the Reagan administration shifted its position to support a phaseout of ozone-depleting substances. Commentators state that the shift in position occurred as a result of individual pressure on the President. Benedick (1991) notes that Reagan was eventually persuaded to support a CFC phaseout partially through the efforts of Secretary of State George Shultz, who wished to prevent damage to the United States' international reputation. Other observers contend that the administration feared the United States would be unilaterally required to regulate CFC production under its own Clean Air Act requirements, thus putting U.S. companies at a competitive disadvantage vis-à-vis companies from nations not affected by the same constraints.

The United States signed the Montreal Protocol at the conclusion of negotiations in September 1987, and ratified the Protocol seven months later. In August 1988 the EPA issued final rules implementing the Protocol that went into effect when the Protocol entered into force in January 1989.

The 1990 London Amendments to the Montreal Protocol accelerated the phaseout of class 1 substances, such as certain CFCs, by the year 2000. Yet Congress's 1990 amendments to the Clean Air Act went even further. They added a new stratospheric ozone title to the Federal Clean Air Act, and provided the EPA with the authority to phase out certain ozone-depleting substances more quickly than the Act required if necessary to protect human health or the environment. The amendments also limited U.S. production of hydrochlorofluorocarbons (HCFCs) as of 2015 and eliminated them as of 2030. Furthermore, the Act required the EPA to bolster the production phaseouts through other means, including recycling and recapture rules, labeling requirements for products either containing class I or II substances or utilizing them in the product's manufacture, policies to promote transition to safe substitutes, and a requirement that federal agencies modify their procurement procedures to make them consistent with the Act. The EPA has subsequently moved forward to promulgate rules implementing these requirements.

In February 1992, President Bush announced that the United States would phase out the production of ozone-depleting substances by the end of 1995, five years earlier than required under the London Amendments. Later that year, at the fourth Meeting of the Parties in Copenhagen, the United States pressured other countries to agree to a similar acceleration (Dumanoski 1992). The U.S. delegation also supported caps and percentage reductions in HCFCs, and advocated a major cutback in the use of methyl bromide. The United States led the industrialized parties on this issue (Rowlands 1993).

President Clinton sent the 1992 Protocol amendments to the Senate for ratification in July 1993, urging the Senate to give them early and favorable consideration. The Senate approved the amendments four months later.

At the fifth Meeting of the Parties in Bangkok in November 1993, the United States joined fifteen other countries in signing a declaration to phase out methyl bromide as soon as technically feasible. The EPA then issued a final rule for phasing out methyl bromide completely by 2001. The United States also agreed to phase out HCFCs by 2015 at the latest.

Compliance. Stratospheric ozone depletion has been a controversial scientific and political issue in the United States since the early 1970s. Since then, the federal government and various nongovernmental and government-funded groups have participated in the ozone debate and influenced U.S. policy. Until the late 1990s, with a few deviations, the United States has leaned heavily toward the side of the Protocol supporters. In 1992, the United States was 42 percent ahead of the schedule mandated by the Montreal Protocol, which ordered a CFC phaseout by the year 2000.

To date, enforcement of the Montreal Protocol by U.S. courts has been strong. For example, in January 1995 a federal grand jury in Miami indicted two persons on charges of smuggling more than 100 tons of ozone-depleting chemicals into the United States. This case was the first known instance in which the Clean Air Act was invoked by a grand jury against individuals importing ozone-depleting chemicals illegally into the United States. In January 1996, a federal judge sentenced a Florida man to fourteen months in prison and ordered him to pay over $3.4 million in tax restitution under a plea agreement related to his smuggling of over 500 tons of CFC-12 into the United States by using cargo containers shipped through the United Kingdom from Eastern Europe.

Several years earlier, in 1990, civil suits were filed against five importers of CFCs, the first suits alleging violations of the Clean Air Act's rule to protect the stratospheric ozone layer. In those cases, the Department of Justice alleged that the defendants imported CFCs into the United States without obtaining the required consumption allowances. The EPA rule requires producers and importers of CFCs to obtain allowances from the EPA and limits the number of allowances available in a twelve-month period. The rule also requires that importers of CFCs submit quarterly reports identifying the quantity of CFCs imported and the total amount of unexpended consumption allowances. The U.S. attorney general's office sought civil penalties and an injunction requiring the companies to comply with the EPA rule.

According to the Department of Justice, a black market has developed to smuggle in untaxed CFCs as the U.S. government tightens controls on the availability of ozone-depleting chemicals and imposes higher excise taxes on the chemicals. Taxes were $4.35 per pound in 1994 and were raised to $5.35 per pound in 1995.

It remains to be seen, however, what effects recent changes in attitude in the U.S. government toward environmental regulation will have on compliance with the Montreal Protocol. The 104th Congress pledged to reexamine international environmental treaties and the science behind them, with special mention of the Montreal Protocol. In January 1996, Clinton administration officials informed Congress that they would seek an amendment to the Clean Air Act to allow the EPA to grant "critical exemptions" to the ban on methyl bromide production and import in the year 2001. Environmental groups characterized this action as an election year strategy to gain votes in California and Florida, two agricultural states with high methyl bromide use.

In addition, in a surprising move, the EPA asked the E.I. Du Pont Corporation to renege on its corporate pledge to stop making ozone-depleting CFCs for sale in the United States by the end of 1994. Having developed technologies capable of using available alternatives to CFCs, Du Pont had announced it would cease CFC production at the end of 1994, a full year before it was required to do so under the Montreal Protocol. The EPA, to head off a projected refrigerant shortage, asked Du Pont to continue to provide adequate supplies of CFCs through the end of 1995. According to the EPA, the request for Du Pont to continue production through 1995 was intended to help the American consumer ease through the transition from vehicle air conditioners using CFCs to those using refrigerants that do not harm the ozone layer.

From the viewpoint of environmentalists, this request represented a serious backtrack in the area of environmental protection. What is more, it came on the heels of an earlier decision on December 10, 1993, to phase out the U.S. manufacture of methyl bromide by January 1, 2001, a full year later than the proposed deadline of January 1, 2000. Despite the environmentalists' disappointment at the later date for phaseout, however, the United States has still done more to control methyl bromide and phase it out faster than any other country.

The idea of providing funding for developing nations under the Montreal Protocol has been a contentious one for the United States since the idea's inception in 1990. At the 1990 Meeting of the Parties, the United States opposed the creation of a $100 million fund, complete with its own secretariat, to help developing nations. At that time, U.S. officials maintained that the World Bank had sufficient funds to meet those needs and should administer such a program using existing monetary resources. According to administration sources, Bush administration Chief of Staff

John Sununu and Office of Management and Budget Director Richard Darman feared that the fund, which involved a U.S. contribution of $25 million, would set a precedent for more costly programs to address the issue of global warming.

After intensive lobbying efforts by U.N. officials and members of Congress, however, Sununu announced in June 1990 that the United States would support additional funding. Those pressuring the White House had warned that U.S. opposition to the fund was jeopardizing the future of the Montreal Protocol. Interestingly, a representative of the Alliance for Responsible CFC Policy, a CFC industry association, noted that the federal government was collecting several hundred million dollars a year from CFC users through the excise tax, and that it would be foolish not to use some small percentage of it to assist developing countries (Dumanoski 1990).

Eventually, the United States also agreed to be represented on a fourteen-member executive committee for the Multilateral Fund, comprised of seven members each from the donors and the recipients. Initially, the United States had desired more authority to influence operation of the Fund. U.S. contributions to the Fund were approximately $38 million in 1994, having risen steadily by $10 million every year since 1991.

Explaining Implementation and Compliance: General Observations and Conclusions

Treaty-Specific Influences on Compliance

As illustrated above, the United States proceeded to implement all five agreements in this study with a high level of consistency. Although generally good, however, the quality of U.S. compliance with the agreements has been less consistent. This study reveals several variables that have affected the United States' compliance record on the five agreements at issue.

Preexisting Legislation Regarding the Same Subject Matter. In four of the five cases, the United States already had legislation in place addressing the subject matter of the treaties. For example, in the case of the WHC, the United States had its own preservation network in place before it signed the agreement. Although World Heritage designation added to the knowledge of the significance of the sites in question, care, accessibility, and interpretive programs had been under way for most of the sites for decades prior to their designation. Indeed, it is difficult to say whether World Heritage Sites in the United States would be managed differently absent the treaty, or if the sites were not World Heritage Sites (R. Cook 1993a).

Similarly, endangered species and ocean dumping legislation emerged in the United States in the early 1970s, when the U.S. environmental movement exercised great influence. The ESA and the MPRSA both followed on the heels of the first Earth Day, the passage of the National Environmental Policy Act, and the creation of the EPA. At that time, intense public interest in the environment was felt in both the legislative and the executive branches. The MPRSA was actually passed after a commissioned report by the CEQ alerted the Nixon administration to the dangers of ocean dumping. U.S. domestic law preceded the obligations of the London Convention of 1972 at almost every step, and virtually all that the Convention required was already required by U.S. domestic law. Thus, U.S. law required minimal amendment to bring it into line with the Convention.

Curtis (1993) observed that, with regard to ocean dumping, the United States generally seeks to influence other nations only with respect to compliance efforts it has already undertaken. A possible exception may have been the U.S. opposition to including a prohibition against dumping low-level radioactive waste in the Convention. Although the United States had already banned such dumping under domestic law, the Bush administration refused to support a Convention ban in the early 1990s. The United States changed its position in November 1993, and actually took the lead in advocating a ban on dumping of all radioactive waste. According to Greenpeace, public outcry stemming from the discovery of the Russian disposal of radioactive waste in the Sea of Japan led to the change in U.S. position.

The United States also took measures to reduce ozone-depleting substances before entering into the Montreal Protocol negotiations. U.S. scientific research in the wake of the 1973 Rowland–Molina report catalyzed a federal government policy response on stratospheric ozone depletion as early as 1975. Formal hearings soon after the publication of the Rowland–Molina study led to the passage of ozone protection legislation in the 1977 Clean Air Act Amendments and the 1978 Toxic Substances Control Act. Later, after diplomatic negotiations in December 1986, the Senate and House endorsed a strong U.S. position on new international controls on ozone-depleting substances. In most cases, U.S. reductions have been premandated by U.S. legislation; only the time schedules have changed.

Although the subject matter of the ITTA was not previously addressed by U.S. domestic legislation, no significant deviations in the strength of compliance between that agreement and the others can be detected. However, it is worth noting that the obligations under the ITTA, unlike those of several of the other agreements, are not difficult for the United States to meet. For example, the United States does not face difficulties in complying with the ITTA's information-sharing requirements because the government already collects the necessary market data (Hicks 1994). Because several U.S. agencies were working on tropical forest projects in other countries before the ITTA, continuing U.S. involvement in special projects is not problematic.

However, funding for the ITTA poses a different issue. In fact, some of the most common instances of U.S. noncompliance with the agreements in this study have occurred when the United States has failed to provide proper funding under the agreements. In the case of the ITTA, this was initally a result of not having specific legislation to authorize funds to the ITTO in place. In several other cases, however, the United States' failure to provide proper funding derived from political motives.

As can be seen by comparing the contribution amounts for the Montreal Protocol Development Fund against the World Heritage Fund, the amount of a contribution does not seem to indicate whether, or even how promptly, the contribution will be paid. Compliance with the central financial obligations of each agreement appears to be more a function of whether or not the contribution is required by statute. Although the United States generally makes its required contributions in the end, the timeliness of those contributions is problematic when domestic law confers discretion.

Public Interest in the Subject Matter

In most instances, the public's awareness of and interest in the subject matter of each of the agreements influenced the U.S. government's activities under the agreements. For example, perhaps the best single explanation for the high level of U.S. compliance with the London Convention of 1972 is a series of widely publicized events that focused public attention on the consequences of ocean pollution. In 1976, a massive fish kill off the coast of New York and New Jersey sparked citizen interest in cleaning up ocean pollution. In 1987, six years after the *City of New York* decision effectively ended efforts by New York and New Jersey municipalities to find alternatives to ocean dumping, "brown tides" or algae blooms were found off Long Island, generating further concern among scientists about ocean pollution. Then, in the summer of 1988, medical waste and dead fish washed ashore on New York and New Jersey beaches, again sparking citizen interest in cleanup. On the West Coast, high pollution levels off southern California beaches continue to generate public concern about pollution.

Public perception of the danger of ozone depletion has also had a tremendous impact on the politics and economics of the issue. The ozone depletion theory was featured prominently in the media and congressional debates immediately following its issuance by the scientific community, and began to influence consumer behavior almost immediately. Even before the aerosol ban of 1978, consumers began to demand alternatives to products containing CFCs. During the Montreal Protocol negotiations, the public reacted loudly and negatively to reports of Secretary of the Interior Donald Hodel's suggestion that U.S. citizens should wear hats and sunglasses instead of the nation's signing on to the treaty. Once the United States signed the Protocol, the media and the public reacted favorably. *Newsweek* characterized it as "An Exemplary Ozone Agreement," and the *Washington Post* called it "an extraordinary achievement." Shortly thereafter, the city of Berkeley, California, banned all fast-food packaging made with CFCs. Other cities enacted similar measures.

The subject matter of CITES, the ITTA, and the WHC has fostered public acceptance of their implementing legislation. As opposed to treaties that deal with esoteric scientific matters remote from popular understanding, CITES protects animals, and Americans love animals. (When a California jogger was killed by a mountain lion, the lion was tracked and killed—its cub received more money in public donations than the jogger's child!) With regard to the ITTA, during the late 1980s the media began to publicize the negative environmental effects that occur when developing countries allow massive timber harvesting to build up their foreign exchange reserves. The American public's reaction has been largely negative. Public support for ecolabeling tropical hardwoods, and for other programs like "debt-for-nature" swaps, may have had some influence on decisions concerning U.S. obligations under the ITTA to support sustainable forest management. Finally, although public awareness of the WHC is low, its subject matter—at least with regard to domestic sites—has already been the object of long-standing and well-organized conservation efforts throughout the United States.

However, it is worth noting that although the public may be aware of and support the broad issues addressed by each of these agreements, they are less aware of the technical requirements of the agreements, such as reporting and funding obligations. These obligations, left out of the policy spotlight, are generally the first to suffer.

Environmental Group Participation. Participation by domestic and international environmental groups in each of the agreements has been an essential component of the United States' compliance efforts. For example, it is clear that the growth of the environmental movement influenced increased U.S. efforts to comply with the WHC in the late 1980s. In the 1970s, when the WHC was signed, the environmental movement was in its infancy. Between 1985 and 1986, however, the environmental movement experienced a major explosion in budget, membership, and impact. Although they had been active before, environmental organizations' influence increased dramatically at that time, perhaps in part because of the public's perception that the Reagan administration was hostile to environmental concerns.

The participation of environmental organizations has been integral to the success of the WHC, as international groups such as the World Wildlife Fund and other domestic organizations have lobbied Congress for the funds the National Park Service needs to implement the WHC fully. The World Wildlife Fund is also increasing its efforts to raise public awareness of the Convention. IUCN monitoring reports on several U.S. World Heritage Sites have resulted in positive follow-up (Thorsell 1993). In addition, under contract with the United States, US/ICOMOS has carried out investigative studies and aided developing countries in the preparation of site nominations. It also has participated in projects to train local officials in conservation and site development.

At a grassroots level, environmental NGOs have played a significant role in CITES's implementation and compliance. Organizations such as the National Wildlife Federation, the Sierra Club, the National Audubon Society, and numerous others provided strong support for CITES and lobbied in favor of implementing legislation. Active NGOs also engage in continuous and rigorous monitoring of U.S. compliance with CITES. One such NGO is TRAFFIC-USA, a program of the World Wildlife Fund, which identifies itself as "the principal U.S. source of objective information on international wildlife trade for the U.S. government, Congress, NGOs and industry" (TRAFFIC-USA 1993). Several other environmental organizations perform monitoring functions as well.

During the original ITTA negotiations, environmental groups were instrumental in ensuring the inclusion of article 1(h) on sustainable forest management in the agreement (Drucker 1993). Viewed at the time as little more than an afterthought, it has provided environmental NGOs with a basis for lobbying in support of greater U.S. involvement in tropical forest conservation. Environmental groups have also lobbied for attendance rights at ITTO meetings; in fact, the opening of ITTO meetings to NGOs represents the first time NGOs have been permitted to observe UNCTAD negotiations (Mankin 1993). Environmental groups have participated vocally in preparatory meetings for the U.S. delegation and have surveiled U.S. funding of forestry projects in other countries.

Organizations in the United States have long been interested in the activities of the London Convention of 1972, and over the years the Secretariat has received numerous requests for information that reflect such interest (Coenan 1993). Although Diehl (1993a) observes that the NGOs' influence on the positions the U.S. delegation takes on issues at London Convention meetings has been limited, environmental organizations have been dominant and influential in the area of ocean dumping itself (Lishman 1994). Thorne-Miller (1993) recounts the role played by NGOs in the mid-1980s, when, because of pressure from environmental groups, the EPA decided not to wait for the results of scientific research but began a search for land-based disposal alternatives for low-level radioactive waste at an early date. NGOs were also influential during the controversy over incineration at sea, and in the past few years they have become active regarding the dumping of dredged materials containing contaminated sediments. These groups have played a vital role in monitoring U.S. compliance with its own domestic law, and in the case of Greenpeace, with international law as well. American environmental NGOs strongly influenced the passage of stricter domestic limitations with regard to ocean dumping, such as the 1988 amendment to the MPRSA on industrial waste dumping.

American environmental groups took the lead in educating the public and Congress about the dangers of ozone depletion by publishing studies, holding press conferences, and funding research. These groups took legal and political action to force change within the United States. In 1984, the Natural Resources Defense Council (NRDC) pressured the EPA to promulgate regulations on CFCs required under the 1977 Clean Air Act amendments by filing a lawsuit against the EPA. In 1991, NRDC, Friends of the Earth, and the Environmental Defense Fund jointly submitted a petition calling on the EPA to accelerate the phaseout of ozone-depleting substances pursuant to its authority under the 1990 Clean Air Act amendments. During the Montreal Protocol negotiations, representatives from NRDC played a strong role by proposing a global CFC phaseout (NRDC 1991). The NRDC pressured the Reagan administration to support a treaty with obligatory CFC reductions and eventual elimination.

Again, however, although these groups have encouraged compliance with the main, most visible obligations of the agreements, they are too far removed from the inner workings of government bureaucracy to exert much influence regarding the agreements' more technical, yet by no means less important, reporting, funding, and staffing aspects. For example, environmental groups have brought no legal actions and have instigated no publicity campaigns to ensure that the reporting requirements under CITES are met or that the FWS inspection program is adequately funded and staffed. Nor have they made efforts to encourage the granting of funding for the ITTO.

Industry Cooperation. With several of the agreements, industry cooperation is shown to have influenced U.S. compliance efforts quite favorably. For example, the views and activities of U.S. industry have had a very important impact on U.S. compliance with the Montreal Protocol. Initially, CFC manufacturers like Du Pont denied that adequate scientific evidence existed to merit a CFC phaseout. However, once the scientific connection between CFCs and ozone depletion was undeniable and the Senate approved ratification of the Protocol in 1988, industry changed its stance. Through the Alliance for Responsible CFC Policy, created after the aerosol/nonessential use ban in the late 1970s to represent both producers and consumers of ozone-depleting substances, industry urged President Reagan to ratify the Protocol (Roan 1989). Du Pont, the world's largest producer of CFCs and also the first large-scale manufacturer to support global controls on the production and consumption of five CFCs and three halons, announced in March 1988 that it would phase out all production of CFCs rather than simply cut production by 50 percent by 1999, as required under the Montreal Protocol. Some analysts suggest that President Reagan's support for the treaty was not fully confirmed until Du Pont announced its decision to phase out the CFCs voluntarily (Shea 1988).

In September 1988, Du Pont announced that in 1990 it would begin to market products that would replace CFCs. At the time of that policy statement, fourteen chemical companies from eight nations were cooperating to bring substitutes for CFCs to the market and to pool efforts on toxicity testing of alternatives.

Thus, after the decision had been made to proceed with CFC reductions under the Protocol, industry made the most of it. In fact, industry moved more rapidly than the Protocol required. This change of heart makes economic sense when one considers that, during the transition period, demand was expected to exceed supply for a product these firms controlled (Roan 1989). In addition, the public's demand for alternatives to ozone-depleting substances may have influenced manufacturers to find CFC substitutes as soon as possible, in order to avoid having to label their products as containing an ozone-depleting substance. Concern about company image and cost of labeling thus probably hastened companies' phaseout efforts.

Despite positions in favor of CFC phaseouts early on, however, industry has not always supported measures to be taken under the Montreal Protocol. The late date for phasing out HCFCs, used in air conditioners, has been credited largely to industry influence at the fourth Meeting of the Parties at Copenhagen in 1992 (Rowlands 1993). U.S. industry lobbied for late phaseout because the United States utilizes more large cooling units for buildings than any other country. The units have a long economic life. By lobbying for an HCFC phaseout date nearly forty years in the future, industry was guaranteeing that the units could continue to be used until their value depreciated (Bryce 1993).

Commercial interests have also had a positive impact on U.S. compliance with the obligations of the ITTA. Industry, represented by groups like the International Hardwood Products Association and the American Forest Products Association, supported the ITTA because they wanted a treaty mechanism that would help make the tropical timber market more transparent (Waffle 1993; Drucker 1993). Market transparency is a desirable goal because U.S. manufacturers that use tropical hardwoods are small in comparison with European and Japanese firms, and greater market information can help them compete internationally (Drucker 1993). Industry support for the ITTA through commercial NGOs may thus have made it easier for the United States to accept the ITTA. Continuing support from industry also has some impact on U.S. positions at meetings of the ITTO and ITTC (Drucker 1993).

On the other hand, interests engaged in the practice of ocean dumping have been less ready to comply with the prohibitions of the London Convention of 1972. It is important to note that land disposal may cost 10 to 100 times more than ocean disposal (Moore 1992). Because of the cost, Thorne-Miller (1993) observes, some domestic actors may have had a negative effect on compliance. Universities and scientific research interests have often sought to use the ocean disposal option because it is cheaper. In addition, the Army Corps of Engineers faces a seeming conflict of interest in its oversight of the regulations on ocean dumping. Under the MPRSA, the Corps of Engineers issues permits for dumping dredged materials. However, almost 90 percent of all dredged material dumped at sea is generated by the Corps. Finally, illegal dumping activities by private interests still continue.

Under CITES, as we have seen above, illegal smuggling continues. However, as with the London Convention of 1972, such illegal smuggling seems to be more the result of individual plotting by malfeasors than of any overall strategy by traders to influence U.S. government policy.

Conclusions. The most discernible variations in U.S. treaty compliance occur with respect to the differing categories of treaty obligations, rather than to the different treaties themselves. Seemingly, U.S. reporting and funding obligations are less stringently complied with than other, more public obligations such as actual bans and, for the most part, enforcement activities. Correspondingly, such obligations are less often mandated by preexisting legislation or affected by public outcry, environmental group activities, or political maneuvering.

Therefore, although we can identify the strongest influences on U.S. behavior under a specific treaty, we cannot stop there. The influences will affect each and every obligation of the treaty, but they may affect some more than others.

General Influences on U.S. Compliance

Despite the variations in compliance identified above, this study has demonstrated that U.S. compliance with its environmental agreements vis-à-vis most other nations must be characterized as high or good. In addition to the variables addressed with respect to the five agreements, a number of factors are together responsible for the relatively high quality of compliance with environmental agreements in general. It is not, in most instances, possible to isolate any one of those factors as a sole cause; they are interrelated. In some instances, it is difficult to imagine one factor working out of context, in the absence of others. It is hard to see, for example, how NGOs could operate with equal effectiveness in a less open political system, or how the media could be a causative factor without the activity of NGOs, or how the American scientific community would operate if the United States had a planned economy. "Counterfactual" analysis has severe limitations, and however unsatisfactory it may be, the conclusion is inescapable that American political, economic, religious, and cultural forces acting together and upon one another have brought about high levels of compliance, as a brief overview of these factors reveals.

Historical Forces. From the earliest days, religion has played a central role in American life, and it has influenced American environmental policies in two different ways. One approach links the historical destruction of nature to Judeo-Christian anthropomorphism. Christianity teaches that it was God's will that man exploit nature for his proper ends. According to Genesis 1:26, man has a God-given dominion "over all the earth and over every creeping thing that creepeth upon the earth" (see McCormick 1989). The other approach focuses on Earth as God's gift to man to utilize and enjoy, not to consume, damage, alter in any way, or waste (see Marsh 1864). Man is not a conqueror of the land-community but a member of it. "We abuse land because we regard it as a commodity belonging to us," Aldo Leopold wrote. "When we see land as a commodity to which we belong, we may begin to use it with love and respect" (Leopold 1949: viii).

Adlai Stevenson used the metaphor of Earth as a spaceship on which humanity travels, dependent on its vulnerable supplies of air and soil. Production should be minimized rather than maximized. The measure of success of the economy is not production and consumption, "but the nature, quality and complexity of the total capital stock" (Boulding 1966). For most of American history, particularly that period focused upon "taming the frontier," environmental policy (if it can be called that) was dominated by the first approach.

Following World War II, state governments became increasingly concerned about air and water pollution and the disposal of toxic waste. It was logical that states would attempt to take the lead in these areas because the problems traditionally had been addressed most directly by state law in fields such as torts and property. The long-standing common-law doctrine of nuisance, for example, provided a base for the statutory codification by states of limits on polluters.

State legislatures, however, turned out not to be credible protectors of the environment. Malapportionment into the 1960s made them too easily controlled by financial interests that had less concern to protect the environment. Their governments too often were filled not with trained civil servants but with patronage workers who owed their jobs to political bosses. Also, the nature of the pollution problems made effective action by even the more progressive states difficult. Pollution frequently occurred in interstate waterways, which meant inadequate state jurisdiction. Scientific analyses of risk were often beyond the resources of state administrative agencies and legislative committees. And the mobility of polluters made it clear that a national approach was required.

As attention turned to the federal government, the two conflicting attitudinal patterns emerged in sharp relief. The first, discussed by Westbrook (1994), focused on entrepreneurial freedom and on America as the land of opportunity. The second viewed America's natural heritage as a valuable resource in and of itself. This latter view incorporated notions of paternalism as well as religion. So, for example, federal lands management reflected an entrepreneurial view, whereas, Sax (1981) states, species protection, the establishment of the national parks, and establishment of wilderness areas protect the environment for reasons of stewardship. However, the push in 1871 to found the first national park at Yellowstone was led by railroad interests, which sought to deliver tourists to the park's natural attractions (Runte 1974).

Concern for the environment in the United States reached its zenith in the late 1960s and early 1970s. Well-publicized environmental catastrophes caught the attention of voters and politicians. Earth Day, April 22, 1970, allowed Americans to focus their fears and passions in a way that attracted congressional attention. Although Congress enacted major statutory regimes to protect against environmental degradation, one of the ironies in American environmental law is that the American public had grown to distrust administrative agencies. As a result, activists sought to police these new policemen. The Environmental Defense Fund was founded in 1967, the Natural Resources Defense Council in 1970, and the Sierra Club Legal Defense Fund in 1971.

Public concern about the environment has remained strong in the United States. In fact, environmental organizations have turned into large enterprises and environmental issues are frequently used to market goods and services. Recent economic scholarship may offer an explanation for contemporary American concern over the environment. Coursey (1992) and Grossman and Krueger (1992) argue that as nations grow wealthier, they become more attentive to environmental quality. Some research indicates that this phenomenon is confined to open, competitive market economies (Bernstam 1991).

The reasons are thus multiple and various, but the conclusion is inescapable: environmental consciousness is central to contemporary American culture.

American environmentalists, like other activists, often are faulted for harping on what's wrong. Yet change is seldom a product of praise, and whatever "greening of America" has come about in recent decades was not caused by applauding the system or even always by working within it. The debate within the United States on environmental protection has been rancorous and at times bitter. Recall Love Canal ... baby harp seals ... the spiking of ancient redwoods ... "Mr. Ozone" ... owls vs. people ... grazing fees. No one who has participated in the debates on recent American environmental controversies will remember only reasonable people, reasoning calmly together. The factiousness of the debate, and the unrelenting and uncompromising approach of some of the environmental reformers, are not unrelated to their effectiveness. In fact, the very relentlessness of the activist critics that brought about so much of their success does sometimes obscure the real change that has occurred—and those who have traveled abroad cannot doubt that change.

Imagine a country where puddles of oil or raw sewage mar the countryside; where dirty air makes running hazardous; where restaurants are filled with smoke;

where tap water is undrinkable; where public sanitation facilities threaten health. *Anyone who has traveled widely has encountered such filth; no one who has traveled widely will picture the United States.*

True, concerns about health and the environment are not universal in the United States. Many Americans are indifferent to lead poisoning. Some believe global warming is a myth. And Americans surely lag behind other societies in a number of areas of environmental concern, such as recycling.

But what other culture is concerned about *lawn mower* pollution? In what other culture would $2.7 million be spent—in one year—for the restoration of Stephen's kangaroo rat?

Whatever problems American society faces, and it faces many, it cannot be said that Americans don't care about the environment. Few, if any, people in the world do more to keep their lakes clean, their air pure, and their wildlife alive.

Why? Causes of cultural patterns of belief and behavior are virtually impossible to isolate. To ask why Americans love animals is like asking why Americans love baseball. Academics surely can come up with a dozen rationales, ranging from frontier nostalgia to the veneration of the Creator's work to compassion for things helpless. Americans like baseball because they like hot dogs, hot dogs because they like beer, and beer because they like baseball. Causative explanations for other social and cultural attitudes are about equally satisfactory.

Nonetheless, it is possible to speculate at a general level as to why American culture is environmentally conscious.

The United States' market economy has had attitudinal implications. For one, it limits the phenomenon known as the "tragedy of the commons." The prevalence of private ownership gives specific, identifiable individuals a stake in keeping property clean and healthy. To be sure, private enterprise has at times been a mixed blessing; the profit motive does not always take purely aesthetic considerations into account, and greed has had more than its share of victims in nature. Still, compared with cultures in which common ownership has been widespread (namely, eastern Europe), the market culture of the United States has engendered a measure of individual responsibility not present in nonmarket cultures.

This sense of individual responsibility has not, in the United States, been limited to one's "own backyard," though it is perhaps most notable there. In western Europe, environmental consciousness is not optional; it is compelled by population density. In the United States, on the other hand, a New Yorker need not necessarily be concerned about preserving old-growth redwoods or restoring wolves in Yellowstone Park. Yet support for such environmental initiatives comes disproportionately from persons outside the target region. Indeed, one of the ironies of American environmentalism is that it often has drawn the least support from those who live where its immediate effects are felt. In part this is because those effects are sometimes detrimental economically. Preservation of old-growth forests, for example, means a loss of logging jobs; restoration of wolves means livestock losses through increased depredation. But the cross-continental concern about the other person's backyard flows from a sense of shared concern that transcends narrow pressures of regionalism.

Common historical challenges, coupled with a derivative mythology, fortify the national sense of stewardship. Frederick Jackson Turner wrote of the centrality of the taming of the frontier to the American character. James Fenimore Cooper's "noble savage" reappears throughout the literature of the West, from *The Lone Ranger* through *Dances with Wolves.* Ralph Waldo Emerson sang the early praises of the restorative powers of a pristine environment in "Nature," and his friend Henry David Thoreau lived out these themes, immortally recorded in *Walden.* Modern nature writers from John Muir to David Brower continue to capture the imagination of millions.

The late 1980s and 1990s have seen not only a rekindling of interest in America's natural heritage but a reinterpretation as well. As late as the 1950s, for example, Indians were "bad guys," bent on slaughtering innocent white settlers. Nature was still depicted as something to be conquered. American children dressed up as "Davey Crockett, Injun killer"—who "kilt him a b'ar when he was only three." (The word "development," as applied to the replacement of natural surroundings with man-made ones, is a product of this era.)

The 1960s, however, saw a reawakening of interest in native American culture and religion, the imagery of which has provided powerful support for present-day environmental stewards. Indian religion reveres nature rather than seeking to subdue it, emphasizing the oneness of all life and the need for harmony between man and other life forms. The Dakota regarded as relatives the "four-legged, the wingeds, the star people of the heavens and all things," because they all "drank the same water and breathed the same air" (Suzuki and Knudtson 1992). Human responsibility for nature flowed from a deep-seated sense of shared origins. So it is not surprising that a generation of Americans raised on television westerns would be receptive to a philosophy advocating respect for nature.

More and more, American attention has focused on preservation of nature. Inexpensive transportation has meant that nature is no longer accessible only to society's elite. The market for second homes today is as hot as it has been in years, reflecting again Americans' desires to get closer to nature. Increasingly, American heroes are those who respect the environment rather than those who destroy it.

Modern Influences. Viewing the five agreements through this cultural lens, several more specific factors emerge as responsible for the generally high quality of U.S. enforcement and compliance of environmental treaties in general.

Timing. The United States became a party to many environmental agreements in the early 1970s. The effects of affluence, the age of atomic testing, Rachel Carson's book *Silent Spring* (1962), well-publicized environmental disasters, advances in scientific knowledge, discontent among young people, and the influence of other social movements transformed the environmental movement in the United States during that time (McCormick 1989). Persons with increased leisure time and greater prosperity shifted attention from accumulation and enjoyment of material security to the environmental consequences of the affluent society (McCormick 1989). *Silent Spring* was on the *New York Times*'s best-seller list for thirty-one weeks. Well-publicized environmental disasters sensitized many to the potential costs of careless economic development. The "back to the earth" 1960s saw a return to wilderness and nature as the only way of retaining earthly values in a materialist world. At the end of the 1960s, as the civil rights and antiwar movements lost momentum, activists turned to the environment (McCormick 1989).

The largest environmental demonstration in history took place in 1970, when 300,000 Americans took part in Earth Day. *Time* magazine called the environment the issue of 1970, and *Life* predicted that the environmental movement would dominate the 1970s. By 1969–1970, membership of the Sierra Club had tripled since 1966. Many new environmental organizations were born, including the Environmental Defense Fund, Friends of the Earth, and Greenpeace. Beyond the old alliances, the new movement included the consumer movement, including corporate reformers, the movement for scientific responsibility, a reinvigorated public health movement; birth control and population stabilization groups; pacifists and those stressing participatory democracy and consensual decisions; young people emphasizing direct action; and a diffuse movement searching for a new focus for politics (Pursell 1973).

Metropolitan decay and expansion of urban areas might also have increased interest in environmental protection (see Schnaiberg 1977). In the past, only a wealthy minority owned cars and had access to national parks, suburban housing, and resort areas. Downs (1972) explores how the environment of this elite began deteriorating once more people gained access to these places. It was during these salad days of mass environmentalism that the United States entered into the World Heritage Convention, the London Convention of 1972, and CITES.

Since that time, the perspective in the United States has changed considerably. Now more pragmatic in its outlook, the United States applies a high level of scrutiny to any obligations it takes on through both domestic legislation and international agreement. Although the environmental attitude in the country has been slowly changing over time, no corresponding shifts in compliance levels with the agreements we have studied have been noted. It is apparent, therefore, that timing is *not* everything.

Existence of legal enforcement mechanisms. Legal enforcement mechanisms generally exist to enforce the agreements the United States signs. This ensures a relatively high level of compliance. For example, in three of the five cases studied here, United States agencies and other groups have acted to enforce the agreements in the face of violation by an agency or an individual. Where a cause of action is provided in the domestic legislation implementing the agreement, the agreement has the force of law in a U.S. court. Where no cause of action is provided, as in the cases of the ITTA and the World Heritage Convention, the incentive for the United States or its agencies to self-police compliance does not exist. For this reason, individuals have been prosecuted for violations of CITES, and agencies have been compelled to comply with the timing requirements of the Montreal Protocol, whereas Everglades National Park—though listed as a World Heritage Natural Site—has lost 50 percent of its wetlands to the state of Florida's development efforts.

Administrative agencies and courts are regarded by the public as legitimate. The American judicial and administrative system, where most of the day-to-day enforcement work is done, is relatively sophisticated. For all their shortcomings—crowded dockets, expense, delay—the courts and administrative agencies are nonetheless capable of dealing with environmental and other scientific matters that would strain a system backed with fewer resources. They generally produce results that the population perceives as legitimate; public actors comply with judicial and administrative mandates not because the views of the courts and agencies are always seen as correct, but because they are generally seen as exercising rule-making and adjudicatory authority by right.

Why this is so raises questions far beyond the scope of this chapter, relating to matters such as the ready presence of the English legal tradition in the newly independent colonies; a need for public adherence to dispute-settlement mechanisms as a matter of physical survival on an expanding frontier; and the desire of business for predictability in regulatory matters. Perhaps most pertinent for our purposes is that judges, administrators, and law enforcement officials normally are seen as competent. Judges in most cases are highly qualified and administrators are well-trained, carefully selected, and dedicated to their work. Though government budgets are limited and nearly all agencies would like greater resources, salary and support levels are generally adequate. The consequence is that administrative agencies and courts are almost always obeyed—even though the results may be disliked.

The U.S. political system is open. Information on governmental operations at virtually every level is readily available to the general public. "Sunshine" laws have caused meetings to be open. Under statutes such as the federal Freedom of Information Act and analogous state legislation, internal governmental documents are frequently available to the public (though often after considerable delay). Openness vastly enhances the effectiveness of NGOs, as discussed below.

Some Americans have been increasingly concerned about a lack of governmental "accountability," that is, about the difficulty in reaching decision makers, and causing decision makers to take responsibility for their actions. These concerns are perhaps most frequently directed at legislatures, where the skill of buck-passing seems to have reached its zenith. With greater and greater frequency, "hot potato" issues are passed by legislatures to commissions and administrative agencies—where job tenure is not related to the outcome—for resolution.

In some parts of the bureaucracy, too—including elements responsible for enforcement of these treaties—accountability is indeed a problem. Voice mail systems give telephone callers the runaround, letters go unanswered, and officials are "too busy" to schedule appointments with private citizens. Strong sentiments have thus arisen in many parts of the country to downsize government by decreasing legislative and administrative budgets, staffs, and responsibilities. The aim, one supposes, is to make government more accessible by making it smaller.

Nonetheless, governmental operation, although not always simple, is normally possible to follow. The "pressure points" are identifiable, and concerned observers can register their views. Indeed, the right to petition government is guaranteed by the First Amendment to the Constitution. More often than not, the problem is not that government is unwilling to listen, but that so many actors speak at once on the same issue that the system is overloaded with competing views.

The press has been free to publish information critical of governmental malfeasance in protecting the environment. The role of the press in the United States has been described above. Suffice it to say here that, but for the links provided by a free press, the causative chains leading to the implementation and enforcement of these treaties would have broken at many points.

The United States can afford environmental protection. There is an economic hierarchy of needs, as obvious as it is important, that accounts for the purposes for which federal and state governmental units spend money. Public health and safety, like other purposes related to immediate survival needs, rank more highly than, say, aesthetics in public buildings. Squad cars are purchased before statues. The environment generally is not top-ranked unless the issue raised presents imminent health concerns, as did Love Canal and Three Mile Island. The availability of more public revenues means that lower-level felt needs can be met; species protection, park preservation, and resource conservation are more likely when hospitals, prisons, and fire departments are thought to be adequate, and they are more often thought adequate in rich countries than in poor ones.

This economic point ought not to be overstressed, however: American environmentalism owes its relative success to much more than prosperity. It is noteworthy that the rise of the "environmental movement" in the United States has corresponded closely with the period, beginning in the early 1970s, that has seen a gradual *decline* in real weekly income per capita.

American political leaders have been both environmentally minded and politically responsive. One need not accept the "great man" theory of history to recognize that certain American political leaders, in the "right place at the right time," have had a clear effect. Widely known, for example, is the role played by Theodore Roosevelt in establishing the national park system; less widely known, perhaps, is the role played by Russell Train in the adoption of the World Heritage Convention and a variety of other key environmental laws and agreements, or the role of Senator John Chaffee in the accelerated phaseout of CFCs.

It would be a mistake, however, to overemphasize the role of a single individual in a society as pluralistic as that of the United States. A public official is not a free agent. Decision-making is influenced by social pressures, cultural traditions, and the myriad factors that account for variations in human behavior in general. For just those reasons, it has been extremely difficult to attribute any changes in compliance with the environmental agreements at issue over time merely to changes in presidential administrations in the United States: the factors influencing presidential and congressional political behavior are simply too diverse.

President Richard Nixon's "good" environmental record is a case in point. It is true that a star-burst of environmental legislation and a number of important treaties—including CITES, the London Ocean Dumping Convention—and the World Heritage Convention—originated during his administration. It is also true, however, that the impetus for these laws traces largely to factors exogenous to the Nixon White House. A man who walked wing-tipped through the Pacific surf did not find sudden inspiration in the writings of John Muir, and in this epiphany conceive of the Endangered Species Act. Nixon was a lifelong politician. He was elected in 1968 by only 43 percent of the popular vote. He wanted to be reelected in 1972. The energy of the environmental movement reached a new high during his first term of office. Many Republicans counted themselves within it. The movement was embraced by the middle class, where pivotal votes lay. One of its leaders in the Senate, Edmund Muskie of Maine, was widely expected to be the Democratic nominee for President in 1972. However great the role Nixon or any other individual may personally have played in a given environmental accomplishment, therefore, it would be wrong to think that he was unprompted by external events and other personalities. Nixon was influenced by the press, by lobbyists, by demonstrators, by polls—by all the factors, again, that determine human behavior in general and political behavior in particular. Nixon, like Muskie, did lead. But he also followed.

The federal system has made U.S. compliance cheaper. Preexisting state and local legislation has reduced the cost of compliance for the federal government in several ways. Already-regulated businesses have been less likely to lobby vigorously against federal laws (or treaties) if their behavior already comports with proposed federal standards. A lower level of federal inspection and enforcement resources has been needed because of the backstop provided by state and local authorities. Administrative costs, such as those incurred by courts, also have been lower. So, all told, federalism probably has enhanced compliance.

This might not have been true had the supremacy of federal authority not been established early in American legal history (or at least by the 1860s). Today, however far from Washington one may go, the jurisdiction of the federal government is unquestioned. Federal supremacy has two practical effects. First, states face federal preemption in the event their laws conflict with laws of the federal government; there can be no "war of laws" within the United States. (The United States fought a civil war over the issue, and the federal government won.) Second, the federal government sets the floor level—it, not states or cities, determines the minimum level of protection to be accorded the environment. The upshot is that local obstruction has not proven to be a significant impediment to compliance.

NGOs have held the government's feet to the fire. The role of American NGOs can be overdrawn. All are not alike. Some favor sensationalism. Others prefer a quiet approach. Some are aligned with business, hunting, or development interests. Others are led by zealots and eschew compromise. Some are led by individuals who pull punches in the hope of getting governmental jobs. Some are foundation-backed; others have membership rolls in the millions. Still others are encumbered by their "501(c) (3) status"—because contributions to them are tax-deductible, they are precluded from engaging in lobbying and other political activities.

Still, the more active ones, such as the Sierra Club, the World Wildlife Fund, Greenpeace, the Wilderness Society, the Defenders of Wildlife, and the Audubon Society, have a dramatic effect on environmental policy at both federal and state levels. Some of these work closely with federal officials and engage in quasi-governmental monitoring and investigatory activities. They plant editorials with newspapers, give awards to friendly lawmakers, telephone committee staff with arguments for or against legislation, and actually draft amendments. TRAFFIC-USA is respected both within and without government as a credible source of information on species endangerment. Others, such as Greenpeace, influence environmental policy less directly by staging highly publicized events aimed at taking advantage of press coverage and thereby marshaling public opinion. Both sets of groups have had undoubted impact on a wide variety of environmental issues ranging from the ivory trade to CFC use, an impact that would have been inconceivable absent a free press and open political institutions.

Conclusion

When law and behavior are isolated and extracted from a larger social context, it is easy to presuppose a causal relationship, to assume that the behavior in question was caused by the law in question, to think that law and behavior are independent variables. The reality, of course, is otherwise. In any analytic framework, variables inevitably are taken out of context—a context that contains many, many more variables than merely law and the specified conduct. In a democracy, at least a democracy like the United States, law not only reflects culture, law is a *part* of culture. No less than songs, law shapes; no less than poems, law is shaped. Its roots are as deep and sinuous as an old oak's. When one asks why a given law or treaty exists, or what difference or contribution it makes, one might as well ask the same of a song or a poem.

Thus we cannot know with certainty what the United States might have done or not done in the absence of these treaties. We do know from this study, though, that the United States does well in both implementing and complying with international environmental agreements when those agreements require conduct that is generated by concurrent or preexisting political processes. We know, too, that the United States compares favorably with other countries in such efforts. And, as this study also reveals, we know that the United States can do more to improve implementation and compliance.

Notes

The authors wish to thank Danae Aitchison, Erin Burke, Allison Hayward, John Leman, and Steven Plesser for their research assistance, and the University of California, Davis, School of Law and Morgan, Lewis & Bockius LLP for their support.

7

The European Union and Compliance: A Story in the Making

Alberta M. Sbragia with Philipp M. Hildebrand

How effectively does the European Union (EU) comply with international environmental accords? The simple answer is that its compliance depends on the agreement in question. Table 7.1 shows the relationship of the EU and its member states to the five treaties included in this study.

The reason for such difference in compliance lies in the Union's sui generis institutional structure and the very nature of its institutional capacity. The Union's political "will" can be easily expressed in the form of legislation, but in the areas of administrative implementation and enforcement, the Union has limited institutional capacity. Its institutional structure, therefore, allows it to enforce very effectively those environmental accords that are dependent on legislation and limited administrative capacity. That same structure hinders the Union's effective compliance with agreements that require strong administrative implementation and enforcement to execute.

The Montreal Protocol and its amendments were more effectively implemented than was the Convention for International Trade in Endangered Species (CITES, known by Europeans as the Washington Convention). The nature of the demands imposed by the Montreal Protocol were far more compatible with the Union's institutional capacity than were the demands associated with the effective enforcement of CITES. Compliance is a complicated issue for Brussels because it highlights the Union's fluid status as an international actor as well as its limited capacity in the areas of administration and enforcement. Compliance involves two related dimensions, both of which are problematic for this entity, which is neither a full-blown polity nor a traditional international organization. The first dimension has to do with the Union's status as an international actor. Do other sovereign states perceive the Union as enough of an international actor to negotiate with it? Do they deem it capable of enforcing an agreement, and thereby view it as a worthy counterpart at the negotiating table?

The second dimension has to do with the relationship between the Union and its member states. Can the Union ensure that its constituent members will in fact comply with the provisions of an international environmental accord? Does it have the authority to adopt legislation implementing an international agreement? Does it have the power—and the capacity—to enforce such legislation?

Table 7.1
Treaty membership of the European Union

Treaty	Year treaty came into effect	Signed by European Union	Ratified by European Union
WHC	1977	Member states are signatories—not EU	Member states are signatories—not EU
CITES	1975	Member states are signatories—not EU	Member states are signatories—not EU
ITTA	1985	1984	1985
LC	1975	Member states are signatories—not EU	Member states are signatories—not EU
MP	1989	1987	1989

The Union's competencies, powers, and reach have evolved over time and across policy sectors. Given the diverse subject matter and the different historical periods represented by the agreements included in this study, it is not surprising that the Union has played a different role across accords. In fact, the varying role it has played in such agreements reflects the "asymmetrical integration" that has characterized the European Community (EC) and its successor, the European Union (Sbragia 1993a).[1] The Union is not involved with the London Convention because the member states, rather than the Union, have competence in the policy areas covered by the Convention; the same holds for the World Heritage Convention. By contrast, it is a party to the Vienna Convention and the Montreal Protocol, as well as to the International Tropical Timber Agreement. Finally, although it is not a party to CITES, it has implemented that convention at the EU level.

In order to explore the complex relationship that has existed between the negotiation, legislative implementation, and administrative enforcement of these international accords within the Union, this chapter deals with each of the relevant environmental agreements individually. The Montreal Protocol and CITES are given the most attention because the Union has been much more concerned with those two treaties than with the International Tropical Timber Agreement. The Union has been very successful at the legislative implementation of both treaties and has improved its record over time, but enforcement of that implementing legislation has differed in the two cases. The Union's institutional structure and administrative capacity are much better suited to the enforcement of the Montreal Protocol than of CITES.

The European Union as an International Environmental Actor

A description of the Union's involvement in international environmental initiatives needs to be cast in the light of two key questions: First, does the Union have the competence to become a party to international environmental agreements? Second, if the Union is able to sign a treaty, either by itself or jointly with the member states, what are the tools and procedures it possesses to obtain compliance? Is the Union's institutional capacity such that it can comply with the agreements it ratifies?

With regard to the first question, in the Treaty of Rome the founders of the European Community did not explicitly envision Community competence in the area of environmental protection. Nevertheless, it soon became apparent that the economic objectives of the Common Market could not be kept entirely separate from environmental issues. The elimination of trade obstacles led to the harmonization of certain technical standards linked to environmental quality. However, only after the 1972 Stockholm Conference on the Human Environment did the Community effectively begin to take some responsibility for environmental protection.

Environmental policy was formulated without being grounded in a formal legal base. Typically, a policy was decided in Brussels only after the member states agreed to include that policy sector in a treaty. The member states, however, agreed to legislate at the Community level in the field of environmental protection but did not amend the Treaty of Rome. The lack of an explicit legal base rendered the Commission somewhat wary of overreaching, since its legal standing in this field was weaker than in those areas explicitly singled out in the Treaty of Rome. The Single European Act in 1987 (which amended the Treaty of Rome) finally gave the Community an explicit mandate to pursue environmental policies (title VII, article 130r-t) (Hildebrand 1992). This mandate was reinforced by the adoption of the Treaty of European Union (known as the Maastricht Treaty) (Hull 1994; Kramer 1995; Sbragia 1996). As the Community's powers became formalized, Brussels adopted an extensive legislative framework for the protection of the environment (Haigh 1992a; Sbragia 1993b).

Parallel to the gradual development of an internal environmental policy, the Community has become active in the international environmental arena. Such activity has raised the controversial issue of the Community's external competencies. The Treaty of Rome did not provide a legal basis for extending external competencies to the area of environmental policy. However, over time, the European Court of Justice, by virtue of its case law, has established that the existence of external powers is in principle dependent on the question of whether or not such powers are necessary to attain Community objectives. Such rulings have made it possible for the Community to undertake important international environmental initiatives since the early 1970s. In general, it has become increasingly accepted that the Community should be an international actor in international environmental negotiations. As Robert Hull points out, "The Maastricht Treaty makes it clear that a basic objective of Community environment policy includes action at global and regional levels" (Hull 1994:154).

Nevertheless, the gradually acquired external competence of the Community for the protection of the environment is not an exclusive one. Member states and the Community are required to complement each other in their environmental efforts. As a result, virtually all international environmental agreements belong to the category of "mixed agreements." That is, both the Community and its member states are parties to the relevant treaty.[2]

The Community is represented in many international negotiations by the European Commission (with the presidency of the Council of Ministers also playing a role in some negotiations). The Commission is the Community's executive arm, and its civil servants are important in the formulation of policy. In negotiating international environmental agreements, the Commission is given instructions by a special committee responsible to the Council of Ministers, a body composed of member-state ministers that must approve all Community legislation (Kramer 1995; Hayes-Renshaw and Wallace 1997). Depending on the position of the member state holding the presidency of the Council of Ministers, the Commission may be allowed considerable negotiating latitude or may be given almost none.

Compliance

Implementation Through Legislation

Once an international agreement is ratified by the Council of Ministers on behalf of the Union, the Council typically implements it by adopting a "regulation." A regulation is a form of legally binding obligation that does not require that it be "transposed" by the member states' parliaments.[3] Whereas most legislation approved in Brussels does require national transposition, regulations do not, and are mainly used for the implementation of international accords.

If an accord can disrupt the Common Market, the Union becomes involved in the legislative implementation of that agreement even if it is not itself a party. The Union adopts internal measures to reconcile Common Market requirements with international environmental objectives.

The first step in implementing an international environmental accord at the Union level involves Directorate General XI (DGXI), the Commission's main administrative unit concerned with environmental issues. The establishment of DGXI in 1981 symbolized the Community's new interest in environmental protection, and it is involved in both the international negotiation of treaties and their implementation. DGXI officials draft the implementing legislation that, after approval by the College of Commissioners, is sent on to the Council of Ministers as a formal legislative proposal for a regulation. The Council of Ministers then adopts the regulation implementing the international accord. The status of treaties that are unaccompanied by implementing legislation within the Community is problematic. Since the mid-1980, DGXI has acted to ensure the simultaneous ratification of an international environmental treaty and the approval of implementing legislation.

Once the legislation is approved by the Council of Ministers in the form of a regulation, the Commission's role diminishes. The actual execution of the regulation "on the ground"—as well as the determination and imposition of criminal penalties for the lack of execution—is the province of the member states.

Postlegislative Implementation

Once the Council of Ministers has adopted a regulation, each member state's administrative apparatus becomes chiefly responsible for executing that regulation "on the ground."

In Nigel Haigh's words, "Although EC policy has the attributes of foreign policy during its formulation, it then becomes integrated with home policy in its implementation" (Haigh 1992b:233). The Commission has not been given the authority to monitor or inspect how member states carry out their duties on the ground.

In the case of noncompliance with the obligations of the regulation by one or several member states, the Commission's only recourse is to initiate infringement proceedings under Article 169 of the Treaty of Rome. Ultimately, such a proceeding can lead to a ruling by the European Court of Justice against the concerned member state or states. However, the Commission cannot, for example, withhold EU funds allocated to the offending member state.

The Commission has fewer "tools" at its disposal than does a traditional state, including federal states (see, e.g., Hood 1983). In brief, it is not a policeman. It can use the European Court of Justice and the infringement procedure to judicially penalize member states that do not carry out the obligations of a regulation—but only if the Commission is able to obtain the kind of evidence necessary for such a judicial ruling. That is often very difficult.

DGXI does not have the tools of law enforcement enjoyed by, for example, the Environmental Protection Agency in the United States (Kramer 1995). The Community relies on legal instruments more exclusively than does a state, which is able to use administrative and law enforcement powers to complement its legal powers. The Commission cannot use nonjudicial tools to ensure compliance—for example, it cannot conduct independent investigations (Commission 1993b).

As we shall see in the case of CITES, the lack of field staff, inspectors, and on-the-ground investigative powers prevents the Commission from actually monitoring compliance. It cannot hire undercover agents, for example (the Fish and Wildlife Service in the United States, by contrast, can and does). It cannot engage in "sting operations," nor can it send lawyers to the member states to collect information independently. The member states control law enforcement and the collection of information—two key aspects of ensuring compliance—as well as the imposition of penalties.

The EU is disproportionately dependent on the legal system to ensure compliance. While that system works well when legislative noncompliance is involved, it is not easily brought into play when noncompliance involves inadequate enforcement by the member state. The pivotal role of the judiciary, however, does mean that an analysis of compliance in the European Union necessarily is more "legalistic" than the analysis of strictly national compliance would be.

It is important to note that the Commission was designed to act as the *animateur* of the Community rather than as its policeman. It is an initiator, a creator, a formulator—it is not an implementer. Given the initial lack of environmental legislation in most of the member states, the Commission has focused on putting into place a fairly comprehensive legislative framework designed to protect the environment. Since implementation is a national rather than a Union responsibility, and since the staff is very small (in comparison with, for example, the Environmental Protection Agency), the implementation and enforcement of environmental policy have understandably not been given as much attention by the Commission as has the formulation of legislation. By 1996, however, it was widely recognized that the lack of compliance with Community environmental legislation was a serious problem for the cause of environmental protection. It was unclear, however, whether the political will would exist to improve compliance significantly (see, e.g., Cutter 1993b, 1994k).

The great disparity among the Union's member states makes compliance with environmental legislation—whether inspired by an international agreement or by the Union—very problematic. In a report assessing the state of implementation in the area of environmental protection, the Commission was blunt. It stated that there exists a "wide disparity in enforcement agencies or mechanisms among the member states, with some putting considerable resources into well-supported inspectorates . . . which monitor the practical application of Community environmental law and others making lesser provision or none at all" (Commission 1996a:8–9). Given the highly decentralized nature of administration in the Union, such a disparity makes uneven compliance with environmental legislation inevitable.

In fact, the role that the Commission should play even in reporting how well member states are complying with EU laws is contested. (Such reporting would have to be based on accounts formulated by the member states themselves.) Environmentalists suspect that political pressures keep the Commission from reporting information that might lead to demands from environmental groups that the Commission bring infringement proceedings against noncomplying member states (Cutter 1994k; Environmental Data Services 1994). Environmental groups are dependent on the Commission for judicial action, for they themselves do not have the legal standing necessary to sue member states. On the other hand, the understaffing of DGXI is undoubtedly an important contributor to the lack of reporting.

Judicial Implementation

Infringement proceedings are the chief instrument used by the Commission to ensure compliance. The ultimate result of such a proceeding is a case tried before the European Court of Justice. Numerous infringement proceedings have been brought against member states, but no infringement proceedings have been brought against a member state for noncompliance with an international treaty as such.[4]

Given the centrality of the judicial process to enforcement within the Community, the legal affairs unit within DGXI has been viewed as critical to the enforcement of EU environmental legislation. That unit proposes infringement proceedings for noncompliance with environmental laws (its proposal then has to be approved by the Legal Affairs Service of the Commission).

Even though the division has employed fewer than a dozen officials, its special status within DGXI, as well as the uncompromising and aggressive enforcement attitude of its director, Ludwig Kramer, gave enforcement of environmental legislation a relatively high profile. Such enforcement also became highly controversial. Critics argued that Kramer insisted on enforcing aspects of legislation that were not at the core of environmental protection, thereby not using the Commission's limited resources most effectively. The critics seemed to include an advocate general of the European Court of Justice. Whatever the merits of the debate, Kramer, after directing enforcement efforts for a decade, was transferred in January 1995 and became head of DGXI's waste division. Environmentalists charged that the legal affairs unit had been downgraded in status, a charge the Commission denied (Cutter 1995a, 1995b). The controversy over Kramer's removal may, however, have obscured efforts undertaken by the Commission to help member states improve their own enforcement efforts (Europe Information Service 1992; see also Macrory 1992; Council 1990).

Given the importance of legal instruments within the EC, it is perhaps not surprising that national courts have become very important in the judicial implementation of European Community law generally. In the case of environmental laws, however, the lack of access to the national courts by environmental interest groups has meant that such groups are less powerful than they are in the United States. (The Netherlands is an exception; see Klik 1995:16.) In general, civil law systems (characteristic of the EU's mainland members) do not allow nongovernmental groups to bring class-action suits. The absence of the judicial remedies extensively used by American groups differentiates European environmental politics from its American counterpart (see, e.g., Sbragia 1992; Rose-Ackerman 1995).

The Commission and Implementation

Given the lack of priority accorded environmental protection in a number of EU member states (Sbragia 1996; Baker et al. 1994; Prendiville 1994), it is predictable that lack of compliance with environmental legislation will continue to be a problem within the Union. It is also predictable that environmentalists will pressure the Commission to increase its commitment to implementation while the Commission carefully weighs the political risks of being viewed by the member states as encroaching on their "turf." Extending the reach of the Commission into the implementation of environmental legislation—legislation that is of relatively low priority (and very controversial) in some of the member states—is fraught with political peril for the Commission.

In brief, the Union's institutional capacity to ensure compliance is significant but limited. Capacity is very robust at the legislative stage but declines steeply in the postlegislative stages of implementation. The Commission, for its part, can be a significant actor to the extent that actions short of law enforcement activities and independent investigations can yield results. It can highlight or minimize the problem of compliance, stress the importance of an international environmental treaty or downplay it, and build an administrative structure and an institutional framework that are more or less helpful in ensuring compliance with a particular treaty.

In turning to the three cases, the Union's experience in obtaining compliance with the Montreal Protocol indicates the Commission's ability to work with a small number of large, reputable firms in an area that is relatively clear-cut. The case of CITES indicates the limits of the Commission's reach, although it is possible that the Commission could be more proactive. The negotiating history of both treaties also affects implementation, with the Community promoting the Montreal Protocol far more aggressively than it has promoted CITES. The case of the ITTA is one in which the Community focused on its trade rather than its environmental dimension, although it changed its position once the treaty was due to be renegotiated.

The Montreal Protocol

The Vienna Convention and the Montreal Protocol are significant in that they represent the "success story" in any evaluation of Union compliance with international accords. The negotiations over ozone protection marked the entrance of the Community as a significant actor in the global environmental arena. In fact, the Commission's concerns about its international status initially seemed to outweigh its concern with protecting the ozone. By the time of the Copenhagen Amendment, however, the Community had taken a leadership role in the negotiation of ozone protection and its international status was firmly institutionalized. Its record on implementation at least partially reflects the Commission's desire to maintain a leadership role in this area.

The evolution of the Community's position in the area of ozone protection has been marked. During the negotiations that led to the Vienna Convention for the Protection of the Ozone Layer (1985), the Community's position was one of skepticism concerning the importance of the problem of ozone depletion. By 1993, the Community had emerged as a strong supporter of stringent regulations on ozone-depleting chemicals. Whatever its substantive position, however, the EC's role as the supplier of nearly 40 percent of the world market meant that "the EC had to play a major role in all attempts to build a regime for the protection of the ozone layer" (Jachtenfuchs 1990:262).

Negotiating About More Than CFCs

The Vienna Convention, from the point of view of the Commission, was not only about negotiating the production and consumption of CFCs. It was also about the Commission's competence in environmental matters within the Community, its role as the Community's negotiator with third parties (Commission 1985), and the Community's international profile. In that sense, the Vienna Convention was about "institution-building" in Brussels as much as it was about CFCs. It represented an opportunity for the Commission to strengthen its own reach and influence vis-à-vis the member states in the field of environmental protection and for the Community as such to gain international standing.

The Commission saw the negotiations over the Vienna Convention as an opportunity because the Community had not yet been given explicit legal competence to deal with environmental matters (that is, there was no such provision in the Treaty of Rome). Such competence was granted only in the Single European Act, which came into force in 1987. The Commission wanted an external negotiating role in order to be able to increase its institutional power within the Community as well as to strengthen the Community's international profile (Jachtenfuchs 1990).

In October 1984, the Council of Ministers approved the Commission's negotiating strategy, which called for the Community to be able to participate in the Convention without any conditions. The Commission did not want to "accept any clause which would make participation by the Community subject to prior participation by one Member State ... or by a majority of Member States" (Commission 1985:1).

The Commission was forced to compromise in the final negotiations, but it did manage to become a signatory. The United States initially opposed the participation of the Community but was finally forced to accept it. The Vienna Convention thus was one of the first "mixed" international agreements that now characterize the Community's participation in international environmental accords (J. Lang 1986).[5]

During the Vienna Convention negotiations, the Community refused to negotiate on the question of CFC use and consumption. The Convention did not limit CFCs primarily because of an irresolvable conflict between the Community, whose national governments protected the interest of their chemical companies, and the so-called Toronto Group (Finland, Norway, Sweden, the United States, and Canada).

The Vienna negotiations could agree only on the need for research and monitoring. "It was clear to the Toronto Group that forcing an agreement to cut CFCs would mean losing the participation of the EEC" (Granda 1990:30). The EC came to the Vienna Convention having been less aggressive in limiting CFCs in aerosols than had the United States and the Scandinavians (Rowlands 1995). In fact, the negotiator for the Community was a relatively junior official who, by his own admission, was inexperienced. Nonetheless, the United States was unable to prevail. One participant argues that "there was hypocrisy on both sides." As Nigel Haigh points out, "Each group proposed that the first protocol to cover CFC's should embody the policies already adopted in their own group of countries" (Haigh 1992b:245).

Industry clearly played a central role during the negotiations for the Vienna Convention.[6] The member states put forward positions determined by their industries. The French, the British, and the Germans represented the interests of their producers, with the British being particularly determined to avoid specific reductions. Further, during the negotiations for both the Vienna Convention and the Montreal Protocol, the firms opposing regulation skillfully exploited the existing scientific uncertainty. As one Community participant analyzed the Vienna negotiations,

> We were operating under conditions of scientific uncertainty. The principle of prevention was not very strong. Industries that were going to be hurt had fertile ground of resistance. We were trying to convince industry that we needed the precautionary principle in legislation. There has been a sea-change in the way the danger is perceived. But back then industry was using uncertainty against us.

Scientific uncertainty played an important role in the Montreal Protocol negotiations as well because the Protocol was negotiated before the importance of the Antarctic ozone hole was recognized. As one participant put it, "The Chair of the Scientific Assessment Panel did not have a smoking gun." (For a scientist's view of that uncertainty, see Albritton 1989.)

Once negotiations began for the Montreal Protocol, the Community's position differed according to which country held the EC presidency. In the second half of 1986, the British presidency did not facilitate movement toward an agreement. By contrast, under the Belgian presidency (in the first half of 1987), the Commission was given more latitude in the negotiations. Furthermore, public opinion in Germany was in favor of reductions, so that Germany "abandoned its reluctant attitude and at least formally adopted a more environmentalist position" (Jachtenfuchs 1990: 265–266).

Echoing the experience of the Vienna Convention negotiations, the Community insisted on becoming a party as a single unit. The Commission viewed the Montreal Protocol as an opportunity to increase the Community's international standing, and thereby its own institutional prestige and influence. That posture caused diplomatic difficulties of such magnitude that only a compromise ending a "nerve-racking midnight standoff over this issue" allowed the Montreal Protocol to be agreed to (Benedick 1991:96). After the Protocol was negotiated, politics in the reluctant member states gradually changed as industry became willing to consider reductions more stringent than those agreed to in the Montreal Protocol. Once ICI, the main British producer of CFCs, and Atochem, the principal French producer, shifted their positions, Britain and France did as well. In September 1988, Prime Minister Margaret Thatcher supported the "green" position, and Britain no longer obstructed reductions in the way it had previously (see Levy 1991; Jachtenfuchs 1990; Benedick 1991). Once industry's opposition eroded, governments followed suit. From then on, the influence of new scientific information on shaping policy options increased significantly.

By mid-1993, the Community's position on protecting the ozone layer bore very little resemblance to its position during the negotiations for the Vienna Convention. Whereas the United States had been the environmental leader during the Vienna Convention and the Montreal Protocol negotiations, the European Community now claimed that role. The Commission proudly stated that "the Community and the Member States played a leading role" in the negotiations leading to the Copenhagen Amendment to the Montreal Protocol (Commission 1993c:9).[7]

In December 1992, the Council of Environment Ministers decided to move faster than provided by the Copenhagen Amendment on most CFCs and to ask the Commission to develop proposals on limiting HCFCs, HBFCs, and methyl bromide (later introduced as COM (93) 202 final). The ministers decided to phase out CFCs completely by January 1, 1995; totally eliminate halons by January 1, 1994; eliminate carbon tetrachloride by January l, 1995; and eliminate trichloroethane by January 1, 1996 (Europe Information Service January 7, 1993). And in July 1994 the European Union (along with Austria, Finland, Norway, and Sweden) called for international agreement to accelerate the phasing out of HCFCs (Bureau of National Affairs 1994c).

Beyond Copenhagen: The Legislative Record Improves

The Community's commitment to phase out CFCs and carbon tetrachloride by the end of 1994 attracted worldwide attention.[8] A high-level official from the U.S. Environmental Protection Agency, referring to the EC's policy, stated, "The challenge is to watch that rapid phase-out, learn from its successes and its mistakes, and learn from the strategies that they would employ" (quoted in "Tensions Between..." 1993:812).

In June 1993, the Commission asked the Council of Ministers (which would be able to use qualified majority voting on this issue once the Treaty of European Union came into force) for controls on HCFCs and methyl bromide more stringent than those agreed to in Copenhagen. It proposed that HCFCs should be eliminated by 2015 and that methyl bromide's production and consumption be cut by 25 percent by 1996 in addition to the freeze for 1995 agreed to in Copenhagen (Council 1993a:5; Commission 1993c:8). The European Parliament asked the Commission to propose a ban on the production of HCFCs from December 31, 2002, rather than allowing manufacturing for export purposes, but its proposal was rejected. The Commission also rejected the Parliament's request for a ban on the manufacture of methyl bromide from December 31, 1999 (Europe Information Service 1994a; see also Commission 1994a; Cutter 1994b).

The Community's willingness to consider measures more stringent than those required by the Copenhagen Amendment was resisted in some quarters. The French chemical industry was especially hostile to regulations more stringent than those adopted at the international level (Cutter 1993a). Elf Atochem was a particularly important firm in the decision-making process because it had invested very heavily in HCFCs, is Europe's largest producer of methyl bromide, and is 51 percent owned by the French state (Europe Information Service 1993d). Furthermore, food and refrigeration industries opposed the Community's adoption of measures stricter than those required by the Copenhagen Amendment (Environmental Data Services 1993).

The Council of Ministers. The complaints of user industries were so loud, and their resistance so fierce, that some within the Commission were worried about an eventual backlash against the Community's desire to move faster than required. Fears of losing competitiveness were very real, and given the state of unemployment within the Union, such fears received a respectful hearing. The Council of Ministers, however, agreed with the Commission that the Community should regulate more stringently than does the Copenhagen Amendment. Although all the member states have ratified the Montreal Protocol, they are now bound to the Community's more stringent regulations.[9]

At a preliminary meeting in December 1993, the Council of Environment Ministers agreed that from January 1, 1995, each producer would face a cap on the volume of HCFCs that it could use or market. The cap would consist of its use of CFCs and HCFCs in 1989 plus 2.6 percent (Cutter 1993d).[10] In 2004, the use of HCFCs would be cut by 35 percent (compared with 1989), and reductions would continue until the end of 2014, when use of HCFCs would be phased out altogether (Cutter 1993d). The use of HCFCs for certain specific purposes also was banned. In

1996, bans would affect household refrigerators (the ban would affect industrial refrigeration in the year 2000) and car air conditioners. France "had managed to delay until 1998 a ban on HCFC use in air conditioning systems aboard trains, a solution that fits in with its program for re-equipping its fleet of TGV high-speed trains" (Cutter 1993e).

On June 8, 1994, the Council of Environment Ministers agreed definitively to approve legislation setting more stringent regulations than those agreed to at Copenhagen: the phaseout of use (but not production) of HCFCs by December 31, 2014, with methyl bromide to be cut by 25 percent from 1991 levels by 1998.[11] Under pressure from German Council of Environment Minister Klaus Topfer, the Council of Environment Ministers agreed to tighten restrictions on imports of new and recycled CFCs (Cutter 1994f:6).[12]

The politics behind the Council of Ministers' decision (which was approved as Regulation 3093/94) involved the differing interests of the member states. In general, Belgium, Germany, Italy, Denmark, Great Britain, the Netherlands, and Luxembourg supported an aggressive response to ozone depletion. Denmark, Germany, Luxembourg, and the Netherlands argued that Community legislation should act as a floor for national action rather than as a ceiling. In the area of methyl bromide, France, Spain, and Great Britain seemed to favor a target date of 2000.

Belgium, holding the presidency of the Council, suggested a freeze beginning in 1995 and a 25 percent cut beginning in 1998. In the area of HCFCs, Denmark and the Netherlands thought the Commission position too weak, whereas Great Britain essentially supported it. France wanted to wait until 2004 before beginning to reduce HCFCs to any significant degree. Spain supported the French position. Belgium decided to fight for 2014 as the target date for complete phaseout of HCFC usage. The Commission assured the member states that production would be allowed so that export markets would be protected (Europe Information Service 1993f:7).

The Council of Ministers' decision represented a floor rather than a ceiling on efforts to protect the ozone layer. Germany, the Netherlands, and Denmark have passed national legislation more stringent than that approved in Brussels (Cutter 1994e:7; see also Dutch CFC Committee 1993; Commission 1993c:8).

Industry Response. As the deadline of January 1, 1995, for the cutoff of CFC production and use approached, the relevant officials within DGXI and the firms involved were cooperating well. DGXI saw no problem with compliance and regarded the confidential data turned in by the companies to an accounting firm in London as trustworthy. The relevant firms were seen as highly reputable and carried a great deal of credibility with Commission officials. Although the Commission had earlier obtained data on aerosols from producers and from aerosol fillers in order to examine whether they matched (they did), they no longer carried out such checks.

Given that the number of firms involved is small, the Commission does not face the kinds of compliance problems it faces in implementing environmental legislation in general.

Industry, however, has been unhappy about selected actions taken by the Commission. In 1994, the German and French chemical producers fiercely criticized the Commission's decision to increase the quotas for imports of used or recycled ozone-depleting compounds while keeping domestic producers to previously agreed to schedules (Europe Information Service 1994b:21–22; Cutter 1994c:1).

While producers complied, the refrigeration and air-conditioning sectors moved slowly to phase out CFC usage. (In contrast to the United States, no member state within the EU except Denmark imposed a CFC tax.) While firms in those sectors may be counting on stockpiles of CFCs to allow operations after the phaseout (Cutter 1994i:8–9), they may also be very receptive customers for the massive quantities of CFCs that customs data indicate are being illegally imported into the EU (Cutter 1994j:9–10). In fact, by the end of 1995, the Spanish government had acknowledged that Spanish companies were illegally importing CFCs, with small firms in the refrigeration and air-conditioning sectors being the most likely buyers (Cutter 1995f:4–5, 1995c:10).

The problem of CFC smuggling became increasingly serious during 1996. In October 1996, the British Environment Minister called for more coordination among law enforcement agencies throughout Europe in order to address the growing black market in CFCs. The fact that CFC prices had increased only fourfold since production was phased out, whereas prices in the United States (where enforcement has been strengthened) had climbed twelvefold, heightened the suspicion that the black market in the EU (probably supplied by Russian firms) is a very significant one (Cutter 1996c).

The Commission hopes to be able to track illegal shipments more effectively in the future, but the size of the illegal imports calls into question whether the phaseout of CFCs will actually occur as planned by the Community (Cutter 1994j: 9–10). The behavior of some of the member states will help the Commission's efforts, while that of others will hinder them. Whereas the government that had acknowledged the problem of CFC smuggling in Spain had prepared a decree establishing penalties for illegally importing and using CFCs, the government that succeeded it changed course. It claimed no new legislation was necessary even though the Spanish association of air conditioner producers pointed out that the lack of sanctions damages the interests of companies that have invested in substitutes. Spain, however, is not alone: Greece, Italy, and Portugal have no national penalties for the illegal use of CFCs (Cutter 1996b). By contrast, Denmark is stiffening its penalties against environmental crime and the Netherlands has established a special police task force to "fight organized environmental crime in The Netherlands" (Cutter 1996a).

The Problem of Methyl Bromide. Although methyl bromide is far more damaging to the ozone layer than are CFCs, controlling its use—important for the southern member states—has proved especially difficult. In the negotiations leading to the Copenhagen Amendment, the Community supported more stringent restrictions on HCFCs than did the United States (MacKenzie 1992a:6). However, the situation was reversed in the case of methyl bromide (MacKenzie 1992b).

The reluctance of Greece, Spain, Italy (the main methyl bromide users), and France (the most important methyl bromide producer) to reduce the use of methyl bromide made it difficult to agree to a phaseout date when the parties to the Montreal Protocol met in Vienna on November 28–December 7, 1995. Although the EU's position at the conference was to support restrictions even more stringent than those already adopted at the EU level, that position had been agreed to over the strong objections of France and Portugal.

The southern countries finally agreed to the phaseout date of 2010 (adopted by the Vienna conference). By contrast, the United States had argued for a target date of 2001 (with exemptions for certain types of agricultural uses). In a similar vein, although eleven of the then fifteen EU member states signed a separate declaration committing them to phase out methyl bromide as soon as possible, France, Greece, Spain, and Portugal did not sign[13] (Cutter 1995e:6–7, 1995f:4).

Overall, the Union has a good record when it comes to implementing the Montreal Protocol and its revisions. Legislatively, the EU has moved faster than the Protocol requires in several areas, so that its record in restricting ozone-depleting compounds has improved over time. The Union has moved from a position of skepticism and reluctant protection to one of environmental leadership in this issue area. Compliance with the Protocol has been relatively smooth, although the smuggling of CFCs may cause problems in the future. The negotiating history of the Protocol and its descendants, as well as the fact that a small number of highly reputable firms are involved in this arena, has rendered compliance relatively unproblematic. The Commission can enforce legislation that is more stringent than the Protocol and the Copenhagen Amendment, and it clearly is committed to doing so. In this area, we find a convergence of the Commission's institutional interest and of its institutional capacity.

CITES

If the Montreal Protocol is a relatively easy treaty for the Community to implement, CITES is the exact opposite.[14] The Union has neither the same institutional self-interest nor the same institutional capacity to enforce compliance that it has had in the case of ozone protection. The issue of institutional capacity is particularly salient. CITES in fact highlights the difficulty the Union, with its extraordinarily decentralized structure, faces in ensuring compliance on the ground with any legislation involving a multitude of actors (as opposed to a few very large reputable firms). (For a reference book on CITES, see Wijnstekers 1992.)

CITES contains all the elements that make an international environmental treaty problematic in Brussels: the Union is not a party, although the member states are; numerous actors (traders and public authorities that issue import and export permits) dispersed throughout the Union are involved; and the acquisition of evidence concerning noncompliance by public authorities that would be required to take a case to the European Court of Justice would be nearly impossible for the Commission to carry out on its own. The decentralized nature of implementation and enforcement within the EU is thrown into relief by CITES.

The Union's role in implementing CITES is complex. The Union is not a party to the Convention, although it very much wants to be. All its member states are parties without reservations, however, a feat for which the Community can claim credit. The Union is not a party largely because of American opposition—opposition based on the argument that the Community/Union does not possess sufficiently strong powers to ensure compliance. To confuse matters further, an EU regulation implements CITES within the Union, and thus even member states that were not parties until recently (Greece became a party on January 6, 1993) were bound to implement CITES through the Union's regulation. Finally, that regulation (no. 3626/82) was more stringent than the Convention, as is the regulation approved by the Council of Ministers on December 9, 1996. Some member states' national legislation is even more stringent than Regulation 3626/82, and criminal penalties are similarly disparate throughout the Union.

Although other reasons are also at work, the fact that the Community has been kept from being a signatory has been resented enough by top officials (at least as of mid-1993) that it may have discouraged the Commission from being more proactive in implementation. The institutional self-interest of the Commission in being recognized as an international actor has been stymied, and it is probable that recognition of the Community's standing internationally would give the Commission new incentives to improve implementation and compliance within the member states. In the case of CITES, the negotiating history of the Convention seems to have significant implications for its subsequent implementation.

Negotiating History

The Community is not a signatory to CITES because of what one high-level official termed "a historical accident." The Convention was signed in 1973—several months before the Community was given legal competence in environmental matters. (The political decision to develop policies for environmental protection at the

Community level had been taken in 1972.) CITES is therefore not a "mixed agreement" like the Montreal Protocol—but neither are the member states opposed to the Community's becoming a party. In fact, the Community has attempted to become a party since 1983.

On April 30, 1983, an amendment (known as the Gaborone Amendment) was proposed that would allow the European Community to accede to the Convention (Vandeputte 1990). During the negotiation of the Gaborone Amendment, the Community pledged, in the words of Caroline Jackson, a member of the European Parliament, that "adequate staff and finances would be allocated in order to ensure full implementation of CITES within the Community" (Council 1985:28). In 1986, however, the World Wildlife Fund presented a critical assessment of the Community's implementation efforts in a background report for the U.S. Department of State's decision on whether to ratify the Gaborone Amendment.

In 1987, the United States deferred a decision on ratification and announced that it wanted an intensive discussion of the role of the Community vis-à-vis the Convention to take place at the sixth meeting of the Conference of the Parties (Thomsen and Brautigam 1987:285). At the meeting, Greenpeace reported that "loopholes and gaps in enforcement ... demonstrate that the aim of a full and uniform enforcement of CITES by the EEC is far from being achieved" (Thomsen and Brautigam 1987:285). As a consequence, the Conference of the Parties adopted a resolution that stated the parties were "concerned that information and reports presented by the CITES Secretariat, both to this meeting and to previous meetings have identified serious enforcement problems in ... Member States of the European Economic Community." It went on to request that the Community, "in view of its abolition of internal border controls, urgently establishes full means of Community supervision of its legislation by means of an adequately staffed Community inspectorate." Further, the resolution recommended that the Community "monitor the movement of CITES specimens within and between Member states in accordance with the mechanisms foreseen in EEC Council Regulation 3626/82 and by use of existing forms available under Community legislation" (Convention on International Trade in Endangered Species 1987).

As of the beginning of 1997, the United States had not ratified the Gaborone Amendment. In fact, as of April 1996, only thirty-two out of the necessary fifty-four states had. The issue of how effective the EU could be as an executor of CITES has been critical in stalling the acceptance of the Amendment (Favre 1989). As of mid-1993, at least, the fact that the Gaborone Amendment had not been ratified was deeply resented by key officials in DGXI. The American position on the Amendment is particularly criticized: given the leadership and prestige of the United States in the area of nature protection, it was strongly believed that American approval of the Amendment would quickly lead other parties to ratify it.

The lack of ratification keeps the Community from being able to play an officially recognized international role in the policy processes connected to the Convention.[15] In general, it frustrates the very strong desire of the Commission to play an international role in the environmental arena, a role that would increase its influence and status within the Community's policy processes more generally. It is important to remember that the Community is still struggling to define and consolidate its international status in areas outside of external trade relations, and that the Commission regards itself as the key "Community" actor in that struggle. It therefore cannot be overemphasized how much the institutional self-interest of the Commission is linked to international recognition and access to international forums. Given that the Commission provides significant funds for the CITES Secretariat, the lack of official status is particularly galling.

The lack of status seems, over time, to have affected the relationship between Commission officials (both mid- and upper-level) and those in the Secretariat. Relations as of mid-1993 were not harmonious, and while the reasons are complex and beyond the scope of this study, it is clear that the structural position of the Community contributed significantly to the tension that existed. One high-level Community official made it clear that the Secretariat has been only too ready to contact the Commission for money and resources even though officially the Secretariat pretended the European Community itself did not exist.

The resentment has been deep enough that it has borne on the organizational context within which implementation of CITES takes place. Even official statements have implied that the international role of the Community and its efforts at implementation were linked (Vandeputte 1990). The fact that the Community has not been allowed to accede to CITES has at the very least shaped the political context within which implementation has taken place. As one very knowledgeable analyst put it, "I feel that the ratification of the Gaborone Amendment is necessary; it may be already too late, but without it, I am afraid the little attention now given to CITES will diminish even further."

The Union's Role in Implementation

The Community/Union's role in implementing CITES has involved requiring member states to drop their reservations to the treaty; encouraging accession of member states; adopting two major pieces of legislation, both more stringent than the Convention; and putting together a loose administrative framework within which the member states work. Over time, the Union's legislative record has improved: the most recent piece of legislation is more stringent than the first and follows up on the Community's commitment given at the ninth Conference of the Parties (Fort Lauderdale, Florida, November 7–18, 1994) to improve implementation (CITES

1994:8). Some of the member states have strengthened their compliance over time, but it is unclear how much credit Brussels can claim for such improvement.

The Commission required that before the entry into force of the Community's main piece of legislation, member states withdraw their reservations (Thomsen and Brautigam 1987:275). Furthermore, the Community has encouraged Greece and Ireland to accede to CITES. Both have done so. Greece, widely viewed as the most serious "weak link" in the Community's implementation efforts, acceded to the Convention on January 6, 1993 (Europe Information Service 1993b). In brief, the Community can be credited with increasing the number of parties to CITES and rendering member states' participation more comprehensive.[16]

Implementation through Legislation. The Community's principal legislative mechanism for regulating wildlife trade has been Regulation 3626/82, which was approved in 1982 by the Council of Ministers and came into effect in 1984.[17] (A new regulation was approved on December 9, 1996, and came into force on June 1, 1997.) Regulation 3626/82 has served as a legislative "floor" for the member states (which are permitted to impose stricter regulations) and sets up an administrative structure that brings the representatives of the member states together in an institutionalized fashion. It was innovative in certain respects. TRAFFIC-Europe concluded that "the principle of using tough import conditions as a means of affecting the export policies of producer countries ... has in many instances proved a successful conservation tool" (Broad and De Meulenaer 1992:2). From a comparative perspective, the fact that an EC regulation is immediately applicable throughout the Community, without the need for national legislation, is important, because the passage of national legislation has been a stumbling block in several countries outside the EU (Forster and Osterwoldt 1992:82). The Union, ironically, is not a party, but its legislative activity has ensured that legislation applying to all its member states is in force.

The regulation was more stringent than the Convention in several respects (Commission 1994d:4–5). Thus, the legislative "floor" that the EC imposes on the member states is more restrictive than the Convention itself and does not preempt even tougher national laws. (The fact that national laws can be more stringent is, however, criticized by the CITES Secretariat.) Essentially, the regulation treats some of the species listed in CITES appendix II as if they were in appendix I (thus practically banning trade) and requires an import as well as an export permit for certain species listed in CITES appendices II and III (Johnson and Corcelle 1989:243). The legislation also regulates the transportation of CITES-listed animals, and generally prohibits trade within the Community of appendix I species (see Broad and De Meulenaer 1992:1; TRAFFIC-Europe 1992:36).

The legislation and its implementation, however, were criticized by both the European Parliament and, in 1988, by TRAFFIC in a wide-ranging report (Barzdo and Broad 1988). The Commission, in response, proposed a new regulation in 1991. This regulation, developed over five years, was much more ambitious than the existing one and moved far beyond CITES (Broad and De Meulenaer 1992:3).

The Commission's proposal was generally met with approval by the European Parliament's Environment Committee (European Parliament 1993b:52). The Council of Ministers, however, took a very different view. The draft regulation—with its expansion of the annexes listing regulated species—was met by a barrage of criticism. In general, its scope was seen as far too ambitious, bringing with it unwanted paperwork and regulatory activity for the member states. The member states wanted a reduction in the number of species regulated and in workload rather than the increased regulation embedded in the Commission's proposal.

There were also disagreements over the legal basis of the Commission's proposal. The Commission wanted the Council of Ministers to use qualified majority voting and to preempt stronger national regulations, whereas some member states wanted to require unanimity, and still others wanted to allow stronger national measures. Germany in general wanted more restrictions than those proposed by the Commission, and the Netherlands took the Commission's position. The other states wanted less than the Commission but differed among themselves. As one Commission official put it, "There is a different coalition on every article in the proposed regulation and seems to be no overall view on the part of the member states."

The Commission put forth an amended proposal that seemed to have a greater chance of obtaining the needed approval (Commission 1994b). It did in fact substantially reduce the number of species included in the legislation's annexes A and B; it also, controversially, repealed existing Community legislation on imports of whales and cetacean products. Although the European Parliament had suggested listing the species of tropical timber in annex B rather than annex D (thus increasing the degree of protection given to tropical timber), the Commission refused to accept that request (Commission 1994b:1). The amended proposal gave the Commission less of a role in implementation than the earlier version and reduced the regulatory burden. The proposal to establish a registration system for the movement of live specimens of selected species was dropped, for example (Commission 1994b:6).

Most important, the Commission changed the legal basis of the proposed legislation (from articles 100A and 113 to article 130s, paragraph 1). That change, insisted on by Germany in particular, allowed member states to impose more stringent national regulations than those agreed to in Brussels.

The French presidency helped move the negotiations on the revised proposal so that political agreement was reached at the Environment Council meeting of June 22 and 23, 1995 (Europe Information Service 1995:25(IV)). It was definitively approved by the Council of Ministers in December 1996. Substantively, the new regulation reinforced the controls at the Union's external borders, thereby addressing one of the major concerns voiced by critics (Agence Europe 1995; Commission 1996b). The regulation also covered infringements that member states will be required to penalize.

The Commission and Implementation. Regulation 3626/82 set up an administrative body known as the CITES Committee (referred to in the regulation as the Committee on the Convention). The CITES Committee is composed of member state representatives and chaired by a Commission representative. In general, each member state is represented by the director of its Management Authority. Critics of the Commission's role argue that it has defined its role as being a secretariat for the member states' management authorities. Some member states do not have a well-institutionalized scientific authority, so the Commission set up the Scientific Working Group shortly after the regulation went into effect. That group meets the day before the CITES Committee meets, and makes recommendations. Such recommendations are particularly valuable for those member state representatives who do not have a national scientific authority upon which to draw.

By establishing the CITES Committee and the Scientific Working Group, the Commission provided an administrative structure that could facilitate CITES implementation. The structure, however, was loosely articulated, and both the regulation and Commission practice have clearly left implementation measures to the member states. As one official put it, "In terms of implementation and compliance, the Community has the CITES Committee, the Scientific Working Group, and TRAFFIC to monitor what goes on.... But, finally, implementation is a member state responsibility."

Given the current level of resources, the Commission cannot do a great deal more than it currently does. Critics do, however, argue that even within the current structure, the Commission could be more proactive. In particular, DGXI could work more closely with DGXXI so as to train customs agents to recognize wildlife covered by CITES. DGXXI does have an anti-fraud unit that could be used to better implement CITES, but DGXI and DGXXI have not coordinated their activities toward that end. Whereas the CITES Secretariat has received funding from DGXXI to train customs officers in several member states, DGXI has not participated.

The fact that the Community is not a party to the Convention serves as a disincentive to a more proactive Commission. Given the lack of staff within DGXI, the explosive growth of environmental legislation at the Community level since the late 1980s, and the increased number of environmental treaties to which the Community is a party, the staff has far more to do than it can possibly accomplish. It is not surprising that the frustration felt at the highest levels over the exclusion of the

Community as a party is also felt at lower levels of officialdom. As one official put it, "There are lots of other things to do."

The Convention's Secretariat. The Convention's Secretariat became increasingly critical of the EC during the 1980s. In 1983, it "singled out two EEC countries, Italy and France, as having major enforcement problems" (Thomsen and Brautigam 1987:283–284). In 1984, it criticized the Community as a whole, concluding that "implementation of CITES in the EEC is extremely poor." This concern with the EC's lack of implementation reappeared very strongly during the Secretariat's attempt to improve compliance in Italy (discussed below).

The NGO community renewed its criticism of the European Union's implementation record as the November 1994 Fort Lauderdale Conference of the Parties approached (Reuters 1994). At that meeting, the Secretariat presented a report on implementation within the European Union. Although the report praised the EU's legislative framework, it was hard-hitting in its criticisms of other aspects of implementation. It concluded that the new resources promised by the EU in 1983 had not been made available. The Commission, for its part, did not coordinate the activities of the member states in the way it should.

Although inadequate resources were recognized as being primarily responsible for the Commission's failings, the report did note that at times a "lack of political will" accounted for the reluctance of the Commission to bring strong pressure on member states whose implementation was below standard. It called for the approval of the Gaborone Amendment in the hope that allowing the Union to become a party would improve implementation. Implementation in the EU as of 1994, however, was judged to be at the level of "the State with the lowest implementation level" (CITES Ninth Meeting 1994:4).

Reporting. The Secretariat treats reporting far more seriously than does the Community. For the Secretariat, reporting is a key component of implementation, whereas it is not for the Commission. The Community often requires member states to report a wide range of data to the Commission. The member states, however, often fail to respond to reporting requirements in any kind of timely fashion. Commission officials do not place a high priority on prosecuting violations of reporting requirements—whether they be for EC legislation or for international treaties. Interviews in the legal services division indicated that a member state would not be brought to the European Court of Justice for infractions of reporting requirements. As one interviewee put it, "We are not likely to bring infringement proceedings because of a lack of reporting. Problems of reporting are '*maigre*'; we want to bring cases of substance." In general, the Commission does not treat its own reporting requirements (set out by legislation) with any more seriousness than do the member states. Reporting is simply not a high priority within the Commission.

Reporting related to CITES is no exception (Commission 7.04.1994). It is not a high priority for the member states. For their part, Commission officials think much of the required reporting is not useful. A markedly contrasting attitude exists between the Commission and the member states, on the one hand, and the Secretariat, on the other.

Implementation in the Member States: An Overview

How effective is the Union at ensuring compliance with the legislation implementing the CITES Convention? Answering such a question is extremely difficult, because the question implicitly carries with it a comparative referent. Does the Union do better or worse than, say, extremely decentralized federal systems? The data do not exist to give even a partly systematic answer to that question. How does it compare with its own member states? Although again data are scarce, it is reasonable to argue that the Union does better than some of its member states and worse than others.

If we ask whether the state of enforcement in general is good across the Union, the answer is most probably "no" (in sharp contrast to the successes identified in the case of the Montreal Protocol). The responsibility for that assessment, however, lies far more with the member states than with the Community's efforts. Nonetheless, as we shall discuss below, over time several of the member states have improved their compliance with the regulation implementing CITES.

The implementation and enforcement of CITES in the various member states vary considerably. Although some member states have greatly improved their performance over the last several years, others still face enormous obstacles in implementation. Belgium is widely considered to have improved its compliance. Germany has improved its legislation and has begun to regulate the port of Hamburg. France has strengthened its controls in French Guiana (Thomsen and Brautigam 1987:286–287). As we shall discuss in the next section, Italy, after a great deal of pressure, also has significantly improved its performance.

Southern members of the Community are particularly troubled by poor compliance.[18] Greece, for example, has an especially bad record of implementation, and did not even accede to CITES until 1993 (TRAFFIC-Europe 1992:30).

While the Greek legislative framework is particularly inadequate, other member states also have significant problems. As one official put it, "Denmark implements CITES best. But it has limited trade and only one airport. Denmark is an easy case. In Germany, someone sold permits which were circulating throughout the EC. The Germans have local management authorities. They give out permits to anyone. In the UK, reporting is lousy. It misses a lot." However, there is a widespread impression that implementation is a problem in other developed countries as well. As one official pointed out, "The TRAFFIC report on implementation in the United States was never published, so it was probably critical as well."

The Secretariat Intervenes: The Case of Italy

Italy, of all the Union's members, has drawn the most intense scrutiny from the CITES Secretariat. As one of the Union's southern members, it is widely viewed as a country with lax environmental enforcement and low environmental awareness among its citizenry. Whereas other southern members are poor, however, Italy is one of the richest members of the Union. Its lack of enforcement infrastructure, therefore, cannot be blamed on a lack of financial resources; rather, it is rooted in a public administration widely viewed as ineffective as well as in a lack of political will.[19]

Given the importance of Italy's leather industry, lack of enforcement of CITES regulations led to the unregulated import of reptile skins. The Italian problem with CITES was widely known in the relevant NGO community (especially in TRAFFIC's office in Italy), and that community helped the Secretariat obtain the evidence that publicized problems of compliance in Italy.

The Secretariat spent a great deal of time and energy trying to improve Italian compliance, but it met with little response. Finally, the Secretariat informed Italy that it was prepared to contact the Standing Committee. The document outlining the course of this dialogue (or lack thereof) pointed out that the Secretariat had contacted numerous officials in Italy's public administration as well as Commission officials—all to no avail (CITES Secretariat 1991).

On September 26, 1991, the Secretary-General sent a fax to all members of the Standing Committee that outlined a number of alleged infractions by Italy for the period 1990–1991. The fax was accompanied by a document outlining what the Secretariat had done to improve Italian compliance, as well as a seven-page draft document entitled "Major Alleged Infractions Registered by the CITES Secretariat Concerning Italy." The fax began with the information that "for several years, the Secretariat has been concerned to receive information about weaknesses in the implementation of CITES in Italy." It also included information about a recent case of chimpanzees that had been brought into Italy, seized, released by the courts because Italy did not penalize CITES violations, and allowed to be slipped into Austria even though the Italian customs authorities had been informed by the Management Authority not to allow the chimpanzees to leave Italy (CITES 1991).

The case of the chimpanzees received considerable publicity within Italy and throughout Europe (World Wide Fund for Nature 1991). A report prepared by TRAFFIC-Europe helped to set the problem in a larger context and used the drama of primates being moved illegally across borders to draw attention to the lack of CITES enforcement in Italy. The report pointed out that Italy lacked legislation allowing for enforcement of the EC regulation. It also criticized in scathing terms the administrative apparatus responsible for implementing CITES (TRAFFIC-Europe 1991:70). In general, the report judged the implementation of CITES to be very poor in Italy.

The CITES Standing Committee considered the problem of Italian implementation at length. At its meeting on January 20, 1992, the Secretariat announced that Italy had not answered any of the diplomatic notes sent to its embassy, and that the Secretariat leaned toward proposing a ban on trade in CITES specimens with Italy (CITES Standing Committee 1992a).[20]

At that same meeting, the Italian representative was asked to speak. Although he informed the Standing Committee that progress was being made in regard to the passage of legislation, strong concerns were expressed. The Secretariat pointed out that two draft resolutions were to be presented at the eighth Meeting of the Parties concerning implementation of CITES in the European Community, and that the Italian problem was a "contributing cause" for those resolutions. Furthermore, several Latin American countries wanted action against Italy by the Standing Committee, and if none were forthcoming, "then countries in South America might take stronger action with respect to Italy at the eighth meeting of the Conference of the Parties."[21]

When the EC representative pointed out that a ban on trade with Italy was unenforceable unless a ban was imposed on the entire Community, the Standing Committee agreed that Italy and the EC would have to be addressed simultaneously. It decided "that the EC should also encourage the passage of ... legislation by Italy." Finally, Italy was given three months to improve its compliance with CITES.

At the eighth meeting of the Parties, held in Kyoto on March 2–13, 1992, a resolution concerning the European Community was approved. It listed problems that, given the discussions going on within the Standing Committee, clearly referred to Italy (live animals and reptile skins were specifically mentioned in the resolution).[22] It urged Community members that were parties to the Convention to improve their legislative framework and to increase the resources allotted to enforcing CITES; it also urged Community members that were not parties to accede to the Convention (Agence Europe 1992). It concluded by urging ratification of the Gaborone Amendment.

At the twenty-eighth Standing Committee meeting on June 22–25, 1992, the Secretariat reported that a mission to Italy had concluded "that very little progress has been achieved. Italy has passed new legislation, but has failed to adopt decrees ... needed for its full implementation" (CITES Standing Committee 1992b). The Secretariat therefore recommended "that the Standing Committee recommend to all Parties to not accept any CITES documents which have been issued by Italy, and to not issue any CITES documents with Italy as the country of destination."[23] The

Secretariat reported that its staff members had been told by Italian civil servants that "the Italian government would do nothing to improve CITES implementation in that country without trade restrictions being imposed." Although the Italian representative informed the Standing Committee about the progress Italy was making, the Committee accepted the Secretariat's recommendation.

Italy, however, continued the progress that its representatives had reported at the various Standing Committee meetings. On January 12, 1993, the Council of Ministers in Italy approved penalties for CITES violations. A new law that came into effect on March 13, 1993, makes funds available for the administration of CITES and regulates the condition of animals being transported internationally. Furthermore, by March 31, 1993, all firms "in possession of crocodilian skins were obliged to register them with the Ministry of Agriculture ... where they would be permanently tagged" (TRAFFIC-International 1993:89). The approval of penalties convinced the Secretariat that Italy was serious about compliance. On February 19, 1993, the Secretariat sent a notification to the parties that the Standing Committee's recommendations were suspended (CITES Standing Committee 1993). In October 1993, the Finance Ministry designated selected customs offices to regulate trade falling under CITES (Guttieres and Aguilar-Lizarralde 1994:40).

The Italian government continued to improve its administrative infrastructure. In March 1995, the Secretariat sent a mission to Italy to examine the improvements in depth. Its report at the thirty-fifth meeting of the Standing Committee later that month was striking in its laudatory tone:

> The Secretariat was very impressed by the quality of the work and the motivation of the staff of the Forest Corps [Corpo Forestale dello Stato]. The result is a very high standard of achievement, which in our opinion, figures among the best in the world. The computer system that has been established is, as far as the Secretariat is concerned, the best system for CITES purposes that it has ever seen. The enforcement section has initiated several investigations with outstanding results.... Now Italy is among the Parties which have the highest level of implementation of the Convention. (CITES Standing Committee 1995:1, 3)

The improvement in Italian compliance with CITES is noteworthy, for it indicates what can be achieved even in countries with an inadequate administrative apparatus.

Why the Secretariat and Not the Commission?

The Community's representative, when queried at the June 1992 Standing Committee meeting about the possible role of the Community in improving Italian compliance, responded in a fashion that indicated the skepticism of some Commission officials about the Secretariat's pursuit of Italy.

Officials within the Commission clearly felt that Italy was being used as a scapegoat and that indeed it had been "pursued." (It should be noted that none of the Commission officials involved were of Italian origin.) They felt that the Secretariat was searching for a rich developed country to punish in order to balance sanctions that had been applied to poor Third World countries. Cooperation between the Secretariat and the Commission on the Italian case seems to have been minimal at best. One official within the Secretariat was identified by a knowledgeable Commission official as "having drafted the resolution against Italy and then saying he was trying to stop it." In general, Italy was viewed as not being significantly more lax in its enforcement than other countries with a strong economic interest in CITES-covered trade. In the words of one Commission official, "If you put the same effort into Belgium or the United States, you would come up with the same problems."

Given the lack of cooperation between the Secretariat and DGXI on operational issues such as those involved in the Italian case, as well as the resentment felt by DGXI's top leadership over the exclusion of the Community as a party to CITES, it is not surprising that implementation on the ground is weak. CITES is an extremely difficult treaty to enforce even under ideal conditions: the tension between the Community's officials and those in the Secretariat have rendered implementation even less effective than it otherwise might have been. The roots of such tensions are beyond the scope of this study. It is clear, however, that many of the EU's member states have significant problems in complying with CITES. Given that the CITES Secretariat is unusually strong and that the EU is weaker than traditional federal systems in its enforcement powers, a working partnership between the two would seem to be in the interests of both if compliance with CITES is an important goal.

Such a partnership would need to go beyond DGXI's funding of Secretariat activities; it would have to involve real administrative coordination between the Secretariat, the Commission, and each member state. Given the challenges presented by CITES, and given the demands placed on both the Commission and the Secretariat, such a partnership is probably the only realistic way of improving compliance at the member state level in a sustained fashion.

ITTA

Although the Montreal Protocol and CITES incorporated trade provisions, the Community interpreted them as primarily concerned with environmental protection because they *restricted* trade in ozone-depleting substances, in the case of the Protocol, and in endangered species, in the case of CITES. In fact, both are widely viewed as multilateral environmental agreements (Brack 1996). The combination of environmental and trade provisions did not dilute the effectiveness of those concerned with environmental protection, and in fact is credited with strengthening that effectiveness (Brack 1996).

The ITTA stands in contrast to the Montreal Protocol and CITES. In its 1983 version (which entered into force on April 1, 1985, but did not become operational until 1987), the International Tropical Timber Agreement was unique in that it was the only commodity agreement in the world with an environmental provision. That provision was concerned with the conservation of tropical forests (Chase 1993:757). Nonetheless, trade in tropical timber was not restricted, thereby differentiating the ITTA from CITES and the Montreal Protocol (Demaret 1993). In fact, as mentioned in chapter 5, the ITTA was designed to facilitate international trade in tropical timber rather than to regulate it.

In spite of the environmental dimension, however, the European Commission, even as it has supported the strengthening of that dimension, has viewed the ITTA primarily as a commodity agreement. Its environmental provisions have been viewed as so nonbinding on producers that interviewees questioned the inclusion of this treaty in a study of international environmental agreements.[24] The International Tropical Timber Organization has been categorized as "part commodity agreement, part development agency, part environment agency," but its potential for environmental protection was downplayed in Brussels until the 1990s (B. Johnson 1991:7).[25]

The ITTA is a "mixed agreement" like the Montreal Protocol, and neither the Commission nor the member states that are parties to it have favored the regulation of trade in tropical timber. However, both have favored strengthening the environmental dimension of the Agreement and argued for much stronger provisions in its successor agreement. In 1990 and 1992, the ITTO adopted Guidelines for Sustainable Management of Tropical Timber, and in 1991 it adopted "Target 2000." By the year 2000, it was hoped that all tropical timber sold internationally would be harvested from sustainable, managed forests. The European Commission strongly supported Target 2000, and in the negotiations for the successor agreement tried to make that goal a binding commitment on producers (Demaret 1993).

The European Parliament has kept the issue of tropical forests on the agenda without, however, having much impact on the Commission's substantive position concerning restrictive trade measures. The Parliament has called for quotas in order to curtail imports, the use of import licenses to assist monitoring, and the actual ban of imports "from countries which do not participate in any forest management and protection programme" (European Parliament 1993a:16).

As late as 1991, the Commission (which spoke on behalf of all twelve member states) sent only one commodity expert as its representative to meetings of the International Tropical Timber Council. As a World Wildlife Fund study put it, "Not surprisingly, the Community takes a minimalist view of the Organization's environment potential" (B. Johnson 1991:7).

The Commission has certainly had doubts about the operational capacity of the ITTO and about its ability to finance projects that make a real difference (Commission 1989:18). It has argued that the ITTO should develop a distinctive role and not try to develop as a funding agency. The Commission laid out its views rather explicitly in a report issued in late 1989:

> ... the Commission has strongly taken the view that rather than try to become yet another funding organization ... ITTO should put agreat deal of emphasis on developing what is broadly termed the "normative approach." This implies the establishment of a set of standards. This is a task that is being undertaken by no other international organization.... A number of areas are of particular interest such as the preparation of model contracts and model laws for tropical forest exploitation, advice on royalties, handling of concessions, replanting conditions etc. and the definition of a standardized approach to the granting of licenses, based on well defined criteria. (Commission 1989:18)

Following up on Rio, the Commission proposed a new regulation that would direct the spending of the 50 million ECUs earmarked for the protection of tropical forests in the 1993 budget (Commission 1993a; Europe Information Service 1993c). The Parliament's response was to criticize the Commission's exclusion of specific measures for regulating trade in tropical timber and for concentrating exclusively on financial aid and technical know-how in the Community's efforts to conserve tropical forests. It proposed the use of ecolabeling and the creation of a "Tropical Forest Unit" within the Commission (European Parliament 1993a:18–19).

The Commission, however, refused to accept the Parliament's request for the use of trade measures. In its amended proposal, the Commission explicitly stated: "Like the first version, the modified proposal concentrates on cooperation activities in the fields of the environment and development, and does not contain provisions for commercial activities" (Commission 1994c:1).

Although the Commission highlights its participation in the ITTO when responding to the Parliament's demands, the Community has provided very little funding for ITTO. Japan has been viewed as the dominant consumer country, and neither the Community nor the member states have provided significant funds. As of 1989, non-Japanese contributions for the funding of projects totaled a mere $2.0 million, to which the Netherlands and the Federal Republic of Germany contributed (Commission 1989:12).

The Role of the GATT

Although the Commission has addressed the issue of tropical forests in a variety of frameworks, it has consistently refused to consider restrictive trade measures. The limitation of tropical timber imports has been viewed as anti-GATT by the Commission. Some involved in this issue view the Commission as "obsessed by GATT," and therefore unwilling to support any measures that would really affect the conservation of tropical forests. Although in 1989 the Commission had indicated a willingness to consider including endangered species of tropical timber in the CITES listing, it rejected such a proposal from the European Parliament while it was drafting its amended CITES proposal in 1993.

The issue of GATT is a sensitive one. Malaysia, the chief tropical timber exporter, has used the GATT to ward off Community intervention. After Austria (not then an EC member state) introduced in 1992 an "ecolabeling" law designed to indicate which imports contained tropical timber as well as which timber had been harvested in a sustainable manner, Malaysia "feared that it was only a matter of time before other European Community members would be spurred on by domestic ecological movements to enact similar legislation ... and lodged a formal protest with the GATT's Committee on Technical Barriers to Trade" (Chase 1993:761–762). Such a fear was not unfounded, given that "200 city councils in Germany and 51% of Dutch municipalities have banned the use of tropical timber" (Chase 1993:765). The situation was defused in December 1992, but the possible intervention of GATT remains a factor in the Community's debate.

The European Parliament has been far less worried about GATT restrictions. It has passed several resolutions asking for a moratorium on imports of hardwood from Sarawak. It has also urged that the Community use an ecolabeling scheme (Europe Information Service 1993e:7).

National Activity

In general, environmental nongovernmental groups have lobbied national governments to ban imports of tropical timber. Within much of the environmental community, the restriction of trade is seen as the only effective way to save tropical forests. The Commission's argument that producer countries needed to adopt binding provisions for the protection of their tropical timber, made during the negotiations that resulted in the 1994 Agreement, needs to be understood within that context (Europe Information Service 1993g:24(1)).

Political activity has been most effective at the national level. Certain member states have seen significant mobilization around this issue. In Great Britain, for example, activists have organized boycotts. Frustrated by the ITTO's decision to postpone its labeling program, in 1994 British environmental groups announced a campaign against the sale of Brazilian mahogany. Roughly half of Brazil's mahogany exports are sold to Great Britain, and environmentalists claim that "at least half of that timber is illegally extracted from biological or indigenous reserves" (Cutter 1994g:12). Major store chains have responded by refusing to deal in mahogany.

The issue of tropical timber has been particularly controversial in the Netherlands, which has the highest per capita demand for tropical timber in western Europe (Cutter 1994h:11). The Dutch government responded to criticism by agreeing to limit imports to those from sustainable, managed forests but seemed unable to meet deadlines to which it had agreed. In fact, the Dutch experience highlights the problems and opportunities intrinsic to protecting tropical timber by changing the behavior of buyers.

In the summer of 1993, the Dutch government, environmental groups, and three associations of tropical wood importers agreed that, beginning December 31, 1995, only timber from areas with sustainable forest management policies would be imported into the Netherlands (Cutter 1993c:17). However, in March 1994 two of the environmental groups withdrew from the agreement over a dispute as to the definition of "sustainably managed forests" (Cutter 1994d:15).

In July 1994, a commission appointed by the Dutch government announced that the government would be unable to meet its December 1995 deadline, partially because the ITTO had decided to defer a labeling program until at least the year 2000 (Cutter 1994h:11). In December 1994, the Dutch government announced that, in cooperation with producer countries, it would develop a voluntary ecolabel for all (rather than just tropical) sustainably produced timber. The scheme might be in place by 1996 (Cutter 1994l:12). However, by December 1995, no agreement had even been reached on the definition of "sustainable." Meanwhile, environmentalists' publicity about tropical timber had significantly hurt the market for such timber. In fact, environmentalists claimed that "over the last two years, imports of tropical wood ha[d] fallen by 40%" (Cutter 1995g:13).

Development Assistance and Tropical Forests

Environmental protection for tropical forests has been of more concern within the framework of environment and development than within the ITTA. In mid-1994, the European Commission published a proposed regulation to promote tropical forests (Commission 1994e, 1994c). However, the proposal was not related to the ITTA, it concerned development policy rather than trade. In a similar vein, a protocol was added in 1995 to the Lomé Convention (a pact on trade and aid between the EU and seventy developing countries known as the African, Caribbean, and Pacific [ACP] countries) that encouraged the ACP countries to sustainably manage their tropical forests and contained measures to promote trade in sustainably produced tropical timber from the countries concerned (Cutter 1995d:11). In fact, the Commission has concretely addressed the environmental aspects of tropical forestry more directly and consistently within the framework of the Lomé Convention than within the ITTA (Commission 1989).

Because the ITTA was viewed as a commodity agreement rather than as an environmental accord, it does not fall under the jurisdiction of DGXI (which is responsible for the Montreal Protocol and CITES). Rather, the Directorate-General responsible for development assistance and the Lomé Convention (DGVIII) represents the Commission within the framework of the ITTA. Just as the ITTA falls under the aegis of UNCTAD at the global level, so it falls within the jurisdiction of the Brussels bureaucracy responsible for development.

The New ITTA

As indicated in chapter 5, the negotiations over the successor agreement to the ITTA were very contentious. In general terms, the European Union, along with other consuming countries, pressed for the adoption by producers of binding guidelines and criteria ensuring sustainable forestry management. However, the Europeans were unwilling to accept binding commitments for the sustainable management of their own forests. The result was an agreement much weaker than that desired by environmentalists.

Some officials were deeply disappointed at the provisions of the 1994 agreement. Officially, the EU and the member states reserved their position so as to evaluate the agreement. In particular, disappointment was expressed at the lack of a strong commitment by the producers to sustainably manage their tropical forests by the year 2000 (Cutter 1994a:4). Nonetheless, on March 29, 1996, the Council of Ministers signed the International Tropical Timber Agreement 1994 on behalf of the European Community (96/493/EC).

The World Wide Fund for Nature (WWF) blamed the northern countries' refusal to follow up on the commitments they had made at Rio for the weakness of the ITTA. By refusing to include temperate and boreal forests in the new ITTA, the northern countries had essentially agreed to weak protections for tropical forests. The EU's position was that expanding the scope of the ITTA would completely eliminate the possibility of agreement on a global convention on forests. The WWF, however, argued that the position taken on temperate forests makes producer countries suspicious about the likely fairness of a global forest convention (Cutter 1994a:5).

In fact, at a meeting in Helsinki in June 1993, the Europeans had refused to accept the proposal that by the year 2000 they harvest timber only under environmentally safe cutting rules. Although the proposal for the target date had been put forth by the Dutch, the European Community, Britain, and Sweden were opposed, and the target date was not adopted. (They did agree on a common definition of what the term "sustainable forestry" means, however.) By contrast, the United States changed its previously held position and agreed to manage its own forests sustainably by the target date of 2000 (MacKenzie 1993:9).

In general, the Community did not treat the ITTA as an agreement with a significant environmental component. Furthermore, since that component applied to producers rather than to consumers, compliance with the Agreement's environmental provisions cannot be analyzed in the way that compliance with the Protocol or CITES can be.

However, the Community did, over time, pay more attention to the environmental dimension. It strongly supported Target 2000, and negotiated with other consumer countries to strengthen environmental protection in producing countries in the 1994 agreement. The Community did not, however, play the kind of leadership role it did in the negotiations for the Copenhagen Amendment to the Montreal Protocol, nor did it show the kind of interest and engagement it has shown in trying to become a party to CITES. Given the hostility of the Commission to imposing unilateral trade restrictions on tropical timber, only a globally negotiated treaty that restricts trade in tropical timber is likely to gain the Community's support of restrictive trade measures.

Conclusion

The European Union is implementing the three agreements analyzed here quite differently. It has done very well in the case of the Montreal Protocol, has done less well in the case of CITES, and has interpreted the ITTA as a commodity agreement and downplayed its environmental dimension while subsequently pressing for stronger environmental protection on the part of producers. It has been actively engaged in a variety of ways with both the Protocol and CITES but has been much more passive in the case of the ITTA.

While both the Montreal Protocol and CITES have been implemented well legislatively, they have been treated differently in the postlegislative phase of implementation. Institutional factors help account for this difference. The Union's institutional capacity is well suited to implement the Montreal Protocol, for it involves a small number of reputable firms that are willing to cooperate with the European Commission and national governments. Furthermore, at the negotiating table, the Union is able to represent member states whose positions on ozone depletion are relatively convergent. As illegal trade in CFCs grows, however, the Union's lack of law enforcement powers may damage Brussels' good implementation record.

CITES presents a more difficult case for the Union. The Union has done well in implementing CITES at the legislative level, and that record has improved over time. The regulation that has been in force—and the one that replaced it in June 1997—provide more protection for endangered species than does the Convention itself. The difficulty in obtaining compliance with the Convention is rooted in the postlegislative phase of implementation. Brussels is not a party in spite of strenuous efforts to become one, and therefore lacks an important incentive for proactive implementation. Second, and more important, CITES requires implementation "on the ground" and involves a very large number of economic actors throughout the Union. The exceptionally decentralized nature of the Union makes achieving high degrees of compliance in that context extraordinarily difficult. Third, the member states diverge sharply on the priority they give the issue of endangered species and the resources they are willing to devote to implementing CITES.

Given that the Union has problems with implementing CITES on the ground, where it has almost no institutional capacity, but has done well in the legislative arena, in which it has the greatest institutional capacity, its record is mixed. Implementation on the ground will improve only to the extent that either the Union is given the authority to intervene more forcefully in the implementation efforts of member states (not likely in the short term) or the member states devote more resources to obtaining compliance on the ground. The case of Italy demonstrates that the latter is possible. External pressure is needed, however, to achieve such an outcome. And here the Commission could probably play a larger role while still honoring the political requirement of not intruding in the implementation processes, for which national governments are responsible.

In the case of the ITTA, the Union did not view the initial agreement as an important environmental accord but did support stronger environmental provisions in the 1994 Agreement. Although the environmental dimension, protection of forests, is acknowledged in the 1994 Agreement, the Union supported its application to producing rather than consuming countries. The Commission has been unwilling to consider any kind of import bans on tropical timber that might contradict GATT requirements and has refused to include tropical timber in the regulation implementing CITES. In brief, the ITTA has not been given high priority within the Commission. The fact that it does not fall within the jurisdiction of DGXI may help account for the Agreement's low profile.

Implications

Although the European Union does not have an impeccable record in the implementation of international environmental accords, its involvement, on balance, has favored environmental protection, and its record has improved over time. When Brussels is involved, environmental protection in many member states benefits. Particularly in the area of legislative implementation, the involvement of Brussels reinforces protection rather than dilutes it.

The involvement of the European Union has been important in convincing some of its member states to take more environmentally friendly positions than they otherwise would have on issues such as the restriction of HCFCs and methyl bromide. In the case of CITES, the Community was important in convincing some of its members to accede to the treaty. It has also provided important financial resources to the CITES Secretariat. Finally, both regulations implementing CITES adopted in Brussels have been more stringent than many of the member states would have adopted if left to their own devices. In brief, at least several of the member states take "greener" positions because they are members of the European Union than they would if acting unilaterally either in the international arena or at home (Sbragia 1996).

Given the Union's institutional structure and the fact that its institutional capacity is still evolving, however, compliance with accords "on the ground" has been far more problematic a goal than implementation of accords in a legislative/legal sense. Third parties, in fact, have often been reluctant to admit or recognize the Community as a negotiating partner with equal legal status precisely because it lacks the enforcement capacities of a traditional sovereign state.

Although the "center" in federal systems exercises significantly less control than it does in unitary states, federal governments typically are able to police, to gather evidence independently of subfederal authorities, to impose federally determined criminal sanctions, and to use federal agencies to monitor and inspect (if only selectively) the activities of agencies "on the ground" (Sbragia 1997).

The Union's capacity to implement is weak if measured against such criteria. The Union does not possess any of those powers, whereas, by contrast, the American federal government possesses all of them. Brussels can nonetheless "add value" to the implementation process because it can bring both formal and informal pressure on its member states to improve implementation. It also provides an institutional framework that is a necessary precondition for future, more effective compliance.

The Union is far from impotent. It can adopt legislation applicable to those of its member states that would ignore the implementation of international environmental treaties altogether (or might not accede to them to begin with). It can also bring informal pressure on such states to improve compliance with the legislation adopted in Brussels. It is the best mechanism available to the international community to influence nation-states that, left to their own devices, would give environmental protection (nature protection especially) a low priority. Although the Union lacks nonjudicial enforcement powers, the international community can use the European Union as an ally to improve compliance within the nation-states that are its members.

The European Union, although not as strong an implementor as a traditional federal state, can be a key actor in two ways: (1) the construction of a consensus among fifteen disparate member states in the negotiation and subsequent legislative implementation of an international treaty and (2) the strengthening of compliance with obligations in those member states that, without the Union's involvement, would do even less.

Notes

1. Legally, the term "European Community," rather than "European Union," is correct to use when discussing the EU's role in environmental affairs. However, the Treaty of European Union (usually known as the Maastricht Treaty), which came into force in 1993, changed the

name of the European Community to European Union in the field of international affairs generally. I shall refer to the Community in the pre-1993 period, and to the European Union in the post-1993 period.

2. "Mixed agreements" are often complex. For example, when the London Amendment to the Montreal Protocol was negotiated in June 1990,

The EC had competence for negotiating the percentage reductions in the quantity of ozone-depleting substances that could be produced, because of the existence of an EC Regulation covering this. However, the EC had no competence for the decision to establish a fund to assist Third World countries to obtain the more expensive alternative substances and technologies, and on this point the Member States acted on their own. (Haigh 1992b:241)

3. However, regulations approved by the Council of Ministers typically need penalties associated with them in order to be effective. Given that criminal penalties cannot be legislated at the Community level but, rather, remain in the hands of national governments, national action will be necessary if the regulation is to have sanctions attached to it.

4. Although a case was brought against France for noncompliance with CITES, one highly placed interviewee stated that the generalization stood "because from the Community's point of view, the regulation on CITES is not implementing CITES in that sense because the EC is not an official contracting party."

5. The first "mixed agreement" was the Geneva Convention on Long-range Transboundary Pollution (1979).

6. Enders and Porges (1992:137) point out that

the game between the EC and the United States has two levels—the national level and the industry level.... For many years, the industry resisted any form of regulation of CFCs. Unilateral government regulation of CFCs in one entity led to the capture of market share by producers of the other entity—as the domestic industry argued after the United States' aerosol ban in 1978. Once regulation was inevitable by the mid-1980s, only coordinated regulatory action taken simultaneously by the governments where major consuming markets are located was acceptable to producers.... Quantitative limits on CFC production are a form of regulation preferred by the industry.

7. The Copenhagen Amendment agreed to phase out CFCs on January 1, 1996, with a 75 percent reduction, based on 1986 levels, by January 1, 1994. The EC had proposed an 85 percent reduction. Regarding carbon tetrachloride and halons, the EC proposed tougher restrictions than were adopted, while the restrictions on methyl chloroform were "in line with EC proposals" (Environmental Data Services 1992:13).

8. ICI, the British producer of CFCs, supported the EC's tight deadline even before Du Pont decided to halt all CFC production in developed countries by January 1, 1995. Thus, the position of key European producers has shifted rather dramatically (Zurer 1993).

9. During the negotiations, the United States insisted that all the member states should become contracting parties. This condition was not included in the final text but is reported by Winfried Lang (1988:106–108).

10. The Copenhagen Amendment to the Montreal Protocol limited HCFCs, beginning on January 1, 1996, to 3.1 percent of the calculated level of consumption of CFCs and HCFCs in 1989, and included a time scale for phasing down from 35 percent on January 1, 2004, to total elimination on January 1, 2030.

11. SAFE (Sustainable Agriculture, Food and Environment) Alliance, an environmental lobby, complained that the decisions about methyl bromide were far from satisfactory. An earlier version of the legislation had proposed cutting methyl bromide usage by 50 percent,

but that proposal was subsequently changed to a 25 percent reduction by 1996. SAFE complained that then "Ministers diluted it further by postponing the 25% cut to 1998." It claimed that the French forced the postponement in order to protect Elf Atochem, the French manufacturer of methyl bromide (Europe Information Service 1993h:3).

The legislation was formally passed (after Parliamentary review) as Regulation 3093/94 (Council of the European Union 1994).

12. The German position was taken to defend the interest of producers in Germany. Hoechst, a major producer of CFCs (it held 40 percent of the German market and 5 percent of the world market), had already developed a substitute coolant known as 134A (and had built the third largest plant in the world to produce it), which is far more costly to manufacture than are CFCs ("Chemicals: Hoescht..." 1994:390). It had come under such intense pressure from the NGO community in Germany that production of CFCs in Germany ended ahead of schedule. Rudolf Staab, former CFC production director at Hoechst, said the company was pressured to pull out of CFC production ahead of its planned timetable, following a Greenpeace anti-CFC campaign that "had put us under tremendous public pressure to quit CFC production" (quoted in "European Union..." 1994:390).

13. Much the same pattern is found vis-à-vis restrictions on HCFCs. France, Greece, Spain, and Italy did not sign a declaration at the 1995 Vienna conference committing them to phase out HCFCs (Cutter 1995f).

14. It should be emphasized that CITES is considered a very difficult treaty to enforce in any country. As Ann Misch has noted, "CITES agreements are notoriously difficult to enforce" (1992:30).

15. Godelieve Vandeputte argues that "accepting the EC as a Party to CITES would be beneficial because it would allow the real decision maker in the European forum to use the communication channels provided by CITES. Presently, communication between the EC and the CITES Secretariat is unsatisfactory" (1990:260).

16. The Bern Convention offers an interesting contrast. The Convention on the Conservation of European Wildlife and Natural Habitats was signed at Bern on September 19, 1979, but has no EC implementing legislation. As of 1989, France and Belgium had not signed the Convention (House of Lords 1989:8).

17. The regulation has been amended several times to reflect the decisions taken by signatories to CITES (see Vandeputte 1990:248 for a listing).

18. A report by TRAFFIC-Europe on wildlife trade in Greece, for example, reported:

Until now, conservationists have virtually overlooked the role of Greece in international wildlife trade, but have suspected that, as in other southern European countries, such as Spain and Italy, there are enforcement problems. (TRAFFIC-Europe 1992:1)

TRAFFIC-Europe forms part of the TRAFFIC Network, which describes itself as the "world's largest wildlife trade monitoring programme with offices covering most parts of the world." TRAFFIC-Europe receives funding from the Commission of the Community and is considered one of the most knowledgeable NGOs in the field of wildlife trade. TRAFFIC Network receives funding from WWF and IUCN.

19. Angela Liberatore, for example, concludes that parliamentary procedures cannot explain Italy's lack of compliance with EC environmental legislation:

Even more important are in fact the low level of effectiveness of the Italian public administration, its scarce technical expertise ... its fragmentation and its lack of credibility in the eyes of target groups used to obtaining postponements of deadlines or even "condono" (remissions) in the case of tax policy. (1992:6)

20. At the meeting, Doc SC 24.7 was discussed. "Implementation of the Convention in Italy" summarized the infractions by Italy that the Secretariat considered to be serious, and pointed out that Italy had failed to respond to diplomatic notes as well as to the draft report on the alleged infractions. The report noted:

The Secretariat recognizes that Italy is not the only Party in which these problems exist but the volume of international trade taking place in this country is particularly significant, mainly concerning reptile skins ... and live animals. Several repeated infractions have been noted. Despite numerous letters sent to the Italian Management Authority, the situation has not improved. (CITES Standing Committee 1992a:1)

21. This information as well as direct quotes are all taken from CITES Standing Committee (1992a).

22. However, the Secretariat had already identified the Community as a major problem from the point of implementation during the fifth Meeting of the Parties in Buenos Aires (European Parliament 1988:25).

23. That position was a shift from the recommendation it had made in a document prepared for the twenty-eighth meeting of the Standing Committee, Doc. SC 28.5. That document was entitled *Implementation of the Convention in Italy* and included the following passage:

Taking into consideration the opinion stated by the representative of the EEC Commission during the 24th meeting of the SC (that it would be impossible to implement a ban because of the removal of border inspections between EEC member States), the Secretariat proposes that, if it is confirmed that the EEC rules do not allow its member States to suspend trade with Italy, exclusively, a suspension in all CITES trade with all EEC countries be recommended.

Nevertheless, if during the discussions at the current SC meeting, the commitments made by representatives of Italy appear to be serious, the Secretariat is ready to reconsider its position. (CITES Standing Committee 1992b:2)

24. Such a view is not surprising. In the scholarly literature, the ITTA is routinely included in analyses of commodity agreements (Tillotson 1989) but not in analyses of environmental agreements with trade provisions (Esty 1994).

25. The European Parliament has written several major reports on tropical forests. See, e.g., European Parliament 1990a, 1990b.

8

Japan: Consensus-Based Compliance

James V. Feinerman and Koichiro Fujikura

With respect to its adherence to international agreements to protect the environment and/or to conserve the natural environment, Japan has demonstrated an interesting pattern of actions. Originally quite reluctant to join in such international undertakings, Japan was stung by widespread foreign criticism and eventually agreed to participate. However, the domestic circumstances that would have to change in order to ensure Japanese compliance with these new commitments have been slow to evolve.

Perhaps the most publicized example has involved whaling; although Japan has finally acceded to the international ban on commercial whaling, the dietary habits of a significant portion of the Japanese population have been slow to change enough to eliminate the demand for whale meat—even though commercial establishments providing whale for consumption have been banned.

In Japan, national implementation of international agreements to which Japan has consented remains a difficult issue. Part of the problem certainly involves legislation required to effect the international obligation, but much also depends upon the attitudinal shifts and increasing public awareness of environmental concerns in those instances where national legislation may not be required to make international environmental agreements effective.

Finally, despite Japan's active participation in some international treaties (with respect to ozone-depleting substances and tropical timber, for example), an annoying reality is the willingness to pay lip service to international concerns by joining international conventions and other arrangements that require a departure from past domestic practices, then simply failing to implement newly enacted national laws or to propagandize the new regime. Such halfhearted implementation efforts elicit complaints from fellow parties that, when backed up by appropriate threats, may be able to compel compliance.

As this chapter will demonstrate, there are numerous factors that influence the national implementation of, and compliance with, Japan's international treaty obligations in the environmental sphere. These have deep roots in the nation's political and legal culture, beginning with the earliest development of Japan's legal order and culminating in a relatively unusual congeries of modern legal institutions that have evolved since World War II.

Historical Foundations of the Japanese Legal System

The early legal system and law of Japan were, like those of other cultures, closely tied to the observance of an indigenous religion. According to Hiroshi Oda (1992), in the archaic period, communities were organized by kinship, forming clans called *uji*. The *uji* chieftains, serving the functions of both king and priest, governed their clans by means of religious rituals. By the end of the fifth century, the *uji* controlling the Yamato region (Kyoto–Nara area) emerged as a dominant political power and exalted the deity of their ancestry by the establishment of a Shinto shrine (Van Wolferen 1989). The emperor became a religious as well as a political leader; law and religion were, in this era, inseparable.

Once contact was made with the Chinese, the Japanese ruling class became eager to learn from China, the most advanced civilization at the time. Chinese influence on the early Japanese legal system and culture was of great importance. In the sixth century, Buddhism was imported from China. The ruling class incorporated it into its religious practice and officially supported it. Such incorporation of a new religion was not perceived to be incompatible with Shintoism; rather, the ruling class used this new religion together with Shintoism as an instrument to endorse the existing order by promoting popular obedience.

In the seventh century, Chinese legal codes (called *ritsuryo* in Japanese) were introduced into Japan from the T'ang dynasty and became the basis of the Japanese political system. The *ritsuryo* consists of penal statutes (*ritsu*) and administrative regulations (*ryo*). These codes were adapted rather selectively and gradually, because some of the Chinese tradition embedded in the *ritsuryo* system was not easily transferable to the existing practices and patterns of political authority in Japan.

One example of such practices was the governance based on kinship ties rather than on Chinese-style merit-based bureaucracy. As a result, the selective adaptation of the *ritsuryo* system served to reinforce the domination of a clan-based aristocracy in Japan. In addition, John Haley (1991) has noted that the Chinese notion of imperial legitimacy derived from the Mandate of Heaven was incompatible with the Japanese imperial institution, because incorporation of the Chinese interpretation would imply political submission by the Japanese imperial family to universal Chinese sovereignty. Further, the ruling class enforced *ritsu* much more leniently, partly due to the influence of the prevailing Buddhist value of mercy and partly as a result of a lesser concern with mandating exact requital for specific crimes.

The *ritsuryo* codes were nominally in effect as Japan's legal foundation until the mid-nineteenth century. However, in Oda's recounting of Japanese legal history, as a new governing group came to power, first the nobles (*kuge*), then the proprietors of large estates (*shoen*), followed by samurai warriors, each promulgated their own laws to supplement the *ritsuryo* or to fill the gaps between the codes and reality.

In 1868, the Meiji government succeeded the Tokugawa regime. The opening of Japan to the West necessitated the establishment of a new legal system in order to ensure its independence and also to have the Western nations eliminate from bilateral treaties the provisions for consular jurisdiction and for control over Japanese tariffs and foreign trade policies. The leaders of the Meiji government added to their agenda the drawing up of a new constitution and other civil and criminal codes. The resulting legislation included the Constitution of the Japanese Empire of 1889, the so-called Meiji Constitution, which was largely based on the Prussian Constitution.

The Meiji Constitution conferred sovereign rights on the Emperor. All governmental organs were controlled by or accountable to the Emperor, except for the elected House of Representatives and the autonomous judiciary. However, it should not be too quickly assumed that the sweeping terms of the Meiji Constitution effectively granted the Emperor absolute powers.

The lawmakers of the Meiji government needed to describe the authority of the Emperor in absolutist terms as exercisable power precisely because the Emperor was not to exercise power nor to be responsible for the consequences of political decisions. However, Haley (1991) maintains that at least the nominal conferral of absolute powers was necessary for the maintenance of the sanctity of the imperial institution as the source of authority and power. In other words, by giving the imperial institution "absolute powers" in the Constitution, the Meiji government ensured the existence of an imperial institution that was incapable of exercising any centralized control. This weakness allowed the military to increase its power in the 1930s, which led to World War II.

After its defeat in World War II, Japan was placed under the control of the Allied occupation forces. Demilitarization and democratization were the primary objectives of the postwar legal reforms. Under the auspices of the Allied occupation forces, radical measures were taken in order to eliminate the perceived causes of the war. Oda (1992) particularly notes the following developments: women were given the right to vote; education in Shintoism and Confucian ethics, which provided justification for Japan's aggression before and during World War II, was abolished; major labor laws were promulgated to enhance the rights of the workers; and business conglomerates (*zaibatsu*) were dissolved.

In 1947, the new constitution drafted by the Allied occupation forces, containing a strong American influence, was promulgated. In article 1 it proclaimed that sovereignty rested with the people of Japan, not with the Emperor. That same article gave the Emperor formal status as a "symbol" of the state, devoid of any political power. The new constitution also provided for Japan's renunciation of the sovereign right to wage war (article 9). Further, Oda (1992) notes that the new constitution incorporated an extensive bill of rights. In 1951, the peace treaty was signed with the Allied nations, allowing Japan to return to the international arena.

Governmental Structure in Postwar Japan

Modeled along the lines of the American political system, except for maintaining the prewar cabinet system, the power under the new Japanese constitution is distributed among three branches of the government: legislative (Diet, comprising the House of Representatives, or Lower House, and the House of Councillors, or Upper House), executive (the cabinet), and judicial.

Article 41 of the new constitution made the Diet the sole lawmaking organ of the state. Although the two houses of the Diet are given an equal formal status, the House of Representatives plays a more significant role in two respects: (1) under article 59 it may override, by a two-thirds vote, the opposition of the House of Councillors, and (2) under article 60 it controls the budget. The cabinet is headed by the Prime Minister and is responsible to the Diet.

Article 67 requires the Prime Minister to be chosen from among the members of the Diet. However, because of the power given to the Lower House in the appointment of the Prime Minister by article 67, the Lower House's decision, in effect, is determinative. Article 67 also provides that the decision of the Lower House prevails if the two houses cannot reach an agreement or if the Upper House fails to vote on the appointment of the Prime Minister within ten days after the Lower House votes. Beneath the cabinet are ministries, commissions, and agencies that plan and draft most of the bills passed by the Diet.

From the time of the Liberal Democratic Party's (LDP)'s establishment through the merger of the Liberal Party and the Democratic Party in 1955, the LDP effectively controlled the Diet until the collapse of one-party rule in the summer of 1993. The LDP's main supporters are from rural districts. LDP candidates projected themselves as the only ones who could effectively obtain funding for their districts from the central government, and usually succeeded in getting elected. Because the LDP had maintained its majority in the Lower House, the Secretary-General of the LDP always served as Prime Minister. Thus, the selection of the Prime Minister ultimately depended on the power struggle among factions (*habatsu*) within the LDP.

The LDP is unique in its special relationships with business management and with the bureaucracy. Many elite bureaucrats who draft bills enacted by the Diet become members of the LDP and enter politics. Others often join the top management of major corporations after their official retirement from the government. As documented by Frank Upham (1987), these LDP members work hand in hand with business and bureaucrats, using the ties developed during their government service.

Some future changes are expected, in part because many of the newest LDP members are not former bureaucrats but sons (or even daughters) of politicians who have inherited their fathers' districts, and in part because the LDP is no longer certain to be the controlling party in the Diet.

However, the LDP still has a continuing relationship with bureaucrats in the process of drafting bills or formulating policy, and still has the largest number of seats in the Diet. Further, the founders of the Shinseito and Japan New Party, the most influential parties forming the new coalition government, came from the LDP. Aside from their views and policies on political reform, the views held by these new parties and those of the LDP may not be very far apart. Thus, the political influence of the LDP cannot be dismissed as negligible; the tripartite structure mentioned above still needs to be probed in order to analyze Japanese politics.

The Bureaucracy: Consensual Governance. The importance of the role the Japanese bureaucracy plays in the political system cannot be overstated. According to Karel van Wolferen (1989), everyday policy-making by the government is mainly controlled by the elite bureaucrats, especially those in the Ministry of Finance, Ministry of International Trade and Industry (MITI), Ministry of Construction, and Ministry of Posts and Telecommunications. They decide on governmental policies through consensus-building among themselves, industry leaders, scholars, and other parties who may be affected by such policies. The power relationships among various ministries and parties are often keys to understanding the implications of bills, regulations, or administrative guidance.

The explanations for this consensus-oriented political process have been provided in terms of Japanese history, culture, social structure, and political control. John Haley (1978, 1991), for example, has argued that Japan's consensual governance is a product of institutional history and cultural environment. First, he claims, the incorporation and replication of the Chinese administrative system (*ritsuryo*), narrowly defined as penal and administrative regulations, helped Japanese customary rules to exist as an "extralegal" system that the adjudicating authorities either recognized as binding and enforceable or disregarded. Second, the Japanese relied on customs and conventions formed through consensus within the society, because there were no universally applicable moral or ethical standards. Further, the lack of formal regulatory enforcement throughout the country helped to establish small communities called *mura*, where villagers avoided interference from the central government and enjoyed relative autonomy by settling conflicts through their submission to consensual community resolution.

Although this system originated in premodern Japan, the consensual type of dispute resolution and its implications for governmental power continue to resonate in present-day Japan. The consequence of such consensus-oriented government is a government with authority to command but without power to coerce the implementation and enforcement of those commands.

Under consensual governance, informal resolution of disputes, which often takes the form of institutional mediation, is given preference. By allowing for particularistic solutions based on consensus, informality in dispute resolution eliminates governance by an abstract formulary of universal rules and minimizes the possibility of one group's challenging the dominant social consciousness. At the same time, this assures at least that the bureaucrats will maintain their policy-making leadership in the face of need for social change, although these elites may be neither monolithic enough to agree on detailed social objectives nor powerful enough to achieve such objectives against the will of society.

The result of consensual governance by bureaucrats, in van Wolferen's (1989) view, is the elusive power structure subsisting among bureaucrats, businessmen, and politicians. Because of the informality in governance, the jurisdiction of each ministry is open-ended. Consequently, intense territorial rivalry among ministries and agencies often leads to political deadlock. For those who would attempt to determine the focal point of responsibility and authority in policy-making, the Japanese political system is extremely frustrating. The truth of matter, in the eyes of many foreign observers, is that there is no particular focal point of power or authority under the Japanese political system.

Administrative Guidance. According to Michael Young's landmark study (1984), the primary distinguishing characteristic of national government regulation of industry in Japan is the use of "administrative guidance" (*gyosei shido*), defined as "non-binding administrative action that encourages regulated parties to act in a way that furthers an administrative aim." Despite the existence of analogous practices in all industrialized countries, Young feels that the overwhelming reliance upon administrative guidance in all areas of regulation has led the Japanese, as well as many foreign commentators, to perceive administrative guidance as a peculiarly Japanese institution. Young cites estimates of commentators that 80 percent of all bureaucratic activity in Japan is administrative guidance. John Haley observed: "In Japan informal enforcement is not a process of governing, but has become *the* process of governing" (1991:163).

Administrative guidance has no formal legal effect on the rights of regulated parties. Because the bureaucratic agency is unable to rely upon the courts to enforce its directives, the regulated parties' compliance with administrative guidance is voluntary, at least in a narrow technical sense. To encourage the cooperation of regulated parties, administrative agencies use rewards in the form of government financing or other aid, and punishments in the form of delayed or withheld permits and services, instead of threats of legal sanctions.

The use of extralegal means to enforce administrative decisions has several important results. First, the availability of extralegal enforcement provides officials who resort to administrative guidance with significant authority in determining which policies should be enforced. Second, in Young's analysis, it extends agency power beyond legal limits. Third, it permits collateral enforcement of agency decisions through unrelated administrative action.

Nevertheless, since compliance with administrative guidance is voluntary in reality as well as in theory, agencies must negotiate and compromise with the regulated parties to ensure compliance. Attempts to implement administrative policies without taking into consideration the affected parties' concerns are likely to be frustrated if the cost borne by the regulated parties is too high. As Richard Samuels has aptly concluded, "The Japanese bureaucracy does not dominate, it negotiates" (quoted in Haley 1991:167).

Michael Young also has described how affected parties contribute to the policy-making process through extensive consultations with the government agencies. Although more casual communications are often used, there is also a formal version of these consultations known as *shingikai* (deliberation councils). *Shingikai* are committees "attached" to various ministries that are composed of scholars, journalists, industry representatives, and other relevant groups.

The *shingikai* have come to play such an essential role in policy formulation that the noted scholar Chalmers Johnson (1982) asserts they have usurped the role of the Diet in deliberating on laws. The Diet has routinely rubber-stamped, without changes, laws on such essential matters as taxes and tariffs, and even the privatization of the nation's railroads. However, the active participation of the regulated parties in the policy-making process minimizes competitive disadvantages, the dislocations of regulatory burdens, and the cost of achieving the regulatory goals.

One criticism of the *shingikai* "attached" to the Environmental Agency and other ministries involved in environmental policy is that their membership remains closed, in all but rare cases, to those outside of business, the academy, and government; thus, as Joseph Badaracco and Susan Pharr (1986) argue, victims of pollution, representatives of environmental and conservation groups, and members of the general public are excluded. Effective administrative guidance requires that the number of active participants in the negotiating process be kept to a minimum, because each participant must be satisfied with and consent to the final outcome. As a result, there is a conscious exclusion of NGOs from both formal and informal policy-making in the environmental protection field.

Neil Gross (1989) provides an example of this in the Japanese government's exclusion of environmental groups from a 1989 conference on global warming, pollution, and Third World development that was hosted by the Japanese government and the U.N. Environment Programme. Upham (1987) describes how MITI and the Atomic Energy Safety Commission exclude experts and critics from explanatory hearings regarding nuclear plant siting that are held for residents of the affected areas.

The discretion of the bureaucrats in creating and implementing environmental policy is further broadened by Japanese law concerning the justiciability of administrative guidance and the standing of litigants. Historically, all direct legal challenges to administrative guidance have been held nonjusticiable by Japanese courts. The courts have relied on the formalistic reasoning that regulated parties cannot challenge administrative guidance because compliance with it is strictly voluntary.

More recent doctrine cited by Michael Young (1984) allows courts to restrain agencies from forcing negotiations between parties in cases where the negotiations do not advance goals approved by societal consensus. This "societal consensus" is of course so vague that it provides only a minimal check upon governmental actions. Nor does this doctrine affect administrative guidance in cases not involving governmentally encouraged mediation.

Standing, the question of who may bring suit, is an even greater barrier to private challenges to administrative guidance than is justiciability. The Japanese Supreme Court limits standing to plaintiffs whose private rights and duties are immediately and directly created or delimited by administrative acts. Young's study of administrative law provides the official rationale: legal rights and duties are created only through final and legally formal acts, so the majority of agency actions, including informal threats and arm-twisting, are not subject to judicial review. The entire policy-making process, which Upham (1987) notes is considered an internal government action, also is unreviewable because it does not directly affect the legal rights or duties of private citizens.

Litigation. Dan Fenno Henderson (1965) has provided a masterful study of the history of formal adjudication in Japan, which stretches back over eight hundred years. By the eighteenth century, one discerns the existence of an incipient legal profession and a rudimentary private law system. However, this system existed not so much to provide redress of grievances and enforce legal obligations as to ensure political order and stability.

The importation of aspects of a Western legal system shortly after the 1868 Meiji Restoration brought rapid change to litigation in Japan. In 1872, the new Department of Justice began to organize local courts and established the nation's first law school. The same year, the government recognized the principle of representation in court by legal professionals. By 1881, Japanese procedural law was fully replaced by a Code of Civil Procedure patterned on French and German models.

Soon after Western models were adopted, Japanese citizens, primarily landlords, began to exercise their recently granted legal rights in court. The number of civil cases filed grew sporadically until 1928, when, as Haley (1991) demonstrates with a detailed table, it reached its peak of almost 400 cases per 100,000 persons. The number of cases, according to Haley (1991), declined rapidly immediately prior to and during World War II, and by 1988 had only regained a level of 60 percent of the prewar peak.

Despite this historical aversion to in-court dispute resolution, litigation during the late 1960s and early 1970s, especially the "Big Four" pollution cases,[1] played the crucial role in mobilizing public support for environmental protection and in reversing the postwar Japanese government's low-profile environmental policy. The "Big Four" pollution cases, filed between 1967 and 1969 and decided by 1973, were suits by residents against local industries to recover for personal injuries resulting from water and air pollution, and to enjoin these companies from continuing polluting activities. When the financially and politically weak victims, afflicted with visible and highly painful diseases caused by pollution, first confronted local authorities and the offending corporations, they were met with denials, token payments to ward off litigation, highly biased mediation, and public ostracism.

The courts, according to Upham (1987), were much more sympathetic after public airing of the grievances in these lawsuits, granting record awards to the victims. In deciding for the plaintiffs in the Big Four cases, the courts made several innovations in Japanese legal doctrine. First, the courts imposed a standard of care on pollution-generating manufacturers, drawn from article 709 of Japan's Civil Code, practically equivalent to strict liability. Second, they recognized joint liability among several polluters. Third, strict scientific standards of causation were relaxed and supplemented with clinical and pathological evidence. These revolutionary changes "rewrote the code of operations" for the chemical industry, establishing standards far more stringent than existing regulations.

The doctrinal innovations, however, were minor compared with the litigation's role as a focus for political and other extrajudicial activity. The Big Four cases received the focused attention of the media, which mobilized public and political support for environmental protection, perhaps best documented in the famous study *Island of Dreams* (Ehrlich 1975). The Big Four cases also encouraged and emboldened other victims of pollution; Badaracco and Pharr (1986) record that by 1973, citizen groups were involved in over 10,000 local disputes concerning pollution.

Yet, the momentum of the Big Four cases has not continued into the present period. Upham observed, "Environmental litigation has largely disappeared as a major political or legal factor in national policy, and the central government has recaptured the initiative in environmental planning" (1987:62). Upham sees the decline of environmental litigation as due in part to the success of the government in alleviating the very severe pollution problems that were the root cause of popular discontent. The channeling of pollution disputes into extrajudicial mediation and conciliation proceedings by the Law for the Resolution of Pollution Disputes also has accounted for a portion of the decline. The remaining factor accounting for this decline, however, is systematic bias against litigation in Japanese society and government.

The cause of the relatively low litigation rate in Japan has been the subject of much debate, with commentators usually taking the position that it is due to unique cultural traits or, alternatively, that it is a result of calculated reasoning by Japanese citizens in response to institutional barriers to litigation. Representative literature in support of the latter position includes works by Sanada (1988), Haley (1978) and van Wolferen (1989). A law professor and former dean of the University of Tokyo's law faculty, Hideo Tanaka (1985), makes a persuasive argument that both cultural and institutional barriers to litigation exist.

The most commonly cited cultural barrier to litigation in Japan is the strong expectation of harmony fostered by government and large institutions. This attitude discourages lawsuits, which emphasize the conflict between parties. Quoting from T. Kawashima's study (1963), Tanaka explains the Japanese conception of harmony as follows: "There is a strong expectation that a dispute should not and will not arise; even when one does occur, it is to be solved by mutual understanding" (1985:379).

A second barrier noted by many commentators is that even if a dispute is acknowledged, most Japanese prefer to retain control over the outcome of the dispute rather than surrender their autonomy to a third party; Haley (1978) argues that a preference in favor of mediation is not peculiar to Japanese society but is found in most societies, including the United States.

Finally, these cultural biases against litigation, whether valid or not (Haley argues that cultural factors are largely irrelevant in explaining Japan's low litigation rates), are strongly held by most Japanese, have been internalized, and have become a self-fulfilling prophecy affecting Japan's legal system. Haley concludes that since most Japanese believe they and their society prefer nonjudicial dispute resolution, there is little pressure from outside or within the government to increase judicial resources.

The most serious institutional barrier to private litigation is the interminably slow trial process. Haley's description (1991:8) represents the norm:

> In 1970 over 75 percent of the civil cases decided by the Supreme Court took longer than 3 years to complete from initial docketing for trial to final disposition by the Court. Fourteen percent were decided between 7 and 10 years after docketing and 11 percent took over 10 years to complete.

The long delays in resolving disputes in court discourage many from filing suit and simultaneously add to the attractiveness of extrajudicial mechanisms such as mediation. These long delays have at least one obvious cause: scarcity of judges and lawyers. First, the number of judges is not sufficient to process the current level of lawsuits. Haley documents that the number of judges did not even double between 1891 and 1986, although the population more than trebled and the amount of commercial activity expanded exponentially. Second, compared with their number in other modern nations, lawyers in Japan are extremely scarce. In 1991, there were only 14,420 lawyers in all of Japan.[2] This comes to 11.7 lawyers per 100,000 persons, compared with 273.8 lawyers per 100,000 persons in the United States in 1985.[3]

The scarcity of judges and lawyers is, Haley argues, a result of the government's strict limits on entry into the legal profession. Although the desire to become a lawyer is quite strong in Japan—Haley (1978) notes that in 1975, more Japanese took the judicial examination per capita than Americans took bar examinations—less than 2 percent have passed the examination since 1974; Haley sees the bureaucracy's policy of limiting admission to the legal profession both as a premeditated effort to reduce or contain judicial activity and as a result of the natural Japanese predisposition against litigation.

Another barrier to environmental litigation in Japan, noted by Upham (1987), is the absence of class-action suits and very restrictive standing-to-sue requirements. The cost of pollution damage to individual people is usually quite small, although the aggregate amount may be enormous. Therefore, individuals are deterred from bringing suit because the cost of litigation almost always outweighs the benefit to the individual from winning a suit. In the United States (and a few other countries), class-action doctrine permits a group with similar claims to be represented by only a few members and for lawyers' fees to be borne by the unsuccessful defendant. If such a device is unavailable, potential plaintiffs are likely to be deterred by the high costs and delay of court action, resulting in random enforcement of laws.

At present, litigation plays a relatively insignificant role in policy-making on environmental issues. As a result of the lack of class-action suits, the energy of judicial and citizen activism in the 1970s has dissipated, and environmental litigation has largely disappeared as a major political or legal factor in national policy. Although this situation is due in part to the implementation of successful pollution control policies, Julian Gresser et al. (1981) see the decline in litigation as largely the result of the government's purposeful channeling of environmental disputes from the courts to government-sponsored mediation.

The government's predisposition to enforce mediation and conciliation, as formally enshrined in the Law for the Resolution of Pollution Disputes (*Kogai funso shori ho*), Law no. 108 of 1970 (partially translated in Gresser et al. 1981), and the Law for the Compensation of Pollution-Related Health Injury (*Kogai kenko higai hosho ho*), Law no. 111 of 1973, diverting most pollution disputes from the jurisdiction of the courts to that of the bureaucracy. The administrators' goal in using mediation, according to Michael Young, is to "reallocate bargaining power between the parties so as to assure serious negotiations, but then distance themselves from the process, thereby allowing the parties themselves to make the difficult determinations" (1984:943).

The resolution of environmental disputes through government mediation has proved to be an effective method of compensating pollution victims without the expense and unpredictability of litigation. However, it has had several consequences adverse to the environmental movement. First, mediation, which need only satisfy the immediate parties, prevents local conflicts from providing the impetus for nationwide measures to combat pollution. Second, government-sponsored mediation is inherently biased: "Because all mediators are appointed, trained, and supported by the government and mediation itself is intrinsically ad hoc, the potential for subtle, hidden, or even unintentional manipulation toward preferred government outcomes and policies is undeniable" (Upham 1987:66).

Japanese Adherence to and Practice Under the Treaties

This section considers the implementation of and compliance with the five international treaties. Table 8.1 shows the years in which the treaties entered into force and the years in which Japan acceded to them.

World Heritage Convention

The Convention for the Protection of the World Culture and Nature Heritage was adopted in November 1972 and became effective in December 1975. Japan ratified the convention in June 1992 (which became effective that September) and quickly recommended four sites (and eight other provisional sites) for placement on the World Heritage List. Table 8.2 shows the years of additions of all of the World Heritage Sites in Japan as of December 1996.

Table 8.1
Treaty membership of Japan

Treaty	Year treaty came into effect	Signed by Japan	Ratified by Japan
WHC	1977	1992	1992
CITES	1975	1973	1980
ITTA	1985	1984	1985
LC	1975	1980	1980
MP	1989	1988	1989

Table 8.2
World Heritage Sites in Japan

Year	Natural site	Cultural site
1993	Yakushima Island Shirakami Mountains	Himeji Castle Buddhist monuments in the Horyuji area
1994		Historic monuments of ancient Kyoto (Kyoto, Uji, and Otsu cities)
1995		Historic villages of Shirakawa-go and Gokayama
1996		Hiroshima Peace Memorial (Genbaku Dome) Itsukushima Shinto shrine
Total	2	6

The two original culturally and historically significant architectural structures were approved and put on the list in 1993: Horyuji Temple, a five-story pagoda that is the oldest wooden structure (built in A.D. 607), and Himeji Castle, known as a "snowy heron," built in the early seventeenth century. Two natural sites were also approved that year. Natural sites under the committee's purview include the Shirakami Mountains with their natural forests of beech trees (*Fagus Sieboldi*) and Yakushima Island, covered with ancient cedar trees.

In October 1993, Japan was elected to a committee consisting of twenty-one countries that decides on World Heritage Sites and manages the World Heritage Fund.

Domestic Actions. The news that four sites made the World Heritage List was enthusiastically greeted in Japan. This new distinction may bring more international as well as domestic tourists to those sites, stimulating the surrounding local economies. The two cultural sites are well known and already frequented by tourists; they would seem to enjoy financial independence through the income from admission fees and other public sources. The increasing number of tourists may be absorbed without drastically altering the existing facilities or requiring new regulations.

However, the two nature sites may present a different situation. There are already tensions between groups of local citizens who favor the preservation of the sites and those who expect an economic boom generated by tourism and outside investment. The local governments have prepared plans for the preservation of the sites, and environmental groups have paid close attention to the development of these plans. The debate over how to preserve the two nature sites may expose inadequacies of existing national conservation policy and regulatory measures.

It was twenty years before Japan joined the steadily growing group of treaty signatories as the 125th member. Initially, when the treaty was under discussion at

UNESCO and the IUCN in the early 1970s, Japan did not show any interest. There are no records of involvement of either Japanese officials or NGO members in the treaty formation. Japan supported and signed the treaty when it was proposed at the seventeenth UNESCO meeting in 1972.

Over the following two decades, the Japanese government took no action and, according to a Foreign Ministry official, maintained a "wait and see" attitude. In 1992, members of the Socialist Party and the Communist Party questioned the status of the treaty in the Diet. The director of the Environment Agency responded positively and obtained approval from the cabinet, proposing the ratification of the treaty to the Diet. There was no opposition to its ratification, and the treaty was swiftly approved. The Rio de Janeiro Earth Summit that June certainly was in part responsible for reviving interest in ratification of the treaty.

There are several reasons why the treaty remained dormant in Japan for such a long time. First, it is generally thought that historical and cultural, or natural, heritages in Japan have been well preserved and protected by existing domestic laws and regulations. Second, several government agencies had to be involved and work together in order to ratify the treaty, and none was willing to take the initiative to do so. Third, the realization that decaying historic and cultural treasures in other countries needed international action to preserve them came late in Japan. Such realization has come as a result of the recent Japanese awareness of other global environmental problems.

The treaty was ratified primarily because it was a good and timely act for Japan to take in view of the Earth Summit. Also, it required no new legislation to implement the treaty and no substantial domestic costs, for there already were several domestic laws for the preservation of historical and cultural sites as well as nature sites. The media helped by frequently reporting on decaying cultural sites in Asia and on needs for international efforts to preserve them.

World heritage preservation was not a high priority in Japan in the early 1970s. By 1992, however, the Environment Agency was willing to take the initiative and played an active role in ratifying the treaty. There was an increasing awareness of the fates of many World Heritage Sites in other countries and of the need for Japan to join the international cooperation to preserve them.

In Japan, there are several laws affording protection to domestic heritage sites. The Law Concerning the Protection of Cultural Properties, the Law Concerning the Preservation of the Natural Environment, and the Law Concerning Natural Parks are examples of such laws. No new and additional costs were incurred domestically because of the treaty, although the government provided a small appropriation in 1993 for gathering more data on possible sites to be recommended to the World Heritage Committee.

Effects of the Treaty on Japan. As a member nation, Japan contributes to the World Heritage Fund. Its contribution is set at the amount corresponding to 1 percent of the membership share of UNESCO's annual budget. Recently, Japan's UNESCO share has been the largest, so it is also the largest contributor to the Fund.

The Law Concerning the Protection of Cultural Properties provides for the protection of important buildings of historical and cultural value. On the basis of architectural heritage, shrines, temples, castles, traditional residences, and late-nineteenth-century buildings are designated and preserved as cultural and historical properties. The most valued sites are designated as national treasures. In this category there are 58 shrines, 159 temples, 16 castles, and 20 residences. Himeji Castle and Horyuji Temple, now on the World Heritage Sites List, are among them.

The law also provides a zoning system for areas where clusters of traditional houses present especially good examples of historical architectural styles. By 1993, thirty-two such areas had been designated. Other special laws protect preservation areas in such ancient cities as Nara, Kyoto, and Asuka-mura. These laws regulate activities affecting existing conditions of the designated site, and provide for restoration and maintenance of the properties and their surroundings. The 1992 national budget allocated Y5 billion for architectural property preservation.

The Agency for Cultural Affairs is responsible for the conservation of cultural properties. It also can designate certain animals, plants, geological features, and minerals of special scientific value as "natural monuments" (165 animals and 248 plants were so named as of 1993). Ancient, giant cedar trees in Yakushima have been designated and protected as "special natural monuments."

There are three major agencies and several laws that work together to preserve nature and natural objects. Local governments, whose jurisdictions are often overlapping and overly divisive, are also involved. Preservation and protection laws are complex and ineffective when applied.

The Nature Conservation Law of 1972, amended in 1992, covers natural sites. Under the law, the Environment Agency can fund basic research, designate preservation areas, and restrict activities within those areas. The agency also uses the Natural Park Law and the Birds and Animals Protection Law to preserve nature. The Ministry of Agriculture, Forestry, and Fisheries may also be involved in conserving natural sites under the Marine Resources Protection Law, the Fisheries Law, and the Forest Law.

The Nature Conservation Law sets, in highly abstract and ideal terms, the basic protection policy and policy goals for nature conservation. The law declares as "a unique cultural tradition of Japan" that "man, nature and man's works of art form an organic unity," and that nature is not only a provider of the resources necessary for human economic activity but also an essential element of human life (OECD

1994:81). It recognizes the need for sustainable development of agriculture, forestry, and fisheries, in view of their roles in the conservation of the natural environment.

These idealistic policy goals do not seem to have been translated into specific quantitative targets for nature conservation, and there seems to be no national strategy or plan for nature conservation that is expressed in terms of specific programs and priority tasks having detailed performance measurements.

Historically, Japanese forest management has from the beginning been preservation-oriented. In 1897 the Forest Law, which established a preservation forest system, was enacted. Under that system, the Minister of Agriculture and Forestry and the prefectural governor may designate a preservation forest for the purposes of conserving water sources, preventing soil erosion and landslides, protecting the shoreline, and so on. The designation can be made for national as well as private forests. The areas for preservation were extended markedly in the 1970s; by 1993, they encompassed one-third of all Japan's forests (totaling 8.9 million hectares).

In 1919, the Law to Preserve Historical, Scenic Sites and Natural Monuments was enacted. In 1931, the National Park Law followed. Even before these laws, there was practice of nature conservation by designating certain areas in national forests for special protection. For the purpose of research and preservation, the designation "protected forest" created sanctuaries for birds, animals, and plants.

By 1931, when the National Park Law formalized the practice, 93 designations had been made; the protected areas extended to about 110,000 hectares. In 1989, the protected forest system was revised in order to cover a wider area essential for the forest ecosystem. By 1994, twenty-one forest ecosystems had received protective designations, including the Shirakami Mountains and Yakushima Island, two natural sites on the World Heritage List. The number of protected forests had increased to 779 (totaling 380,000 hectares).

The Environment Agency has overall policy responsibility and administers the national parks, wilderness areas, and nature conservation areas. The agency also oversees, in cooperation with local governments, quasi-national parks and some other protected natural areas. The Forest Agency administers the national forests, especially the protected reserves within the national forests, for the purposes of soil and water conservation, health preservation, and scenic zone protection.

Japan has a system of twenty-eight national parks covering 5.4 percent of its territory. Most national parks were designated for their scenic qualities. By the late 1960s, the area of national parks began to level off at around 2 million hectares. Since the late 1970s, only two national parks have been designated.

A system of 55 quasi-national parks has been designated by the government under the Natural Parks Law, and 299 prefectural nature parks have been designated at the local level. These parks are managed by local governments. Natural areas in suburban and rural areas are subject to national, and possibly local, land planning. However, "Natural areas, particularly in areas on the periphery of cities, tend to be under pressures that are often poorly controlled by fragmented physical planning mechanisms" (OECD 1994:84).

What has been happening in Yakushima Island and the Shirakami Mountains, the two World Heritage nature sites, since their designations may be instructive. Even before Yakushima Island was placed on the World Nature Heritage List, tourism had hit the island. In 1992, 240,000 persons visited this remote island. The number of visitors has doubled since 1989, mostly due to improved ferry services. Ancient cedar trees along the forest passes had been damaged by the increased tourist traffic.

Local people's reaction to the World Nature Heritage designation has been mixed. The island has no strong local industry, and its economy had previously depended chiefly on public works spending. Some welcomed the economic boom brought by tourism, while others feared the disruption of previously untouched nature.

The local government prepared a preservation plan for the nature site and enacted an ordinance to protect the environment while accommodating tourism, and considered imposing an environmental charge on tourists. Some local residents fear that new measures and regulations to preserve the site may become too restrictive of local economic activities. A similar tension between protecting nature and promoting tourism exists in the Shirakami Mountains.

The Environment Agency decided to institute by 1994 a certification system for nature education specialists to serve at national parks and nature preservation areas. There are about 100 guides at the visitor centers of various public parks, but most of them are not professionally trained. The new system is expected to provide human resources for nature appreciation and preservation.

CITES

Japan signed the Convention on International Trade in Endangered Species of Flora and Fauna (Washington Treaty) on April 30, 1973. The treaty became effective in 1975 when ten countries ratified it. It was five more years before the Diet ratified the treaty on April 26, 1980. During this period, the government tried to persuade affected industries, many of which were small manufacturers and traders engaged in old and traditional crafts and trades, to bring their practices into conformity with the Treaty. After making necessary adjustments and realigning those industries, the cabinet decided to accept the treaty obligations. Thus, the treaty came into effect on November 4, 1980.

At that time, Japan enacted no new laws to implement the treaty. Instead of directly regulating domestic transactions involving endangered species, a regulatory target was set to stop imports at their ports of entry. Customs officers under the Ministry of Finance have been responsible for regulating the listed species subject to export/import control like any other products. Japan's export/import control system is based on the Foreign Exchange and Foreign Trade Control Law.

In October 1984, a round-table administrative conference was held among the agencies involved to discuss and coordinate treaty implementation. Attending the conference were delegates from the Environment Agency; the Foreign Ministry; the Ministry of International Trade and Industry (MITI); the Ministry of Agriculture, Forestry, and Fisheries (MAFF); the Ministry of Finance; the Ministry of Health; and the cabinet secretariat.

Given its international reach, MITI has become the chief agency regulating international transactions in endangered species through its control over licensing procedures. MAFF exerts similar control over marine life. The Environment Agency and MAFF serve together as a scientific council to the control agencies, giving advice on whether or not particular transactions may affect the protected species.

Domestic Actions. Japan imports a large volume of wildlife, second only to the United States. In spite of—or because of—that fact, Japan's record of protecting wildlife has been weak and unimpressive.

On June 2, 1987, the law regulating domestic transactions such as the sale and transfer of possession of "rare wildlife" was enacted; it took effect on December 1 of that year. The law is intended to prohibit domestic transactions involving designated rare wildlife. The wildlife species named in the law are the same as the endangered species listed in appendix I of the treaty. It should be noted that transactions in endangered migratory birds have been regulated by the Law Concerning Special Birds since 1972.

On June 5, 1992, the Species Preservation Law was enacted; it took effect on April 1, 1993. This law incorporates the 1987 and 1972 laws, and strengthens regulatory measures. Endangered species listed in the treaty's appendix I, with eleven specific exceptions, are subject to regulation. The regulatory scheme based on export/import control is retained, and is used for regulation of the species listed in the treaty's appendices I–III.

Issues concerning the administration of CITES, illegal imports, and the quality of border control are the responsibility of MITI, which consults with the Environment Agency as a "scientific authority." NGOs such as the World Wide Fund for Nature Japan also play a role in monitoring trade in parrots, cacti, crocodile skins, rhino horns, and sea turtles, surveying pet shops for illegal wildlife and investigating whether zoos abide by the provisions of CITES.

Implementation in Action and Attendant Problems. The international community has criticized Japan's lax implementation of CITES. Many of the factors discussed below may have contributed to Japan's slow pace in treaty implementation.

First, it was politically difficult for Japan to coordinate the affected parties, many of whom were small in size and engaged in diverse trades. It took quite a while for those parties to phase out operations or to make the necessary transitions. Under these circumstances, it was not possible for the government to enact laws imposing a total ban on domestic transactions of endangered species listed in the treaty's appendix I. The government could only utilize the export/import control mechanism to stop endangered species from flowing in.

From the beginning, there was also an ideological split, internationally as well as domestically, over the basic policy goals and appropriate methods of implementing the treaty. One group held that protection efforts are best promoted through a total prohibition on trafficking. The other group, however, believed that a strategy of "sustainable use," with proceeds to be used for the protection and preservation of species, is the best route to achieving the treaty obligations. Apparently, Japan has taken the latter stance as its policy guide, and has stuck to a program that results in the minimum implementation of treaty obligations. Japan has maintained several exceptions to the treaty's appendix I, and at times has increased the number of exceptions, although the number was lowered to six in 1992.

Another factor has been the fact that before 1987, there was no method for regulating endangered species brought into Japan, legally or otherwise. Under the Species Preservation Law, transactions in endangered species are prohibited. Yet, transactions in these species are permitted in the following cases: (1) species obtained before the treaty became effective in Japan; (2) species artificially bred or farmed for commercial purposes; (3) one of the parties to the transaction is a government agency.

In addition, the protected species were not clearly defined under the Species Preservation Law. The law uses the term "rare species" and lists about 800 to be regulated. The list of designated rare species corresponds to the treaty's appendix I. However, some endangered species are not protected under the law because Japan has taken exceptions to the treaty's appendix I. Also, the law narrowly limits its scope of regulation to an individually identifiable wildlife species.

At the same time, it has been difficult for Japanese officials to distinguish between legal and illegal transactions in endangered species. For practical purposes of enforcement, the law regulates only living wildlife and such easily identifiable items as stuffed birds and animals, eggs, etc. The law covers neither the use of portions or derivatives of endangered species nor products using them. Shop owners dealing in wildlife are generally aware of the law, but many have confessed that they are confused about which species are regulated.

In addition, the legal framework for implementing the treaty may be too complicated and not well integrated. Two different regulatory schemes have operated. One has aimed at controlling export and import, while the other has regulated domestic transactions. Also, several government agencies have been involved in various aspects of treaty implementation. They have not been well coordinated. The implementation of the treaty has been put largely in the hands of customs officers, MITI officials (export/import control), and the Environment Agency (domestic transactions, registration, and certification).

It is interesting to note that in 1992 the General Affairs Agency, which oversees the workings of administrative agencies, found it appropriate to investigate the overall administration of protection for endangered wildlife in Japan. One of the foci of investigation was treaty implementation. The report, *The Problems of Protection Measures for Endangered Species* (1993), found several problems.

First, according to the General Affairs Agency, MITI's supervision of export/import control has failed to collect sufficient data concerning the numbers and types of imported species, the countries of origin, and any numerical shifts in export/import patterns. Those statistics are essential for policy planning as well as for the evaluation of the effectiveness of the export/import control measures.

Second, basic information about the treaty and protected species has not been widely disseminated. Japanese tourists returning from overseas trips are ill-informed about endangered species, and leaflets supplied are not at all helpful for identifying prohibited species.

Third, the report notes that customs officers were provided with manuals for identifying the regulated species under the treaty. However, the manuals were written in English, and the pictures of wildlife were printed in black and white. (As of 1994, customs officers had received a multivolume reference manual in Japanese, written by the Environment Agency.)

Fourth, the number of endangered wildlife stopped at customs offices for lack of proper permits or documentation jumped from 1,329 in 1988 to 2,355 in 1990. An importer of such wildlife was supposed to return it at his own expense to the exporting country (Species Preservation Law, art. 4). However, when, as in many cases, it was not feasible to do so, the importer could voluntarily abandon the contraband. MITI would take custody of the surrendered wildlife, and ask university farms, zoos, and aquariums to care for it. MITI would pay the costs of caring for animals and fish.

Finally, the General Affairs Agency report noted that the Environment Agency was responsible for reviewing applications for registration, issuing certification, and on-site inspection of shops dealing in regulated rare wildlife. The agency, however, had few enforcement officers and no branch office outside Tokyo.

The General Affairs Agency made several recommendations. For those species listed in the treaty's appendix I, but not regulated under the 1992 law, it suggested that regulatory measures be devised in consultation with fifteen other agencies. It also suggested that registration and certification responsibilities be delegated to private organizations. Finally, it concluded that more extensive dissemination of information about the contents of the treaty and the 1992 law was needed.

Summary. Japan's performance in complying with the CITES obligations has not been satisfactory. There was virtually no implementation of the treaty in relation to domestic transactions involving endangered species before 1987. Until then, the treaty's implementation was largely left to customs officers, who tried hard to intercept the species listed in the treaty's appendices without adequate manuals to identify them. With the enactments of the 1987 and 1992 laws, Japan started to regulate domestic transactions involving rare species.

As a result of exceptions and differing interpretations of rare species under these laws, it is difficult to judge what species are prohibited, protected, or permitted, and what species are to be registered or certified.

MITI's export/import control involves a complicated process of going through several offices in other agencies, including an office of the exporting country that gives a prior verification. The Environment Agency is overburdened with registration and certification. It has virtually no enforcement staff to do the on-site inspection that the law authorizes it to do. One solution is to delegate more power to local governments and get them involved in overseeing transactions in wildlife. Private-sector organizations can be utilized to carry out the registration and certification of regulated species.

The public has not been well informed about the need for protecting endangered or rare species. The government agencies simply cannot control numerous transactions involving diverse species. Channeling essential information to raise public awareness is important. International meetings on wildlife held in Japan undoubtedly have made a considerable impact on public awareness. The eighth CITES convention, held at Kyoto in 1992, was a good example. It was attended by 1,000 people from 105 member countries and a number of observer countries, as well as numerous NGOs. The convention enjoyed extensive media coverage, largely because of the heated discussion over the fate of whales and black tuna, which are important to many Japanese.

NGOs can play a significant role in informing the public and defining issues that the government may ignore. It is worthwhile to note that the TRAFFIC Japan office was established in 1982, and the Japanese version of the Red Data Books (detailing trade in endangered species) was published in 1989 through a cooperative effort between environmental NGOs and the Environment Agency.

After an initial period of halting compliance and confusion, Japan has taken a step toward the protection of domestic rare species by enacting legislation. It is also a much-needed step toward a fuller compliance with the treaty.

ITTA

Domestic Actions. The International Tropical Timber Organization (ITTO) was established in 1980 for the purpose of developing tropical forests for timber products while preserving forest resources. By 1994, the International Tropical Timber Agreement (ITTA) had been signed by twenty-seven producing/exporting countries and twenty-five consuming/importing countries. When the ITTO was created to implement the agreement, Japan actively sought, and in 1986 succeeded in getting, the administrative headquarters of the ITTO established in Yokohama. The political and diplomatic story behind convincing the ITTO to come to Japan has yet to be revealed. Japan was active in the U.N. Conference on Tropical Timber. When the ITTO was formed, Japan saw an opportunity to take a leading role in an international organization that would deal with trade and environmental problems close to its shores. Undoubtedly, there was every advantage in hosting the headquarters, because Japan was keenly interested in securing continuing supplies of tropical timber to meet its ever-growing domestic consumption, and in facilitating smooth transactions between producing and consuming countries.

Japan sought to watch over her own interests and to succeed in these endeavors in the face of deepening international concern with overexploitation of tropical forests. Yokohama, a major port city, offered a good location near Southeast Asia, where a vast expanse of tropical forests existed and where active trading in tropical timber was under way. The city provided the ITTO with spacious modern offices in one of its international convention facilities. The ITTA has served Japan's need to satisfy the domestic demand for tropical timber. The construction industry consumes a large amount of plywood, and paper manufacturers depend on imported wood pulp.

International trading companies in Japan have taken full advantage of the terms of the ITTA, freely investing in and buying up tropical timber. The ITTA imposes few obligations on its signatories to perform domestically, except for reporting of the amount of tropical timber exported/imported. The Japanese government has duly filed reports and has exercised virtually no control over imports of tropical timber.

Japan imported a modest amount of timber (2.48 million cu. m.) in 1955 and completed liberalizing the import of timber and wood products by 1964. In 1993, the total amount of imports was thirty-three times the amount in 1955. Since 1993, the annual total domestic consumption of wood has remained in the range of 100–110 million cubic meters. About half of the total amount of tropical timber imports is used for producing lumber, about 40 percent for pulp chips, and the rest for plywood and other products.

Of all the tropical timber exported from developing countries in 1990, Japan purchased 32 percent. Japan is by far the major importer of tropical timber on a per capita basis. In 1993, more than three-quarters of domestic timber demand (76.4 percent) was met by imported foreign timber. Japan imported softwood from North America (38.3 percent) and Russia (4.4 percent), and tropical timber mostly from Malaysia and Indonesia (16.2 percent).

The ITTA did not require any legal measures for its domestic implementation. Importing of tropical timber is subject to well-established export/import control. Japan's Forestry Agency keeps a record of the amount of tropical timber imported and reports the results to the ITTO as required.

Under the ITTA, Japan has succeeded in securing sufficient supplies of tropical timber while contributing the largest share of funds annually to the ITTO's operations and activities. The organization expects its member countries to pay membership shares as well as to make voluntary contributions. Japan's membership share and its contributions are the largest of the ITTO's members. In 1994, Japan paid Y93 million as its membership share (34 percent) and made a Y1.652 million voluntary contribution (61 percent). Ironically, however, Japan's success in securing enough imported timber at low cost has made its domestic forest industry structurally and chronically depressed.

By the end of World War II, Japanese forests were nearly exhausted and could not meet the growing domestic demand for wood. Afforestation efforts have been instituted, and Japan now has forest covering more than two-thirds (67 percent) of the total land area. More than 40 percent of the forest land consists of man-made forests. Most of the trees planted (80 percent) in forests since the 1950s have not matured enough to be harvested. Roughly speaking, 60 percent of Japan's forested land is privately owned, 30 percent is nationally owned, and the other 10 percent is under some form of public ownership. Most forests grow on steep mountainsides. Almost all privately owned forests are very small (less than five hectares) and are difficult to manage efficiently.

While domestic timber prices generally have remained low since the late 1970s, forest management costs have increased dramatically. Using 1965 as the base year (index = 100 in 1965), the price indices show the following trend: afforestation costs are 929 and harvesting costs are 502, while the price index for cedar timber, the most standard and common tree in Japan, is 70.

The Japanese forest industry faces a difficult situation caused by high labor and forest management costs, and low prices for domestic wood products. The forest labor force is rapidly aging and quickly dwindling in numbers (77,000 persons in 1994, one-fifth of the number in 1965). Domestically produced timber cannot compete with imported wood cut from naturally grown forests with virtually no management costs.

The Japanese distribution system for domestic timber is complex and inefficient. Multiple parties—wholesalers, intermediaries, and distributors—are involved in the distribution process from forest owners to end users of timber products. Parties involved in business transactions are mostly small or medium-sized entrepreneurs. Most products are made for diverse, specified uses, and usually in limited quantities. Generally, the parties involved are not innovative in developing and marketing their products. All these factors add to high distribution costs of domestic wood products.

In contrast, imported timber can be traded in large lots and distributed widely in great quantities. A lumber company can buy large lots of imported timber directly from a trading company. The quality of imported timber is fairly consistent, and the timber can be used to make large numbers of standardized products. These products can be distributed through dealers belonging to the same consortium as the importing trading company. Large-scale transactions and a simple distribution process further reduce the prices of imported wood products.

National and public forests are in no better condition than privately owned ones in terms of being productive and efficient. National forests have been ill-managed, with a huge deficit accumulating in each successive fiscal year since 1975. The government insists that the national forest management should be financially independent and self-supporting, but has not developed an effective policy to restructure and promote the domestic forest industry. Private forest owners have so far failed to organize as strong a lobby as rice farmers have against imports.

Ironically, the Japanese forests have remained sustainable. Japan's overall forest resources have increased since the 1950s. Annually, the total amount of forest cuttings remains smaller than overall forest growth: "Forests are sustained for future productive use" (Forestry White Paper, 1994:28). Of course, this is made possible at the expense of foreign forest resources and through a lack of government action.

Strengthening the ITTA. Throughout the 1980s, the market played a major role under the ITTA scheme, and exploitation of tropical timber continued. In Japan, it was essentially left to the market to dictate the amount of imported wood needed to satisfy domestic consumption. No meaningful overall policy for coordinating sustainable forest development, either internationally or domestically, has yet been established.

Notwithstanding the existence of the ITTA and creation of the ITTO, tropical forests in Asia have been decreasing rapidly and steadily since the late 1980s. In 1990, the eighth ITTO Council meeting set a goal of trading only tropical timber produced from sustainably managed forests by the year 2000. The ITTO sent a team to Sarawak, Malaysia, to investigate the possibility of sustainable forest management. The team found that tropical forests in Sarawak would last only eleven more years if exploitation continued at the current rate.

Yet even as of 1997, Japan has no national policy to limit the use of tropical timber despite the scale of domestic consumption, current and predicted. The news media have occasionally covered the worsening situation of tropical forests. Several groups of concerned citizens and lawyers have visited affected forests; all reported on their critical condition and the need for action. However, no strong citizen movement for conserving tropical forests has appeared, though some attempts have been made to reuse plywood in the construction industry and some local governments have set a goal of reducing the use of plywood by 30 percent in public works and construction.

The current ITTA expired at the end of March 1994. The ITTO members negotiated through 1993 to renew and to strengthen the agreement. Consuming/importing countries proposed to formalize the year 2000 goal in the new agreement. Producing/exporting countries opposed it, insisting that it would be discriminatory to regulate only tropical forests. The member countries finally reached an agreement at the meeting held in Geneva in January 1994.

The goal of trading only tropical timber produced from sustainably managed forests was incorporated into the 1994 formal agreement. The renewed agreement also contains a provision imposing a duty on exporting countries to replant and revive their forests. The preservation side of the original agreement is reemphasized. The consuming/importing countries have agreed to establish a fund to finance the costs of sustainable management of tropical forests. Japan ratified the agreement at the Diet session of the summer of 1994.

Apparently, ITTO now places more emphasis on preservation-oriented projects than on development-oriented ones. The organization funded 220 such projects in 1994, compared with 11 in 1985.

Japan adopted the 1994 ITTA and participated in negotiating as well as formulating standards and indicators for sustainable forest management. She became an active member of the Commission of Sustainable Development of the United Nations and other intergovernmental conferences on forest management. Japan has hosted several conferences in coordination with the ITTO.

In 1991, a conference of senior foresters was held in Yokohama to discuss the action plan for sustainable forest management. It was attended by tropical forest specialists and management officials from forty-two countries and seventeen international organizations. The conference adopted the Yokohama Declaration on Forests and Forestry. Similar international conferences of senior foresters followed at Hiroshima in 1992, at Hokkaido in 1993, and at Miyazaki in 1994. These conferences, it is hoped, will contribute to forming an international consensus on dealing effectively with tropical forest management issues.

Summary. Under the original agreement, the productivity of tropical forests, rather than their preservation, was emphasized. Securing an abundant supply of tropical timber seems to be a main concern of importing countries. Producing countries are more interested in exporting timber than in preserving their forests. The market mechanism is at work, and excessive exploitation has resulted. The ITTA has proved ineffective in coping with market forces.

Japan could take a more active role in the ITTO by proposing positive measures to preserve tropical forests. For example, a standard manual (or guidelines) for sustainable forest management is needed. Environmental impact studies should precede the granting of overseas development aid for any project involving the development of tropical forest areas. More technical assistance is called for, and more bilateral and multilateral cooperation is necessary. Even before the year 2000, importer/buyers could agree among themselves to buy only from sustainably managed forests. A charge might be imposed on imported tropical timbers in order to generate funds specifically for the preservation of tropical forests.

In sum, sustainable forest management is as much an international issue as a domestic one. The Japanese government has never initiated any policy to stem its domestic overconsumption of tropical timber. Now, Japan needs to develop a domestic as well as an international forest management policy to coordinate more preservation-oriented measures with the market-oriented commodity demands.

London Convention

Japan is a contracting party to the London Convention. Hazardous material is dumped by Japan in conformity with this convention in five areas: one in the Sea of Japan and four in the Pacific Ocean. Surveys of the dumping sites have indicated no significant degradation of the marine environment.

A consultative meeting of the Convention agreed that the dumping of industrial waste should cease by December 31, 1995. Japan has been studying alternative disposal methods with a view to adopting stricter regulations on dumping industrial waste (4.36 million tons in 1992, of which 2.2 million tons was inorganic sludge). The key concern is to ensure that the environmental effects of alternative methods are not worse than effects avoided by a ban on ocean dumping. In addition, 10.6 million tons of dredged sand and 3.2 million tons of municipal waste have been dumped yearly in the ocean.

Domestic Actions Prior to the Treaty. Japan promulgated the Marine Protection Law in 1970. It was one of several major pieces of legislation enacted in the "Pollution Diet," in response to a crisis of worsening air and water pollution. In single-minded pursuit of industrial growth, many factories carelessly discharged hazardous wastes, such as mercury, cadmium, and oil, into streams and bays. A large number of people living in heavily polluted coastal areas became afflicted with strange illnesses. Their health was impaired by eating contaminated fish and shellfish.

Victims brought their lawsuits beginning in the 1960s, and the courts awarded them damages. The public demanded that the government impose strict liability on industries discharging waste and take stricter measures to protect the environment. During the United Nations Conference on the Human Environment, held in Stockholm in 1972, the appearance of the group of Japanese afflicted by mercury poisoning made a profound impression on the participants. It was the victims' first experience in learning that action by NGOs could be effective internationally as well as domestically.

Compliance with and Implementation of the London Convention. The Convention on Marine Pollution by Dumping Wastes and Other Matter (London Treaty) was adopted in 1972 in London. That same year, the Marine Protection Law took effect in Japan. In June 1973, Japan signed the London Treaty, which came into effect in 1975. In 1978, at the third convention of signatory countries, the treaty's appendix was amended. Consequently, Japan amended the Marine Protection Law in accordance with the London Treaty in 1980. The Diet ratified the treaty and the cabinet decided on implementation measures. In November 1980, the treaty became effective in Japan.

In 1986, the government agencies responsible for implementing the treaty were mobilized. The Environment Agency established the Ocean Pollution and Dumping Countermeasures Section. In accordance with an implementation plan devised under the amended Marine Preservation Law, administrative agencies began to make recommendations and plans for implementation based on the London Treaty. A delegation including representatives from the Foreign Ministry, MITI, the Ministry of Transportation, the Ministry of Health, and the Environment Agency discussed jurisdictional issues regarding vessels from coastal countries.

The Marine Protection Law is the basic Japanese domestic law that provides measures necessary to implement the London Treaty and the MARPOL Treaty of 1973. It adopts the basic policy and the regulatory approach of these two treaties and provides for their domestic implementation. The law has been amended several times to implement specific measures adopted in the treaties' appendices.

In accordance with the treaties, the law prohibits, in principle, dumping of any regulated substances into the ocean. It permits dumping of certain regulated substances if they meet standards provided in the law or in case of an emergency. The 1980 amendments to the law extended the scope of regulation to aircraft and marine facilities engaged in the burning of hazardous substances. It also established a registration system for vessels used to discharge substances and a prior notification system for seagoing vessels that discharge substances into the sea.

The Maritime Safety Agency oversees enforcement of the law. There were 846 instances of illegal dumping in 1992, the lowest number since the Maritime Safety Agency started keeping records in 1971. Most of the instances of illegal dumping were negligent oil spills from vessels, although about 20 percent of the oil spills are considered intentional. Of all the instances of illegal dumping reported in 1992, 176 cases involved substances brought from land, virtually all of which were intentional dumping. The total number of instances of illegal dumping has been steadily decreasing every year.

Every year, the Maritime Safety Agency and the Environment Agency investigate and monitor levels of pollutants such as oil, PCBs, and heavy metals in the seawater and seabed in designated dumping areas. They report that the level of pollution has not noticeably changed or worsened.

Nuclear Waste Dumped by Russian Navy. A shocking event took place in the Sea of Japan, literally Japan's backyard, in the fall of 1993. On October 16, the international environmental group Greenpeace reported that a Russian Navy tanker was on its way to dump a shipload of radioactive waste (900 tons) in the Sea of Japan. The next morning, the tanker began dumping liquid radioactive waste. The Greenpeace ship at the scene filmed the dumping operation, which was broadcast repeatedly in Japan. The Japanese government lodged a strong protest and renewed calls for an immediate halt to the dumping. It was learned that Russia had engaged in dumping of radioactive waste into the Sea of Japan since 1959.

Russia's Environment and Natural Resources Ministry confirmed that it authorized a Navy plan to dump 1,700 tons of radioactive waste in the sea between mid-October and November 1993. Under the London Treaty, Russia maintained, it was allowed to dump low-level radioactive waste into the ocean until the year 2018. The former Soviet Union had abstained from signing the 1985 moratorium on the dumping of radioactive waste.

The London Convention did not require prior notice to neighboring countries, although the International Atomic Energy Agency (IAEA) recommended such a practice, along with notice to the Agency and the Convention Secretariat. An IAEA spokesperson confirmed that Russia had given notice of its intention to dump 1,700 tons of radioactive waste into the Sea of Japan.

In April 1993, Japan pledged a $1.82 billion Russian aid package; $100 million of that sum was designed to help dismantle Russia's nuclear weaponry. When President Boris Yeltsin visited Japan only a week before the dumping took place, he signed an agreement with Prime Minister Hosokawa of Japan regarding nuclear dumping. The pact called for a joint investigation of radiation levels in the Sea of Japan.

The Japanese public reacted. Prefectural governments, politicians, civic leaders, and fishermen's unions joined in the chorus demanding an immediate and complete halt to Russia's dumping of radioactive waste. They picketed the Russian Embassy, delivered letters of protest to Russian consulates, and sent telegrams directly to President Yeltsin. The Director General of the Science and Technology Agency convened a meeting to discuss measures to be taken, and decided to send survey vessels from the Maritime Safety Agency, Meteorological Agency, and Fisheries Agency to the dumping site to take water samples.

On October 21, 1993, Russia announced that it would suspend dumping radioactive waste into the Sea of Japan, but added that it might be forced to resume dumping if it was unable to build a new processing plant within eighteen months. Japan agreed to hold a working-level meeting in Moscow that November to discuss a better way of disposing of radioactive waste.

The October 1993 event had great impact on Japan's policy on nuclear and industrial waste disposal. Japan had opposed a temporary moratorium on ocean dumping proposed in 1983 and had abstained when the same resolution was made in 1985. It seems that Japan then wanted to keep the ocean dumping of low-level radioactive waste as a policy option. These events also affected the meeting of signatory nations of the Convention held at London in November 1993. A total ban on radioactive waste was approved by the majority of member nations, Japan included. No country opposed the ban, and only five signatories abstained from the vote. Japan made it clear that it would abide by the Convention's ban on the dumping of radioactive waste.

The meeting also adopted an amendment to the London Treaty concerning a ban on the dumping of industrial wastes by 1996. When a similar resolution had been made in 1990, Japan abstained. The 1993 amendment prohibits dumping of industrial waste in principle. Only certain nonhazardous industrial wastes are permitted to be dumped, and these exceptions are specified.

Most likely, Japan was dumping what accounts for the largest amount of industrial wastes into several designated zones in the Sea of Japan and in the Pacific Ocean. The Environment Agency estimates that 4.55 million tons of industrial wastes were dumped in 1991. These industrial wastes were largely by-products of the construction and smelting industries. If Japan decides to accept the 1993 amendment, the government must persuade the powerful construction and metal industries to change their disposal practices. Japan's Marine Protection Law needs to be amended in order to implement the obligations under the Convention.

Summary. The London Convention has provided a framework and the impetus for developing Japan's domestic regulatory policy and measures for its implementation with respect to ocean dumping. Japan has absorbed most of what the Convention requires into domestic laws and regulations. It remains a constant and continuing battle between the regulators and the regulated to determine the appropriate extent and standards of applicable laws.

Japanese regulators may be able to use the Convention to their advantage, or Japan could renege on its obligations. Active participation of citizen organizations and NGOs is necessary to change the behavior of both bureaucrats and industrial actors. Enforcement is difficult because of Japan's extensive coastal zones and the need to involve in the regulatory process many actors in addition to the officials responsible for monitoring illegal dumping. The Russian Navy's dumping of radioactive waste in the Sea of Japan could not have been discovered without the effort of Greenpeace. It gave Japan a good example of NGOs having a significant impact on treaty implementation.

Montreal Protocol

Japan has successfully implemented the Montreal Protocol ozone layer protection measures concerning the control of production, import, and export of CFCs, but has so far failed to contain CFC emissions and to promote systematic retrieval of CFCs. The former set of measures involves a limited number of major producers and users of CFCs, a setting where a consensus-based implementation was feasible, while the latter set of measures relates to a large number of users in different trades and of diverse consumer products, a setting where no consensus-forming among affected parties is possible. The administrative agencies in charge of implementing emission control and retrieval have depended mainly on providing affected parties with financial and technical assistance and on public education.

During the 1980 meeting of the OECD Commission on the Environment, Japan promised to freeze the production of certain CFCs and to curtail the use of spray-type products. The Ministry of International Trade and Industry (MITI) implemented this initial promise by means of administrative guidance. Private industry began to explore the possibility of using substitutes in view of impending restriction on certain CFCs. However, the initial reaction of industry was slow because of scientific uncertainty regarding environmental harms and the disputed safety of possible substitutes.

In 1982, a Japanese research team stationed at the South Pole first observed and reported an unusual depletion of the ozone layer. The matter caught public attention and accelerated international efforts to arrest the depletion.

Domestic Actions. After joining the Vienna Convention and the Montreal Protocol, Japan enacted the Ozone Layer Protection Law (May 22, 1988), becoming the first county to fulfill the Protocol's domestic requirements. The law set forth procedures and standards for regulating certain CFCs. It incorporated precisely the measures required by the Protocol, such as issuing production licenses and import approval mechanisms. The law also adopted emission reduction and rationalization of the CFC use as policy goals.

These two policy goals and the guidelines issued for their implementation appeared, at that time, to be a step ahead of the measures required under either the Protocol or the Vienna Convention. The law was amended in 1991 and 1994 in response to the 1990 London Amendment and the subsequent agreements. The 1992 Montreal Protocol (fourth conference) undertook to promote retrieval, recycling, and destruction of CFCs. Japan funded research projects on CFC retrieval and treatment technology, and initiated a model undertaking by certain local governments for retrieval of CFCs.

Japan has taken other actions agreed upon by the parties to the treaty. The MITI has banned exports of CFCs to countries not party to the Montreal Protocol and trade in methyl chloroform. Japan contributed U.S.$9.6 million in 1992 to the Montreal Protocol Multilateral Fund. It has helped organize training courses and conferences to assist developing countries in introducing measures to reduce the use of ozone-depleting substances.

By 1994, production of specified CFCs (11, 12, 113, 114, and 115) had been reduced to 19.8 percent of the amount in the base year of 1986. Consumption of CFCs was down to 18.5 percent of the base year. In 1996, CFCs were banned in conformity with the agreement at Copenhagen (November 1992).

While major firms are progressively abandoning their use of CFCs as solvents, smaller firms are finding it difficult to stop their use or switch to other substances. The Environment Agency and the MITI have issued guidelines for restricting discharge and for rationalizing the use of controlled CFCs. Special measures have been taken to recover CFCs from mobile air conditioners and new measures have been introduced for refrigerators. Tax incentives and low-interest loans have been introduced to encourage changes in cleaning equipment. In 1992, over Y1 billion was available for loans.

However, small firms have little use for special tax exemptions or low-interest loans, since they cannot afford to make the initial capital investment for changing their equipment. The Japan Development Bank and the Japan Environmental Corporation make loans available for small firms, but there have been few applications.

Implementation Measures. The Montreal Protocol was implemented through the adoption of a specific regulatory scheme that began with the issuance of official proclamations to explain the basic provisions of the treaty (January 1, 1989). The scheme also included production regulations, such as production quantity licenses (granted by the MITI); designated limits for export production (set by the MITI); approvals for imports (Foreign Exchange Law, article 52, administered by the

MITI); and verification of the amounts of CFCs destroyed (administered by the MITI).

To aid the implementation of emission controls and use rationalizations, the scheme relied on announcement of targets (January 4, 1989), administrative guidance, and advice for private industry (various cabinet ministers in charge of certain sectors, the director of the Environment Agency, and the Minister of MITI). In addition, private industry received governmental assistance, including preferential tax treatment and financial assistance, to aid implementation.

Another part of the scheme has been carried out by the director of the Meteorological Agency, who is responsible for monitoring and announcing results. The director of the Environment Agency has been responsible both for inspections and for announcing results of inspections. Japan's implementation scheme for the Montreal Protocol has also relied on the promotion of research studies within Japan, as well as the promulgation of reporting requirements from industries and on-site inspection.

Factors Favoring Domestic Implementation and Compliance. Successful implementation of the Montreal Protocol in Japan during the initial phase was based on several factors. MITI, in cooperation with the Environment Agency, took initial actions to implement the treaty obligations. Official proclamations were issued to explain the basic provisions of the treaty. Based on the Ozone Layer Protection Law, MITI formulated regulations controlling CFC production, employing all the familiar tools used for implementing industrial policy. The MITI was responsible for granting production quantity licenses, setting limits on export production and the amount of imports, and verifying the amounts of CFCs destroyed. MITI could exercise direct control over these matters. The Ozone Layer Protection Law imposes no affirmative duty on the part of regulatory agencies to control emission of CFCs. The law simply provides that users shall make efforts to restrain emission and to rationalize the use of CFCs (article 19), and that relevant agencies may issue guidelines, if necessary (article 20).

The MITI and the Environment Agency jointly announced targets for emission controls and use rationalizations. They issued the guidelines for CFC users to follow in order to achieve targets. Other cabinet ministries were to give administrative guidance and advice to private industries involved in their respective fields. Preferential tax treatment and financial assistance were promised to firms complying with the CFC regulations. The Environment Agency became responsible for inspecting and announcing the results of overall compliance. Regulated industries were required to report their compliance and to accept on-site government inspection. The rapid success of these programs was aided by the factors discussed below.

The officials responsible for implementing the treaty were involved early on in its negotiation. Generally, a team of officials responsible for the domestic implementation of the treaty from MITI and the Environment Agency, together with diplomatic officials from the Foreign Ministry, who attend international conferences, took an active part in the negotiations.

In Japan, the drafting of major legislation originates in a particular ministry, not in the Diet. A typical law drafted by bureaucrats delegates general power to certain ministries, but usually omits specifics of enforcement, thereby giving the administrative officials ample room to work out the details. The Ozone Layer Protection Law, however, was drafted by officials who were closely involved in the international negotiations. They could draft the law as they saw fit to carry out their policy objectives. The law as drafted contains very general terms in order to give the lead ministry maximum latitude to maneuver when implementing treaty obligations.

Within the general framework of the Ozone Layer Protection Law, the lead ministry, MITI, has made administrative guidance a major tool of implementation (as compared with a command-and-control approach). Administrative guidance is the method most often used by Japanese bureaucrats, although it lacks any legal basis or force. It is often used in an informal meeting or negotiation between the regulators and the regulated. It is usually communicated orally and is offered as a piece of information, suggestion, advice, guidance, or informal direction. The party who is given administrative guidance is free to accept or to ignore it, in theory, but in most cases the party obeys it out of fear of unfavorable treatment by the ministry officials in other matters. In a web of extensive regulatory actions, most parties in business are likely to have continuing relationships with the ministry and may appear before its officials for a permit or license application. In such a situation, administrative guidance works effectively even without legal force.

Policy implementation has relied widely on ministries establishing informal councils (*shingikai*) consisting of twenty to thirty representatives of various sectors. Such councils are usually organized to provide forums for interested parties to express their concerns on the problems involved. They also have served as sounding boards before a ministry proceeds with policy measures. A council may be used to lay the ground for forming and facilitating the issuance of effective administrative guidance.

The Committee on the Ozone Layer Protection Measures was created in October 1987, as a part of the existing Council on Chemical Substances. One section, established later, consists of twenty-eight members who represent CFC manufacturers, users, components makers, consumers, labor unions, academics, journalists, and scientists. The members are appointed by the MITI. Meetings are held four or five times a year when MITI needs to discuss protection targets and measures, often after the major international conferences. The responsible officials of MITI, as well as officials from the Environment Agency, attend the Council meetings.

Data-gathering and information-channeling have been important functions of the Council. Often, representatives from major manufacturers and industry associations have been invited to submit their reports on technical data and production information. Their reporting and participation usually have been on a voluntary basis. These visitors prepare and present factual information to the Council to help form regulatory policies. At the same time, the Council gathers information through regular contacts with the regulated parties.

The Council on Chemical Substances meetings have provided an arena for lobbying for and against the protection policy and specific measures. Industry representatives certainly have used the *shingikai* meetings as a forum for airing the practical difficulties and burdens that they will face if the proposed measures are implemented. In turn, the regulators may adjust and redefine the policy objectives, and the obligations under the treaty and domestic legislation. Regulators also may float, in these meetings, "trial balloons" consisting of possible regulatory targets and measures. Other parties have asked probing questions and offered commentary from a variety of perspectives; the ensuing discussion and exchanges have helped to promote an understanding of the problems involved and measures needed to address them.

These Council on Chemical Substances discussions have been transparent much of the time; the meetings have been public, except when otherwise agreed. Some journalists (editorial staffs of national newspapers and television networks) have been among the members, and they have written articles and editorials based on information obtained at the Council meetings. Occasionally, MITI has prepared its position papers and reports on the measures taken for the protection of the ozone layer. They have been circulated in draft form and discussed at the Council meeting before their publication.

The Council on Chemical Substances, at its best, has served as a forum or process for building a consensus among the participating parties regarding the necessary and appropriate regulatory measures to be taken. Participants have expressed their opinions freely and informally. The meetings have not been rigidly structured or confrontational, and diverse interests and views have been represented. Often, the need for providing financial incentives and assistance has been discussed, and some informal assurance has been made in order for the regulated industry to comply with certain measures. Once policy targets have been clearly defined and necessary measures have been agreed upon, the regulated parties have been bound to carry out their obligations on a largely voluntary basis. The implementation of policy based on consensus has been almost self-executing, with little regulatory cost.

Summary. Japanese regulatory actions in response to the Montreal Protocol were quick and effective in reducing CFC production and consumption. Several factors may account for the success of this portion of regulations.

First, MITI has had almost sole responsibility, and can exercise exclusive jurisdiction over the matters it is regulating. As a result, MITI has developed a consistent policy and has taken necessary measures without being hindered by intragovernmental turf fights. It has also helped that the harm of certain CFCs has been scientifically established, and the need for regulating certain CFCs has been understood by the parties involved in the regulatory process. In addition, MITI's administrative guidance approach, coupled with an effective use of the Council on Chemical Substances as a consensus-forming forum, is suited to deal with a limited number of well-informed industry organizations and parties. Consensus-based regulation has been cost-effective, saving MITI from directing large amounts of resources toward legal enforcement of the treaty obligations. (Only a few officials are assigned to enforcement-related work.)

However, MITI's consensus-based approach may not be as effective when regulatory needs extend into retrieval, reuse, and destruction of certain CFCs. In view of the recent international decision to move the target year for total ban on the production to 1996, CFC retrieval and reuse became immediate policy concerns. The substances are used in many forms and products. Here regulation will be different from supervising a small number of CFC manufacturers. In devising regulations for the retrieval and reuse of certain CFCs, MITI must deal with a great diversity of parties with numerous and conflicting interests.

The process of retrieving certain CFCs from consumer goods like air conditioners and refrigerators must be carried out at individual user sites, which can be a very costly operation. Major home appliance makers started efforts to retrieve CFCs from discarded refrigerators, and set up 1200 stations. They decided to use portable retrieval machines made by General Electric. Retrieval stations and storage spaces for retrieved CFCs are not readily available. Technology for the destruction of CFCs is still in the experimental stage. It is necessary to get local governments involved in the retrieval process because the operation is closely tied to local waste management.

There may be demand for retrieved CFCs, if their quality can be assured, and there is also a certain window of time in which the market mechanisms deem retrieval an economically viable operation. The demand for reuse, however, may quickly disappear if a less costly substitute becomes available. The destruction of certain CFCs and the disposal of products containing CFCs in a nonretrievable form may require major investments in the development of new facilities and technologies.

These problems involve far broader issues of restructuring the existing socioeconomic system into one of conserving resources and recycling products, while allocating the costs of doing so fairly among all parties involved. The task may go well beyond MITI's traditional mission, but it is a challenge that MITI will face when its regulatory emphasis shifts to use rationalization and direct control (destruction and disposal). A MITI *shingikai* discussed the possibility of imposing some of the retrieval costs on consumers of products using CFCs. The central government must delegate more power to local governments, especially in controlling CFC emission and retrieval. Japan now faces the challenge of developing a comprehensive system to conserve resources and recycle reusable products.

Conclusions

Since World War II, Japan has made remarkable progress with its prompt economic recovery and growth; at the same time, serious disruptions occurred in the natural environment, incurring costs with respect to human health. In many ways, Japan's recent environmental policy can be said to follow—and to apply—the industrial policy model that proved so successful in bringing economic prosperity. Japanese industrial policy has been made by the central government, with the bureaucracy taking an active part in national economic policy-making. Major ministries worked jointly to set goals for national economic development, using a regulatory approach that—on the whole—has been based upon a consensus of the regulated.

Once national goals have been set, officials usually implement their policies without resorting to legal coercion or overt enforcement. Implementation of national policies relies largely upon voluntary compliance by the regulated industries. The central government often provides tax incentives and financial assistance in order to achieve defined goals and to assure compliance by regulated entities, but it also often works closely with them, gathering information and providing necessary guidance and protection. Perhaps this working relationship has enabled Japan to meet the data-filing requirements of the three treaties in this study that require annual reports. Table 8.3 compares Japan's success in complying with the reporting requirements of CITES, the London Convention, and the Montreal Protocol between 1991 and 1996.

This industrial policy model has served the government regulators well and has proved effective where the regulated industries are well defined and the number of players is small. The Japanese government seems to have adopted a regulatory approach similar to its domestic implementation of international environmental treaties; however, the applicability of this industrial policy model in that arena is questionable, at best.

The Japanese "industrial policy" regulatory approach seems to have worked best in the case of ozone protection, quickly achieving the treaty goals regarding production and use of certain CFCs. Major producers of CFCs are few in number and well defined. Yet, this model has proved ineffective in controlling CFC emissions

Table 8.3
Japan's compliance with treaty reporting requirements, 1991–1996: Report transmission/receipt dates

Treaty	Entry into force	1991	1992	1993	1994	1995	1996
CITES	11/4/80	3/8/94	imports: 7/21/94 exports: 3/7/95	imports: 3/7/95 exports: 5/1/95 (extension granted)	imports: 12/8/95 (extension granted) exports/ ocean products: 12/20/95	1/16/97 ocean products: 1/8/97	N/R
LC	11/14/80	9/10/92	9/2/93	9/13/94	10/4/95	10/18/96	N/R
MP	1/1/89	R	R	R	R	R	R

Notes: For all treaties, N/R indicates that no report was received by July 1997. R indicates that the report was received sometime prior to July 1997 but the date is unknown. Montreal Protocol reports received include those filed on time, late, or subsequently revised; in some cases a country may have met two or more years' reporting requirements with the submission of a single report.

and retrieval, where implementation necessarily involves controlling many users of diverse products.

The ITTA may be regarded as an example of the success of the Japanese regulatory approach because the domestic policy goal of securing an adequate lumber supply has been achieved. However, Japan has undertaken a policy of nonintervention, leaving the matter of tropical timber imports to the market, without any effective control over wasteful domestic consumption. This has brought about the destruction of many tropical forests while leaving the domestic forest industry chronically depressed.

CITES implementation clearly shows a failure of the industrial policy model where the regime has a large number of regulated items and many confusing exceptions. Overlapping and conflicting regulatory responsibilities of several ministries have not aided the treaty's effective implementation and enforcement, particularly in light of a large traffic in regulated items among numerous parties.

Japan has apparently complied with the London Convention on ocean dumping. Official control of dumping operations by permits and resulting statistics show permissible amounts of waste materials legally dumped in designated areas around Japan. Yet Japan faces a future total ban on ocean dumping, along with the almost impossible task of disposing of huge amounts of industrial waste on land. Due to its extremely limited land area, Japan faces a severe lack of disposal sites. Thus far, waste disposal has been essentially left in the hands of local governments and generators of industrial waste. No clear and comprehensive national waste disposal policy has yet been developed. No workable definitions of industrial waste, hazardous or otherwise, seem to exist. No manifest system has been adopted. Illegal dumping is expected to increase, and effective legal controls over the generation and disposal of industrial waste are needed.

While the World Heritage Convention may seem, at first, to require no major implementation efforts, it has already created tension between developers and conservationists. The effectiveness of national policy on nature conservation and regulatory structures may be questioned. A new process of policy-making and of implementation, in place of the traditional consensus-based approach, is clearly called for, on a local as well as a national basis.

Due to the impact of international environmental accords and the requirements for their implementation, Japan is facing the need to change traditional regulatory approaches. Japan may resort to a more active use of law and legal means to implement international and national environmental policy objectives more effectively, calling for more citizen participation and greater involvement of local governments. More positive use of law is an effective means of making the official process of environmental policy-making more transparent, accountable, and efficient.

Notes

The authors would like to acknowledge the significant contributions of their respective research assistants, over the life of this project, to the completion of their manuscript. Professor Fujikura wishes to thank his research assistants who compiled materials in the early stage of the project: Takesada Iwahashi (former graduate student, Faculty of Law, University of Tokyo); Shinji Matsuka (former graduate student, Faculty of Law, University of Tokyo); and Jon Richard (former student, University of Wisconsin Law School). Professor Feinerman would like to thank his research assistants, all now law school graduates but formerly students at Georgetown University Law Center: Ken Lebrun; Hiromi Maruyama; and Rebecca Rejtman.

1. The "Big Four" cases are the Itai-Itai Case (*Komatsu* v. *Mitsui Kinzoku Kogyo K.K.*, 22 Kakyu Minshu (Nos. 5 and 6) 1 (Toyama Dist. Ct., June 30, 1971) (*aff'd.* 674 *Hanrei Jiho* 25 (Nagoya High Ct., Kanazawa Branch, Aug. 9, 1972)); the Niigata Mercury Poisoning Case (*Ono* v. *Showa Denko K.K.*, 22 Kakyu Minshu (Sept.) 1 (Niigata Dist. Ct., Sept. 29, 1971)); the Kumamoto Mercury Poisoning Case (*Watanabe* v. *Chisso K.K.*, 696 *Hanrei Jiho* 15 (Kumamoto Dist. Ct., March 20, 1973)); and the Yokkaichi Air Pollution Case (*Shiono* v. *Showa Yokkaichi Sekiyu*, 672 *Hanrei Jiho* 30 (Tsu Dist. Ct., Yokkaichi Branch, July 24, 1972). For a detailed description of each case, see Gresser et al. 1981: chapter 3.

2. JETRO (1991:33).

3. Ibid.

9

The Soviet Union and the Russian Federation: A Natural Experiment in Environmental Compliance

William Zimmerman, Elena Nikitina, and James Clem

This chapter analyzes the behavior of two states, one—the Russian Federation—the legal successor to the other—the Soviet Union. As a result, we are reporting in this chapter on a kind of natural experiment in which there existed a prior system, the Soviet Union, that experienced an enormous internal and external shock, and that was succeeded in December 1991 by what is to a considerable degree a new state, the Russian Federation. As one would expect, there have been dramatic changes in the pattern of compliance with international environmental accords that have paralleled the other enormous changes that have occurred. What is surprising, however, is that the effects on environmental treaty compliance have been decidedly mixed. We expected to find that the differences between the Soviet and Russian politico-economic systems would largely account for the differences in compliance patterns observed. This in fact proved to be a central finding of this chapter. What we did not expect was our other central finding: that in some instances compliance has improved in post-Soviet Russia, whereas in others it has worsened, and that these differences across issues are largely understandable in terms of the differences between the Soviet and Russian political and economic systems.

A Tale of at Least Two Countries and Five Treaties

This is initially a simple story of a now defunct country, the Soviet Union, and its ratification, implementation of, and compliance with five environmental treaties: the London Dumping Convention, the Convention on International Trade in Endangered Species of Wild Fauna and Flora (CITES), the Vienna Convention and the Montreal Protocol on substances that deplete the ozone layer (the Ozone Treaty), the International Tropical Timber Agreement (ITTA), and the World Heritage Agreement. Of the five treaties, the USSR[1] was an active participant in the negotiations leading up to the signing of the London Dumping Convention, CITES, the Ozone Treaty, and ITTA; it ratified the World Heritage Convention thirteen years after that convention had entered into effect.

The dates of signing and ratification by the Soviet Union are given in table 9.1.

Table 9.1
Treaty membership of the Russian Federation

Treaty	Year treaty came into effect	Signed by Russian Federation	Ratified by Russian Federation
WHC	1977	1988	1988
CITES	1975	1974	1992
ITTA	1985	1985	1986
LC	1975	1972	1976
MP	1989	1987	1989

The London Dumping Convention was in part an outgrowth of the Nixon–Brezhnev détente, and the Soviet Union was actively involved in its negotiation. It was one of the first signers and ratified the treaty exactly three years after signing.

CITES was even more directly linked with détente. It logically and chronologically followed not only the 1972 U.S.–Soviet Agreement on Cooperation in Environmental Protection but the 1972 Stockholm conference on the environment as well, which opened up a wide range of Soviet–American cooperative efforts. CITES was but one of many international agreements signed by the Soviet Union during the mid-1970s. It was a counterpart to many bilateral U.S.–Soviet treaties signed at approximately the same time; these in turn had parallels in bilateral and multilateral treaties the USSR signed with other arctic states (see Pryde 1991; Osharenko 1989) in a wide range of conservation areas, including the preservation of endangered species.

There are some grounds for cynicism about much Soviet activity in this vein during the 1970s. At the same time, however, the growth in Soviet attention to international agreements dealing with the environment should also be seen as flowing directly from a tradition, dating back to the 1950s, of thoroughgoing conservation and concern for species preservation. Prior to the signing of CITES, the Soviet Union had adopted national norms and laws directly pertinent to the obligations undertaken in that treaty. A partial list of such agreements would include those providing for strict hunting regulations; animal breeding in captivity; conservation of wildlife in nature preserves and reserves; control, protection, and reproduction of several key species of wildlife; and provisions for preserving flora species. In any event, preservation of flora and fauna was a more pleasant international conversational topic for Moscow in the mid-1970s than, for instance, human rights (see Ziegler 1987).

For the Soviet Union, ITTA was a tertiary concern. It was neither a producer nor a major consumer of tropical timber. It was, however, a member of the Legal Drafting Committee and the Credentials Committee of the founding conference, and thus would have to be identified as an active participant in the negotiation of the treaty, a treaty it signed and ratified in 1985 (see *International Tropical* 1984).

The Ozone Treaty was another matter entirely. It must be seen in the larger context of the improvement in East–West relations in the mid-1980s with the rise of Mikhail Gorbachev. Ozone depletion vividly exemplified the kind of global human problem threatening all states, regardless of social system, that called for new thinking (*novoe myshlenie*). The Soviet Union was active in the negotiations from the outset. It and its successor, the Russian Federation, have continued to play a major role in the international organization. As Elena Nikitina (1991) has remarked, the Soviet Union was among the first to elaborate and implement international regimes for the protection of the ozone layer, just as it was similarly involved in efforts to address the prevention of transboundary air pollution, and at the cutting edge of global change broadly.

Of the five treaties, the only one in which the USSR did not play a negotiating role was the World Heritage Convention. Just why thirteen years elapsed before the USSR acceded to that convention is not clear to us. It may not have been willing to accept the obligations implied by designating sites as "global treasures."

There is a hint of an explanation in two speeches by Gorbachev in the summer of 1988—note the juxtaposition of historical monuments and environmental protection, which makes sense only against the backdrop of the World Heritage Convention—just months before the Soviet Union ratified the Convention in October. Initially there may have been some fear that non-Russian nationalists might use the designation of sites for their own purposes rather than as a way of celebrating the Soviet Union, fears that were seen as less troublesome in the new climate of glasnost. On June 29, Gorbachev observed, "In recent years, the process of democratization and openness has also shed light ... some problems ... connected with language, culture, literature and the arts, historical monuments and environmental protection."[2] A month later, he was more explicit: "We must be [attentive] to the development of native languages and national cultures, *to environmental protection and historical monuments*, and to everything that defines ... each nation's [unique] inimitable contribution to the general treasure house of Soviet culture."[3]

Country Characteristics

By December 1991, the Soviet Union was no more. Russia considers itself the legitimate successor state of the former Soviet Union and has agreed to honor all international obligations undertaken by its predecessor.

The slimmest vestige of a continuing Soviet-wide organizational structure was provided by the creation of the Commonwealth of Independent States (CIS). In practice, however, the CIS has played a very modest role since 1991. The triple redundancy contained in its name (Commonwealth, Independent, States) brings home the point that the CIS is an international organization. Its activities have been largely limited to agreements to coordinate environmental activities—in the case of the ozone treaties, including coordination of domestic policies of the several states and the establishment in 1994 of a working group.

Moreover, most of the former republics other than Russia have not signed the treaties in question, though they have agreed in general to observe preexistent treaty obligations. As a result, much of Russia's current difficulties with respect to the observance of, for instance, CITES arise because much of what were domestic relations in the USSR are now international relations. At the same time, Russia has open borders with nonsignatory states and is in the distinctive position of being obligated to issue permits for endangered species in former parts of the USSR that are not a part of Russia.

A second distinctive feature of the post-Soviet Russian situation derives from the unique place Russia held within the Soviet Union. Viewed from the Soviet periphery, Moscow meant Russia. For Russians, however, there were, paradoxically, ways that Russia was disadvantaged vis-à-vis other republics. In the Soviet Union, no specifically Russian Academy of Sciences existed. Not until the very last years of the Soviet Union was there a Communist Party organization for Russia, in contrast to all the other republics. Similarly, according to N. N. Vorontsov (former chair of Goskompriroda, the Soviet State Committee on the Environment), environmental committees "were created in other republics at the end of the fifties. Only the unfortunate Russian Federation did not receive this right until 1988—after the organization of the USSR Goskompriroda." The result, he maintained, was that "we were several decades behind" (see FBIS 1989a).

The Traditional Soviet Model. The traditional Soviet model existed in 1972 (the year in which the first of the five treaties was signed) and persisted until the mid-1980s.[4] It had many distinctive features, but basically the phrase "command-administrative system" captures it well. It was a strong state that was intolerant of autonomous internal political players, an attitude that was manifested in spades for public and private external actors, such as NGOs. It was centralized; its capacity to penetrate the society was substantial and relatively homogeneous across the country.

Another important aspect of the traditional Soviet system was its high level of militarization, resulting in a military–industrial complex operating in the near absence of controls or monitoring. Western estimates of the size of the defense sector in the early and mid-1980s range from 15 percent to 25 percent of the economy and even higher—compared with about 5 percent in the United States at that time (see Peterson 1993b). Cloaking itself in the high level of secrecy that usually accompanies defense-related industries everywhere, the military–industrial complex successfully shielded itself from efforts to bring its activities into accord with existing environmental regulations.

Finally, the old Soviet model was a low-information system. Strenuous and often successful efforts were made by the state to prevent authentic political participation by the citizenry and to prevent them from acquiring basic information. Borders were secure and virtually impenetrable. The same "Iron Curtain" that prevented the free flow of people and ideas also virtually precluded the import of drugs and the export of Siberian tiger skins.

Within the traditional Soviet system, environmental management was plagued by organizational and attitudinal difficulties, as well as by the economic incentive system. The environmental administrative bureaucracy of the pre-Gorbachev era had two key organizational disadvantages. Bureaucratic compartmentalization was legendary. Responsibility for the planning and implementation of environmental protection directives was divided among a host of state committees and ministries; this institutional fragmentation made coordination, implementation, and monitoring of environmental protection initiatives extremely problematic (see Peterson 1993b).

Even during the Brezhnev era, some measures were taken to alleviate the situation. Steps toward centralizing environmental controls were taken in 1978 when the Hydrometeorological Service (established in 1974) was elevated to a state committee, granting it autonomy from the ministerial system, and renamed the State Committee on Hydrometeorology and the Environment (Goskomgidromet or simply Gidromet). The new state committee was charged with setting and monitoring enterprise compliance with air and water pollution standards, and it had the power to submit proposals to suspend operations of violators, but it had no enforcement functions.

Supervisory powers were also held by other ministries. For example, responsibility for supervising water pollution was shared by the Sanitary Epidemiological Service of the Ministry of Health, as well as the ministries of Land Reclamation and Water Resources, Agriculture, and Fish Industry, among others (see Ziegler 1987).

This points to the second fundamental administrative problem of environmental protection in the prereform Soviet Union: individual ministries were charged with both the "rational use" of resources and the responsibility for conserving them. Stanislav Tsurikov (1990), then deputy chairman of Goskompriroda, stated that this was akin, in the Russian metaphor, to "having the goat guard the cabbage." The Ministry of Fish Industry (Minrybkhoz), for example, was responsible both for harvesting of fish and for meeting its centrally planned targets, as well as for the preservation of commercial and sport fish stocks (see Tsurikov 1990; Pryde 1991). In like fashion, the State Committee for Forestry regulated both the harvesting and the planting of trees (see Pryde 1991). Wresting control from these powerful, well-established ministries quickly proved easier said than done. Thus, Goskomgidromet was a small step toward centralization of the environmental protection bureaucracy, but what was needed was a central protection agency with powers across the board.

Underlying and reinforcing the organizational weaknesses of environmental protection efforts in the prereform USSR were core attitudes expressed in traditional Marxist–Leninist ideology. One ideological impediment was the labor theory of value; another, the overall revolutionary optimism. As interpreted in the Soviet Union, the labor theory of value implied that goods derive their value from the amount of human labor involved in their production.[5] The products derived from natural resources had value equal only to the value of the labor required to process them. Consequently, natural resources were not priced, and the state, the formal owner, provided state enterprises access to them without charge. No fines for pollution were levied on those responsible for it. Thus, there was no incentive or rationale for the users of natural resources to invest in environmental protection equipment or to restrain their demands on those resources.

Indeed, official Soviet images of the environment generally promoted extensive economic development and downplayed caution. The socialist project and its primary goal of industrialization were portrayed as a struggle between man and nature; a zealous belief in the power of scientific development, fueled by notions that there were no barricades that Bolsheviks could not storm, was used as the cure-all for potential disruptions to the environment.[6]

The Soviet legal tradition also hindered attempts to change environmentally destructive behavior. Marxism–Leninism denounced bourgeois law as an instrument promoting the interests of the ruling class at the expense of the workers; nevertheless, practical issues demanded the development of a new system of law, "socialist legality," that could be used as an instrument of social engineering, as well as a guide to rational decision-making within the socialist system. While the Communist Party and the Soviet government enacted a series of resolutions and decrees ostensibly aimed at protecting the environment, most experts contend that Soviet environmental law was "not so much a mechanism for resolving conflicts among contending parties as a set of idealistic and often unattainable principles epitomizing the regime's professed commitment to environmental protection" (Ziegler 1987:79).

A central principle of socialist legality in the traditional Soviet model was its educational function. The extent to which that notion of law was internalized by Soviet elites is illustrated by an interview with N. N. Vorontsov in 1989. Vorontsov was a scientist, the head of Goskompriroda in the late 1980s, the first non-Communist of ministerial rank in the entire Soviet period, and someone quite willing to inveigh against the old "technocratic ideology"—as close to a "new thinker" as one could come in the Soviet administrative hierarchy (see FBIS 1989b). Nevertheless, we find him arguing that "We must also think about regulation by law, for a good law can have tremendous educational significance" (see FBIS 1989b).

Law existed not only as a mechanism for regulating behavior but also as a propaganda tool. This educational role of socialist legality helps to explain the apparent contradiction of a state in which the legal record was thick with environmental protection laws while the air was thick with smog. Normative environmental protection measures repeatedly lost out to economic growth and bureaucratic self-interest. Legal enactments themselves sometimes indicated the gaps in Soviet compliance with international law. In other instances, the gap between government resolutions and decrees, and reality served to give the appearance of compliance, masking the underlying environmental degradation.

In addition to the weakness of the environmental protection administration and the overarching ideological/legal framework of the Soviet state, the centrally planned economy's developmental imperative created strong incentives for enterprises to produce, regardless of the environmental price. Successful fulfillment of the centrally assigned production quotas was the overwhelming priority; it was understood that violations of environmental protection laws and guidelines would be tolerated as long as the plan was fulfilled. Natural resources were allocated to enterprises essentially for free, resulting in a system in which resource conservation made little economic sense[7] (see Peterson 1993b). When the Union republics tried to adopt legislation to curtail the activities of the central ministries and their enterprises, they found it virtually impossible to subordinate all-Union ministries within a system of highly centralized political control and economic planning. This system of vertically organized ministries, essentially autonomous self-regulators of their subordinate enterprises, emphasized the "branch" priority of productive output over the "territorial" concerns of the local social and territorial infrastructure (see DeBardeleben 1990).

The Gorbachev Era. Mikhail Gorbachev became General Secretary of the Soviet Communist Party in 1985, intent on reform. In his six-year tenure, the institutional forms of the previous system were by and large retained. Indeed, many of these institutions, such as the Supreme Soviet, persisted for the first few years after the collapse of Soviet power. But the changes in behavior were astonishing. In a phenomenon first noted in the environmental domain but that diffused rapidly, the number of domestic and external actors participating in the political process grew exponentially. Gorbachev's first years in power were characterized behaviorally by greater political openness (glasnost), limited democratization, a largely unsuccessful program for economic restructuring (perestroika) to move toward a market economy, and a dramatic decline in the central government's ability to govern its constituent republics.

These tendencies persisted after the dissolution of the USSR in December 1991 and the emergence of the Russian Federation headed by an elected President, Boris

Yeltsin. The latest round in a continuing struggle to establish an accepted constitutional order ended with the ratification of a new constitution in December 1993. Economic reforms continue apace as many, but scarcely all, aspects of the old command system have given way to private ownership and marketization. Moscow continues to face a major challenge with respect to reestablishing central administrative authority. A once mighty military has witnessed its reputation, status, and budget share drop precipitously. Russia's borders with states outside the old Soviet Union have become far more open, both to the inflow and outflow of people and ideas, and to smuggling. At the same time, its borders with the "near abroad," especially in Central Asia, remain almost nonexistent.

In the domain of environmental administration, the situation also has changed, largely for the better. Two major efforts at restructuring the treatment of the environment are especially notable. In 1988, Gorbachev organized the first all-Union organization for environmental protection, Goskompriroda, the main goal of which was to eliminate the overlapping responsibilities that had previously characterized environmental administration.[8] The subsequent transformation of Goskompriroda into the Ministry of Environmental Protection (Minpriroda) indicated greater formal capacity to enforce environmental rules directly on enterprises.

To its credit, Goskompriroda accomplished a revolution regarding the public availability of data; it revolutionized the attentive public's discourse by publishing voluminous data concerning the environmental situation and environmental activities in the USSR, a practice continued by its Russian Federation successor (see, e.g., *Sostoianie* 1990; "O sostoianii" 1992). Moreover, it was organized in such a way as to imply a systems, and systematic, approach to environmental matters. In addition, as Tsurikov points out, the creation of Goskompriroda produced a gradual change in the focus of environmental compliance monitoring. No longer were the ministries self-policing. "Monitoring ... compliance with ecological demands is becoming extra-departmental. And in international matters these functions have now been transferred to our committee" (1990:61).

Nevertheless, there was much to regret about the new agency. As Peterson notes tersely, "Goskompriroda remained fiscally impoverished, institutionally weak, and perpetually embroiled in bureaucratic struggles" (1993b:182). Its personnel were often people seconded from other agencies who were poor performers in their former place of employ; many apparatchiks were dumped on the agency; its relation with Gidromet, the State Meteorological Agency, was imprecisely defined and a source of constant bureaucratic battle; its enforcement capacity was limited; and it lacked ministerial status.

After the collapse of the Soviet Union, efforts were made to overcome the organizational problems that had been typical of environmental oversight in the Gorbachev era. The new Russian government created a kind of superagency; several agencies from the Soviet government and four from the Russian government were amalgamated into the omnibus Environmental Ministry (Minpriroda). (By the mid-1990s the official title had been changed to the Ministry of Environmental Protection and Natural Resources.) Thus far, as shown by Peterson (1993a), the evidence suggests that the committees have by and large remained in practice autonomous.

In addition, the new government passed a comprehensive law, "On Environmental Protection," in December 1991. (The law was signed in December 1991 but published by some media sources only in March 1992.) To some extent the law represents an effort to codify new norms concerning the environment. Thus, it attempts to enact rules that take into consideration the shift from plan to market, and it provides explicit protection for environmental groups that may target particular enterprises or industries (see Bond 1992). It recognizes the uniquely insulated place of the military in the old Soviet system by making specific mention of military and defense installations in a specific article (55)—thus explicitly indicating the intention to subordinate the military to the civil law.

Another noteworthy effort to alter traditional Soviet attitudes was the development of a natural resource licensing system. This implied the emergence of an economic mechanism to regulate resource property rights and constituted an explicit break with the Soviet tradition of free natural resources.

In brief, at the macro level the central theme for the time period commencing with the signing of the first environmental treaty under examination and extending to the mid-1990s was one of fundamental change in almost every respect: new states, radically altered political systems, dramatic restructuring of the economy, and an altered political culture in which largely independent media, extensive availability of scientific information, and autonomous social mobilization were viewed as normal and legitimate. Despite the heavy weight of Soviet bureaucratic culture, these macro changes have been paralleled in new institutions, attitudes, and incentives relevant to the environment.

The relationship between changes in these country characteristics and levels of compliance will be explored below. It suffices here to note that the kind of natural experiment we are witnessing permits us to address the effects of domestic political and economic regime change on the propensity to implement international environmental treaties.

Extent of Implementation and Compliance with the Treaties

Actions Taken Over Time

A caveat at the outset: obtaining data on the level of Soviet and subsequent Russian compliance with these treaties is extremely difficult. One reason is the closed nature

Table 9.2
Russian Federation's compliance with treaty reporting requirements, 1991–1996: Report transmission/receipt date

Treaty	Entry into force	1991	1992	1993	1994	1995	1996
CITES	1/13/92	7/27/93	11/27/93	12/26/94	1/29/96	1/17/97	N/R
LC	1/29/76	N/R	N/R	10/5/93	N/R	N/R	N/R
MP	1/1/89	R	R	R	R	R	R

Notes: For all treaties, N/R indicates that no report was received by July 1997. R indicates that a report was filed, but the date of receipt is unavailable. Montreal Protocol reports received include those filed on time, late, or subsequently revised; in some cases a country may have met two or more years' reporting requirements with the submission of a single report.

of the Soviet political system until the late 1980s and the lack of environmental data in general until very recently. (The lack of data is in itself an indicator of the general Soviet record of compliance.) The Russian Federation's record of reporting shown in table 9.2 reveals some of the problems.

A second reason has been the turmoil in post-Soviet Russia, especially with respect to center–periphery relations, and the short period of time since the December 1991 collapse of the Soviet Union. Data for 1992–1995 are quite limited.

ITTA and the World Heritage Convention

Soviet, and subsequent Russian, involvement in the International Tropical Timber Agreement (ITTA) has been modest at best, though the role Russia plays in the future is likely to increase to the extent that ITTA is increasingly conceived of as focusing on sustainable growth rather than on tropical timber per se. As noted above, while the Soviet Union signed the ITTA, neither it nor any of its successors—pending global warming—are producers of tropical timber. The Soviet Union contributed slightly more than $15,000 to the administrative budget, and Russia contributed almost $24,000 in 1993.

The Soviet Union finally ratified the World Heritage Convention in 1988. Soviet compliance was largely ceremonial and limited to the passing of resolutions. Russia, by contrast, has become more actively involved in the designating of cultural regions. Two reasons may be advanced for this change. First, the Russian leadership—as did Gorbachev to a limited extent—has increasingly seen World Cultural Heritage designation as an opportunity to celebrate traditional ethnic *Russian* values and symbolism in the designation process.

Table 9.3 presents the World Heritage Sites that have been listed.

Among the sites that have been listed with the since 1989 are Red Square, the Kremlin, and Kolomenskoye in Moscow; the historic center of St. Petersburg (for-

Table 9.3
World Heritage Sites in the Russian Federation

Year	Natural site	Cultural site
1990		historic center of St. Petersburg and related group of monuments
		Khizi Pogost
		Kremlin and Red Square
		historic monuments of Novgorod and surroundings
1992		cultural and historic ensemble of the Solovetsky Islands
		white monuments of Vladimir and Suzdal
1993		architectural ensemble of the Trinity Sergius Lavra in Sergiev–Posad
1994		Church of the Ascension, Kolomenskoye
1995	Virgin Komi forests	
1996	Lake Baikal	
	volcanoes of Kamchatka	
Total	3	8

merly Leningrad); Novgorod; Vladimir and Suzdal; Kizhi; Trinity-Sergius Lavra in Sergiev–Posad (formerly Zagorsk); the Solovetskiy Islands; and Lake Baikal. (See figure 9.1 for the location of the cultural sites.) Similarly, the creation of an advisory council in March 1993 to protect "especially valued cultural objects of the Russian Republic," without mentioning the World Heritage Convention, seems by virtue of its composition designed primarily to protect especially valued *Russian* cultural objects[9] (see *Sobranie* 1993).

Second, in 1993–1994 Russian officials seem to have become aware that World Heritage status brings *benefits* as well as responsibilities: UNESCO's Director General, Frederico Mayor, announced that UNESCO would supervise the reconstruction of the Bolshoi Theater in Moscow, help finance renovation of the Hermitage in St. Petersburg, and fund the computerization of the Russian State Library (see FBIS 1993b). In this respect the Russian government has overcome its predecessor's hostility to external penetration and has recognized that international organizations can offset the Russian government's diminished capacity to extract resources from its own citizenry.

In general, however, accounts of the three other treaties—the Ozone Treaty, CITES, and the London Dumping Treaty—are far more interesting and relevant to

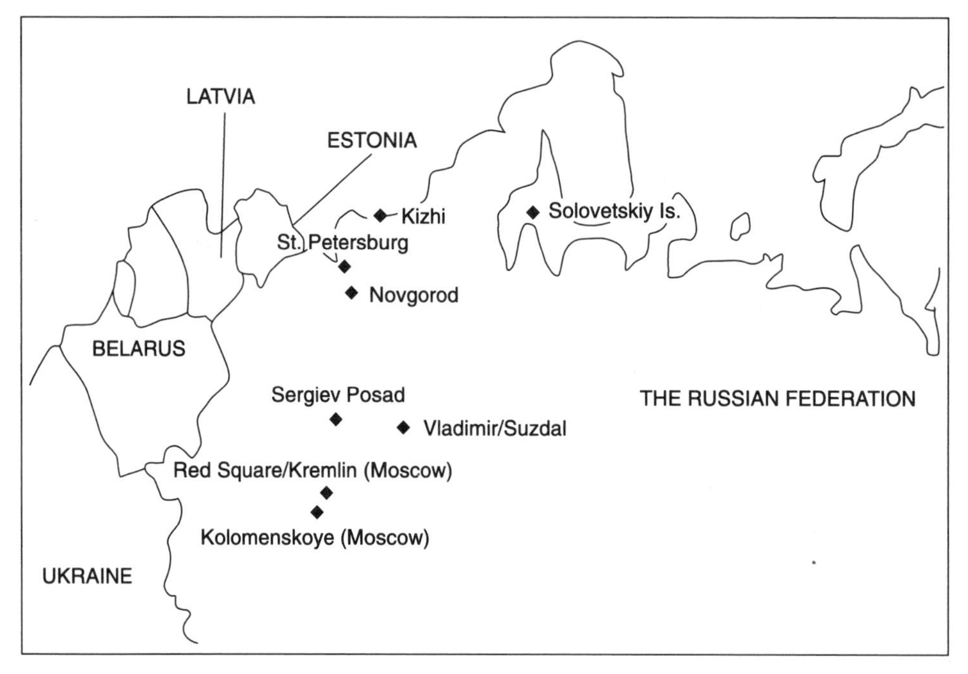

Figure 9.1
Cultural sites designated by the World Heritage Convention

an understanding of Soviet and Russian environmental compliance. They are not the same stories.

The Vienna Convention and the Montreal Protocol

The Soviet Union manufactured approximately one-eighth of the world total of ozone-depleting substances. There is no doubt that Soviet participation in the Montreal Protocol was directly related to Gorbachev's East–West policy. The joint statement at the 1987 summit explicitly mentioned American–Soviet cooperation regarding the environment. The timing of these international treaties coincided with the new initiative of the Gorbachev regime to establish better relations with the West.

At the same time, there is a history to Soviet ozone research that bears telling (see Nikitina 1991 for an elaboration). Long before the discovery of the ozone hole over the Antarctic, the ozone layer was a major area of inquiry for Soviet science. Indeed, in the early 1970s some Soviet scholars were discussing the possibilities of ozone depletion, but no consensus existed among Soviet scientists in that decade. In the late 1970s highly visible Soviet specialists took the position that ozone depletion was a minor threat to northern climes. Indeed, some writing targeted at the mass public was quite dismissive of the phenomenon. In one account published in 1980, it was asserted that ozone destruction as a result of nuclear explosions had not been proved; that the data on ultraviolet radiation risk were to a large extent exaggerated; and that the human body had mechanisms that were highly adaptive—and hence there was little reason for concern.

It was only in the early 1980s, and especially in the late 1980s—after the appearance of ozone holes in the Antarctic in 1985 and 1989—that the anthropogenic contribution to ozone pollution received widespread coverage in the major media sources. In contrast to Chernobyl, public opinion played a limited role in engendering action. Rather, Soviet press coverage and media attention came with the elaboration and signing of the Montreal Protocol, not before. Moreover, the scientific community remained divided. Some continued to minimize the dangers of ozone depletion and others endorsed strategies that overdramatized the situation as a way of mobilizing public opinion. The government, meanwhile, decided to emphasize ozone depletion as an exemplar of the kind of global human problem that East–West collaboration could address.

Similarly, the decision to participate in the international ozone regime occasioned overt governmental commitments to bring the production of CFCs under control. Both the Soviet Union and the Russian Federation have publicly pledged to comply with the targets of the treaties.

Very initial steps toward converting enterprises to the production of nonactive ozone substances were taken in 1988–1989. The production of propane–butane-based spray cans began in Byelorussia in 1988. Three aerosol enterprises shifted their production to nonactive propellants based on ozone-saving technologies in 1988–1989—an event deemed sufficiently newsworthy that it was announced by Minister of Foreign Affairs Eduard Shevardnadze (see *Literaturnaya Gazeta* 1989).

The Russian Federation's 1992 Law on Environmental Protection was the first enactment concerning ozone-depleting chemicals that had any real teeth. It provided (in article 56) that industries were bound to reduce, and later phase out completely, production and consumption of CFCs—though the act contains no specific deadlines or quantitative targets. (These were supposed to be elaborated separately.) This was followed in August 1993 by another Council of Ministers' resolution creating an interagency council specifically charged with coordinating national efforts to implement the Montreal Protocol (see "Ob utverzhedenii" 1993).

Despite claims in 1989 by a former director of Gidromet that the Soviet Union was in compliance with the Montreal Protocol and the production of CFCs was not increasing, the actual compliance record is very likely mixed. In 1993 the Ministry of the Environment stated officially (in Gosudarstvennyi 1993) that Russia was not in complete compliance with the Vienna Convention and the Montreal Protocol. Russia has, the Ministry claims, fulfilled its obligations on information exchange and scientific research, and has met its obligation to reduce production and consumption of ozone-depleting substances.

The former was doubtless a result of government policy, and by and large something of a success story. The 1988 Council of Ministers' resolution established a large-scale program of monitoring and research on the state of the ozonosphere. A scientific commission was established to address various facets of the ozone depletion problem, including evaluation of the long-term impact on human health, ecosystems, climate, and agricultural productivity. It also was designed to address the development and use of alternative substances and ozone-saving technologies for industry. An important part of this project was realized with the creation of the Soviet–American project Meteor 3-TOMS (Total Ozone Mapping Spectrometer), involving the Soviet (now Russian) Committee for Hydrometeorology (Gidromet) and NASA; this joint effort focuses on monitoring the ozone layer and linking changes with human-induced activities, notably the production and consumption of CFCs.

In the early 1990s, Russia began the process of elaborating a national program titled "The Production of Ozone-Benign Refrigerants in Compliance with the International Obligations of Russia to Protect the Ozone Layer" (see *Zelenyi Mir* 1993). The program envisaged continued research on the ozone layer, including the modeling of the interactions between man, the ozone layer, and the biosphere, along with regular monitoring of the ozone layer.

As for the production and consumption of ozone-depleting substances, these are likely to be primarily a happenstance of the huge social, political, and economic transformations that occurred in the former Soviet Union and in the Russian Federation, largely independent of governmental policies.

One part of this explanation concerns the collapse of the Soviet Union. In the words of a U.N. Environment Programme document, obviously paraphrasing the remarks of the Russian representative:

> His country had accepted the position of successor State to a previous multinational State that had fissured into its component parts. Although that had meant a considerable reduction in its territory and population, there had been no recalculation of its assessed contribution. A further complication was that most of the other successors to the previous multinational State were not Parties to the Protocol, with the result that what had been previously transactions in controlled substances had become trade with non-Parties and would henceforth be governed by Article 4. (UNEP 1992a:11)

What was domestic politics in the Soviet Union had become international relations for the Russian Federation.

The other part was, of course, the situation in the economy. Comments made by environmental officials of the post-Soviet Russian Federation in the early 1990s suggested that little progress had been made in efforts to comply with the Montreal

Protocol. In an interview in 1992, Alexei Yablokov, the State Counselor for Ecological and Health Policy of the Russian Federation, stated:

> We ... signed this Protocol ... but aren't doing anything to put its provisions into practice. The whole world is now changing over from chlorofluorocarbons, which destroy the ozone layer, to their substitutes.... The President assured us that a state programme would be worked out for reducing the discharge of ozone-destroying substances into the atmosphere. None of the programmes adopted in the Soviet Union before worked, but who cared? (FBIS 1992a:6)

The situation evidently remained little changed in the mid-1990s. Consumption of CFCs had clearly declined. In its quarterly report "Facing the Global Environment Challenge" (September 1995–January 1996), the World Bank reported a drop in consumption from 100,000 to 40,000 tons. More problematic are the facts about production, imports, and exports. These difficulties leave us somewhat up in the air in our efforts to establish precise levels of Soviet and Russian compliance. This is a continuing story.

At the fourth Meeting of the Parties to the Montreal Protocol in November 1992, the Secretariat observed dryly, "The former USSR has reported the same data for the years 1986, 1989, and 1990" (UNEP 1992d:12). Nikitina observed:

> It was rather difficult to obtain the data of monitoring the emissions of [ozone-depleting substances] as well as CFC production data. It was not widely published and it was only for internal use. According to ... international experts on ozone layer issues, before [the] Vienna Convention entered into force [the] USSR ... refused to provide CFC production data to the international community. One [reason was the] involvement in ozone research of interests of the Ministry of Defense, that resulted in substantial secrecy in this field. (Nikitina 1992:16)

More recently an American NGO, Ozone Action, has been sharply critical of the role of India, Russia, and China in the illegal trade of CFCs.

Nevertheless, we can report some interesting developments in the mid-1990s that flesh out the story of Russia's compliance with the Montreal Protocol. In May 1995, Russia, Belarus, Ukraine, and Poland jointly submitted a request that they receive a five-year extension of their phaseout schedules, exemption from contributions to the Multilateral Fund, and international assistance in phasing out ozone-depleting substances. For some experts, this represented a step toward noncompliance and a harbinger of Russian noncompliance beginning in 1996 with respect to the consumption of certain CFCs, carbon tetrachloride, and methyl chloroform.[10] On another perspective, it is to the Russian government's credit that it has reversed the traditional Soviet practice of avoiding statements of noncompliance.

And this change in behavior seems to have been effective. After much tough bargaining involving Russia and the Implementation Committee of the Montreal Protocol, in collaboration with a bevy of international organizations, Russia accepted most of the Committee's recommendations. It has reported baseline production and consumption data and has agreed to submit a detailed phaseout plan. The World

Bank launched a project in March 1996 to replace CFC propellant with hydrocarbon aerosol propellant in two enterprises that consumed slightly more than 10 percent of the Russian total in 1992. As in the case of the World Heritage Treaty, Russian cooperation with international organizations has resulted in support from those institutions that offsets the Russian government's diminished political capacity.

CITES

The progress made in the 1970s and 1980s that resulted in part from a change in Soviet attitudes on the value of endangered species, away from the purely exploitative attitudes that characterized the earlier Soviet period, has recently been threatened by changing circumstances in the former Soviet Union. The traditional Soviet attitude toward species protection was quite utilitarian. Thus, a 1960 Russian conservation law stated, "It is forbidden to destroy noncommercial wild animals *if they do not harm the economy or public health*" (see "Zakon ob okhrane ..." as quoted in Pryde 1991:185).[11] Evidence of change in that attitude began to appear in the 1970s. First in international agreements and then in national legislation, a more genuinely conservationist ethic began to take hold throughout the 1970s and 1980s. CITES was just one of the many international agreements signed by the Soviet Union in these years pertaining to endangered species, including polar bears, Siberian cranes, snow geese, and many varieties of fish and crustaceans. Article 18 of the 1977 "Brezhnev" Constitution referred to the "scientific, rational use of the ... plant and animal kingdoms" (see Pryde 1991:185).

The USSR Red Book (describing declining and endangered flora and fauna species) was published in 1974 and revised in 1978 and 1983. By the late 1970s, most republics had followed suit with comparable books. As shown by Pryde (1991), the Red Books follow the categorizations employed by the International Union for the Conservation of Nature and Natural Resources (IUCN).

Two years after signing the treaty, the Soviet Union proceeded to adopt national legislation to implement CITES as a precursor to Supreme Soviet ratification. A Council of Ministers' resolution prohibiting the taking of wild animals and plants whose trade is regulated by CITES was adopted on August 8, 1976. It forbids the sale, purchase, or exchange of such species of wildlife and of their parts and derivatives, except in cases provided for by CITES. Moreover, such species can neither be imported into nor be exported from the USSR, other than by a permit issued by specially empowered Management and Scientific Authorities.

Initially, the Management Authority was the Department of Nature Protection, Nature Reserves, and Hunting of the Ministry of Agriculture. When Goskompriroda was established, it took over that task. Subsequently, the monitoring of the fulfillment of international obligations became the task of Minpriroda. The Department of Nature Protection's functions included issuing permits for the shipment of wildlife species as well as their export and import operations. Similarly, the Ministry of Agriculture's Research Laboratory on Nature Protection was initially designated as the Scientific Authority provided for in the treaty, a task subsequently assumed by the Research Institute for Nature Protection, housed in the Ministry of the Environment.

An annex to the resolution details a list of species found in the USSR that are subject to trade control as a result of CITES. It has been regularly revised and updated in the interim.

A set of resolutions paralleling the August 1976 Council of Ministers' resolution was introduced in most Soviet republics. For instance, a month (September 8) after the adoption of the Soviet decree, the Russian Federation passed a resolution that virtually repeated the provisions of the Soviet act.

Several governmental agencies at the all-Union level augmented the implementing legislation by issuing a number of administrative regulations. These explicitly regulated the export of wild animals and plants covered by CITES, provided for the inspection of trade traffic, and made provision for the functioning of the Management and Scientific Authorities provided for in the convention.

Both Soviet authorities and, after them, the authorities of the Russian Federation have been particularly attentive to articulating a detailed system of issuing trade permits. In order to obtain a trade permit, a declaration must be sent to the Management Authority, with a copy to the designated Scientific Authority. The declaration must contain the following information: the purpose and provisional dates of the export or import, the name of the species, the quantity, a description, and the origin, plus additional documentation describing breeding conditions, date and place of birth if born and bred in captivity, and so on.

For imports from a state not a party to CITES (an important category, given that the Central Asian republics are not members), documents from the designated Scientific Authority *in Russia* are necessary.

The national legislation often was at least as strict and, in some instances, envisaged stricter controls than the Convention norms. The August 8, 1976, Council of Ministers' resolution implementing the CITES treaty called for stricter controls than the obligations assumed under CITES. It was not limited to the international trade of endangered species of fauna and flora. Rather, the resolution prohibited (article 2) even the acquisition (or allowed it only by special permit issued by responsible government organs) of certain types of wild animals and plants included in the national Red Book and covered by the annexes to the Convention. Although these norms were written into a document implementing CITES they apparently reflected the content of prior domestic legislation.

It is clear that enforcing compliance with broad legislation designed to protect endangered species was never an easy task because of the attractiveness of poaching. In the Soviet preserves and reserves it was often open season on game wardens, who appear to have enjoyed a popularity and legitimacy akin to "revenooers" in Kentucky during the 1920s. Former Goskompriroda chief Vorontsov complained in 1989 that there were "still instances of rangers and fish protection workers [being] killed on duty," and observed that "In almost every issue of the Goskompriroda magazine ... examples are given of poaching by high officials" (see FBIS 1989a:81)

With respect to CITES per se, which focuses only on international trade, the consensus of persons we interviewed, both in the relevant American agencies and in the United Nations Secretariat, is that the Soviet Union complied well with CITES. In this respect there is nothing like a good old-fashioned authoritarian and bureaucratic regime. The Soviet Union had tight borders. There was little smuggling of icons or Siberian tiger skins. It reported its trade statistics regularly to the CITES Secretariat. (It reported hunting permit totals less regularly.) The trade statistics were extensive and detailed. They included export, import, and reexport operations; specific numbers and destinations; and names of the species. They indicated where the animals originated, the purposes (see table 9.4), and numbers traded illegally. Regular reports on legal trade in animals were submitted. The reporting has been sufficient to make it possible for the World Trade Monitoring Unit in Great Britain to compile comprehensive trade reports for many years.

The granting of permits was limited to a small group of regular players in the export of exotic animals. Noncommercial permits for exchanges were granted to zoos and a few scientific institutes, and to the state circus for the animals to take part in circus performances abroad. Units within the Ministry of Foreign Trade had monopoly privileges over commercial permits. Similarly, the selling of fur pelts at the Leningrad fur auction was strictly regulated, and representatives of the Management Authority (currently Minpriroda) visited the auction regularly.

The authoritarian nature of the Soviet system in the Brezhnev era was not, however, the whole story. By the 1970s, protection of wildlife had become an accepted norm in the USSR. There were some important achievements as a result. One of the specific features of the national parks in the Soviet Union was scientific research and maintenance of breeding, reproduction, and reintroduction of endangered and rare species of wildlife. Ninety percent of the amphibians, 55 percent of the birds and fish, and 40 percent of the mammals listed in the USSR Red Book were conserved and reproduced in the national parks.

The result was that, coupled with the establishment of strictly controlled hunting areas, certain endangered species were revived. The populations of many valuable types of wild animals, including sable, river beaver, elk, and wild boar, increased during the 1980s.

Table 9.4
Russian trade in live endangered/vulnerable species

				Exports			
Species	1985	1986	1987	1988	1989	1990	1991
Snow leopard	3	1	0	1	2	0	2
Markhor	0	0	0	2	6	0	0
Polar bear	1	3	4[1]	0	1	12	37
Siberian tiger	10	2	0	0	7	3	25
Cheetah	0	5	7	9	0	0	1
Mountain sheep	0	0	0	0	0	25	0
Bottle-nosed whale	0	0	0	0	3	0	11
Red wolf	0	0	0	0	0	0	0
Totals	14	11	11	12	19	40	76
				Imports			
Species	1985	1986	1987	1988	1989	1990	1991
Snow leopard	1	0	0	1	0	0	1
Markhor	7	1	0	0	5	0	0
Polar bear	0	0	0	0	0	0	19
Siberian tiger	0	0	9	8	6	3	18[2]
Cheetah	0	0	1	7	0	2	2
Mountain sheep	0	0	0	0	0	0	0
Bottle-nosed whale	0	0	0	0	6	0	4
Red wolf	0	0	0	0	4	2	4[3]
Totals	8	1	10	16	21	7	48

[1] Importing countries claim they received only 2 live polar bears.
[2] The data show that these 18 animals were originally part of an export arrangement that originated in Russia. It is unclear whether these animals were returned to Russia after exportation or whether they represent some sort of verification process of the original export arrangement.
[3] Country of origin claims it exported only 2 red wolves to Russia.

The situation began to unravel in the late 1980s and deteriorated drastically in the early 1990s. There were three obvious factors at work, and a fourth proved quite important as well. One obvious factor was the collapse of the Soviet Union. This resulted in a situation where key areas of the former Soviet Union, the Central Asian republics primarily, are not formal parties to CITES and where governmental agencies are in fact actively engaged in the hunting and export of highly endangered species. Even within Russia, moreover, the central government's capacity to affect behavior in the far reaches of the country and to control its borders has diminished substantially. Third, the shift in the direction of a market economy has had profound effects. Finally, the growing affluence of East Asian states has enormously increased the demand for exotic pelts, horns, and parts of endangered animals as well as endangered plants. This has contributed to the rapid growth of illegal trade in flora and fauna between Russia and China and Russia and Taiwan.

The legal situation with respect to the relevance of CITES, the Soviet Red Book, and Soviet environmental laws designed to implement international agreements for the Central Asian Republics is at best ambiguous. (Russia issues hunting permits for all CIS states.) In practice, preexisting agreements have, at least until quite recently,[12] been irrelevant to state behavior. In Turkmenistan, local authorities have allowed hunting in the Gasan-Kuliski section of the Krasnovodsk Nature Reserve. Special catalogs advertising hunting options are published in Europe and Asia; they list species covered by various CITES annexes and/or listed in the Soviet Red Book and the Red Books of the former republics. In Tadjikistan, a government organization advertises hunting tours and the precise prices ("prices may vary depending on sex, age, and trophy values") for endangered (urial, $5,000; mouflon, $18,000) and threatened (Tien Shan bear, $2,000) species.

The situation has deteriorated drastically in Russia as well. The shift to participation in the global market has had the long-range effect of greatly increasing the practice of clear-cutting the Siberian taiga. Hyundai, for instance, has permission from the regional government in Ternei (the domain of the Siberian tiger) to cut 437,000 cubic yards of forest every year for 30 years. In the long run, the practice of clear-cutting represents an obvious danger to such wide-ranging species as the Siberian tiger.

In addition, marketization has resulted in the rapid growth of legal commercial sales of exotic fauna and flora. The 1989 abolition of the state monopoly on foreign trade, including that in species covered by CITES, resulted in a vast proliferation of state and nonstate organizations having the right to trade with foreign countries.

With the turn to the market, the volume of trade in species covered by CITES and the number of permits issued by the Management Authority of CITES grew substantially in the last years of Soviet power (see figure 9.2). There was a threefold increase in the number of permits issued for commercial and other operations between 1985 and 1992. During the period November 1, 1990–November 1, 1991, approximately 10,000 fur skins were sold at the Leningrad fur auction for $3 million; 350 skins were sold as hunting trophies ($2 million) and roughly $1 million worth of plants, primarily ginseng, were sold commercially. Figure 9.3 shows the reported pattern of Russian trade in live, endangered or vulnerable species in the last years of the Soviet period. Both it and figure 9.2 should be read with much caution; the large number of exports *and* imports of Siberian tigers, for instance, reported in 1991 probably reflects the serious effort to encourage the breeding of Siberian tigers in zoos around the world. Nevertheless, the overall impression is that the numbers

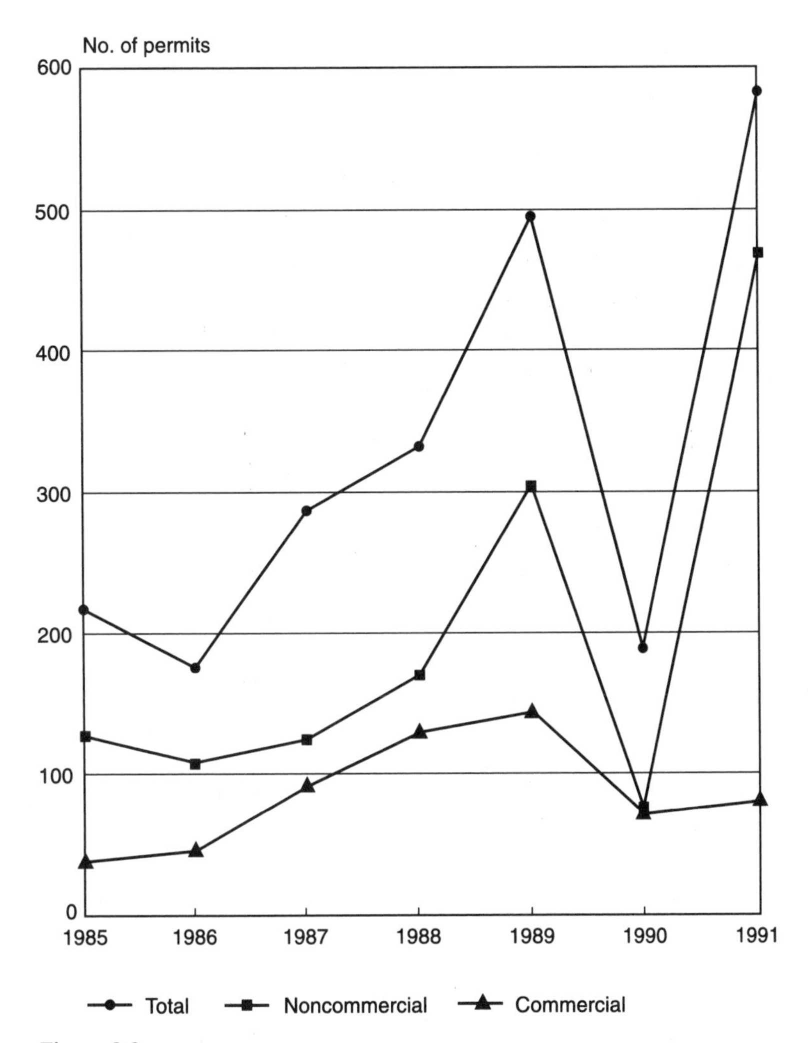

Figure 9.2
Total CITES export/import permits issued in USSR/Russia, 1985–1991

have increased. The growth of legal noncommercial exports and commercial sales, while substantial, is not the crux of the problem.

Legal sales and exchange segue into the far more dubious. We interviewed people in American and U.N. agencies responsible for CITES and consistently heard the theme that Russian administration of permits per se is not, by and large, a problem. The problem, they believe, is that the correlation between reality and the permits is low and diminishing. Putatively scientific permits often cover illegal commercial activities.[13]

What is far more distressing is that poaching and the illegal export of hides, horns, and parts of endangered species are out of control. The incentives to smuggle

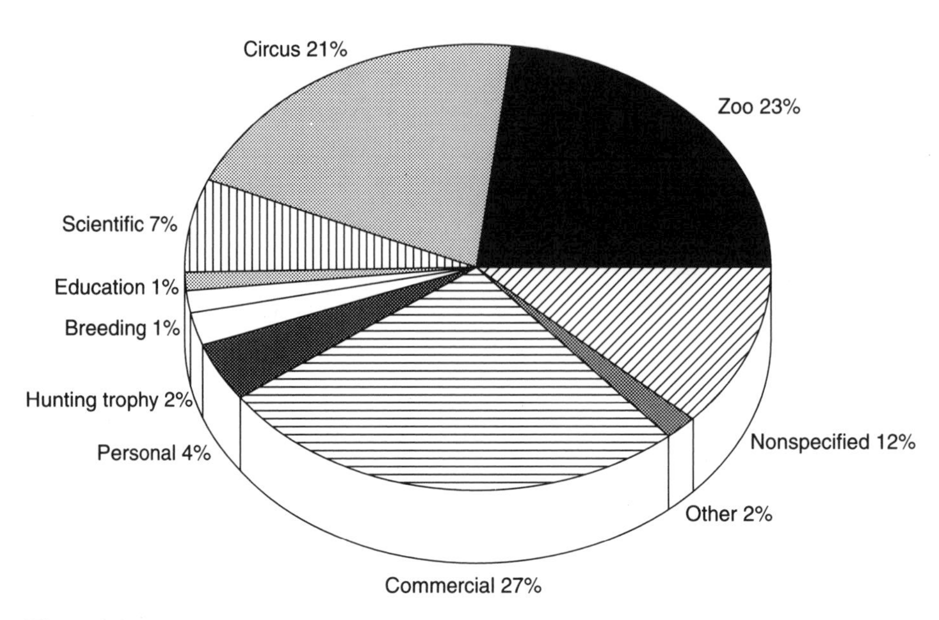

Figure 9.3
Total CITES export/import permits issued in USSR/Russia, 1985–1992, by purpose

and the opportunities across transparent borders are immense. The situation is akin to that for South American dealers in cocaine. In 1992, a game warden made the equivalent of $15 a month. He might be offered as much as $15,000 for a tiger skin. Poachers pay $2,000 to hunt an animal listed in the Red Book. If caught, they pay a fine that is still based on 1977 prices, when the ruble and the dollar were trading at roughly par. "Bones [of Siberian tigers], pulverized and used in 'tigerwine,' bring about a hundred dollars a pound.[11] Saiga horn sells for between $350 and $500 a kilogram. Arctic falcons bring $50,000 [*sic*]. Only about 400 markhor goats still exist; they have 150 centimeter [5-foot] horns that sell for $10,000" (Quigley 1993:43).

The results are predictable. By 1993, the Russian mafia had become involved (see LeDuc 1993). Corruption in state organs was flourishing. The chief prosecutor of the Siberian region was personally involved in shooting and selling the pelts and derivatives of the Ussuri tiger abroad. In a sting operation in February 1993 (see *New York Times* 1993b), the KGB confiscated an Amur tiger skin from a game warden in Terne, where the Sikhote Alinsk Reserve has its office. (Tigers range primarily throughout the Sikhote-Alin Range in the Primorski Krai.) The warden claimed that he had been attacked by the tiger. Instead of reporting the incident, he skinned the animal. In the event, he accepted an offer of $5,000 from an undercover agent. In his defense, he asserted that "most game wardens ... make ends meet by hunting the animals they are supposed to protect" (see *New York Times* 1993b).

Likewise, export permits are issued but nothing else is done. The situation has even affected polar bears, the hunting of which was made illegal in the 1950s, and which are not threatened as a species. Harold Garner (1993), a member of the U.S.–Russian Joint Cooperative Effort, says that "no one really enforces the treaty beyond the issuance of permits." Unofficial reports of poaching have skyrocketed in the last few years. Japanese and Germans are the major hunters. In fact, hunting tours have been promoted by agencies—for instance, Glavokhota—that are allegedly responsible for overseeing the preservation of the polar bear. "In the capitalist tradition, the exploitation of polar bears is the new get rich quick scheme of the immediate future" (Garner 1993).

Other native species are in far worse straits. (One great advantage the polar bear has is that it spends much of its life on ice floes.) The situation with respect to the Siberian tiger has received extensive publicity. It is estimated that only 200–350 remain in the wild. Other animals have been acutely affected as well. Among the mammals in special trouble are the East Siberian leopard, the Amur goral, the saiga, markhors, and bears. For instance, in the Far East taiga (Verkhnebuireinsky region), there has been a fivefold decrease in the musk deer population, a threefold decrease in the numbers of elk, and a twofold decrease in the numbers of sable. The stocks of European saiga are close to depletion, and by the mid-1990s there were no more than 700 goral in all of Russia.[14]

Finally, the permeability of Russia's borders has contributed to a further problem. In addition to the export of species found in Russia and the transshipment of flora and fauna found in other parts of the ex-Soviet Union, Russia has become a transit point for exporting nonnative species from its former key allies, particularly Cuba, Vietnam, and Angola.

In short, while there were a few bright spots on the horizon (discussed below), it is difficult not to conclude that the collapse of the Soviet Union, the permeability of Russia's borders, and the shift toward the market in all the regions of the former Soviet Union have had grievous consequences for compliance with the CITES regime.

The London Dumping Convention

The consequences of the collapse of the Soviet Union for compliance with the London Dumping Convention have been almost completely salutary. The Soviet Union exercised effective control over its boundaries. It seems to have had no control over the activities of the Navy as they pertained to the observance of international agreements or Soviet legislative enactments of international agreements.

The first hints of pollution by the Navy—and its almost complete independence of civilian control—began to appear in the Soviet press during the later Gorbachev years. The following incident, described by Vorontsov in 1989, seems to have been typical. The Vladivostok division of Goskompriroda had collected a lot of material on the oil pollution of the sea and on dumping by the ships at sea. Apparently everybody is on our side—the [regional Communist] party committee and the [regional] executive committee. However, the commander of the Red-Banner Pacific Fleet does not take orders from anyone. I have sent a letter to the minister of defense. We are waiting for a response and, mainly, for results. After all, it is the same nature *priroda* for both the civilians and the military. (FBIS 1989c:114).

It was only after the collapse of the Soviet Union that the most gruesome details of the story were revealed. The Soviet Union turns out to have been an egregious violator of the London Dumping Convention, principally, though not exclusively, through the illegal disposal of high-level radioactive materials at sea. (In 1983, all parties, including the USSR, agreed to a complete ban on the dumping of all radioactive materials.) One can quarrel with whether the Navy's dumping was illegal; as shown in Weiss, Magraw, and Szasz (1992), the Soviet Foreign Ministry took the view that article VII of the Convention exempted "those vessels and aircraft entitled to sovereign immunity under international law." What is not at issue is whether the Soviet government had an obligation to report all dumping. It did. The subsequent clause of the Convention requires that a state's immune vessels nevertheless "act in a manner consistent with the object and purpose of this Convention, and shall *inform*[15] the Organisation accordingly."

The sources of the documentation of these treaty violations are twofold. The major source is a comprehensive report, released in March 1993, compiled by a team of forty-six environmental experts, headed by Dr. Alexei Yablokov, former chief counselor to Yeltsin for the environment. That report followed on the heels of a major report on Soviet nuclear dumping prepared by Greenpeace, which in turn followed pressure on Moscow from the Dumping Convention Secretariat to reveal the real stuation (see *Nuclear Free Seas* 1993). (Yablokov had been head of Soviet Greenpeace prior to becoming counselor to Yeltsin, the President of Russia.) According to an article in *Nezavisimaia gazeta*,

The world community's patience ran out in 1991. A consultative conference of the convention's signatory states demanded that the USSR provide information on all the dumping of radioactive wastes it had done. Greenpeace got involved in the matter, preparing a report and a rough map of radioactive burials in the northern seas in September 1991, based on its own information. The facts cited in the report shocked the world public, and an international furor broke out. It was no longer possible to pursue a policy of silence, especially since in 1992 the next conference of member-countries of the London Convention repeated its demand this time in the form of a thinly veiled ultimatum. ("Russia still..." 1993:21–22)

The Greenpeace report was extremely extensive and differed from the White Paper issued by Yablokov in 1993 only in the details.

With regard to commercial vessels, the Soviet Union took steps to bring its practice into line with the Dumping Convention. Four years after the Convention had been signed and ratified, the Soviet government officially took note of its obligations when the Council of Ministers adopted, in 1979, Decree 222, "On Measures for the Fulfillment of Soviet Obligations Under the London Convention." Over the ensuing five years, commercial dumping of radioactive materials was "reduced and then completely prohibited," with the result that, according to the Yablokov report (1993),[16] dumping by the Murmansk Steamship Line was "discontinued" in 1984.

The chief problem during the Soviet period, however, was the Soviet Navy. No Soviet laws were adopted to address the dumping of RAM by the Navy. Moreover, the Soviet government did not even conform to the reporting requirements of the Convention. "In violation of the stipulations of the London Convention and the Decree by the USSR Council of Ministers, Goskomgidromet did not provide information about the dumping of RAM in the ocean to [either] the IMO or the IAEA" (Yablokov 1993).

Robert Axelrod and William Zimmerman (1981) examined the issue of lying in Soviet foreign policy. They concluded that while the Soviet government's statements about its foreign policy behavior often amounted to dissimulation, official Moscow rarely actually lied. When the Soviet government did lie, however, it was a whopper—as in the Cuban missile crisis.

Such was the case with the dumping of radioactive materials in the Arctic and Pacific oceans. The Soviet government informed the Secretariat of the London Convention in 1989—during the height of glasnost—that "the USSR has not dumped, is not dumping, and does not plan to dump radioactive wastes in the ocean" (Yablokov 1993:26).

This was a whopper: in the Arctic Ocean alone, the Soviet Union at that time (1989) had dumped more than twice the high-level radioactive materials dumped by the rest of the world *combined*. The dumping incidents were proportionately greater in the Arctic Ocean before the signing of the Convention. All the dumping in the Arctic Ocean, and much of it in the Pacific Ocean, took place above the fiftieth parallel in contravention of the London Convention. Moreover, much of the dumping of high-level radioactive material took place in relatively shallow waters—often at depths not exceeding twenty meters—especially in the Arctic Ocean, and the dumping of low-level and medium-level radioactive materials actually intensified during the Gorbachev period, despite the fact that the Soviet government had endorsed the resolution prohibiting all dumping adopted in 1983 (see figures 9.4 and 9.5).

The Russian Federation has undertaken efforts to bring the military within the framework of the law. For the first time, the fleet is now, in theory, subject to Russian domestic environmental law, which specifies that Russia obligates itself to implement its commitments under international treaties such as the London Convention. To give teeth to the law, Gosatomnadzor (the State Atomic Energy Control Committee) was specifically charged with oversight of both civilian and military

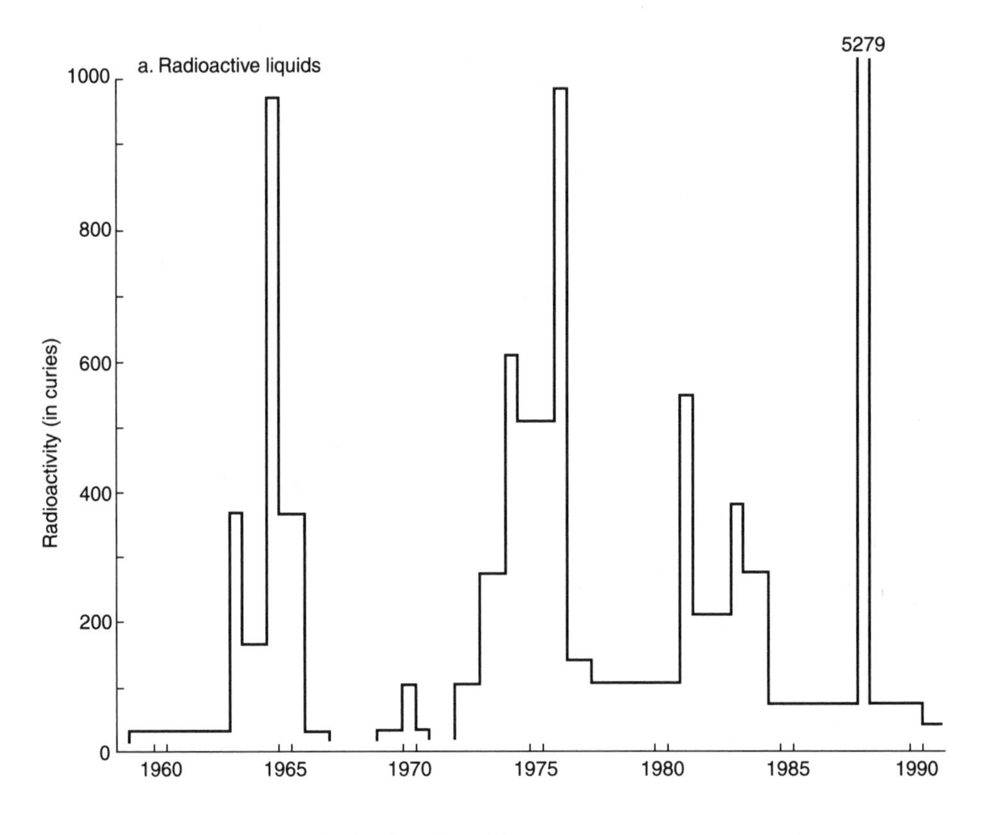

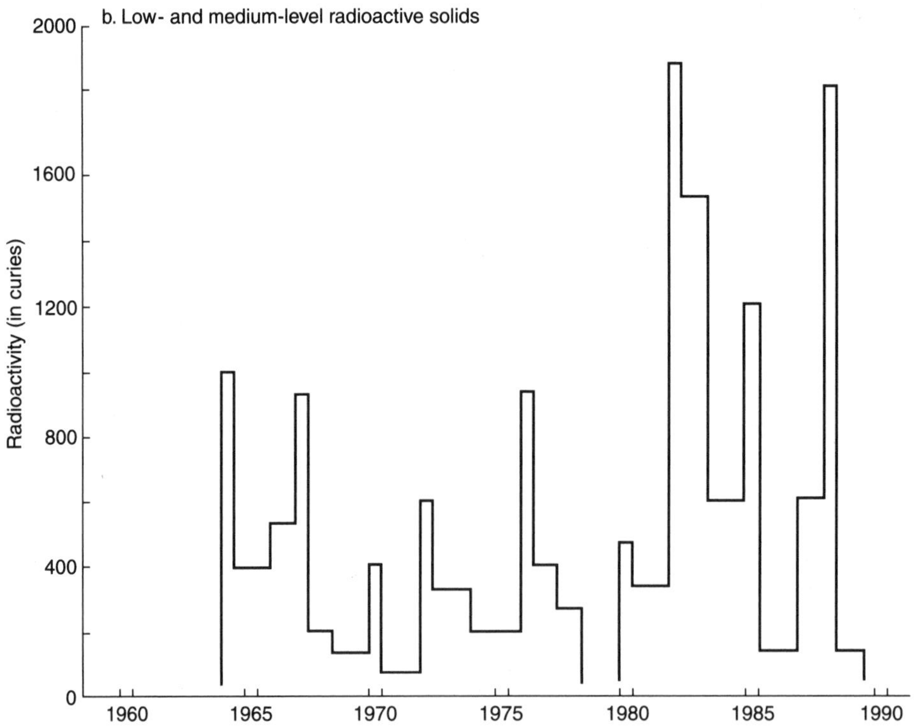

Figure 9.4
Soviet dumping of radioactive materials in the Arctic Ocean, by years

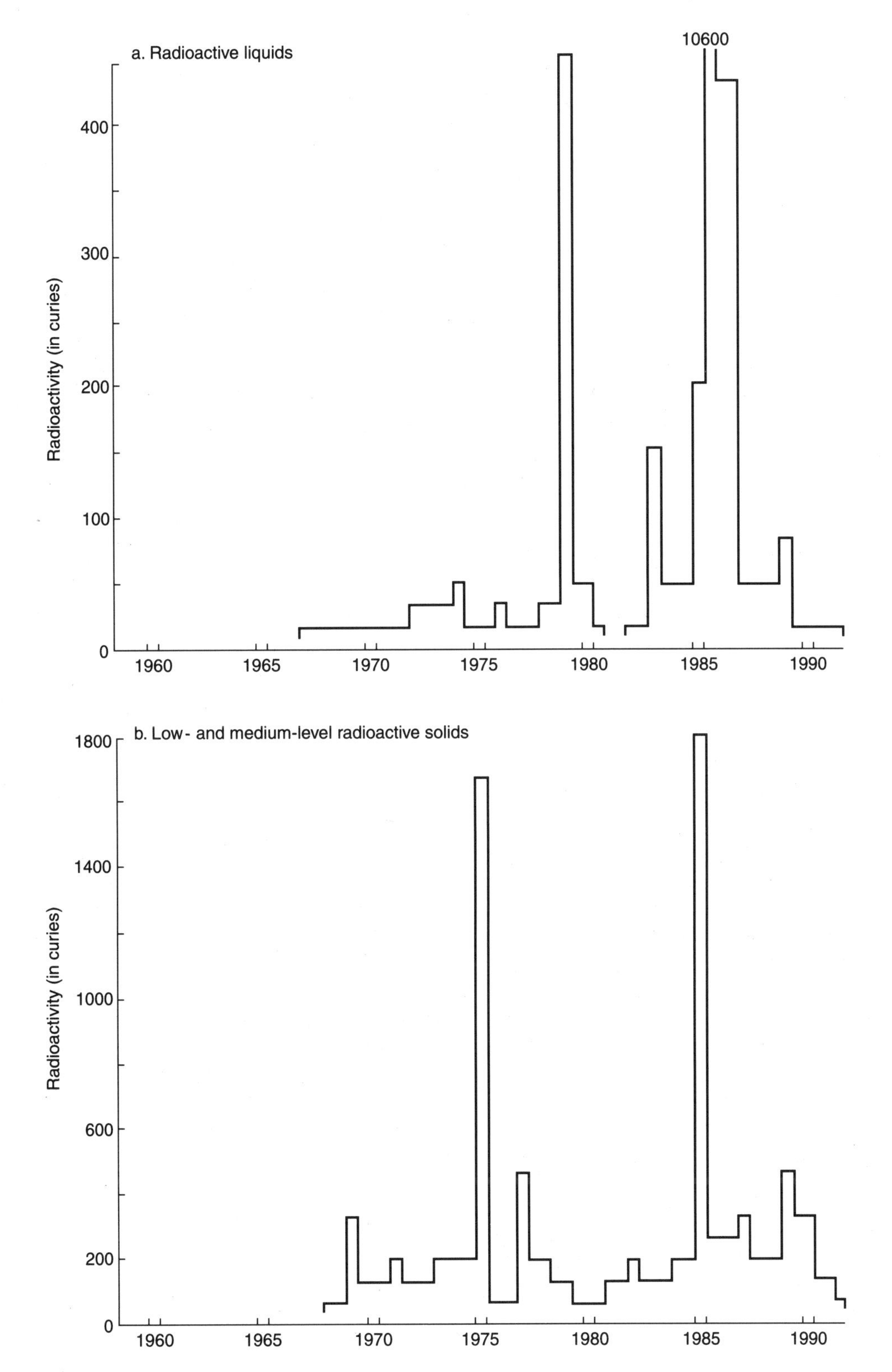

Figure 9.5
Soviet dumping of radioactive materials in the Far East, by year

dumping—and this seems to have had real consequences. The Naval High Command was informed on February 10, 1992, after Gosatomnadzor had received opinions from Minpriroda, the Foreign Ministry, and the Russian Scientific Commission for Defense Against Radioactivity, that oceanic dumping of radioactive materials must be "governed by international norms"—which meant, in effect, a rejection of the fleet's request to temporarily preserve the old order (Yablokov 1993).

And there is no doubt that dumping has decreased. The Yablokov report (1993) indicates that roughly fifty-five curies of radioactive material, down considerably over previous years, were dumped in 1992.

From the Russian Navy's perspective, the source of the problem has not been addressed. The Navy maintains it no longer has space to store the radioactive materials produced by its ships and submarines (see FBIS 1993c). An unpleasant incident occurred in the fall of 1993 when the Navy, rather ostentatiously and provocatively, dumped atomic waste in the Sea of Japan immediately after Yeltsin's visit to Japan. The Japanese doubtless were particularly exercised about this act, inasmuch as in the spring it was reported that they had decided to put up "100 million dollars in grants to dismantle nuclear weapons and prevent radioactive waste being dumped into the sea" (FBIS 1993a).

While the incident created an international furor, there are several hopeful signs for the future. The IAEA and the International Maritime Organization were notified in advance. The action was undertaken with the permission of the Environmental Protection Ministry (Ministry officials were on board at the time of the dumping). The Ministry required the Navy to work with regional fishing authorities and scientists to minimize the effects of the dumping. A joint Russian–Japanese–South Korean research team studied the area in question several months after the incident and reported normal radiation levels (see FBIS 1994). Finally, the Japanese government has pledged to provide the Russian Navy with additional storage facilities to avert future dumping in the region.

The Behavior of the Target Groups and the Achievement of Treaty Goals

We have three quite different patterns of implementation and compliance in the case of the three major environmental treaties involving the Soviet Union and its major successor, the Russian Federation. How did the conventions alter the behavior of the targets of the agreements? It is perhaps most difficult to pin down an answer to that question in the case of the Ozone Convention. Clearly, the Convention has facilitated scientific exchange of data and collaborative research. The major objective was to reduce the production and consumption of fluorocarbons. A few highly publicized changes in production technique were reported. It is very likely that the chief contributing factor to the goals of the Convention being approximated was the collapse of the Soviet, and now the Russian, economy. This may well have occurred without any relevant change by the CFC producers. At the same time, the international commitments undertaken in 1992, followed by efforts to develop a detailed plan for their implementation by 1996, suggests some seriousness about meeting future obligations, even if the realities of the Russian economy probably preclude restructuring and modification at a pace that would bring Russian reality in 1996 in line with Russia's treaty commitments under Vienna and Montreal.

The targets of the CITES treaty, by contrast, were presumably institutions and individuals who would, in the absence of constraint, transmit endangered species across borders. Initially, those targeted seem to have been rather well constrained as the Soviet government, despite some manpower limitations, controlled its borders. The years since 1991 have been a quite different story: the number of target groups in noncompliance grew steadily throughout the early 1990s. Whereas the players in the mid-1980s were largely a handful of state agencies, by the mid-1990s, domestic and foreign individuals, criminal gangs, state agencies, former Soviet republics, and smuggling operations from Angola, Vietnam, and Cuba were all involved and, it would seem, operating almost at will.

As for the London Dumping Treaty, here an important distinction exists between the civilian and military targets. In the case of civilian agencies such as the Murmansk Steamship Line, the evidence seems to be that implementing the convention produced changed behavior and compliance. The Soviet government may never have thought the Navy was a target of the Convention, at least with respect to inducing different behavior. In the event, the Soviet Navy's behavior did not change: it kept on dumping. Indeed, it was only in late 1991 that the Navy began to report on its dumping activities *to the government*, information that became a matter of public knowledge only after the Yablokov report was commissioned by Yeltsin in the fall 1992.

Given the behavior of the targeted groups, therefore, a mixed record must be reported with respect to the achievement of the purposes of the treaties. Since 1993 was the first year the Ozone Convention actually mandated a reduction in CFCs, it may be appropriate to defer judgment on whether its purposes have been achieved—and there remains the nagging issue of whether compliance, if observed, stems from behavior change or a dismal economy. If we take the targets of the London Dumping Convention to be the dumpers, then it seems obvious that it did lead to compliance by civilian agents, but that the Convention might well have had no effect at all on the Navy absent fundamental changes in the political system. Finally, CITES has achieved some of its purposes: a system of permits and monitoring is definitely in place. The situation in the former Central Asian republics suggests that were Russia not a party to the convention, the situation in Russia with regard to compliance would be marginally worse than it is currently, but that the treaty had genuinely induced altered behavior during its first decade, when the Soviet Union had an effective government enforcing norms concerning conservation that resonated reasonably well throughout the society.

Explaining Implementation and Compliance Patterns

To understand the decisions to sign and implement the three treaties to which the Soviet Union, and subsequently the Russian Federation, were major parties requires giving due attention to the international environment, and particularly changes in East–West relations. The signature and implementation of CITES were directly connected with the Brezhnev–Nixon era of détente, as was the London Dumping Convention. Likewise, the signing and implementation of the Ozone Treaty cannot be understood without an appreciation of Gorbachev's effort to transform the U.S.–Soviet relationship.

Some role must be allotted to international factors in order to understand several aspects of Soviet and Russian treaty compliance. Certainly, the growing compliance with the London Dumping Treaty has to be seen in the context of the successful transformation of American–Russian relations that followed the end of the Cold War and the demise of the Soviet Union. Similarly, the diminished compliance with CITES stems in part from external factors: the CITES disaster is not just a story of the collapse of the Soviet Union and its internal transformation. It also illustrates the old adage that nothing fails like success: growing prosperity in China and Taiwan has created an enormous demand for animal parts that are available primarily in the former Soviet Union.

Nevertheless, explaining compliance largely requires attention to domestic variables. We began our research with the assumption that the differences between the conventional Soviet command-administrative system, on the one hand, and the more open Gorbachev period in the Soviet Union and the post-Soviet Russian Federation, on the other, would prove central to an understanding of implementation and compliance with regard to the treaties under consideration. Surely, that assumption has been borne out in broad terms by the Soviet and Russian treaty experiences, especially in the cases of the London Dumping Convention and CITES.

Perhaps the most striking way the change in political systems has been manifested pertains to the involvement of external actors and the emergence of transnational coalitions that affect environmental policies. Sometimes, as in the case of the Ozone Convention, these players are largely limited to scientists and their government agencies, both within Russia and elsewhere, though the enhanced role for domestic *and* external actors needs to be emphasized even in this instance[17] (see Pryde 1991). An example is Ecological International. It is a nongovernmental center for business cooperation with foreign countries established in Moscow, the goal of which is to establish an international fund to protect the ozone layer and to generate ozone in the stratosphere. Significantly, support for the idea came from the Moscow Patriarchate, the industrial association Noosphere, and enterprises of the Ministry of Defense.

Likewise, the role of NGOs in the implementation of CITES, and especially in preventing illegal trade in endangered wildlife species, appears to be increasingly important in the new political conditions. For instance, there is a growing NGO movement in defense of wildlife in Russia's Far East. That movement is trying to establish contacts with foreign NGOs in order to obtain their support. Indeed, early in the First Clinton administration, several ecological and nature conservation organizations in the Russian Far East[18] appealed to President Clinton, urging him to impose economic sanctions on China and Taiwan in order to press them to stop international poaching of endangered wildlife in Russia, in violation of the CITES agreement.

Similarly, a meeting on the illegal trade in endangered species in the former USSR that took place in Moscow at the end of November 1993 illustrates how international organizations and NGOs have become involved in the Russian policy process. Participants included people from the Wildlife Trade Monitoring Unit, TRAFFIC-Europe, the CITES Secretariat, the EEC, and the United Nations' East European Office. Moreover, much of the funding to sustain such species as the Siberian tiger results from joint projects supported by grants from such American organizations as the National Geographic Society, the National Fish and Wildlife Foundation, and the Exxon Corporation.

In the case of the London Dumping Convention, we find the Secretariat, the parties as a whole, Greenpeace International, and Greenpeace in Russia conspiring to induce the Russian government to reveal data about Soviet oceanic dumping—and the President's Counselor for the Environment a willing co-conspirator. In the days of the command-administrative system, one would not have seen the headline that appeared in *Izvestiia* in 1992—"Greenpeace Invades Russia."

Certainly, therefore, the policy process has become more complex. Outputs were affected as well. It is difficult to imagine the Soviet/Russian Navy's dumping of radioactive materials ever being limited had there not been a radical transformation in the nature of the political system, which in turn resulted in a dramatic change in civil–military relations in Russia. The dumping of radioactive materials, the political climate of the post-Chernobyl period, the activation of mass publics and the media, and the increased transparency of the political system to the outside seem to have had considerable effect on Soviet, and subsequently Russian, implementation and compliance. In particular, the military has gradually become a part of the domestic legal order in principle, and to some extent in practice.

Nevertheless, the effects on outputs prove to have been more complicated than we assumed they would be. In addition, other domestic variables need to be given somewhat greater attention. In the case of ozone depletion, the changes in the political system did not have enormous effects on behavior. Given the lack of tangible immediacy of the ozone danger, mass publics were not mobilized. Domestic and foreign actors were limited largely to scientists and scientific institutions. The new Soviet leadership under Gorbachev did press hard for international scientific collaboration, with the U.S.–Soviet Total Ozone Mapping Spectrometer being one example of many achievements, and it did regard ozone depletion as a global human problem. The outcome stemmed from changes in thinking by the leadership and by the Soviet scholarly community, and could easily have occurred independent of regime change.

Similarly, the manifestly changed behavior in the case of CITES is not largely explained by the increased openness of the political system. Rather, the relevant variables seem more to have been the collapse of the Soviet internal empire, the shift to a market economy, and Moscow's diminished political capacity.

Basically, what Russia has gained by way of control over the military, in comparison with the Soviet Union, it has lost with respect to the regions. The enormous devolution of power to localities, coupled with the inability of the central government to provide regional units with resources for environmental monitoring, has resulted in a situation where local governments are encouraging foreigners to hunt and trap endangered species as a way of generating revenue. Some of this growing threat is prompted by crude and unregulated market behavior; much of it is a function of the Russian central government's inability to exert effective political power in the provinces and the changed nature of Moscow's relations with Dushanbe and Almaty (Alma-Ata). Distinctions between open and closed systems were not central, wheseas differences between states with effective and ineffective political capacity were.

Moreover, some of what may have seemed to be a product of political institutions may instead be related to organizational cultures that have persisted across regime change. When Russian officials profess to be shocked (shocked!) that scientific teams and tour groups are reporting an increasing number of dead bears in the Arctic when the number of hunting permits has not changed substantially, they are behaving in ways that were quite familiar in the Soviet period. Indeed, the "they" may not have changed either; given the scarcity of resources for environmental controls in the Russian Far East and Moscow's diminished political capacity, personnel changes in the lower reaches of the Ministry of the Environment may have been quite modest.

Perhaps one reason why there were differences across the treaties as well as across the regimes is that the treaties should not be thought of as having all occurred within a single issue area but, rather, as coming from different issue areas. In principle, CITES, the London Dumping Convention, and the Ozone Depletion Treaty all address preservation of the global commons. In practice, though, key players behaved as though the games were quite different. Naval dumping, at least in the context of the Cold War, was not an environmental issue but a security issue. Security issues generally involve quite closed political processes everywhere, regardless of political system; on reflection, the revelation of widespread dumping and the disjunction between treaty and reality in this respect may be of a piece with the 1993 revelations in the United States of widespread nuclear weapons-related experimentation on human subjects in the early years after World War II.

Similarly, it may be most appropriate to view the lack of CITES compliance in the 1990s as a kind of distributive politics wherein, in the absence of effective sanctions, everyone who counts politically in various Russian localities—entrepreneurs, local politicians, game wardens—are cutting themselves in at the expense of a general interest. The political systems differ substantially between the Central Asian republics and the Russian Far East, with the consequence that the cast of characters who count politically may differ. But in each case, almost everyone who counts politically is getting a piece of the action. The analogy with cocaine production in Latin America warrants repeating: those who are selling hides for a hundred times their monthly salary are behaving rationally.

In general, therefore, implementation of, and compliance with, environmental treaties are perhaps most likely in circumstances where other problems traditionally encountered when trying to regulate the commons are dealt with effectively. Collaboration in environmental treaties occurs when states are intent on improving relations for other reasons. Effective compliance with treaties takes place when states have high political capacity, more or less regardless of political system. Political system matters, however, because the nature of the political system strongly influences a country's transparency—whether by "transparency" one means permeable borders or the active role of external actors in the domestic political processes that shape compliance.

Some policy prescriptions flow from these conclusions that will affect outcomes at the margin. Building bureaucratic capacity in the Russian Ministry of the Environment should be a high priority (see Peterson 1993a). Here the prospects improved in the mid-1990s as the Russian government sought to increase its ability to affect the behavior of persons in other former Soviet republics and to strengthen laws dealing with the import and export of endangered species across Russian borders.

These steps should be encouraged. Interestingly, one of the quickest ways to accomplish this is by the greater involvement of IGOs and NGOs in the policy process. Joint Russian–U.N. efforts to involve those ex-Soviet republics not a party to CITES or the ozone depletion regime are in order, even when such steps produce cries of the restoration of the Soviet/Russian Empire. Restoration of an effective customs regime in Russia seems similarly important—indeed, there were some indications by 1994 that this was under way (see *Tamozhennyi Kodex* 1994).

We make these suggestions, knowing full well that in the short run the Russian government is to some extent utilizing the highly credible threat of collapse to achieve something of a free ride in terms of the monitoring of treaties that it should, in principle, be administering itself. Only steps such as these that combine efforts to enhance governmental capacity (or provide surrogates for governmental capacity) with steps that increase transparency are likely to engender substantial increases in Russian environmental treaty compliance.

Notes

Thanks to Denise DeGarmo and Alexei Roginko for useful research assistance, including the preparation of the CITES tables.

1. As part of the agreement at the end of World War II that resulted in the Soviet Union's participation in the United Nations, the Ukrainian SSR and the Belorussian SSR were also members. The Ukrainian SSR ratified the Vienna Convention on June 18, 1986; the Montreal Protocol on September 20, 1988; the World Heritage Convention on October 12, 1988; and the London Dumping Convention on March 6, 1976. The Belorussian SSR ratified the Vienna Convention on June 20, 1986; the Montreal Protocol on October 31, 1988; the World Heritage Convention on October 12, 1988; and the London Dumping Convention on February 28, 1976. Neither ratified the ITTA.

2. *Pravda* June 29, 1988, as quoted in DeBardeleben (1990:239).

3. *Pravda*, July 30, 1988, as quoted in DeBardeleben (1990:239); emphasis added.

4. While legally the distinction throughout this chapter is between the Soviet Union and Russia, we believe that for the purposes of comparison it makes more sense to distinguish between post-Soviet Russia and the Gorbachev era in the Soviet Union, on the one hand, and the remainder of the Soviet period, on the other.

5. Official attachment to the labor theory of value, however, did not preclude conservationists from having an awareness of the need to conserve and perpetuate important animals and plants.

6. Persons in the Marxist–Leninist tradition are scarcely the only ones who have assumed that natural resources are unlimited. Bruce Catton presents the history of Michigan as being determined by people who believed the supply of beaver was inexhaustible, followed by people who thought the white pine was inexhaustible, followed by people who thought the copper would last forever, followed by people who thought the automobile....

7. For example, oil was sold to enterprises for approximately nine rubles a barrel, less than the price of a liter of vodka.

8. In Soviet administrative law, there is a difference between a ministry and a state committee. A ministry administers things directly, whereas a state committee oversees the implementation of policies by ministries.

9. The creation of the council was in response to an *ukaz* of November 30, 1992, "Ob osobotsennykh ob'ektakh kul'turnogo naslediia narodov...."

10. A view encountered in the Montreal Protocol Secretariat is that in 1995, Russia was in compliance with the Protocol. The Russian Federation's statement that it will not be able to

comply with the Protocol's standards in this construction represents an interesting illustration of how compliance can be said to occur; by informing the Secretariat of noncompliance, the Russian government actually in a sense remains in compliance while continuing to be eligible for international assistance.

11. His italics.

12. In November 1993, there was a meeting held in Moscow, jointly sponsored by Minpriroda and various U.N. agencies and NGOs. It was attended by representatives of former Soviet republics, which suggested some interest in (re)establishing links with the CITES regime. The problem of coordination of environmental agreements was one of the items on the agenda of the 1993 session of the CIS Intergovernmental Environmental Council as well.

13. One vivid example as reported in the *Moscow News* (1993): in 1993, 4,500 tons of crabs were taken for "scientific purposes" by Okhotskuisan, a joint venture, fifteen times the allowable national quota, 300 tons.

14. Estimate provided to us by S. Golubchikov.

15. Italics added.

16. This is the unpublished report to the President issued in March 1993. For the published version, see *Rossiiskie vesti* 1993.

17. Pryde observes:

A new form of bilateral cooperation emerged in the late 1980s when the Soviet Union began working with private conservation organizations on certain specific projects.... [The] Soviet Academy of Sciences signed ... [an accord with the Natural Resources Defense Council] to conduct research on energy efficiency in buildings and appliances, with the goal of reducing the release of greenhouse effect and ozone-depleting gases. (1991:269)

18. These included the Center for the Defense of the Wild "Zov Taigi," the Far Eastern Leopard Fund, the Tiger Protection Society (all in Primorye); the Ecological Club "Putnik" (Vladivostok); the Khabarovsk Regional Wild Life Foundation; the Socioecological Union; the Association of Indigenous Peoples of the Russian Far East; and the International Clearing House of the Russian Far East.

10

Hungary: Political Interest, Bureaucratic Will

Ellen Comisso and Peter Hardi with Laszlo Bencze

Perhaps the most obvious feature of Hungary that strikes one when comparing it with states like Russia, Brazil, China, India, and the United States is its size: in both area (93,000 sq. km.) and population (10,335,000 in 1991), Hungary is a small country. While its level of development is above that of its neighbors to the east and south, it lags significantly behind the states to its west in the European Union. Both its size and its level of development make Hungary neither a major contributor to global environmental degradation nor a major factor in improving the quality of the international environment. Hence, the logic behind its involvement in international environmental accords (IEAs) presents something of a puzzle. Why would a small state whose activities have scant global impact bother to accede to four of the five agreements under study in this project, let alone carry out their provisions relatively efficaciously?

Our answer to this question forms the focus of the first half of this analysis. Two factors were central to Hungary's decisions to join IEAs and to its compliance with them. The first is strategic/political. Political leaders were willing to sign environmental accords because they seemed to represent a relatively uncontroversial and low-cost means of pursuing broader foreign policy objectives that had little to do with the environment.

The second factor explaining Hungarian behavior is a bureaucratic/administrative one. The initiative for Hungarian involvement in the treaties under study came from mid-to-high-level civil servants, who saw accession to the treaty as a means of solving a domestic problem with which they were either administratively or personally concerned.

That involvement in and compliance with IEAs was largely a product of interaction between high-level bureaucratic and political elites in a Leninist political order is hardly surprising. The more interesting question, of course, is whether anything changed after Leninism collapsed in 1989 and a freely elected government came to power in 1990. Accordingly, the second half of this analysis focuses on the impact that the establishment of a competitive political order in Hungary had on compliance with the IEAs negotiated and ratified prior to 1990.

To examine postsocialist compliance in Hungary, we disaggregate the collapse of socialism into three sets of independent variables: political, economic, and international changes. We then explore the impact of each on the pattern of compliance. Somewhat to our surprise, we found the rather radical political and legal changes that occurred had had only a marginal impact on the existing structure of compliance and its effectiveness. Environmental policy—including treaty compliance—thus rests today pretty much in the same place it rested before 1989: in the efforts of mid- and high-level civil servants in the intact state administration. Thus, many of the hypotheses that informed this study regarding the beneficial effects of greater transparency and democratization were not confirmed in our analysis, at least for the initial postsocialist period we examined.

The impact of the economic changes associated with the collapse of socialism were more significant, in that problems that had been relatively minor under socialism began to take on larger dimensions as market forces began to penetrate the economy more fully. At the same time, economic recession and large budget deficits meant an absence of fresh state revenues with which to address them.

The most important change following the collapse of socialism in Hungary and the Soviet bloc was not so much domestic as international. The war in Yugoslavia and the ensuing trade embargo meant that traffic in illicit materials—including endangered species—shifted northward; at the same time, the weakening of state capacities in the successor states of the Soviet Union increased the amount of traffic. Not surprisingly, the existing administrative machinery in Hungary is finding it increasingly difficult to deal with the greater burden that compliance now involves.

Finally, it is important to note that the treaties in this study touch only tangentially on the environmental problems most central in Hungary today. These issues concern air and water pollution, hazardous waste disposal, the impact of the Gabčikovo Dam on the Danube River, and the preservation of the country's remaining natural reserves. Thus, although we argue that the absence of significant public and NGO involvement in the issues addressed by the treaties under study is predominantly due to the general political and economic changes that accompanied the regime change, we wish to note that this absence is also due to the lack of salience of these particular treaties in domestic ecological awareness.

This suggests that when treaties address problems that are relatively marginal in a state that signs them, simply the presence and cooperation of a well-trained bureaucracy may be sufficient to ensure compliance. When treaties seek to regulate resources central to a state's economic development and quality of life, however, purely administrative solutions may well be inadequate. Yet, as we note in Hungary, it may be precisely the strong and influential state administration needed for compliance with less politically catchy IEAs that blocks the ability of popular groups to make themselves heard on local environmental issues that are more problematic domestically.

Signing and Implementing International Agreements

Why Not Free-Ride?

Hungary's size and its level of development mean it is hardly a major contributor to environmental degradation on a global scale. Thus, Hungarian participation or absence from IEAs that aim to correct problems of wide international concern would not appear to make much difference. Equally important, the problems addressed by the treaties in this study are not major environmental causes célèbres within Hungary itself. Given that the interest of the international community in having Hungary "on board" is small, and that the benefits obtained by Hungary from being "on board" are equally minimal, how does one account for Hungary's being a party to four of the five treaties included in this study and compiling a fairly credible record in carrying out their requirements? Why, instead, did Hungary not choose to free-ride on the efforts of larger states that were far more significant contributors to the problem? Or, to phrase the question paradigmatically, how does it happen that among the first IEAs joined by a landlocked country with no significant fleet to speak of is the London Convention on Ocean Dumping?

The explanation rests on two key factors. The first is strategic/political, and explains the willingness of political leaders to buy into international environmental agreements by the late 1970s. Among the political values of any treaty commitment is that it serves as a means of identifying oneself with a particular "community of states" in the international system. Hence, it is a means of signaling both to other states and to one's own population the larger international community in which one has standing.

Hungary's initial treaty commitments after World War II signaled its membership in the Soviet bloc to other states as well as to its own citizens. With détente, however, it became possible for states within the Soviet bloc to become members of other international groupings, and doing so was one way in which the political elite of a given state could signal its desires to increase contacts with non-Soviet bloc partners. Different states in eastern Europe took advantage of these opportunities to different degrees, Hungary being among the most active.

For Hungarian political leaders, joining and complying with broad, multilateral international agreements allowed Hungary both to show its interest in being included in non-Soviet bloc international accords and to demonstrate that Hungary could be a reliable partner in such agreements. Joining existing IEAs and participating in the elaboration of new ones was thus part of a larger strategy designed to "rehabilitate" Hungary's international political image, bringing it into closer contact with west European states without threatening its status in the Soviet alliance. The movement to increase interchange with non-Soviet bloc states gathered steam in the 1980s, the

Table 10.1
Treaty membership of Hungary

Treaty	Year treaty came into effect	Signed by Hungary	Ratified by Hungary
WHC	1977	1985	1985
CITES	1975	1985	1985
ITTA	1985	not a member	not a member
LC	1975	1973	1976
MP	1989	1989	1989

decade in which Hungary formally became a party to all but one of the treaties in our study. Table 10.1 shows the dates when Hungary joined the treaties.

In the eyes of the Hungarian government, then, becoming a party to broad environmental treaties was not a sign of heightened ecological concern. "There [was] only a vague political instruction and there was no explicit intention to go and see what exactly those treaties were," stated a participant in several negotiations (Hungarian Interview Series 1994: #3). On the contrary, accession was simply a relatively inexpensive means of becoming a more independent player in the international system. As such, the rather minor adjustments that had to be made to comply with them was an important factor recommending accession to treaties like the London, Washington, and World Heritage accords at the time: retrofitting a handful of ships to comply with MARPOL or documenting a few sites for the World Heritage Convention appeared to be a small price to pay for international respectability and inclusion.

Moreover, by the end of the decade, this general strategy produced a momentum of its own. Hungary began to be routinely expected by states within and outside the Soviet bloc to take part in multilateral negotiations on agreements dealing with environmental questions. Hence, whereas Hungary became a party to CITES and the World Heritage Convention long after the treaties were formulated, it was involved in the drafting of accords negotiated after 1987, the London and Copenhagen amendments to the Montreal Protocol being the examples in our sample. Indeed, to have rejected such participation by this date would have signaled just the opposite of what was desired by overall foreign policy objectives. As one interviewee commented, "[This happened] also because among the former COMECON countries Hungary was the most interested in the practice of Western countries and was the most willing and able to follow suit. [I]t's better to follow than to be excluded" (Hungarian Interview Series 1994: #2).

In sum, the Hungarian government was willing to join environmental treaties as part of a broader rapprochement with the West. In that process, environmental concerns were a distinctly low priority, as evidenced by the lengthy delays between accession and implementing regulations in several cases and by the fact that accession to treaties was by no means accompanied by major alterations in the harmful environmental policies affecting the many areas not covered by these accords.

Ironically, then, the fact that the treaties in our sample dealt with issues of peripheral concern to Hungary was probably a factor in favor of acceding to them, since it appeared they could be complied with at minimal cost to the state. Thus, for the most part, it was the intangible benefit of inclusion in a non-Soviet bloc community of states at a price it could easily pay that explains Hungary's accession to the treaties in our sample.

As we shall see below, the assumption of minimal costs of compliance for several treaties turned out to be less valid once socialism collapsed in 1989. Nevertheless, insofar as Hungarian foreign policy objectives remained fairly constant—or, more precisely, the goal of integration with western Europe acquired an ever-increasing urgency—exit from the agreements was not an option the post-1990 leaders wished to entertain. On the contrary, Hungary became a party to several new international agreements in the early 1990s, and continued work on implementing the provisions of regional agreements covering air pollution concluded in the 1980s.[1] Nevertheless, environmental concerns as such were no more central to the newly elected government's priorities than they had been to the outgoing socialist government (Okolicsányi 1992). As a result, effective compliance became ever more dependent on funds from and actions of states surrounding Hungary and on the resourcefulness of Hungary's administrative cadres.

This brings us to the second key factor explaining Hungary's pattern of participation in the IEAs considered in this study: the role of civil servants in the state administration. What we described as political leaders' "willingness" to accede to IEAs by no means implies that political figures took the initiative as far as Hungarian involvement went. The lack of political interest in the treaties is evidenced by the almost total absence of publicity given to Hungary's accession to any of the IEAs in our sample. Indeed, the socialist government did not even use its relatively good record in these agreements to defend itself when it came under increasing political attack from the environmentalist movement in 1988–1989.

Rather, initiatives to join the agreements normally originated in the civil service, from medium- and high-ranking career bureaucrats who either had a particular problem to solve or were personally concerned about the resource a treaty sought to regulate. Without a general policy of involvement in environmentally oriented treaties as such, which treaty happened to come to the attention of which bureaucrat inevitably involved a considerable element of randomness. In fact, the specific agendas of individual civil servants and bureaucratic agencies are critical in accounting for why Hungary signed the World Heritage, Ocean Dumping, Ozone Depletion, and CITES agreements while ignoring ITTA.

The story of Hungary's accession to the World Heritage Convention in 1985 is illustrative. The accession was prepared primarily by civil servants in the office concerned with architecture and monument preservation in the Ministry of Construction and Urban Planning. They were joined by colleagues in the separate National Authority for Environment and Nature Conservation. Efforts originated in the early 1980s, when the above-named office sought to restore a large synagogue in central Budapest that had been in major disrepair since the end of World War II. The Jewish community in Hungary—a shadow of its pre-Holocaust self—was simply too small and financially strapped to restore the site. Efforts to reconstruct the synagogue were thus spearheaded by Jewish organizations outside Hungary, which brought the Convention to the attention of the ministry officials and the government as a means of facilitating their fund-raising efforts.

It then took several years until the Ministry of Foreign Affairs was ready to move, responding as much to prodding from UNESCO as from the domestic agencies interested in site preservation. Financially, the only major stumbling block appeared to be the 1 percent membership fee; since Hungary already had law decrees dealing with monument and nature preservation, civil servants favoring accession could argue cogently that no major new expenditures would be required. Indeed, one interviewee noted, "A bit we deceived the Government and Parliament as we wrote in our proposal that if we have got World Heritage sites, then we can get subsidies although we knew it was not quite true" (Hungarian Interview Series 1994: #7).

Several additional points about Hungary's accession can be made here. First, consistent with our remarks above regarding foreign policy objectives, it is important to note that even though Hungary did not join this accord until 1985, it was one of the first East bloc states to do so. Treaty accession was neither part of an East bloc bandwagon nor related to the start of the Gorbachev era, since most of the preparatory work had been completed earlier. The Soviet Union became a party to the convention somewhat later, and then primarily because the British and American exit from UNESCO made accession to World Heritage a means of maintaining contacts with them on cultural issues.[2]

Second, our initial expectation that a socialist government would be reluctant to nominate sites associated with religious activities was obviously incorrect, since it was efforts to restore a place of worship that were the catalyst behind Hungary's decision to join the Convention. Ironically, the Dohany Street synagogue's failure to make the list of World Heritage Sites in Hungary was due to the choices of the international World Heritage Commission, not to the failure of the Hungarian commission to nominate it.[3]

Third, it is worth emphasizing that in the years immediately prior to Hungary's accession to the convention, the ball was carried by civil servants in the same agencies that would be concerned with its implementation. The main actor was the Agency for Monument Preservation in the Ministry of Construction, whose bureaucratic interests would be enhanced by having the aura of an international treaty attached to them. Finally, both the legal infrastructure and the administrative machinery for implementing the treaty were pretty much in place before the accord was actually signed. As a result, the main action that signing the treaty entailed was formally establishing the Hungarian World Heritage Commission to coordinate nominations and documentation. Not surprisingly, it was composed of representatives from the bureaucratic coalition that had pushed for accession to the convention.

The CITES and London Convention followed similar processes. With CITES, experts from the Nature Conservation Authority had been attending meetings of the parties to the Convention as observers since the late 1970s. Undoubtedly, these individuals initiated the bureaucratic lobbying effort for Hungary to join the Convention. Those efforts came to fruition only in 1985, when the CITES Secretariat, noting the lack of action, made inquiries to the Hungarian government, which pulled the accession documents off the shelf where they had lain since 1981 and ratified the agreement without paying very much attention to it. The four-year "disappearance" of the matter suggests that the particular experts initially concerned with CITES had probably moved or been moved on to other matters, and the treaty, lacking individual sponsors, was simply "lost" in the administrative machinery.

Lacking a bureaucratic actor whose career depended on it, and of utter disinterest to political leaders, the treaty languished for another five years without an implementation decree. Meanwhile, the elevation and reorganization of the Nature Conservation Authority into the Ministry of the Environment in the 1980s led to the treaty's "rediscovery" as Ministry officials cast about for matters to bring under their jurisdiction. The pattern of bureaucratic initiative and political acquiescence and disinterest is thus basically similar to the World Heritage Convention, the important difference being that with CITES, bureaucratic proponents were sidetracked to other matters, causing the treaty to lie dormant until administrative interest revived.

Our information on the origins of Hungary's accession to the London Conventions on ocean dumping and on pollution from ships is extremely sparse. As Hungary is a landlocked country, the issue of direct dumping has no significance, but it operates a small marine fleet, so the dumping and pollution from ships have relevance. We suspect that the impetus for accession here probably came from either the state-owned shipping company operating Hungary's minuscule merchant marine or from the officials in the Ministry of Transport who supervised it. The motive may well have been the desire to guarantee credit for modernizing ships in the fleet or the fact that ships were already having problems at various ports of call due to inadequate facilities for waste and oil disposal. The need to avoid fines explains the willingness to install the required equipment and accede to the agreement, and accounts for why only oil filters, and not grinders or incinerators, were installed (the lack of the latter was not subject to financial penalties).

The history of the Vienna and Montreal protocols departs somewhat from the general pattern of bureaucratic initiative followed belatedly by political endorsement. Accession to both appeared to be part of a general interest that emerged in the late 1980s in widening cooperation on environmental issues with west European states.[4] Hungary thus was present at the negotiations for the amendment of the Montreal Protocol, and the Ministry of Foreign Affairs became involved with the treaty much earlier than had been the case with the CITES, World Heritage, and Ocean Dumping accords. The international pressure that arose after the discovery of the ozone hole in the Antarctic also helps account for the relatively short time that elapsed between ratification, promulgation, and implementation decrees of the Montreal Protocol.

The newly created Ministry of Environment and Water Management provided the expert staff for the negotiations as well as additional support within the government for the treaty. Or, as one of our informants observed, "The Ministry of Foreign Affairs and the Ministry of Environment incited each other" (Hungarian Interview Series 1994: #2). In contrast, the Ministry of Industry was not involved at all. As a result, although a study of the treaty's technical feasibility was made prior to accession, there was no preliminary study of the treaty's potential economic impact. In fact, apparently the only economic issue raised prior to ratification was the Finance Ministry's routine query about a membership fee. The absence of consultation with affected companies is quite different from the process followed by the major Western producers of CFCs.[5]

Compliance Issues

The factors accounting for Hungary's accession to the four agreements in our sample also explain why its compliance record is relatively good—but far from perfect. Although environment has not been a major political concern of any Hungarian government, being a reliable partner in international agreements has been. As a result, the environmental concerns that are subjects of international conventions tend, at the margin, to receive somewhat more administrative attention than environmental issues of comparable importance that have purely domestic ramifications. We suspect that this is a key motivation for administrators to bring IEAs to the attention of political leaders in the first place. And when environmental and economic interests conflict, adherence to an IEA by no means guarantees that environmental concerns will triumph, but it does at least provide an avenue for them to be voiced and seriously considered.

Nevertheless, the resources devoted to environmental problems of any sort have always been limited, and adherence to an international treaty is rarely an occasion for increasing them. In fact, our sense of the general process of treaty accession and compliance in Hungary is that governments are willing to let bureaucrats have a treaty as long as the treaty is in line with general foreign policy orientations and is couched as an action that won't cost anything. Should treaty compliance turn out to be expensive, administrative officials are expected to find nonbudgetary resources (especially international) to cover the costs.

Civil servants thus play a central role in initiating accession to IEAs, and their resourcefulness is primarily responsible for the level of compliance with IEAs' provisions. Effective compliance in Hungary's case thus rests less on the level of active political support for a treaty in the society, government, or legislature than on the skills and professionalism of the administrators entrusted with devising appropriate procedures and monitoring their effectiveness.

Although the specific compliance issues involved in the treaties in our sample defy easy generalization, a few preliminary remarks can inform our more detailed discussion. First, as suggested above, there was no designated state budget item for implementing any of the conventions we examined. Rather, implementation costs had to be covered by the regular budget of the responsible ministry or external funding had to be sought. Nevertheless, this was not as draconian as it might appear, insofar as it normally meant that the same staff that had previously prepared the accession documents now used their time to prepare and monitor the compliance procedures. Continuity was thus preserved, although problems arose in the case of treaties whose costs of compliance turned out to be higher than originally expected.

Second, both the heavy bureaucratic involvement with the treaties and the traditions of Hungarian state administration made "command and control" the method of regulation most heavily relied upon. The extensive involvement of professional yet insulated state administrators in treaty preparation and implementation was partly responsible for a third feature of Hungarian compliance efforts: the virtual absence of nonadministrative actors—from NGOs to the legislature—in the process. The only exception here concerned the CITES treaty, where an unusually enterprising civil servant sought on her own to involve a wider public in protecting endangered species. As for the legislature, its passivity and subordination to the government and Communist Party in the socialist regime is well known, but even after 1990, its failure to pass a general law on the environment for five years or take an active role in monitoring state bureaucrats concerned with environmental affairs meant the new Parliament had little impact on compliance. Treaty implementation remained very much the province of ministerial regulations and decrees in which "outsiders" had little input and of which they had little knowledge.

Continuity in the functioning of the state administration is the main factor explaining why compliance mechanisms have been relatively effective to date. Nevertheless, where compliance rests on nondomestic factors—whether international funds or the cooperation of other states—the record has been weaker. Thus, the impact of acceding to IEAs in Hungary is to reinforce or continue a domestic process to comply with their provisions, but not to facilitate international coordination or support for them.

Let us now consider compliance measures in greater detail. Where problems emerged due to the regime change of 1990, they will be examined more carefully in the analysis exploring the impact of democratization.

World Heritage Convention

Compliance with the World Heritage Convention involves nomination and documentation of proposed sites as well as preservation of the sites that ultimately reach the World Heritage List. In Hungary, a committee was established to nominate sites immediately after the ratification of the Convention. Although it was chaired by staff from the National Authority for Monument Preservation and the National Authority for Nature Conservation, the latter apparently "didn't take the Convention seriously" (Hungarian Interview Series 1994: #11), perhaps reflecting the dominant role of the Ministry of Construction in pushing the accord through the government initially. Consequently, most of the sites considered were of architectural and cultural interest, to the subsequent dismay of experts in nature conservation, who wound up with what they today consider a rather subordinate role in the initial 1986–1987 selection process.

Ten locations were suggested for domestic review; only seven were seriously considered, and the cost of documentation limited the number of sites actually proposed to six. Following its review and inspection, the World Heritage Commission suggested that only two sites be officially nominated: Buda Castle and the Danube riverside in Budapest, and Hollókö, a traditional village in the northern hills. Both sites were accepted onto the World Heritage List. Both were named in the documentations prepared by the monument preservation experts, with the result that the nature preserve in which Hollókö is situated is not part of the site. Rivalries between the monument preservationists and nature conservationists continued to complicate site nomination.

Meanwhile, the committee that chose the sites ceased operating in 1989. Our interviews, however, revealed that plans were being made to nominate several new

Table 10.2
World Heritage Sites in Hungary

Year	Natural site	Cultural site
1987		Budapest, incuding the banks of the Danube with the district of Buda Castle
		Hollókö
1995	caves of Aggtelek and Slovak karst*	
1996		Millennary Benedictine abbey of Pannonhalma and its natural environment
Total	1	3

* Site shared by Hungary and Slovakia.

sites. Most of these were natural, probably because responsibility for implementing the Convention had been moved out of the Ministry of Construction and into the Ministry of the Environment, whose core staff consisted of the civil servants who previously had worked in the Nature Conservation Authority. Their hope is to use the Convention to impart international prestige to their domestic nature conservation efforts, although how willing the World Heritage Commission will be to oblige them is open to question.

In addition to the nature versus culture rivalry on the Hungarian commission, geopolitical factors affected site nominations. There was some interest in 1986–1987 in nominating the Fertö Lake area and Aggtelek cave system as World Heritage Sites. However, the former extends into Austria and the latter into Slovakia—neither of them a party to the Convention at the time—so the sites were dropped from the list. As with the other treaties to which Hungary is a party, when compliance rested on international cooperation and not simply on domestic action, it was much more difficult to obtain. In 1995 these sites were added. In 1996, another cultural site was added, the Millennary Benedictine Abbey of Pannonhalma. Table 10.2 lists all of Hungary's sites.

Preservation of the selected sites falls largely to the local governments. As far as Buda Castle is concerned, there is considerable pressure from local developers to build in the area, and it is unclear what action the city and district authorities will ultimately take. The village of Hollókö is relatively remote and virtually unknown as a tourist attraction, and local authorities have done little to publicize it. Its location within a national nature preserve means development is necessarily limited, thereby protecting the village's architectural cohesion and deterring residents from modernizing their traditional cottages. Thus, the village retains its "authentic" status thanks to benign neglect. Ironically, Hollókö is preserved as part of the world's cultural heritage in large part because the world is uninformed of its existence.

CITES

According to the Ministry of Environment staff member responsible for CITES,

> It seems that in Hungary, the CITES is the one nature conservation treaty which really works now. The customs offices are ready to pay attention to it already. It has got publicity and people start to be aware of it. The press deals with it a lot, last year about 200 articles, radio, and TV programs were about CITES. It is not [a] result of the Convention, but of some sensational cases, e.g., there were the Mongolian wolves, then the chimpanzees, and so it has become generally known. (Hungarian Interview Series 1994: #9)

Nevertheless, further inquiries suggest that such an optimistic assessment may well be premature. Certainly, protecting endangered species is today a popular cause in Hungary, but at least part of this enthusiasm is related to the absence of species native to Hungary on the list; hence, it is chiefly a question of protecting animals and plants coming from abroad. Consequently, where other nature conservation efforts (e.g., reserving land for parks or protecting wetlands) encounter strenuous opposition from domestic interests, there are no such barriers to protecting species from other countries. Equally important, the nature of CITES is such that only the violations that have been caught are known, and one can only speculate on the extent of the traffic that evades inspectors. Estimates vary widely, with some interviewees suggesting that as little as 1–5 percent of infractions are actually uncovered (Hungarian Interview Series 1994: #9, #14).

Enforcing the provisions of CITES is the responsibility of a small staff—only one individual is engaged full-time—at the Ministry of the Environment's National Authority for Nature Conservation. It relies heavily on the efforts of the customs service. Unlike the rivalry between monument preservation and nature conservation that characterized the World Heritage Convention, cooperation between the two state agencies responsible for enforcing CITES is quite good.

At the time the Convention was ratified, Hungary was a socialist country. There are no data on how frequently attempts were made to bring endangered species into the country at that time, but the generally high trade barriers that prevailed suggest they were minimal. After 1990, however, foreign trade regulations were greatly liberalized; combined with the war in Yugoslavia, the effect was to make Hungary a fairly major transit route for traffic in endangered species.

The primary burden of compliance thus falls on customs officers, who notify the Ministry of the Environment in cases of suspected CITES violations. The Ministry will then investigate and, if necessary, seize and confiscate the items. The Ministry's authority to take such actions was upheld in 1994, in a widely publicized court case involving the seizure of 100 Mongolian wolves.

The Ministry staff does its best to keep abreast of domestic violations of the treaty, relying on a wide variety of informal sources of information, ranging from telephone calls made by concerned citizens to published advertisements. Often a phone call to a potential offender is sufficient to produce compliance. The relevant Ministry official stated:

> There were many new hotels which wanted to serve turtle soup or stewed otter in their restaurants. I visited them, we discussed [it], and they gave it up.... A travel company advertised tours to South America for collecting parrots. I called the Advertising Association and asked them to stop it. They thanked me and the next day the advertisement disappeared. (Hungarian Interview Series 1994: #9)

Clearly, there is an ad hoc quality to such efforts, which resemble the personal campaign of one or two dedicated individuals. Local police are quite overwhelmed by other pressing matters, and rarely uncover infractions. Inspections at the border are, of course, more systematic. Generally, traffic in live animals is better controlled than trade in prohibited products; the publicity accompanying the recent "sensational" cases noted above may act as a deterrent as well.

Penalties for violations of CITES are not great, and they are difficult to enforce, particularly when violators are citizens of another country. Moreover, since several states bordering Hungary are not signatories of CITES, the volume of the traffic across the Hungarian frontier is increased.[6] Indeed, a customs officer argued, "If all the states were parties, there would be little problem. It would be easy to enforce it in Hungary" (Hungarian Interview Series 1994: #14). While this view is perhaps overconfident, the basic argument is probably valid. In addition, all officials involved with CITES emphasized the need for greater controls in the markets for endangered species and their products. The markets, of course, are in Europe, North America, and Japan, not in Hungary. Thus, to the degree that compliance rests on international coordination, ensuring it remains problematic.

London Ocean Dumping Convention

Hungary signed the 1972 London Convention in 1973 and promulgated it in 1976; it also joined the 1973 Convention on ship pollution in 1985, promulgating it with a decree of the Ministry of Transport in the same year. The Ministry, thanks to its supervision of the shipping company (MAHART) that operates Hungary's minuscule merchant marine, is also responsible for carrying out the treaty's provisions.

Even in its heyday, the Hungarian fleet numbered only twenty-three oceangoing vessels; by 1994, the number had dropped to nine. With a tiny and aging fleet of no military significance and utterly marginal economic importance, it is hardly surprising that major sums have not been spent to modernize waste disposal systems. Oil-cleaning equipment apparently was installed on all ships immediately following accession to the treaty.

According to our interviews, the major control on compliance occurs less through the Ministry of Transport than in foreign ports of call, where it is compulsory for captains to sign the bill for unloading of accumulated waste. We

Table 10.3
Hungary's compliance with treaty reporting requirements, 1991–1996: Report transmission/receipt date

Treaty	Entry into force	1991	1992	1993	1994	1995	1996
CITES	8/27/85	9/13/94	9/7/94	9/7/94	10/23/96	6/3/97	N/R
LC	3/6/76	N/R	N/R	6/30/97	6/30/97	N/R	N/R
MP	7/19/89	R	R	R	R	R	R

Notes: For all treaties, N/R indicates that no report was received by July 1997. R indicates that a report was filed, but the date of receipt is unavailable. Montreal Protocol reports received include those filed on time, late, or subsequently revised; in some cases a country may have met two or more years' reporting requirements with the submission of a single report.

hypothesize that as such requirements became routine at ports visited by Hungarian ships, pressure mounted to sign the treaty because compliance with its provisions was becoming mandatory in any case.

The responsible official, himself a former captain, suggested that Hungarian ships generally comply with the Convention, particularly once one makes allowances for the age of the equipment (Hungarian Interview Series 1994: #1). At the same time, one must acknowledge that even continual violation of the accord by all nine Hungarian ships would scarcely contribute much to ocean pollution. Probably for this reason, there is little systematic monitoring of ships by the Ministry of Transport, which frequently does not even bother to file an annual report, in contrast to the situation with respect to the other treaties. The first reports on the London Convention on Dumping were filed in 1997, covering the years 1993 and 1994 (see table 10.3).

There was, however, quite a lot of complaining about the expense involved when unloading waste at ports (perhaps the best evidence that ships were not dumping it at sea!). Because the treaty grants ports a natural monopoly over a captive market, charges for waste disposal can be high. Moreover, the Hungarian experience is that several ports require vessel captains to sign the bill for the delivery of waste and then leave the waste on the ship. In these cases, even good-faith efforts of one state to comply with a convention become frustrated by the failure of the same accord to deal with abuses by others.

Vienna/Montreal Protocols

Hungary ratified the Vienna Convention in 1988 but promulgated it only in 1990. The time lapse between ratification (1989) and promulgation (1990) was somewhat shorter for the Montreal Protocol, although the implementation decree has undergone a lengthy revision to reflect the changes of the Copenhagen Amendment.

Meanwhile, reducing the time period within which levels of ozone-depleting emissions must be lowered and increasing the number of substances regulated, as prescribed in the Copenhagen Amendment, are the sources of some difficulty. The best guess is that Hungary will be able to meet the requirements of the original Montreal Protocol, but is unlikely to satisfy the more stringent provisions adopted in Copenhagen within the required time period.

Compliance measures, supervised by a small staff in the Ministry of the Environment, follow traditional formats. On the one hand, there are prohibitions and fines. For example, the use and sale of products with aerosol gas were banned after July 1993, immediately reducing emissions to half of the maximum achieved in 1989. Economic recession and the decline in industrial output undoubtedly contributed to this. On the other hand, Ministry staff members work directly with individual firms to facilitate their adaptation to the new requirements. Such efforts follow a cooperative, nonadversarial format involving temporary exemptions, help in preparing applications for support from the Global Environment Facility (GEF), compensation for losses, and the like.

Implementation problems take several forms. First, as one would expect, compliance is facilitated to the degree that the use of CFCs in Hungary was not very great to start with, and most of it is concentrated in a handful of firms with very specific needs. Environment Ministry staff members work directly with such enterprises; while they may (and do) grant postponements, firms acquire these exemptions only by demonstrating plans to meet new deadlines in the near future. Thus, the impact of the Protocol is not so much to produce immediate compliance as it is to start a process that, after a certain number of years, is likely to lead to a significant reduction in the size of the problem.

Second, coordination between the ministries affected by treaty requirements is often lacking. This has positive as well as negative consequences. That is, regulations formulated by the Ministry of the Environment tend to directly reflect treaty requirements; they are not vetted by the Ministry of Industry, which would normally seek to water them down. For example, the use of CFCs in some fields was banned earlier in Hungary than in the European Union: "The opinion of the Ministry of Industry arrived late, and it didn't argue strongly that it was a needless eagerness to ban the use of CFCs in plastic foams from the middle of 1993," states one of our interviewees (Hungarian Interview Series 1994: #2). But precisely because regulation is vigorous, enterprises frequently cannot meet their requirements, leading to firm-specific concessions. Thus, in the case of plastic foam, the impact on the affected company caused "a modification of that decree [to be] drafted where it can be compensated" (Hungarian Interview Series 1994: #2).

Nevertheless, insofar as the bargaining is between the Environment Ministry and individual firms without the intermediation of the Ministry of Industry, general compliance is probably better. In contrast, trade regulations to prevent Western companies from dumping CFC-emitting refrigerators on the Hungarian market were omitted from the 1994 list of import restrictions. The primary cause was the failure of the Ministry of Industry to propose such limits to the Ministry of Foreign Trade.

Third, the bulk of ministry attention is on helping larger enterprises adapt to the new requirements; as a result, smaller violators are quite likely to escape detection. The extent of the "black" economy in Hungary is quite large, but given the country's size and contribution to ozone depletion, even a large number of small violators may not make a major contribution to the global problem. Hence, the decision not to devote limited resources to monitoring small offenders is probably justified.

Last but not least, a major factor slowing the rate of compliance is financial. It is complicated by what one informant characterized as a "weakness of the intellectual infrastructure that we cannot reach the available resources in the world with well prepared projects" (Hungarian Interview Series 1994: #2). For example, plans to create a servicing network for the collection of used CFCs were at a very preliminary stage, too late to meet a January 1996 deadline. Although the World Bank was expected to support the project, money and personnel to work out the program were lacking. The problem was less the economic impact of CFC reduction on enterprises than the simple budgetary cost of planning an infrastructure that would allow firms to meet requirements. While the Environment Ministry is supporting applications from individual companies for subsidies to convert equipment only international sources can supply such funds, but it is unclear whether the Hungarian firms will have access to them.

Outside the few individuals in the Environment Ministry, virtually no one deals with ozone-depleting emissions in Hungary. Unlike the publicity surrounding CITES, efforts to meet the requirements of the Montreal Protocol have been a quiet affair, involving only managers of affected companies and Ministry officials. Such behavior means that a domestic constituency for strict enforcement is nonexistent, probably an outcome welcomed by the officials who are involved, because it provides them with a degree of flexibility a public spotlight would rapidly erase.

The Impact of Regime Change on Accession and Compliance

Impact of Altered Political Institutions

In 1989, Leninism collapsed in Hungary and, indeed, everywhere in eastern Europe. The most dramatic and immediate change that ensued was, of course, political: competitive elections regulated accession to political office, legal constraints replaced party constraints on elected and nonelected officials, courts—especially the new Constitutional Court—began to play a central and independent role in defining law, and policy debates moved to open forums—ranging from the Parliament to the press—from the party backrooms in which they had previously been secreted.

Nevertheless, like other ex-Leninist states in east central Europe—but in direct contrast with the Soviet/Russian experience—the introduction of competitive politics in Hungary did not trigger the collapse of the central state apparatus or even much of a reduction in the influential role the state played in the society. Indeed, in Hungary, there was almost no interruption in the functioning of the state, the bureaucratic institutions, or the structure of administration, despite deep and decisive changes in the political system and leadership governing them.

The question, of course, is what impact the establishment of an open and competitive political order in Hungary had on environmental policy in general and on participation in IEAs in particular. While it is frequently assumed that an elected government is likely to be more responsive to environmental concerns (see Mann 1991), the Hungarian experience illustrates very nicely that this is not necessarily the case. In fact, the centrality of environmental issues depends far more heavily on a government's priorities and the strength of environmental advocates within its coalition than simply on whether or not it is elected.

In Hungary, the importance of the environmentalist mobilization against the Gabcikovo–Nagymaros Dam in bringing down the socialist regime was by no means reflected in an elevation of environmental issues in the newly elected coalition government. In foreign policy, accession to IEAs continued to play a marginal and highly instrumental role, very much as it had in the past. Thus, the pattern of joining environmental accords as a relatively costless means of realizing general foreign policy objectives continued; only the foreign policy objectives themselves were changed in the light of more global changes in the international order. That is, whereas environmental treaties were a means of creating some maneuvering space for Hungary within the Soviet alliance before 1990, they were now a tool of creating and confirming Hungary's dependence on western Europe.

Accordingly, participation in IEAs not only was a means of establishing a reputation as a "reliable partner" but also was perceived as a requirement for acquiring support (especially economic) from Western powers. In such a situation, it is hardly surprising that Hungary made no attempt to resist the Copenhagen Amendment to the Montreal Protocol—even though it entailed an unexpected and in some cases unrealistic burden on local firms. There was simply no political will to resist an accord once the major powers had agreed to it.[7]

While foreign policy concerns mirrored the new power realities in Europe, the internal political logic of accession to IEAs in Hungary was in many ways unchanged. Nor did compliance patterns change very much. While it is certainly true that new political and legal institutions have created real opportunities for members of Parliament, citizens, and lobbyists to hold state officials accountable for their actions and to demand stricter environmental legislation and better compliance, in fact these opportunities have been realized to only a minimal extent.

Part of the story here is the absence of change in the centralized character of the state administration. The benefit of this continuity for the treaties in our sample is that to the degree compliance with their provisions was relatively complete before 1989, it simply continued on the same track. Where new problems arose, they remained unsolved.

For example, the National Authority for Nature Conservation—a key actor in implementing the World Heritage Convention—was formally absorbed by the Ministry of the Environment and Regional Planning. Yet it continued to operate the same way it had in the past: its biggest problem was not elaborating any new policy but struggling to find resources to hire a sufficient number of staff to work properly.

Indeed, as far as the World Heritage Convention is concerned, the regime change may have had negative effects. It is not a coincidence that the national commission responsible for site nomination simply ceased operating in 1989. Lack of adequate documentation and transparency is still a problem, and at this stage, there are no new procedures for better and more reliable data collection, data processing, and documentation.

That bureaucratic traditions dating from the Austro-Hungarian monarchy die hard is amply demonstrated by measures taken in connection with the Montreal Protocol. As mentioned earlier, there was a serious lack of harmonization between the legal and technological/industrial issues. Problems were "resolved" by what had been the standard operating procedure within the socialist system: a general rule was promulgated, and then enterprise-specific exceptions were made to accommodate the individual needs of particular companies. (On enterprise–ministry bargaining in socialist Hungary, see Granick 1975; Laky 1979; Antal 1979; Comisso and Marer 1986.)

The actions taken by the ministry official supervising CITES enforcement were the only example we could find of changing bureaucratic practices that were directly related to the regime change. As noted above, there was heavy reliance on personal persuasion, negotiation, providing information, and mobilizing public opinion through the mass media. Whether such a pattern will spread to other enforcement agencies or even whether it will continue where CITES is concerned is unclear at this point, since such practices are highly idiosyncratic and uninstitutionalized, depending purely on the one-person layer of implementing officers.

The essential pattern of bureaucratic continuity in Hungary is also reflected in personnel, although one less fortunate effect of the regime change was the departure of some of the more qualified staff for positions in the private sector. Most of those we interviewed who were currently working in ministries dealing with the environment had been doing the same or similar work prior to 1989. Only at the very top of the administrative structure, where appointments were made on a partisan basis, was there a change in personnel. Nevertheless, since neither environmental issues nor enforcement of the particular treaties we are examining were high on the new government's list of priorities, these figures proved quite uninfluential in general policy-making and usually opted to avoid rocking the administrative boat.

This brings us to the second factor behind the administrative continuity: the lack of political will to change anything. As already noted, the issues addressed by the treaties in this study are marginal concerns in Hungary. As a former top official in the Ministry of the Environment observed about the Montreal Protocol:

> I maintain that the majority of officials working [in the Ministry], 80–90 percent of them, are qualified to do their job. I don't think it is naiveté [to say this]. They are qualified and informed, they know what to do, but the political intention is missing from the leaders ... conditions, responsibilities are not clear within the Ministry. (Hungarian Interview Series 1994: #5)

The failure to allocate a separate budget line for treaty enforcement is one indicator of this basic indifference. The failure to press for the passage of any new environmental legislation is another (see Bowman and Hunter 1992; Bell 1992; Sajó 1994).

The explanation of such political disinterest is also related to the impact of the regime change. During the Leninist era, environmental issues raised from below easily became attractive as a means of manifesting opposition to the regime. For this reason, many individuals normally not interested in the environment as such joined loosely structured "green" organizations, converging into a major movement to block the construction of the Nagymaros Dam (see Szabó 1991; Galambos 1994). Yet once legal channels opened for "regular" political opposition and organization, many of its most effective leaders abandoned the movement for openly partisan political activities. Environmental causes of all types lost their organizers, and with them, their mass constituency. What remained were a few rather narrow NGOs that could easily be ignored by the government.

As for those elevated to high political or administrative positions, they either found themselves with little practical influence or rapidly became overburdened with general political and administrative affairs. Others were forced by partisan loyalties to defer pressing environmental issues, and in one case—a particularly persistent deputy state secretary—the official was simply removed from office (see Hardi 1992; Waller and Millard 1992; Lipschutz 1993).

The one important exception to this pattern was the President of the Constitutional Court, an individual who had been the leading "green" lawyer and defense attorney of several fledgling environmental NGOs before 1989. Although his status as a sitting judge necessarily limits his field of activity, his influence could be clearly felt when Hungary's restitution act came up for constitutional review (see below).

Meanwhile, one of the consequences of administrative continuity is that the bureaucracy's relationship with the larger society persisted. The CITES experience is most unusual in this respect; typically, civil servants prefer to keep out of the public eye in order to maximize their own discretion, as the examples of the Montreal and London conventions illustrate. The lack of publicity accompanying the acceptance of Hollókö to the World Heritage List is another example. Consequently, high-level civil servants continue to have the same limited but specific influence on decision-making they have always had—being a key channel through which issues are brought to the attention of political elites—and the quality of enforcement continues to depend heavily on their professionalism, rather than on the support of a mass public or even of NGOs. Ironically, however, it is partially this very insularity of the civil service and its adherence to professional norms that makes it suspicious of, and impenetrable by, popular pressures from outside.

Changes in the legal system have made it possible for citizens and NGOs to challenge the actions of state administrators on environmental as well as, of course, other matters. For example, in 1994 an important case regarding a land restitution statute was brought before the Hungarian Constitutional Court by an environmental NGO. The issues in the case concerned whether or not land in areas protected by international and/or domestic laws could be subject to reprivatization. The court ruled it could not, and issued a very broad opinion that may ultimately prove quite relevant to the enforcement of IEAs. In that opinion, the court described at great length the state's responsibility to guarantee the constitutional rights of citizens to a safe and clean environment. From our perspective, the most important element in the opinion was its declaration that it is unconstitutional to decrease the level of protection of "natural values" if they are already regulated by legal means, including international treaties (see Az Alkotmanybíróság 1994). The thrust of the opinion was thus to incorporate the provisions of IEAs into domestic law, a potentially significant change.

The use of the legal system and the newly independent judiciary to enforce treaty provisions was demonstrated in an earlier, widely publicized case in 1991. Here, a Hungarian firm imported over a hundred Mongolian wolves, a protected species under CITES. Customs officers caught the infraction and called in the Authority for Nature Conservation officials. The wolves were confiscated, put in temporary shelters, and, since there was no place to house them in Hungary, an international action to save them was launched. Finally the Brigitte Bardot Foundation took care of them, amid a great deal of favorable publicity for the Authority for Nature Conservation.

This was not the end of the matter, however. The importing firm filed a suit against the Ministry of the Environment for illegally confiscating its property. After several appeals, the case ended at the Hungarian Supreme Court, which ruled the seizure legal because of Hungary's obligations to observe a ratified and promulgated international treaty. The decision, anticipating the ruling of the Constitutional Court three years later, was widely viewed as an important precedent-setting case (Hungarian Interview Series 1994: #8). Meanwhile, the case is also a reminder that the law is a double-edged sword: if environmental NGOs can use it to challenge the government, so can other private interests with far less benevolent intentions.

In conclusion, we see little evidence in Hungary that the establishment of an open and competitive political order had much impact on participation in environmental treaties. Accession to treaties rests on general foreign policy objectives now very much as it did before 1989, with the major change being that west European preferences have acquired greater influence on Hungarian actions. Likewise, compliance with IEAs in Hungary rests on the professionalism of the civil service now very much the way it had earlier. Certainly, Hungary still remains an "emerging" polyarchy rather than a well-institutionalized competitive order, and thus future developments may well alter this assessment. As of the second half of the 1990s, however, we cannot confirm a hypothesis that relates adherence to democratic norms to treaty implementation or accession in the initial post-1989 period.

The Economics of "Transition" and Compliance

It was not only Leninism as a political order that collapsed in Hungary in 1989; socialism as a distinctive economic model fell with it. The result was not so much that assets were immediately divested into private hands (the collapse of socialism as a model should not be confused with an overnight elimination of state ownership in practice) as that the overall framework for managing assets—be they public or private—was radically altered. Price controls and trade barriers were immediately relaxed, the currency became quasi-convertible, credit allocation began to reflect market forces, budget constraints on enterprises "hardened" considerably, and competition on the domestic market increased substantially. At the same time, traditional markets for Hungarian exports, especially in the Soviet Union, contracted sharply. Nevertheless, the substantial foreign debts Hungary had assumed over the previous two decades remained and even increased, while what had been a hidden state budget deficit became a rather large open one.

The immediate effect was a major economic recession from which the country has yet to recover (for details, see Koves 1992; Hanel 1992; Marer 1993). The impacts of the recession on treaty implementation were several. First, state moneys allocated to treaty implementation were severely limited even prior to 1989. This point was well illustrated in our analysis of treaty accession, where we saw how political leaders, and especially the Finance Ministry, had always been asking the same questions prior to ratifying a treaty: How much will it cost to accede? Does the

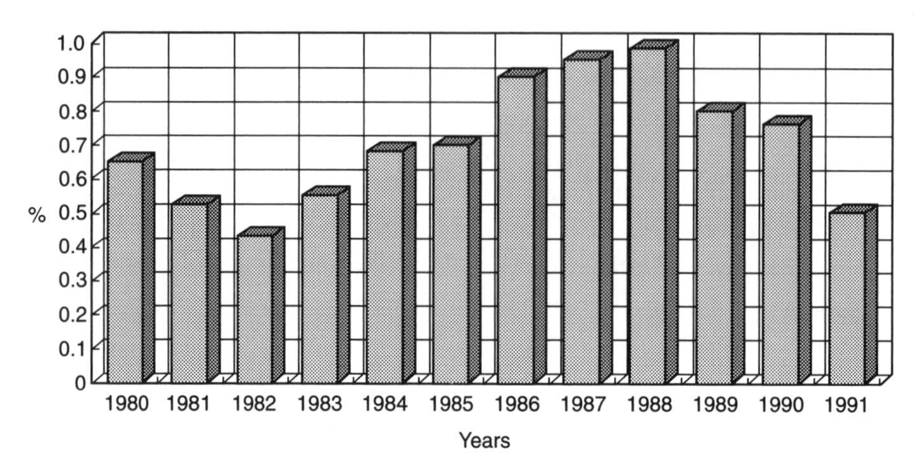

Figure 10.1
Environmental investment in Hungary as a percentage of GDP

country need extra resources to implement the treaty? Will accession bring additional financial resources into Hungary? After 1989, the financial situation only worsened.

To illustrate this point, the changes in the ratio of investments in environmental protection to GDP between 1980 and 1991 are presented in figure 10.1. The data substantiate our earlier observation that environmental protection has not been a government priority. In addition, while the ratio of environmental investment to GDP was always well below the comparable ratio in western Europe (not to mention that Western GDPs were much higher), the doubling of this ratio between 1982 and 1988—from about 0.5 percent to almost 1.0 percent—was followed by a quick decline back to 0.5 percent in 1991, after two years of the political transformation (Hardi 1994:40–41). As of 1994, this downward trend had not been reversed. Equally significant, the economic adjustment process that followed the collapse of the old economic order saw GDP fall in absolute terms as well; accordingly, the amount of environmental investment declined even more precipitously than its proportion of GDP did.

Consequently, the main factor that saved the compliance processes already in effect was the fact that their continuance did not require fresh budgetary outlays and could basically be accomplished with existing personnel. But to the degree that extra expenditure was needed in the light of new conditions, it was not forthcoming from the Hungarian state. For example, the fate of the Montreal Protocol in Hungary depends heavily on whether the applications of individual firms to the GEF are successful.

The second consequence of the economic recession was political. The impact of direct economic hardship upon a large portion of the population was inevitably to shift both popular and elite attention away from environmental issues (Hardi 1992). As a result, both environmental concerns generally and the environmental problems addressed by the treaties we are considering in particular remain the focus of policy makers only to the extent that they have been pushed by international partners in the West or present immediate local threats to public health. While international pressures played a role in Hungarian approval of the Copenhagen Amendment and the favorable international response to the Mongolian wolves episode undoubtedly encouraged enforcement of CITES, neither was sufficient to generate sufficient support to pass a general statute on environmental protection, which remains a poor second cousin to the economic and narrowly partisan concerns of political elites.

At the same time, the overall deterioration in the economy quite understandably has made the government reluctant to enforce environmental regulations too strictly, lest it jeopardize the economic situation of major companies and their employees. The enterprise-specific bargaining that characterized enforcement of the Montreal Protocol is a case in point. Ownership (public or private) did not seem to affect enterprise compliance, except insofar as it was correlated with size. That is, infractions by small firms are far more difficult to monitor, and these are typically in the private sector.

Nevertheless, the economic restructuring under way in Hungary during the 1990s has reduced the amount of heavy industry and the major pollution it entails to a significant degree. As we noted earlier, MAHART, the Hungarian shipping company, has reduced its fleet to roughly one-third of its previous size, presumably scrapping or mothballing the oldest vessels. With respect to the Montreal Protocol, there is also some evidence that the stronger firms affected by the accord were already moving to newer and cleaner technologies on their own. As one of our interviewees observed, "If the Protocol has an adverse effect on the microelectronics or precision machinery industry, it is for different reasons: they would decline or be closing down anyway, as there is more than one factor contributing [to their difficulties]" (Hungarian Interview Series 1994: #2). Thus, in some cases, the impact of economic change has been to accelerate compliance with treaty objectives.[8]

Opening the borders has meant that a far greater volume of individuals and commodities is moving across them, forcing customs inspections to be increasingly spotty. Further, the end of constraints on private enterprise has meant better satisfaction of consumer demand even when it entails violating—knowingly or unintentionally—the provisions of IEAs. Whether it means a tourist agency offering parrot-hunting trips in Latin America or a small restaurant importing a used, CFC-emitting refrigerator from western Europe to keep its costs down, to the degree that there is a market, firms will arise to exploit it. Meanwhile, the basic continuity that characterizes the state administration has meant that officials perfectly capable of controlling such traffic under socialism are badly in need of reinforcement.

Impact of International Changes on Compliance

Although changes in the domestic political, legal, and economic orders have had minimal effects on Hungarian compliance with the treaties in our sample, changes in the international political order have been more important. The dissolution of the Soviet bloc was accompanied by the collapse of three socialist federations: the Soviet Union, Czechoslovakia, and Yugoslavia. All three of these border on Hungary.

Not surprisingly, the major effect of this development has been on treaties whose implementation is affected by border controls, CITES being the leading example. Due to the war in Yugoslavia and the embargo on rump Yugoslavia, Hungary became one of the major transit states for goods moving from East to West. Needless to say, illegal traffic (drugs and weapons in addition to endangered species) increased together with legitimate commerce, and customs officers are hard pressed to conduct timely inspections. That the disintegration of the Soviet Union has resulted in less effective border controls in the successor states has only increased the burden on Hungary to catch violations, whether they be the transport of protected species under CITES or the importation of equipment using substances banned under the Montreal Protocol.

That several of the successor states have not ratified some of the international treaties (including those under study here) to which the Soviet Union was a party hardly helps matters. Though of the countries of the former Soviet Union, Hungary has common borders only with the Ukraine, the Central Asian republics are actively participating in traffic affecting CITES, and trading routes are less controlled by authorities from the eastem border to the western border of the former Soviet Union. Moreover, transport of endangered species and products often occurs via air, in which case the physical proximity of the state of origin is irrelevant. Yet the fact that the Central Asian republics are linked to Budapest through the formerly socialist air connections makes Hungary attractive as a transit point.

With regard to the newly independent Slovakia, we initially expected that a nationalist government there would be anxious to nominate sites for the World Heritage Convention, even if it was not clear that it would collaborate with Hungary in doing so. But so many political and economic problems surfaced in Slovakia following independence that it took a full two years for Slovakia even to join the Convention, and site nomination appears to be the least of any government's concerns there.

The dramatic shift in the international order also had a major impact on Hungary's accession to treaties, as noted above. Although membership in the Soviet bloc had not prevented Hungary from acceding to IEAs, the collapse of Soviet hegemony made Hungary heavily dependent on its relations with western Europe. Consequently, its political leaders have perceived themselves as being under considerable pressure to adapt to west European norms, including environmental ones. To the degree that willingness to adhere to such norms is accompanied by financial assistance, compliance is greatly facilitated, as our interviewees stressed. Western assistance also has contributed to the financial support of environmental NGOs and to the training of specialists in environmental issues. Thus, although one can hardly claim that an "epistemic community" has had any influence on treaty accession or compliance, it is possible that such a group may play more of a role in the future. Nevertheless, external assistance to date has been quite limited—indeed, Western aid has equaled only about 5 percent of overall environmental spending in the formerly socialist countries.

Conclusion

All in all, the great political landslide of 1989 had a surprisingly small impact on Hungary's pattern of accession to and compliance with international environmental treaties. This is not to say that there was no change at all, but it is to say that positive changes balanced negative developments sufficiently well to support the observation that "democracy" and "the market" by themselves are by no means a guarantee of either effective environmental policy or even of compliance with international agreements. In fact, there is very little support for the hypothesized relations between compliance and active NGOs, adherence to democratic norms, greater transparency, size and strength of an "epistemic community," or even economic growth and level of development.

From our point of view, the key factor in Hungary's accession to IEAs seems to have been the general foreign policy objective of successive governments: the desire to expand relations of all sorts with Western states. Since this objective is hardly unique to Hungary, it suggests that the advanced capitalist states have a special burden in the elaboration of international agreements affecting the environment. That burden is not necessarily due to the often-heard claim that they consume the lion's share of the world's natural resources, but to the fact that the desire to be included among them gives them a leverage on less-developed states that few other international communities have. That leverage is particularly pronounced among smaller states like Hungary, but exercising it is likely to be effective only if incentives for compliance—such as financial assistance—are included in accords.

As far as compliance goes, Hungary's relatively good record did not begin with "democracy" or economic liberalization. Rather, it was part and parcel of socialism, suggesting that the key to good compliance is that all-too-nebulous variable "state capacities." A professional, knowledgeable, and relatively noncorrupt civil service, interested in adhering to and complying with IEAs as a way of advancing its own bureaucratic interests, is the critical explanation for Hungarian enforcement both before and after the collapse of Leninism. Certainly, had there been a more pronounced "political will" (as opposed to mere political willingness) in favor of environmental legislation and treaty enforcement, compliance would have been better, but it is equally important to note that had there been environmental enthusiasm on the part of the public and the political elite without a competent administrative cadre to channel it into concrete enforcement measures, compliance would have been substantially worse.

Notes

1. See Hungary's accession to the Basel Convention on the Control of Transboundary Movements of Hazardous Wastes in 1990, the UN Framework Convention on Climate Change in 1994, the Rio Convention on Biological.

2. On Soviet/Russian accession, see the study by Zimmerman, Nikitina, and Clem, this volume.

3. The Dohany Street Synagogue and war memorial were rejected on the grounds that they would "duplicate" Auschwitz, which was already on the WH list.

4. The other treaties include the Helsinki accord on acid rain, the Sofia protocols on nitrous oxide emissions, and the Basel convention on hazardous waste disposal. Hungary had ratified the Geneva convention on transboundary air pollution in 1980.

5. On U.S. accession to the Montreal Protocol, see the study by Michael Glennon and Allison Steward, this volume; on Europe, see the study by Alberta Sbragia and Phillip Hildebrand, this volume.

6. Yugoslavia never acceded to CITES prior to its dissolution; consequently, none of its successor states is a signatory either. Independent Slovakia only began the accession process in 1992–3, while the Ukraine has not yet joined the treaty.

7. A similar dynamic characterized Hungary's recent accession to the convention on Global Climate Change. In that case, Hungarian leaders initially sought to postpone accession, as the costs of compliance appeared prohibitive, especially as it would have affected the energy sector. External political pressure from the European Community and western ministers of environment, however, forced them to change their position. In effect, the Hungarian political leadership had become a captive of its own more general foreign policy goals.

8. It is important to note that although economic restructuring has had a positive effect on efforts to clean up the environment in some ways, it has also had negative consequences as well. For example, there has been a tremendous increase in automobile ownership; since many of the vehicles purchased are used, they do not contain catalytic converters and are a major source of air pollution in the cities.

11

China: Implementation Under Economic Growth and Market Reform

Michel Oksenberg and Elizabeth Economy

Since the advent of political and economic reform in 1978, the Chinese have become increasingly active in international environmental affairs. They now play prominent roles in several international environmental organizations and have signed more than two dozen international environmental agreements in rapid succession. Among these are the five protocols that this book investigates: the Convention in Trade in Endangered Species (CITES) in January 1981; the London Convention against Ocean Dumping in October 1985; the World Heritage Convention in December 1985; the International Tropical Timber Agreement (ITTA) in July 1986; and the Montreal Protocol in June 1992. Table 11.1 shows when China signed the treaties and when they were ratified.

China's accession to these treaties and the nature of its participation in the treaty negotiations are part of a broader effort by Chinese leaders to expand their ties internationally, to appear to be cooperative players in the international arena, and to increase their capacity to develop the Chinese economy and protect the environment (see Jacobson and Oksenberg 1990; Economy 1994; Lardy 1994; Kim 1989; Shirk 1994).

This chapter first summarizes our principal findings. Especially noteworthy is that China's rapid economic growth (averaging 8–10 percent per year for nearly a decade), the increased role of the marketplace, and the extensive administrative decentralization pose severe difficulties for environmental protection measures. Even though Chinese authorities have attempted to meet their international obligations, and progress has been made in developing institutions, laws, and personnel, the situation in many arenas continues to erode. The chapter then notes some of the common characteristics governing China's accession to and implementation of the five agreements. These patterns and procedures are evident in China's involvement in other international environmental regimes as well. Indeed, these characteristics might be called the Chinese style of international engagement. Next, we detail the accession to and implementation of each treaty: the origins of China's interest, the process of engagement, the institutional arrangements, and the implementational

Table 11.1
Treaty membership of China

Treaty	Year treaty came into effect	Signed by China	Ratified by China
WHC	1977	1985	1985
CITES	1975	1981	1981
ITTA	1985	1986	1986
LC	1975	1985	1985
MP	1989	1991	1991

record. The concluding section analyzes the policy implications of our research, both for China and for the international community.

Principal Findings

Seven major conclusions emerge from this study. First, China appears to have met the obligations of each agreement. While its implementation of ITTA and CITES is under scrutiny at the international level and has not satisfied some of the more vigilant international supporters of these agreements, the national agencies responsible for enforcing China's obligations have taken their duties seriously. Further, in each case, an implementing agency at the national level has been established that has increasing contact with the international community. In addition, in each case, a community of individuals and agencies has emerged that believes in the spirit and broader purpose of the existing original agreement. However, China does not appear to be developing the capacity to meet the increasingly stiffer requirements of the agreements; adherence to the letter of the accord is better than to its spirit.

Second, China's performance in each of the five areas is clearly better than if China had not acceded to the accord. The existence of these agreements makes a difference, but the effect varies.

Third, huge problems remain in each of the areas that the five agreements address. In each instance, despite the beneficial effect of the agreement, conditions are worse in the mid-1990s than when China entered the international regime.

This leads to the fourth and most important conclusion. China's rapid economic growth and economic reforms are overwhelming even the most serious efforts in environmental protection (Smil 1993). Increased per capita income, the emergence of wealthy Chinese, and ostentatious consumption have created new demands for environmentally damaging products. According to officials in the Beijing office of the World Bank, the Chinese government at a maximum devotes only $8 out of $1000 of gross national product—0.8 percent of GNP—to environmental protection measures, whereas simply preventing further deterioration would require more than doubling the national effort to 1.7 percent of GNP.

Meanwhile, the political reforms of the Deng era thus far have not enabled the implementational capabilities of the central government on environmental matters to keep pace with the increasing problems (Jahiel 1994). To be sure, new agencies have been established, especially the National Environmental Protection Agency (NEPA), and environmental regulations have been promulgated not just to enforce the five agreements but to address a wide range of other problems as well. But the emphasis on rapid economic growth, administrative decentralization, expansion of the marketplace, and deliberately weakened propaganda apparatus have complicated the tasks confronting the environmental protection agencies. Moreover, as Lieberthal and Lampton (1992), among others, suggest, Beijing authorities find it more difficult to secure compliance from the provinces on issues where control over budgets and personnel assignments resides largely at lower levels.

No single administrative hierarchy exists that can easily convey an edict from Beijing. Enforcement mechanisms exist, but they are cumbersome. If large profits are at stake, the edicts are ignored. Nor can public opinion be easily mobilized. Strapped for funds, the environmental protection agencies cannot afford to convey messages that previously the state propaganda apparatus disseminated for free. Development of norms to support an environmental protection measure, therefore, appears more difficult than in the past.

Because the political reforms have not altered the fundamentally authoritarian nature of the regime, robust nongovernmental organizations (NGOs) and a free press do not exist to stimulate debate and draw attention to environmental issues. Although environmental regulations exist, they are not embedded in a strong legal system. Power in China ultimately resides in the hands of individuals, not institutions, and the judiciary is not autonomous. Enforcement of environmental regulations is problematic and subject to corruption.

Fifth, it should be noted that the agreements under consideration do not deal with the nation's most pressing environmental problems. The five accords are viewed as peripheral to the more elemental agenda threatening the nation: poor air quality, polluted inland and coastal waters, water shortages, soil erosion, loss of arable land, and inadequate toxic-waste management.

Sixth, the considerations that China's leaders and negotiators take into account when deciding to enter an international environmental protocol have changed considerably. In the 1970s and early 1980s, the desire to participate in international activities was overwhelming. After twenty years of isolation from the West and often protracted political persecution, internationally oriented, cosmopolitan Chinese seized upon opportunities they had long awaited. Enjoying the support of Premier Zhou Enlai and then Deng Xiaoping, they sought to ensure a Chinese representation in the world community. Full of optimism and idealism, they wished to contribute to the solution of humanity's problems. Another crucial goal was to drive Taiwan from international organizations. The Chinese also had to balance conflicting desires to win the confidence of the developed world and to retain a leading voice in the developing world.

By the late 1980s and early 1990s, the calculus had changed somewhat. While the earlier motivations had not weakened, economic concerns had come to the fore. In the intervening years, China's membership in the World Bank, UNESCO, U.N. Development Program, and the Asian Development Bank made Chinese more conscious of their eligibility for foreign assistance. And the surge in Chinese exports, with the accompanying greater dependence upon access to foreign markets to sustain the high growth rates, made the Chinese more sensitive and vulnerable to foreign pressures. Increased awareness of environmental problems and the creation of an environmental bureaucracy also have altered the bureaucratic lay of the land. These considerations, plus the accumulated experience at the negotiating table, have made the Chinese more sophisticated, knowledgeable, and tough-minded participants. They have a clearer sense of what they seek to achieve and the impact they wish to have upon the international regime.

Finally, we must note that China's involvement in each of these agreements is critically important to their worldwide success. China has become one of the largest consumers of ozone-depleting substances (ODS) and (with Taiwan) the second largest importer of tropical timber, a portion of which is processed for reexport. Its behavior affects several animals on the Endangered Species List. The cultural treasures of China are among the most prized legacies of all humanity to be bequeathed to future generations. China's coastal waters provide livelihood and fish protein to tens of millions who reside along the coast. As those waters become contaminated and depleted of fish, and with the populace demanding an improved diet, Chinese fishermen are going farther into the Pacific. Fishermen from mainland China are now joining the fleets from North America, Japan, and Korea in pursuit of tuna in the rich waters of several small Pacific island nations. But it now appears possible that a fate awaits these fishing beds similar to those of the North Atlantic and North Pacific. The sheer size of China and the rapidity of its economic rise, in short, mean that in almost every sphere, China's constructive engagement is essential for a global environmental regime to be effective.

Patterns of Entry and Implementation

Accession

Several common patterns characterize Chinese accession to and implementation of the five agreements. The Chinese joined each of the regimes only after careful consideration of the costs, benefits, and responsibilities involved. The initiative for China's involvement in most cases came from the international community. In some cases, a visit to China generated a relationship between foreigners who were interested in the environmental problem and Chinese who were predisposed to work on the issue (perhaps as a result of reading about the issue in foreign publications). In other instances, members of the relevant agency became aware that Chinese participation was essential in addressing the issues, and an invitation was extended to either a Chinese agency or an individual to consider joining the international agreement. The Chinese response and the Chinese policy process in acceding to the international environmental protocol involved protracted deliberations. A new procedural law has codified the internal processes that guide Chinese accession, but these procedures were followed even before the law was promulgated.

The nature of China's participation in the negotiations leading up to ratification of these treaties offers clues to some of the difficulties in implementing the treaties. The agency that assumed the lead role in coordinating the Chinese stance during the negotiations usually became the main implementational body, and this has not been easily changed. This agency has garnered international funds, invitations to attend international meetings, and some enhanced authority over other central and provincial bodies. But the lead agency also has confronted increased administrative burdens and pressures that usually cannot be met with existing financial and human resources.

During the negotiating and implementation stages, the relevant domestic environmental actors in implementational agencies, research institutes, and universities established linkages with international actors. Gradually, a community has assembled in China, especially in Beijing, that has come to believe in the advantages that the agreement offers for China, the environmental purposes, their particular organization, and/or their own personal interests. At the same time, others in China have concluded that the particular agreement offers their cause very little, and may even inhibit the attainment of their objectives and responsibilities. Implementation therefore has become a political issue in China's authoritarian, bureaucratically fragmented system.

Implementation

Several factors determine the success of implementation. The informal status of the implementing agency and of its head affects its ability to guide the interagency effort. Support by one or more high-level political leaders—Politburo members, the Premier, and/or Vice Premiers—is very helpful and even necessary. Lodging the responsibility for the agreement in an agency that has a major stake in its implementation has been important. Agreements that have high international visibility and affect China's stature and access to foreign currency have been more likely to elicit the requisite high-level support. The access, influence, and connections (*guanxi*) of the policy community that gathers around the accord obviously have been crucial, as has the support this community has been able to garner from the international community. Training seminars, transfers of funds through loans and grants, and provision of equipment and technology have enhanced China's implementational capacity in all five agreements. But another important factor determining China's adherence to the agreement has been whether its requirements are congruent and converge with the path China was pursuing prior to signing the agreement.

Provincial and local conditions are pivotal in determining implementation. The agreements fall unevenly across China's highly diverse, vast territory. Some cities, provinces, and regions are much more affected than others.

Moreover, the implementational capacity of different provinces varies considerably, as does the willingness of their leaders to respond to Beijing's directives and pleadings. The balance among conflicting interests in observing or ignoring the terms of the treaty varies even among those provinces affected by it. Put simply, as Beijing bureaucrats state, the farther south and west from the capital, the more difficult it becomes for the central government to secure the desired compliance from the locality. Thus, in the cases of ocean dumping of toxic wastes, removing medicines containing material from endangered species, and encouraging use of substitutes for ozone-depleting substances, Beijing officials believe that Shanghai's performance is better than that of Guangdong.

In both Beijing and the provinces, several factors inhibit effective implementation of the agreements: bureaucratic infighting, mismanagement and abuse of international funds, a weak scientific and technical ability in the managing agencies, an insufficient incentive structure for the monitoring and implementing agents, and a general lack of funds for the specific environmental protection effort. In this, as in most administrative areas, the national government and its ministries issue edicts to the provinces, which then transmit the directives to lower levels. The national level may even mandate the creation of an agency at a lower level to implement the directive. But the actual funding for the mission and the assignment of personnel are the responsibility of the local level of government, which typically has received many more directives than it has the funds and manpower to carry out. As a result, the local leaders must establish priorities and select which of the often conflicting changes they will enforce. Moreover, the top officials in Beijing usually have given some guidance regarding their priorities—for example, that economic growth, maintenance of social order, earning foreign currency, and keeping birthrates low are the main indicators for evaluating performance of provincial and local officials.

In short, in the five cases under review, the national government agencies responsible for implementing the agreements exercise guidance over the provinces, but the actual administration in terms of finance and personnel is in the hands of provinces. The burden on the authorities in Beijing has been to create an awareness and understanding, and hence a commitment, among provincial bureaucrats to alleviate the problem at hand. Beijing's hand has been strengthened because the international agreements now involve the credibility, prestige, and financial interests of the national government. From the vantage of distant Guangdong, however, the interests of the national government are not always of immediate concern.

CITES

Accession

For centuries, the Chinese people have eliminated or endangered flora and fauna that thrive on their land and in their waters. A thousand years ago, much land—especially in the south—was forested; marshlands were abundant. But the pressure upon the environment increased substantially with the growth in population from the 1500s on. The political turmoil of the past 150 years, with the absence of authority to enforce environmental protection measures, intensified the problem, as did the industrialization of recent decades. For example, the Chinese Emperor designated vast tracts of land as imperial reserves and hunting grounds, but with the collapse of the imperial rule, these natural parks were denuded of their forests, and their game animals—including many rare species—were destroyed.

Chinese scientists, historians, officials, and even peasants aware of their local histories did not require Western environmentalists to remind them of the changes to the natural habitat through the centuries. This longer-term trend had become particularly acute during the Mao Zedong era, and especially the Cultural Revolution and its aftermath (1966–1976). It was a period of poaching and major illegal exports of wildlife.

The 1972 Stockholm environmental meeting called for a convention against trade in endangered species. Chinese Premier Zhou Enlai attended that meeting and delivered a speech to it. Through its participation in this international conference, China became convinced that it should accede to such an agreement. Zhou, his colleagues, and advisers from both the environmental and the cultural domains thought CITES offered a vehicle for assisting China to adopt the strong measures the country needed to bring its disastrous situation to an end. However, international diplomatic obstacles prevented an immediate accession. Taiwan's cooperation was vitally important to the success of CITES. For example, it was a major importer of elephant tusks for its ivory-processing industry. As the Republic of China and claiming to represent all of China, Taiwan attended the 1973 Washington meeting at which CITES was signed. In light of the policy of the People's Republic of China's (PRC) not to join international organizations where Taiwan is present as the Republic of China, the PRC absented itself from the Washington meeting and subsequent activities of CITES.

However, the international community (including the nascent CITES Secretariat) retained contact with Beijing and encouraged it to adhere to CITES. Finally, in the late 1970s, U.N. Secretary-General Waldheim specifically asked Chinese Foreign Minister Huang Hua to consider joining the Convention. Upon the Foreign Minister's return to Beijing, he informed the State Council Environmental Protection Leading Group about the U.N. invitation. A coordinating group of seven agencies discussed the matter and prepared a report for the State Council; the group consisted of the ministries of Forestry, Foreign Affairs, Foreign Trade, and Agriculture, the Chinese Academy of Science, the Export–Import Commission, and the Environmental Protection Agency under the Capital Construction Commission.

He Liliang, the head of the Foreign Ministry's International Organization Bureau, was the wife of Foreign Minister Huang Hua. A very capable and intelligent woman, she had spent several years in New York, where her husband was China's ambassador to the United Nations. This experience had exposed her to the wide range of concerns then circulating in Manhattan's various international communities. Among the top ranks of the leadership, CITES also attracted backers. Vice Premier Gu Mu, head of the Capital Construction Commission, where the Environmental Protection Agency was located, became the nominal head of the interagency CITES study group and expressed an interest in its purpose. Song Jian, the rising star in science administration, also became a backer of the effort.

Drafting of the recommendation report to the State Council was assigned to Qing Jianhua, an official in the Ministry of Forestry. Forestry drew this assignment, which affected the location of the implementational agency, for a simple reason. Since 1958, Forestry had been responsible for wildlife conservation; China's remaining pandas lived in preserves under its jurisdiction. And since protection of pandas clearly would be a principal beneficiary, Qing's report foresaw only benefits from participation in CITES. With strong backing from higher officials, the report did not refer to more cautious views of the Foreign Trade, Public Health, and Defense ministries. Indeed, the latter two were not even represented in the working group. The Export–Import Commission argued that responsibility for implementing CITES should be given to the Ministry of Foreign Trade (MOFERT) as the most effective way to meet the obligation to halt export and import of endangered species. Qing's report stressed the need for the implementing agency to have sufficient authority to supervise enforcement. The solution was to place the implementation agency—the Endangered Species Import–Export Office (ESIEO)—directly under the State Council while physically locating it in the Ministry of Forestry building. Qing Jianhua became its head.

However, it remained awkward for this new body to regulate trade in endangered species, since according to State Council regulations, only MOFERT could issue permits or licenses to engage in foreign trade, a prerogative it guarded jealously. Eventually a solution was found. The State Council issued the directive enumerating the banned species under CITES, and the ESIEO issued certificates (not permits) for the exceptions to the ban (such as to send a panda to a foreign zoo). In the Chinese lexicon of licensing documents, a certificate ranks below a permit, thereby assuaging MOFERT concerns. This history remains relevant today, as we note below, as the ESIEO asserts that MOFERT, the People's Liberation Army (PLA), and Public Health have no role in implementing the treaty.

According to ESIEO officials, "We do not have to include MOFERT, the PLA, or Public Health in our decisions. They have no role in our deciding how to implement the treaty. They cannot influence policy-making." But moments later, the same officials complained, "We have problems of coordination. For example, after the ban was issued in 1991 against import of rhino horns and tiger bones, we had difficulties [securing compliance]. Even calls on the relevant ministries failed to achieve a consensus." Not until Politburo member Song Jian called an interagency meeting in May 1993 was the matter settled. This situation is probably related to the bureaucratic dynamics of China's accession to CITES: what agencies were and were not included in the initial working group, the bureaucratic status of the lead agency, and the distribution of benefits from participation.

The Management Structure

China has established a CITES office that, although located physically within the Ministry of Forestry compound, reports directly to the State Council. Contracting parties abroad and within the PRC are permitted to accept permits only from the CITES office.

The CITES office in Beijing is the principal coordinating point; it has ten staff members and is petitioning for twenty more. There are also nine branch offices—in Beijing, Shanghai, Tianjin, Chengdu, Fuzhou, Shenyang, Guangdong, Nanning, Harbin, and a check point in Shenzhen—which the central office hopes to increase to twenty. The budget for these offices is provided by the provinces, and each typically has a staff of three to five people. Although Vice Premier Song Jian is very supportive of China's CITES efforts, the State Council Personnel Management Commission will make the decision concerning its expansion.

Actual enforcement of CITES in forests under control of the Ministry of Forestry and in provinces where departments of forestry exist is the responsibility of China's 46,000 forest rangers. These rangers constitute a dual lead agency that appears on the Ministry of Forestry table of organization as the Bureau of Public Security and in the Ministry of Public Security as Bureau Number 16. The salaries of the rangers are on the budget of the Ministry of Forestry. The rangers have many enforcement responsibilities beyond CITES, especially prevention of illegal logging. The bulk of the 46,000 rangers are deployed in the national, provincial, and local forests, which means that enforcement officers are stretched thin outside of forest lands, where many of the violations occur.

Moreover, in some locales the Ministry of Forestry does not have subordinate units. In those areas, the local Public Security Bureau is responsible for CITES enforcement. Several other ministries also play roles in the management and enforcement of CITES: the Ministry of Foreign Trade and Economic Cooperation; the Ministry of Agriculture, which is responsible for aquaculture; the Ministry of Public Health; and the Endangered Species Research Agency in the Chinese Academy of Science, whose experts ESIEO must consult.

Interaction with the International Community

Chinese efforts to implement CITES have involved significant assistance from foreign experts and officials. Typically, the foreign specialists have found the Chinese management structure for CITES frustrating, often ineffective, and weakened by the existence of competing bureaucratic and institutional interests. One Wildlife Conservation Society analyst noted that the "Ministry of Exports" (the Ministry of Foreign Trade and Economic Cooperation) appeared to have more power than the Ministry of Forestry, which as the lead agency should be coordinating and directing policy. Therefore, this analyst commented, it has been difficult for the PRC to control exports of species.

Provincial rivalries are also keen: reportedly, in 1992, China submitted a large Global Environment Facility (GEF) proposal for CITES enforcement. Incorporated in this proposal was a bid for every province to have its own breeding facilities for tigers, in order to generate tiger bones for traditional medicine. Similarly, there is poor coordination between the central authorities and municipalities. During the late 1980s and early 1990s, when the central government was attempting to regulate international panda loans, it tried to coordinate all the loans of pandas. However, the municipalities frequently bypassed the regulations, often with the complicity of American zoos and the U.S. Fish and Wildlife Service, which was responsible for issuing permits.

Occasionally, even when the U.S. zoos opposed the transfer of pandas, they were overruled by local politicians who wanted them for the prestige and prospects of tourism they brought (Schaller 1993). In December 1993, the U.S. Fish and Wildlife Service began a review of its permit issuance policies and temporarily barred the importation of pandas from China. It turned down the San Diego Zoo's request to import two pandas (*New York Times* 1993c).

Within the PRC, two separate Chinese organizations that lend pandas, the Ministry of Forestry and the Chinese Association of Zoological Gardens, which is administered by the Ministry of Construction (which also controls pandas in the wild), actively competed for control over the pandas (*New York Times* 1993c). According to one U.S. expert, this competition prevented the PRC from developing a comprehensive conservation plan for breeding pandas; further, he suggested that the lack of cooperation extended to the point that training seminars conducted by one organization might not be attended by the other. One key official in CITES's China office has stated that sometimes that office has sent certificate requirements to the provinces but has failed to receive responses. As Schaller notes, "Enthusiasm and goodwill count for little when the enemy is a vast bureaucracy of local officials who myopically use obstruction, evasion, outdated concepts, activity without insight, and other tragic traits to avoid central-government guidelines and create ecological mismanagement on a dismaying scale" (1993:234).

The center itself has participated in illegal trade in endangered species. In the mid-1980s, a state-run organization, China National Nature Produce and Animal By-products Import and Export Corporation, sent its agents into the countryside to stimulate the killing of wildlife. According to the noted conservationist George Schaller, who worked during the 1980s with the Chinese at the Wolong Panda Reserve in Sichuan province, "Musk, bear paws, skins, antlers, bones and other animal parts, often from endangered species, were blatantly sold on the open market" (1993:225). In addition, the PLA has been implicated in protecting the smugglers of tiger bones and rhino horns.

Finally, the management infrastructure does not promote sharing of information and ideas. Even at the most basic level, there is poor coordination of conservation efforts. Reserves and zoos, for instance, do not share specialists, resources, or even good breeding pandas. Also, according to one U.S. environmental specialist, the Chinese government had significant difficulty in setting up a coordinated effort to manage CITES enforcement and biodiversity. The Chinese bureaucracies were unwilling to share data and information. A key Chinese CITES official agrees that there are serious interministerial coordination programs. In attempting to resolve the problem with the use of tiger bone and rhino horn in the manufacture and sale of traditional Chinese medicine, this official called for a joint effort involving several ministries. However, these ministries, which included the Ministry of Forestry, the Ministry of Customs, the Ministry of Agriculture, the Public Security Bureau, the Ministry of Foreign Affairs, the Ministry of Finance, the State Planning Commission, NEPA, the State Oceanographic Administration, the Ministry of Industry and Commerce, and the Ministry of Foreign Trade and Economic Cooperation, failed to reach consensus until Song Jian held a meeting in May 1993. The total estimated cost of closing down the factories that produced these medicines was $2 billion.

Compliance with CITES also appears to have been hampered by the interest of ministries in the potential monetary rewards that could be realized both from overt noncompliance and from abusing the funds made available by external actors to support compliance. According to a Wildlife Conservation Society (WCS) analyst, WWF (originally World Wildlife Fund; now World Wide Fund for Nature) money ended up being spent on cars for people in the Ministry of Forestry. In addition, according to wildlife specialist Alan Rabinowitz, China has a formal list of the prices it charges foreigners to study particular species (Begley 1993).

In some cases, this abuse of funds may be advanced by International Nongovernmental Organizations' (INGO) willingness to support any demand that Beijing makes for them to become involved in conservation efforts. In recounting a statement by Prince Philip, President of WWF, Schaller comments:

> ... the Chinese demanded every single item on that agreement had to be met, in spite of the fact that they knew perfectly well that a lot of the items, like the elaborate equipment for instance, was nonsense. It was mortally embarrassing for them when they couldn't get people to use it. (1993:222–223)

This failure to appropriately utilize international funds and equipment was also evidenced in the fact that, according to one U.S. expert, some of the $4 million that WWF gave to China for the Wolong reserve went to build a hotel and school for the children of resident Chinese scientists. In another case, the World Bank did not approve a rural credit project (a subloan from the World Bank that goes through the Agricultural Bank of China), which was meant to increase the number of breeding facilities, specifically because the project was not perceived as being motivated entirely by environmental concerns.

There have been extensive efforts undertaken jointly by the PRC and external actors in training, transfer of funds, and development of joint projects. CITES is perhaps the treaty that traditionally has engendered the most international interest and assistance for the PRC. The WWF has been involved in the PRC since 1980 and has been one of the most active participants. It has had projects on panda conservation and the south China tiger, among others (World Wide Fund for Nature 1992b). Other efforts have involved cooperation between the U.S. Department of the Interior and the PRC Ministry of Forestry: the two agencies have a joint protocol for wildlife studies, database development, and training. Chinese CITES officials cite a three-month undercover TRAFFIC sting as especially helpful in rooting out the illegal trade in tiger bone. China also has become active in CITES. In the spring of 1994, it hosted the biennial CITES Animals Committee meeting as a prelude to the annual CITES meeting that was held in the fall of 1994 in Florida.

The impact of direct international pressure on China's compliance efforts cannot be measured precisely. In fact, the lead PRC CITES official denies that external pressure plays any role in PRC efforts to halt trade in endangered species. However, there is strong evidence that when the PRC's economic interests are threatened, the leaders respond with tougher laws and promises to adhere to the treaty.

Cases of continued Chinese trade in rhino horn and tiger bone and products made from them illuminates this point. During the summer of 1993, the United States threatened the PRC with trade sanctions for continuing to traffic in rhino horn and tiger bone. The PRC managed to avoid the sanctions by taking a number of well-publicized and definitive steps to address the problem. Immediately prior to the formal U.S. warning, on May 29, 1993, the State Council issued the "Circular on Forbidding All Trade in Rhinoceros Horns and Tiger Bones." In June 1993, however, the U.S. Department of the Interior issued a warning to the PRC, under the Pelly Amendment (which calls for levying trade sanctions against countries trading endangered fish or wildlife species) on tiger bone and rhino horn (TRAFFIC-USA 1993). The PRC responded with a State Council order that rhino horns and tiger bones would no longer be legal in Chinese traditional medicine. Further, the order made illegal the purchase, sale, trade, and transport of these substances (TRAFFIC-USA 1993).

China's rhino horn-based medicines are consumed domestically and in other Asian states. Until the threat of the Pelly Amendment, the PRC had refused to become a party to the 1987 CITES resolution to ban internal trade in rhino products (TRAFFIC-USA 1993). It argued that its internal stocks of such products were acquired prior to 1981. It also continued to import rhino products. In 1989, China had a recorded stock of ten metric tons of rhino horn, possibly the largest in the world. When the Pelly Amendment was invoked, however, the PRC stated that manufacturers would have six months to sell medicines already produced with parts or products of rhinos or tigers (*New York Times* 1993a).

It is difficult to ascertain how effective this new State Council regulation will be, although a number of statements from top-level officials have followed the issuance of the circular. A ban on the hunting of the Siberian tiger and the south China tiger, for example, has been in existence since the 1960s. However, it has yet to be implemented thoroughly (*New York Times* 1993a).

On September 8, 1993, CITES itself issued a directive to "consider stricter domestic measures up to and including the prohibition of trade in wildlife" to China and Taiwan. Following this announcement, Vice Minister of Forestry Shen Maocheng stated that the implementation of China's decision to prohibit trade in rhino horn and tiger bone would be part of a major three-year program to ensure the enforcement of wildlife protection legislation. He called for coordination with the international community to "crack down on illegal trafficking of endangered wild species of fauna and flora" (AFP 1993).

In September 1993, Vice Premier Song Jian called a meeting to undertake a special study on further augmenting protection and administration of endangered species. At the meeting, it was decided to set a deadline for determining which shops were dealing in rhino horn and tiger bone, registering the substances when they were found, and affixing them with a special seal to remove them from circulation. All medicines that contained these substances were to be withdrawn from the market before November 30, 1993. The participants further called for the development of substitutes for these substances in medicines. In addition, they condemned "weakness, incompetence, and corruption" in administrative and law enforcement departments (FBIS China Daily Report 1993). Moreover, at the 1994 CITES Conference of Parties, China and India spearheaded the formulation of an agreement to stop domestic trade in tiger parts.

Record of Compliance

The Chinese have faced significant difficulties in implementing CITES despite their continued efforts. In addition to the bureaucratic infighting and mismanagement by officials, a number of other cultural and economic factors undermine the ability of proactive officials to implement the treaty effectively.

Poachers and dealers are clearly motivated by the possibility of making significant amounts of money. Dealers offer poachers $3000 or more for each panda pelt; the dealers then sell them in Hong Kong, Taiwan, and Japan for upward of $10,000 (Schaller 1993). In spite of fines and even the possibility of death—five men have been sentenced to death thus far, and over 200 imprisoned—poachers appear willing to risk their lives for the chance of making significant amounts of money (*New York Times* 1993a). CITES officials, however, believe that the threat of the death penalty, along with extensive public awareness campaigns, has been a very effective deterrent. They have exerted substantial efforts to go to the rural and remote mountainous regions to educate the peasants. One CITES official admitted, however, that it is far more difficult to educate the people in these areas and that there is over 1,000 years of tradition in the use of wildlife. He further noted an example of one mountain dweller who made his thumbprint on a statement of commitment not to poach pandas and was caught the very next day with a panda skin.

While money is an important motivation for poachers, cultural and technical issues are additional sources of China's poor compliance record on CITES. According to a U.S. wildlife expert, patrols are not undertaken in the reserve parks, only outside them. Chinese forestry officials argue that the village people are poaching for food; therefore, the officials are reluctant to interfere. According to this expert, however, poaching is increasingly the province of outsiders. In addition, he noted that the guards at the reserve parks are reluctant to report incidents of poaching because they will be blamed. While he did point out that experimental patrols are being conducted in Sichuan at the Wolong panda preserve, Schaller found that neither media nor international pressure had any effect on the virtually complete lack of law enforcement (Schaller 1993). In addition, despite the harsh penalties for poaching and trading, according to a foreign expert, Beijing typically exerts little pressure on the provinces to cut down snares in the bamboo forests or to search the panda reserves for poachers.

In addition to the inability of Beijing officials to effectively manage the trading of endangered species, occasionally measures taken to respond to the problem may in fact contribute to its continuation. According to Schaller, the Chinese are dedicated to captive breeding as a means of protecting endangered species. However, another U.S. wildlife expert argues that the breeding programs themselves could be encouraging the growth of endangered species trade. Poor technical skills were manifested in the case in which baby pandas taken to the Wolong center fell ill and died because of inexperienced and incompetent vets. More recently, the Chinese refused to release results of a panda census. According to one expert, it is unclear whether this signifies that they are underreporting to get additional funds, or overreporting in order not to be blamed for failure (*New York Times* 1994).

Chinese efforts to stem the tide of trade in endangered species appear to be woefully understated and underfunded. Moreover, the right incentive and/or penalty system for the implementing agents apparently remains to be developed. A CITES official specifically noted the impact of the economic reforms on hindering enforcement efforts. He stated that previously a wildlife protection poster in the airport was free; now, it costs between 30,000 and 50,000 yuan.

At the same time, however, the repeated calls by Chinese elites for action to control trade in endangered species show no sign of abating. Wang Bingqian, a vice president of the National People's Congress and one of the heads of the provincial inspection effort initiated in August 1993, stated in a speech before the Congress's Environment and Resources Protection Committee in April 1994 that "rampant poaching, killing, selling, and buying of endangered and rare animals which are under national protection are still going on in some places" (FBIS China Daily Report 1994). NEPA's director, Xie Zhenhua, commented, "illegal poaching, killing, smuggling, selling, and buying of protected wildlife animals have survived repeated government sweeps" (FBIS China Daily Report 1994:19).

There are extensive efforts to educate the public through television programs and reports in newspapers. One potential boost to CITES enforcement may emerge from the inspection efforts that were initiated in 1993. These inspections have included, and will include, an assessment of wildlife protection in ten provinces and autonomous regions: Hebei, Jilin, Zhejiang, Fujian, Henan, Hunan, Sichuan, Guizhou, Inner Mongolia, and Guangxi Zhuang Autonomous Region. One thirteen-day investigation of enforcement of environmental protection in Guangdong led to the exposure of a pharmaceutical company speculating in rhino horn (FBIS China Daily Report 1993).

London Convention

Accession

Prior to acceding to the London Convention in September 1985, China had passed the Marine Environmental Protection Law (1983). This law, which was modeled on the London Convention, was designed to address the large amounts of industrial waste residue and household refuse that were being "piled at the tideland" and "deliberately discarded at sea" (Fang et al. 1993). It covered numerous activities related to marine management, including construction of coastal engineering projects; exploration and exploitation of marine petroleum, fisheries, and breeding; industry transport services; military operations; scientific research; and, most important for the London Convention, discharge and dumping of wastes into the sea. In addition, of the thirteen International Maritime Organization (IMO) treaties focused specifically on preventing pollution from ships, China has ratified five (Huang 1993). China also has developed twenty-five regulations on the dumping of waste at sea.

Dumping is regulated through a permit system that is managed by the State Oceanic Administration (SOA) and its Division of Marine Environmental Protection. Any entity that wants to dump any kind of waste into the sea is required to file an application permit with the Department of Marine Affairs; the dumping has to be done at a designated site and within a designated period of time (Huang 1993). The SOA selects the dumping sites and is responsible for closing them if it so desires. The sites must be approved by the State Council (Huang 1993). There are three lists that categorize the types of waste that various ministries or other actors desire to dump. These lists are based on those developed in the original 1972 London Convention: the Black List consists of substances whose dumping is prohibited; those on the Grey List can be dumped with special permits (valid for six months); and those on the White List require only general permits (valid for one year) to be dumped. There are also special one-time emergency permits. If the permit is granted, its recipient is required to pay five cents per cubic meter of waste.

The SOA has the power to penalize violators of the Law. If found guilty, violators are required to abate pollution within a specified period of time, pay a pollution discharge fee, defray expenses for pollution control, and compensate for damage sustained by the state. The amounts of the fines are set by law, although the Ministry of Finance is involved in determining the fee schedule. In addition, for dumping of waste at sea, there are specific criminal punishments—for instance, not more than three years of fixed-term imprisonment and, for especially serious violations, not less than three and not more than seven years. There are also regulations that provide for payment of compensation for ecological damage; in practice, however, such liability remains unenforced (Huei 1989).

The Management Structure

A complex infrastructure has evolved to manage ocean dumping. At the national level, there is one waste-dumping administration, one monitoring service, and one information service. At the marine district level, there are three management agencies and three monitoring institutions. The district-level authorities are responsible for issuing the general and special permits (the SOA must issue the emergency permits). They are supported by five provincial management agencies, fifty-seven offshore inspection stations, two surrveillance aircraft, twenty-seven supervisory vessels, and eleven scientific investigation ships. There are approximately 280 people who manage waste-dumping and an additional 620 who are responsible for research, monitoring, and investigation of ocean dumping. The budget of these agencies is 4,480,000 yuan.

Although the SOA is the lead agency for marine dumping and for implementation of the London Convention, other ministries and agencies hold significant power in verification and monitoring. If the dumping of waste is from ships or boats, the Harbor Superintendency Administration is responsible for examining the waste and ensuring that it does not fall outside the classification of the permit. Representatives of the Administration also may accompany the vessel when it dumps its waste (Fang et al. 1993). According to one U.S. expert, the ministries that deal with marine affairs have "drawn distinctive turf lines." He states that if a marine area has a fishery, then it falls under the jurisdiction of the Ministry of Agriculture. An area with military facilities is under the Navy's control, and transport lanes and areas with transport facilities are the "property of the Transportation Ministry." Land-based sources of pollution are the responsibility of NEPA. An SOA official has commented that, in general, the Ministry of Transportation wants dumping sites to be permitted closer to land, while the SOA pushes them farther out. Often these disagreements necessitate serious negotiations.

For some of these ministries, pollution control conflicts with their economic interests. A U.S. expert has claimed, "The SOA cannot monitor everywhere, and if they find evidence of a ship polluting, they will say something. However, if they go into an offshore oil rig which 'belongs' to the Petroleum Ministry and find excessive pollution from dumping, the Petroleum Ministry would bring forward their own researchers who would claim that 'we pollute very little.'" Even NEPA and SOA sometimes have some conflicting interests over coastal areas/land-based sources of marine pollution. Although NEPA's primary interest is in land-based pollution, it is seeking jurisdiction over coastal management. However, SOA is reluctant to cede this control because it would lose international funds.

Interaction with the International Community

As the necessity of ocean dumping increases in the PRC, the role of external actors appears to be critical in advancing effective Chinese implementation of the London

Convention. There are training courses on marine waste management in China that are directed by the IMO and sometimes are conducted in conjunction with, and funded by, other international offices such as the U.N. Environment Programme. According to a London Convention official, the Chinese SOA researchers and scientific officials are not very experienced or knowledgeable, but they are eager to learn. According to this same official, China is developing the infrastructure to implement China's domestic laws related to the London Convention with its laws and training. The Chinese need specific guidance on what to do because they are "not very good with broad guidelines." They sponsored an ocean dumping symposium in the fall of 1993.

According to the official who used to chair the scientific group for the London Convention, the lead Chinese representative on the scientific group, Zhou Jiayi, is "very aware of techniques and technology related to the Treaty as well as the compliance of China with the Treaty." He was very active in the negotiating and planning of the Waste Assessment Framework, which provides technical implementing guidelines for the Convention. This same official noted that for a long time, China was simply an observer. However, since accession the Chinese have become very active, particularly in the scientific groups. They have made a number of submissions to the scientific and political groups on what they are doing.

According to one U.S. expert, unlike many developing states, the PRC is taking an active role in the scientific negotiations on the London Convention. It is especially active on the Intergovernmental Panel on Radioactive Waste Disposal. China's participation is fair compared with all states but excellent in comparison with developing states. The PRC also has been very active in the International Oceanic Commission of UNESCO. According to a second U.S. expert, the Chinese want a leadership role and have offered to host one of the World Data Centers, which entails developing archival oceanic data on a global scale. The centers are very expensive to develop and maintain.

The SOA also has begun to experiment with new methods of managing marine pollution with assistance from the U.N. Development Programme. In Xiamen, a four-year project is under way in which the municipal government is in charge of coordinating hydro development, industrial enterprises, and agriculture. The fee system for obtaining permits and for dumping is regulated through this system of local governance. In other regions, according to SOA officials, the fees are not well regulated.

SOA officials have subtly suggested that the London Convention needs to be strengthened in order for it to be implemented more effectively. In an informal paper prepared for an international conference, several SOA officials noted two weaknesses in the London Convention: first, it has no system of penalties for parties that violate the terms of the treaty; second, the Convention does not provide a quantitative standard for "significant amounts," and "trace contaminants," which makes it "very difficult to make decisions in many cases on whether the wastes can be disposed at sea or should be prohibited and on whether the general permits or special permits should be issued." These comments suggest an appeal to the international community for a stronger convention that could buttress the efforts of the SOA officials in preventing ocean dumping.

The Record of Compliance

In general, there has not been significant evidence of illegal ocean dumping by the PRC. According to one published study, the failure to strictly regulate waste disposal on land has had the effect of permitting factories and enterprises to dispose of their wastes on shore or to release them directly into coastal waters. Therefore, there have been no real demands for ocean dumping. Coastal water pollution is the result of "land-based discharges rather than dumping" (Jian 1993). Only an estimated 2.8 percent of marine pollution in the PRC is from ports or ships; in contrast, approximately 84 percent is from river discharge (Huang 1993).

However, there is evidence that the central authorities are having difficulty controlling dumping in waters off Guangdong. One SOA official, for instance, has commented that Guangdong "doesn't listen." He claims that superficially Guangdong pays attention to center directives, but local authorities are really in control. The SOA did close a dumping site in Daya Bay in Guangdong in 1992;[1] however, in general, according to this official, the SOA lacks the funds to monitor dumping in the region. In addition, there is concern that with the continued economic development of the coastal region, oil drilling, and marine-based industry, there will be increasing numbers of industries that wish to dump.

There is also evidence that the high level of Chinese compliance is expected to change. According to one expert, with the increased burning of coal and consequent increasing quantities of fly ash requiring disposal, China is considering disposal of fly ash, as well as of calcium carbonate residue from fertilizer production in the Bo Hai (Valencia 1992). If the Chinese are to begin ocean dumping, criteria for dumping site selection will need to be identified (Valencia 1992). As of the early 1990s, the PRC has demarcated at least fifty-two marine dumping areas (Huang 1993).

The enforcement mechanisms developed by the SOA for ocean dumping are constantly being upgraded. The SOA has designated thirty-eight offshore dumping areas that have been approved by the State Council. In 1992, 686 general permits were issued for dumping of 3.48 million cubic meters of dredged substances; this represented a decrease of 1.41 million cubic meters from the previous year. It is unclear that this decrease represents a positive change, however; it may be that the number of permit applications is declining due to an increased ability to dump illegally. Although remote sensing capabilities are being developed by a number of agencies (the Petroleum Ministry, the Geology Ministry, and the Ministry of Transportation), enforcement is hampered by poor communication among these bureaucracies.

The level of China's compliance with the London Convention appears to be fairly high. China has submitted reports on issued permits annually and is actively participating in the scientific and management training sessions offered through the Secretariat and other international organizations. There have been cases of illegal dumping, but they did not lead to offshore pollution. The fines range from 20,000 to 200,000 yuan; in the last few years, however, there have been fewer and fewer cases. Nonetheless, as one U.S. expert noted, the main obstacles to China's complying with the treaty are lack of adequate knowledge about technical matters, overcentralization of administration, and a short-term perspective concerning the importance of development first and environmental protection second. In addition, future compliance may be jeopardized by changes in the mandate of the treaty. As one Ministry of Foreign Affairs official noted, if the mandate of the London Convention is expanded to include coastal waters (as is currently being discussed), China will protest on the grounds of infringement of state sovereignty.

The World Heritage Convention

Accession: A Policy Entrepreneur and the Garbage Can Model at Work

The "garbage can" model of politics emphasizes the randomness and serendipity of the policy process. Politicians in search of answers to their problems latch onto policy innovators, communities, and proposals that have been around for a while. The policies that the innovators advocate may or may not be related to the political concerns the politicians have—to attack a rival, to placate or cultivate a constituency, to garner funds. Thus, the policy process frequently involves the inadvertent wedding of different policy streams. That vision of politics, developed by Michael Cohen, Jack Walker, and John Kingdon in their cumulative work on organizational and American politics, captures China's accession to the World Heritage Convention.

Three policy streams came together. The first was the policy community, long interested in preserving China's cultural relics. The Bureau of Cultural Relics, established in the 1950s, can trace its origins at least to the Republican era (1911–1949), when various archaeologists, architects, historians, artists, and their patrons in the governmental and commercial worlds feared the destruction of China's extraordinary heritage of temples, palaces, bridges, tombs, archaeological sites, and artistic works. After 1949, this community found political patrons in Premier Zhou Enlai and Politburo member Chen Yi. After the Bureau was established in the 1950s, a system developed for identifying and preserving historical sites. This system collapsed during the Cultural Revolution, when much of China's cultural heritage was pillaged by marauding Red Guards. Further damage was done when many of the urban sites were turned into military bases, where soldiers previously located in rural bases were encamped as they restored order in the cities. While not deliberately damaged, the sites decayed under military protection.

By 1978, the Bureau was back in business, with Zhou's backing. Directly under the State Council, it began to reclaim and restore the sites. In subsequent years, its nominal status changed; it retained its independent status until 1982 and from 1987 to 1993, while from 1982 to 1987 and since 1993 it has been part of the Ministry of Culture. With Zhou Enlai's death in 1976, it lost its patron, and with the exception of Vice Premier Gu Mu, now retired, its officials have felt that they have not had a strong patron at higher levels.

The second policy stream involved those concerned with preserving China's natural and scenic spots. In the 1950s, the powerful State Capital Construction Commission (SCC) acquired responsibility for this mission, presumably because its core mission was to guide China's investment projects and therefore also decided what to preserve, what was off limits. The Cultural Revolution wreaked havoc in this area as well. The SCC was abolished. For nearly a decade, China's natural sites fell under local or military control. Much devastation occurred. By 1978, the prior structure had been resurrected, and the reestablished SSC convened a landmark meeting to assess the damage and to develop a strategy for preserving natural and cultural heritage. Vice Premier Gu Mu, the head of the SCC, lent his full weight to the effort. Staffing the effort fell to the Bureau of Landscape in the SSC, and a remarkable, talented, and energetic policy innovator—Division Chief Liu within the Bureau—stepped forward to organize a survey of the national situation.

The investigation lasted three years, bringing together teams from universities such as Qinghua, ministries such as Forestry, and landscape architects. The teams included archaeologists, biologists, and other scientists. In 1981, the SCC, in coordination with the Bureau of Cultural Relics, submitted a report to the State Council detailing the urgent need to preserve scenic and cultural spots. The SSC approved the report, but in 1982, the SCC was abolished, its constituent bureaus were distributed among other agencies, and Gu Mu suffered a political setback. Responsibility for preservation of the natural heritage was assigned to the General Administration of Urban and Rural Construction, which acquired ministerial status in 1988.

Planning went forward under the new setup, and in 1982, the Law on Protection of Cultural Relics was promulgated. The preservationists needed new patrons, and they found them in the Chinese People's Political Consultative Conference (CPPCC), an organ that brings together distinguished personages from various walks of life, including the leaders of China's very small remnant of non-Communist political parties that date back to the Republican era. In 1982, the CPPCC helped select the forty-four national scenic spots that merited preservation. This activity culminated in the June 1985 Provisional Regulations on Management of Scenic Spots and Historical Sites.

The third policy stream was now ready to appear on the scene. After China entered the United Nations, various U.N. agencies began to set up shop in China, such as the United Nations Development Programme and the United Nations International Children's Emergency Fund (UNICEF). Among these, of course, was UNESCO, and in keeping with that organization's methods, the Chinese National Committee of UNESCO was formed. After the 1985 Provisional Regulations were passed, this committee approached the National People's Congress for approval to join the World Heritage Convention. It was, in effect, seeking activities to promote and serving its linkage function between UNESCO and Chinese agencies. However, it had not played any role in drafting the 1985 Provisional Regulations or even, apparently, in encouraging the agencies involved. The initiative was from within. But a domestic system had been developed that could almost inundate the international community with sites worthy for inclusion as part of the world heritage.

The Management Structure

The management structure of the Chinese for this treaty is complex, involving not only the central government but also provincial departments and municipalities. It also has evolved significantly over the years.

While the China National Committee of UNESCO has oversight responsibility, the State Cultural Relics Bureau and the Ministry of Construction are in charge of concrete implementation. The State Cultural Relics Bureau, which has about 110 administrative workers, handles international agreements and the management of work related to cultural heritage. It also formulates policies, promulgates decrees, coordinates relations with other departments, provides funds for protection projects, and offers training. In addition, the Bureau organizes the investigations of sites, registers items of cultural heritage, and submits them to the State Council for approval. From this list of State Council-approved sites, the Bureau recommends sites for inclusion on the World Heritage List. Other agencies also nominate sites for World Heritage recognition.

The Ministry of Construction has, under its auspices, the Scenic Spots and Historical Sites Management Department, which manages scenic historical sites. In addition, this department "drafts regulations on strengthening management regarding protection, examines the overall plan for the state's scenic spots and historical sites, and urges local authorities to solve existing problems." Some Chinese officials have implicitly suggested that the Ministry of Construction traditionally has not been proactive in its protection efforts. As one official commented, "Since China's five national scenic spots were included in the World Heritage List, the Ministry's

[Management] department has been gradually strengthening its conservation efforts and administration of heritage projects and organizing activities conducive to conservation work." The local regions in which the scenic spots are located also have been active in protection efforts. They hold publicity drives to enhance the environmental protection awareness of local people and tourists. In addition, at Mount Huangshan, tourists were banned at some sites in order to promote vegetation growth.

At least one Chinese official has noted that the inability of the management structure to clearly define the lines of responsibility among the Ministry of Forestry, the Ministry of Construction, the Bureau of Cultural Relics, and the National Environmental Protection Agency (NEPA) makes coordination very difficult. Some national parks fall within the jurisdiction of NEPA, while others are the responsibility of the Ministry of Forestry. Any of the four agencies may initiate a nomination for a scenic site. These proposals are submitted to the Chinese National Committee of UNESCO and a number of expert working groups that review the proposals. The Secretariat, which consists of thirty people, selects the list of sites to submit to the international body. Further, many provinces nominate sites within their boundaries.

One Chinese official has noted that the "separation of ownership and use rights to different departments" has brought about a situation in which some actors are interested in reducing tourism to protect the cultural sites while others are more interested in increasing the number of visitors for greater profit. According to a U.S. National Parks Service official, there is intense competition among the Chinese bureaucracies involved in nominating sites. It is the Ministry of Construction, however, that sends a representative to the UNESCO Committee meetings; NEPA and Ministry of Forestry representatives are frequently absent.

Funding for protection of scenic and cultural heritage sites comes from several sources: the national government supplies about 140 million yuan annually, only twice the amount budgeted in 1982. The majority of the funds is derived from ticket sales and other sources, such as fund-raising drives. For example, a significant public effort to raise money for the preservation of the Great Wall yielded 60 million yuan.

Interaction with the International Community

In 1991, China became a member of the World Heritage Commission and, in December 1992, was elected its vice chair. According to a World Heritage Secretariat official, China is very active in the World Heritage Commission and aggressively pursues efforts to get its sites on the list, producing elaborate presentations for each site proposal. An International Union for the Conservation of Nature (IUCN) official has been impressed by the officials involved (from governors to local mayors to park staff) and by the high level of interest evidenced in the IUCN visits. For instance, the Chinese have put together an eight-hour video detailing the visits of the

IUCN officials to evaluate the sites. Nineteen sites in China have been accepted for the list (see table 11.2).

China also has pursued training and funding opportunities from the World Heritage Commission and other sources. The State Bureau of Cultural Relics has established over ten labs to develop improved preservation capabilities and has sent technicians to International Center for the Study of the Preservation and Restoration of Cultural Property (ICCROM) to attend training courses. In addition to the IUCN missions to review the proposed sites, other interactions with the Secretariat and its representatives have taken place: a training seminar on natural areas in Beijing (1989); a training seminar for tourism management in Huangshan (1992); and, in the fall of 1993, a management training meeting in Beijing, a follow-up mission to Huangshan, and an IUCN Commission regional meeting on national parks.

The World Bank and the Getty Museum have been assisting the Chinese in restoring the Mogao Grottoes of Dunhuang. They have coordinated efforts to reduce sandstorms in the area and to conduct environmental monitoring inside the caves. The Getty supplies instruments and trains workers in protection techniques. The PRC, in turn, has begun to export its newly trained experts; it will provide experts for the international effort to protect Angkor Wat in Kampuchea. Direct financial aid from abroad has also been forthcoming for China's protection efforts. In 1991, for example, the Venice and Great Wall Preservation Committee donated U.S.$146,000 to the PRC to rehabilitate part of the Great Wall at Mutianyu in Beijing. Also, $20,000 in external aid was provided to help manage Huangshan.

The Chinese have been responsive to the suggestions of the international community for making changes in sites that appear to be in danger of noncompliance. In 1988, a UNESCO-sponsored delegation inspected the five historical sites and noted at least two problems: construction was destroying traditional urban structures in historic cities; and the PRC needed to develop its science and technology capabilities to protect the sites. In February 1989, a meeting was held in China by the newly formed National Committee of Cultural Relics to address the comments and proposals in the report. The Committee stressed that the PRC should "not allow these places of historic interest to be rated as 'world heritages in imminent danger.'"

The threat of international censure or embarrassment before the international community apparently spurred the State Council and some municipal governments to take action on several trouble spots. China suspended the joint-venture cableway in the Badaling section of the Great Wall, which was responsible for significant pollution, dismantled illegal structures in the area, and assigned blame to the responsible personnel. In addition, the PRC took action at the Imperial Palace and Mogao Grottoes to limit the number of daily visitors. Some Chinese officials have suggested that the treaty has been a great impetus to "clean up" the sites by forcing the provinces to undertake a moral commitment to fulfill the treaty. At the same

Table 11.2
World Heritage Sites in China

Year	Natural site	Cultural site	Mixed site
1987		The Great Wall	Mount Taishan
		Imperial Palace of the Ming and Qing Dynasties	
		Mogao Caves	
		Mausoleum of First Qin Emperor	
		Peking Man Site at Zhoukoudian	
1990			Mount Hungshan
1992	Jiuzhaigou Valley Scenic and Historic Interest Area		
	Huanglong Scenic and Historic Interest Area		
	Wulingyuan Scenic and Historic Interest Area		
1994		Mountain Resort and Outlying Temples, Chengde	
		Temple and Cemetery of Confucius and Kong Family Mansion in Qufu	
		Ancient Building Complex in Wudang Mountains	
		The Potala Palace, Lhasa	
1996		Lushan National Park	Mt. Emai and Leshan Giant Buddha
1997		Old Town of Lijang	
		Ancient City of Ping Yao	
		Classical Gardens of Suzhou	
Total	3	13	3

time, as noted before, public embarrassment appears to be an even greater catalyst. At Wulingyuan, for example, peasants and merchants had taken over many of the scenic spots. However, when a delegation of international experts announced its intention to visit, officials at all levels cooperated to clear them out.

For the management of these sites, the involvement of local level officials is critical. Each site has a local custodial commission, which consists of representatives from the Ministry of Construction and the National Scenic Area Office, to implement management of the site. The Ministry of Construction is responsible for legislation, technical assistance, training of personnel, and approval of plans for the region. Although the center, specifically the Ministry of Finance, may provide funds in the form of a special allocation to help develop the surrounding infrastructure, most of the funding comes from local sources. Local officials may use taxes, ticket revenues, and other resources (as long as they do not damage the site) to finance site management. Of these, the primary source of income is ticket sales. Several provinces are especially noteworthy for the efforts of their leaders to protect scenic and cultural sites. Ministry of Construction officials cite several provincial government officials, specifically those in Anhui, Sichuan, Shandong, and Hunan, as leaders in such efforts.

The Record of Compliance

In spite of these activities and generally favorable reviews by outside experts concerning the nature of Chinese participation, there are suggestions of problems in Chinese compliance efforts.

First, according to one World Heritage Secretariat official, some sites are suspected of poor management. Taishan, for instance, reportedly has problems with flooding and tourism; however, the PRC officials have refused to limit the number of tourists. Also, the situation at Wulingyuan and Jiuzhaigou has engendered concern by the Committee about the "growing human impact and tourism." The Chinese have acknowledged that they are unable to protect many of their Cultural Heritages Sites. There is, for instance, a new express highway crossing the Great Wall. In addition, there are factories near the Peking Man Site at Zhoukoudian that are causing serious environmental pollution. The Chinese tend to attribute their failings in protection efforts to a lack of funds and the developing status of the country. Foreign experts, however, offer additional reasons. A World Heritage Secretariat official commented, "In general, China is committed to maintaining its sites. By Western standards, the physical planning and delivery of visitor services could use improvement." He further has noted that the Chinese have often stressed research rather than management.

In addition, several international organization officials have suggested that the PRC is interested in the World Heritage listing only for reasons of international recognition, access to bilateral funding and expert assistance, and prestige. This has led, in some cases, to abuse of funds (e.g., unnecessary purchases of jeeps and walkie-talkies). The U.S. National Parks Service invested in initial bilateral exchanges but concluded that they were essentially a waste of time. Perhaps a more serious problem is that, according to a U.S. official, the PRC has indicated that it will not invite World Heritage Commission monitoring groups to its sites; it considers such visits an infringement on Chinese sovereignty. While IUCN and International Council on Monuments and Sites (ICOMOS) have made initial exploratory visits to the PRC sites, further follow-up, which often depends on NGO assistance, has not been possible, given the Chinese stance.

Moreover, some Chinese officials have expressed concern that the rapid rate of economic development, the modernization of historic cities, and an ever-increasing number of tourists will continue to "exert enormous pressure on and even destroy natural legacies." These officials also argue that the 15 million yuan (roughly U.S.$3 million) earned from tourists who visit the five World Cultural Heritage Sites is far from enough to protect, conserve, display, and study the sites. They also note that the low level of public awareness has led to frequent incidents of "wanton carving and painting on the Great Wall and stealing of cultural relics from the Palace Museum." In addition, they comment that the national tourist agencies oppose increases in ticket prices to help conserve the Palace Museum ground bricks from the Ming dynasty because they desire increased tourism. These officials are supported by political leaders who fear a negative response from the press and public if ticket prices are increased. At Dunhuang, there were severe problems with degradation of the site from tourism, to which the local commission responded by selling "A" and "B" tickets to see "A" or "B" sites. Frequently, construction is desired where archaeological sites are located. In Datong, for example, consideration is being given to rerouting a major highway to preserve the grottoes, although damage has already been done by the shaking caused by highway trucks. In Zhoukoudian, small lime kilns have been removed.

Finally, one of the most pernicious problems plaguing the protection of cultural relics is theft and smuggling. Illegal excavation was previously considered a theft. Since 1991, however, this has been considered a criminal act punishable by anywhere from four years in prison to death.

The International Tropical Timber Agreement

Accession: A Triumph for MOFERT Turned Burden

China ranks sixth or seventh in the world in consumption of tropical timber. The majority of the timber imports goes to plywood used in construction. Some is processed into furniture or parts of furniture for export or domestic use. At the same time, China does have tropical forests, although they are limited to Yunnan, southern Guangdong, and Hainan. These resources are being exploited, and some illegal timbering occurs as well. China therefore is both an important consumer and a modest producer of tropical timber. Several countries that are important in China's overall foreign policy—Malaysia, Thailand, Indonesia, Myanmar, and Laos—have a major stake in their exports of tropical timber; logs and processed timber are major sources of foreign currency earnings for these countries.

At the time of accession to ITTA, management of China's external and domestic tropical timber policies involved three principal ministries: Foreign Economic Relations and Trade (MOFERT), Forestry, and Foreign Affairs. In 1986 MOFERT still enjoyed monopoly control over trade in timber. All import of timber had to pass through one of the trading corporations in its ministerial domain. MOFERT had an interest in assuring that supply of this commodity would remain steady and sufficient. As an importer of 10 million cubic meters of tropical timber a year, China has to be concerned that it will become more difficult to secure this resource. Some of its suppliers are exhausting their reserves, and others are imposing limits on log exports.

The Ministry of Forestry owned enterprises that processed tropical timber; its corporations, however, were not the only consumers. More important from the Ministry's perspective was its responsibility to exploit existing reserves wisely, to prevent illegal timbering, and to increase the domestic supply through expansion of tropical timber acreage and improvement of species. That latter concern fell particularly in the domain of the Chinese Academy of Forestry and its subordinate research agencies. Finally, the Ministry of Foreign Affairs was concerned that China's policies as a consumer—maintaining low prices and encouraging sustainable harvest to ensure long-term supply—did not alienate countries in Southeast Asia that China was courting for other purposes. China inadvertently could find itself aligned with the developed world—the consumer countries—on this issue, thereby undermining the broader interest to champion developing countries' causes.

But accession to and implementation of ITTA became at least initially a captive of MOFERT. The reasons for this outcome were straightforward. First, the international community presented ITTA to China as strictly a commercial and trade-promoting agreement. It lacked the obvious environmental dimensions it has subsequently acquired. And, as Ministry of Foreign Affairs officials now explain it, the head of MOFERT in 1986 was Politburo member Chen Muhua. She was much more powerful than the Foreign Minister at the time, Wu Xueqian, and therefore her subordinates were emboldened to proceed on ITTA without extensive consultations with Foreign Affairs. MOFERT did not assign high priority to developing capabilities to implement this agreement, however, because it could not benefit from the financial assistance that flowed from the agreement. That assistance went to the research and production side, that is, the concerns of the Ministry of Forestry. Indeed, MOFERT was responsible for paying a fee of roughly $50,000–$60,000 a year to ITTO, calculated on the basis of the volume of tropical timber exports and imports, while ITTO has contributed $3 million to Forestry for research and development of tropical timber.

In light of this history, the perspectives of both the Ministry of Forestry and the Chinese Academy of Forestry are closer to the current focus of the ITTO than that of the Ministry of Foreign Trade and Economic Cooperation (MOFTEC). The Ministry of Forestry has complained that MOFTEC will not grant it increased access to ITTO activities; however, as some officials have noted, the Ministry of Forestry lacks the appropriate combination of technical and diplomatic expertise to manage the issue in international negotiations. Within the Ministry, the technical specialist on tropical timber does not speak English, while the liaison personnel in the International Cooperation Department who speak English and are responsible for ITTA within the Ministry do not understand the technical details. Yet those proficient in English enjoy a privileged position in representing the Ministry on issues involving international matters. In addition, neither the Chinese Academy of Forestry nor the Ministry of Forestry can afford to attend the ITTO meetings without external funding.

According to a Ministry of Foreign Affairs official, both his Ministry and NEPA are attempting to gain entry into this issue through international conferences on forests. This official believes that the only hope of snatching implementation of ITTA away from MOFTEC would be for the United Nations to initiate a new international negotiation on forestry or tropical timber that would be explicitly environmental and would necessitate a reorganized Chinese working group with NEPA, Forestry, or the Academy of Forestry as the lead agency. Until then, implementation of this agreement will not be a priority.

The Management Infrastructure

Not surprisingly, MOFTEC was reluctant to provide information on its implementation of ITTA, and hence we cannot illuminate its efforts to monitor the import of tropical timber. In any case, it has lost its monopoly over trade in tropical timber through the state trading companies it controlled. Administrative decentralization, the growth of private enterprise, and the increased porousness of China's borders have transformed the lumber business, as the economic reforms have changed other aspects of commerce and trade. The Ministry of Forestry now imports tropical timber without reference to MOFTEC, as do many government agencies, trading companies, and manufacturers at the provincial level and below. As an indication of MOFTEC's noninvolvement, the ITTO requests statistics on China's annual production, imports, and exports of tropical timber. The data are very hard to obtain, and MOFTEC evidently plays no role in generating them. The statistics are generated by the State Customs Administration, a body independent of MOFTEC, and by the CAF.

The de facto implementing agencies for ITTA are the International Cooperation Division in the Ministry of Forestry and CAF, not on the export–import side but for preservation and sustainable harvest of China's tropical timber forests. A number of functional bureaus within the central ministry and at the provincial levels in Yunnan, Hainan, and Guangdong have responsibilities on tropical timber. For example, the Forest Security Bureau—the agency responsible for protection of endangered species—is responsible for preventing illegal logging of tropical forests. While news reports detail the vigilance of these bureaus—such as numerous arrests and successful prosecution of illegal loggers—American experts believe the legislation remains largely unenforced. CAF also has a research program on tropical timber, and the ITTO financial contributions for enhanced management go through the Ministry to CAF.

Interaction with the International Community

The Chinese have been fairly active participants in ITTO negotiations and meetings, and have asked for project funding from ITTO for sustainable management projects in tropical forests. The ITTO has provided $3 million to the Ministry of Forestry for tropical timber management.

In 1991, Jiang Jianjun, China's representative to the International Tropical Timber Commission, requested that China be included in the training programs for producing members, even though China is a member of the consuming group. The Chinese received $33,000 to learn how to identify, process, and utilize tropical timber from Africa and Southeast Asia. The three-year program will develop materials to help Chinese importers and users understand tropical timbers; the literature will be supplemented by work on wood anatomy that is to be done by the CAF's Research Institute of Wood Industry (ITTC 1991a).

A second project, also under the CAF, was initiated in 1992. This was an effort to develop a database of wood species and substitute species (ITTC 1991b). This software has been distributed to agencies in the Netherlands and Austria. In addition, a two-year project was initiated under the auspices of the CAF, Nanjing University of Forestry, and Zhongnan Forestry College to study the use of bamboo as a substitute for tropical timber. This project involved a massive effort of about 1200 researchers from the CAF and its Bamboo Information Center, with extensive cooperation with experts from the United States, Great Britain, Germany, Indonesia, and India. In addition, in 1993, the ITTO contributed $3 million to Hainan through the

Ministry of Forestry and CAF to establish a demonstration model for sustainable utilization and management of Hainan's tropical resources (ITTO 1993a).

The World Bank has provided significant loans for the development of forest plantations: in 1991–1996, the Bank invested $500 million for China to develop 16 plantations in 240 counties. The tropical forests that will be supported by this money are located in Hainan, south Guangdong, and Yunnan. In 1994, the World Bank approved a second loan of $200 million for a forest resource development and protection program.

According to a U.S. State Department official, China may be "skewing the usual meeting of the minds among the consumer countries by making decisions more favorable to developing countries." In this vein, in a statement to the U.N. Conference for the Negotiating of a Successor Agreement to the International Tropical Timber Agreement in 1993, the Chinese delegation stressed the necessity for developed consumer countries to provide sufficient financial and technical support for developing countries to realize the goals of sustainable management and utilization of the tropical forest resources. They further stated, "No country should use environmental issues as an excuse to create obstacles for the development of timber industries and the international trade" (Chinese Delegation 1993). China also has noted that while a separate agreement on temperate forests should be concluded, the ITTO should remember that China is a developing country and cannot have sustainable development in its temperate forests.

Record of Compliance

Despite all of the training programs and program assessments, as of August 1993, the PRC had not submitted a progress report on its efforts to meet Target 2000, the commitment by ITTO member countries to have all tropical timber entering trade from sustainably managed sources by the end of the twentieth century. Moreover, China is the only consumer country that planned to substantially increase its tropical log imports during 1992–1993 (ITTC 1993a). China is the second largest ITTO tropical log importer. (Taiwan is included as a province of China in these calculations and is the largest importer within the PRC. (ITTC 1993a).)

In addition, illegal logging domestically has been supplemented by illegal logging internationally; the Chinese have become active in the logging industry in Myanmar. In 1989, after logging was banned in Thailand, Myanmar permitted foreign logging companies to do business. According to a TRAFFIC report, although most of the companies are Thai, in the north they are being replaced by Chinese interests. Most of the illegal operations center on smuggling into Thailand; there is no direct evidence that the Chinese are participating in this smuggling. According to the director of one Thai company, most of the illegal loggers are Thais, some of whom have Chinese and Japanese backing (ITTC 1993a).

The Montreal Protocol

Accession

In 1984, Beijing University Professor Tang Xiaoyan wrote a paper on the depletion of the ozone layer. An atmospheric scientist, Professor Tang sought to draw the global dimensions of this emerging issue to the attention of her professional colleagues. Her paper attracted attention, and a few colleagues began to discuss both the possible effects of depletion on China and China's potential role in accelerating the depletion. Professor Tang, a policy-oriented research scholar, succeeded in eliciting the interest of the State Science and Technology Commission and of Politburo member Song Jian. In 1987, China sent an observer to attend the Montreal meetings. The subsequent report circulated among the relevant community of technicians and scientists, but no recommendations were made as to whether China should join the agreement.

After the initial signatories ratified the Montreal Protocol, international efforts began to be made to encourage Chinese participation. For example, Ministry of Foreign Affairs officials recall such an effort by the United Kingdom. The Ministry began to realize that China would be subject to increasing pressure to join this international agreement; it also saw that China had the opportunity to help lead an effort among potential allies in the developing world, all of whom thought that without compensation from the developed world, the treaty would impose inequitable and heavy costs upon them. The Ministry therefore developed a strategy to seek modifications in the accord prior to Chinese accession.

Meanwhile, NEPA asserted an interest in the issue and organized briefings for Chinese officials. Professor Tang then organized a research project and assigned a graduate student to write a thesis on consumption of ozone-depleting substances (ODS) in China, the first time this topic had been studied. As a result of these efforts, top Chinese officials—with the encouragement of Song Jian—decided to break new policy ground. Until that time, China's officials were only willing to consider making commitments on local and regional environmental issues. They were unwilling to address global environmental problems having only a limited impact on China. They decided that policy should be reconsidered, though self-interest rather than a global consciousness was the motivating factor. They realized that failure to join the protocol could adversely affect China's trade, since the State Planning Commission (SPC) hoped to make China a major exporter of refrigerators and other products that used ODS. But the top leaders also decided that China would not join unless it derived clear benefits.

NEPA then organized six ministries to evaluate the precise costs and benefits of signing the protocol: Light Industry (manufacturers of refrigerators), Chemical

Industry (manufacturers of ODS), Public Security (manufacturers and users of halons), Electronics (users of ODS), Commerce, and Aeronautics. The State Administration of Tobacco and Cigarettes, a major user of ODS and an important source of government revenue, joined later, as did the State Science and Technology Commission, SPC, and the Ministry of Foreign Affairs. In the course of these discussions, the Chinese government concluded both that it should sign the agreement and that it would insist on modifications before doing so. The Chinese calculated that their involvement was necessary to the ozone regime's success, and that they therefore enjoyed considerable bargaining leverage.

Three arguments enabled Chinese proponents of the agreement to convince their leaders to sign it. First, failure to sign would harm imports and exports. Second, financial support could be derived from the Multilateral Fund. Third, China had an international duty to reduce ODS production. At the time, China had two ozone measurement stations, and they recorded the decrease. This information and the consequences were explained to top leaders. One Chinese participant in the working group recalls:

> The environmental issue at stake was based on excellent scientific research, in contrast to many other proposed international environmental treaties that are not based on as much scientific evidence. Moreover, the Chinese scientists were involved early in the deliberations and were able to make suggestions regarding the terms of China's entry. This helped the Chinese side to explain to their leaders the purpose of the treaty. The existence of a worldwide scientific community on the ozone issue and China's participation in it helped sell the issue.

As a result of the internal deliberations, during the Montreal Protocol negotiations and discussions leading to the London Amendments, the Chinese delegation indicated that the Montreal Protocol did not adequately address the financial and technological needs of the developing states. Therefore the Chinese did not accede to the treaty until after the 1990 London Amendments, in which the developed nations agreed to establish the Multilateral Fund and offered stronger guarantees for the transfer of technology. Articles 5.2 and 5.3 of the Montreal Protocol call on parties to "facilitate access to environmentally safe alternative substances and technology" and to "facilitate bilaterally or multilaterally the provision of subsidies, aid, credits, guarantees or insurance programs." Article 10 further calls upon parties to "cooperate in promoting technical assistance" in the developing states.

The Chinese delegates, along with the representatives from India, adopted two tacks during the negotiations. First, they argued that the developed countries were responsible for most of the damage to the ozone layer; hence, they should bear most of the cost of solving the problem. Second, the Chinese and Indian delegations claimed that given the need for developing countries to address more pressing issues, such as poverty, hunger, and disease, they could not afford the costs of CFC abatement (Rosencranz and Milligan 1990:313). A member of the Chinese delegation also raised an important point of contention before the U.N. General Assembly in 1989: "Patent rights and copyrights should not become obstacles to the efforts of the developing countries to promote scientific, technological, and economic progress" (U.N. General Assembly 1989). Because of the pressure exerted by the PRC and India—two relatively low-producing CFC countries with estimated very high future production levels—the Multilateral Fund was established under the auspices of the World Bank. It provides $240 million to article 5 countries on the basis of the principle of additionality (Stammer 1990:A8).

As a result of the modifications to the Montreal Protocol adopted in the London Amendments, China committed to sign the agreement and to prepare an implementation plan. The Chinese draft of the implementation plan sufficiently impressed international authorities that the UNDP hired Tang Xiaoyan, the architect of China's entry into the regime, to counsel India in the preparation of its plan.

The Management Structure

China has established a vast bureaucracy to manage its ozone protection effort. In July 1991, the Chinese organized a group consisting of NEPA (the lead agency), the Ministry of Foreign Affairs, the State Planning Commission, the State Science and Technology Commission, and the Ministry of Finance (deputy agencies). The ministries of Light Industry, Chemical Industry, Public Security, Machine-building, Electronics, Commerce, Aeronautics and Aeronautic Industry, and Foreign Trade and Economic Cooperation and the General Customs Administration were member agencies. This group opened an office and formed a coordinating team to manage the routine work on ozone protection. It was responsible for developing China's program to implement the treaty and for managing the daily responsibilities that the treaty required. The Chinese funded the office with U.S.$100,000.

By 1992, the office had seven full-time staff members and thirty-four part-time workers; currently there are ten full-time staffers. NEPA formed a project management office that provided the Convention Secretariat with data and managed the Multilateral Fund moneys. This office had a staff of ten and was funded primarily by Multilateral Fund money.

Interaction with the International Community

In 1991, China applied for funding for numerous projects from the Multilateral Fund. It is launching 173 projects on ozone depletion with the help of $105.6 million from the Multilateral Fund of the Montreal Protocol.

Du Pont and the U.S. Environmental Protection Agency have been active in Chinese efforts to develop CFC substitutes, although each is proposing a different substitute. Moreover, the Chinese themselves are attempting to develop substitutes for ozone-depleting substances. According to one Chinese report, the Ministry of

Chemical Industry's research center in Zhejiang (Zhejiang accounts for one-third of China's reserves, production, and exports of fluorine raw material, an ODS) has developed "dozens" of substitutes for Freons. Some of them are under test use and are supplied to major refrigeration enterprises. The research center is also studying a potential substitute for halon (FBIS China Daily Report 1993a).

During 1992–1993, China received money from the Multilateral Fund to form a working group to draft the program to eliminate ODS. In 1993, the Chinese submitted their country report to the Multilateral Fund Executive Committee. In it, they outlined the structure of China's ODS industries. Approximately forty enterprises produced CFCs, with a total production capacity of 47,000 tons. The existing CFC production capacity in the PRC cannot meet the domestic demand; therefore, the PRC imports 15,000–20,000 tons of CFCs per year from Japan, France, and the United States. It exports approximately 100–200 tons annually to Southeast Asia (UNEP 1993a).

In the country report, the Chinese set out 140 projects requiring an estimated U.S.$1.4 billion to implement. According to one U.S. expert, China's report was quite impressive; however, he noted that the request for the phaseout of ODS was more money than the entire Multilateral Fund contained. Xia Kunbao, the head of NEPA's Foreign Affairs Department, has stated that the Executive Committee of the Multilateral Fund has approved a dozen projects to transform China's "refrigeration, washing and cleaning and electronics industries so as to phase out chlorofluorocarbons." He has further stated that some of the projects have already been implemented and that China is planning to implement the rest (*Xinhua* 1993).

Record of Compliance

Although China has the highest levels of ODS consumption and production among the developing countries, it was not in 1997 in danger of falling into noncompliance with the Montreal Protocol or the London Amendments. Under article 5.1 of the Montreal Protocol, developing countries were to be granted a ten-year grace period in delaying their compliance, in order to meet their "basic domestic needs." During this time, developing states would be allowed to expand their CFC consumption level to 0.3 kilogram per capita. China's potential CFC consumption during the 1990–2000 period was estimated to reach only .05 kilogram (Rosencranz and Milligan 1990). Thus, what is important in terms of China's compliance with the Protocol is that it submits its progress reports and develops the appropriate legal, managerial, technical, and economic infrastructure to fulfill its future commitments.

China's obligation to report its ODS production and consumption figures began in 1993. Its report was complete, with the exception of data for 1992. Its 1994 report also was complete. Both a Du Pont representative and a U.S. State Department official consider the PRC to be "proactive," for a developing country, in its attempts to phase out CFCs. The Chinese are actively pursuing measures to manage the problem of small-scale factories that emit ODS, to control imports,[2] and to levy taxes and penalties. NEPA officials believe that they have the capacity to halt new projects that use ODS, but not those that currently consume them.

Nonetheless, international actors are already citing potential challenges to future implementation. First, there is an unconfirmed claim that the PRC is in the process of building a CFC plant with the assistance of a German firm. Second, many of the Chinese enterprises that produce and consume ODS are small and spread out; according to a U.S. State Department official, it is therefore more difficult to change over the production system. Third, at least one multinational corporation—Du Pont, which has become involved in the Chinese effort to develop CFC substitutes—has found the Chinese bureaucracy a complex entity. One Du Pont official has commented that it is "very complicated" to deal with ministries in China. He notes that sometimes one ministry "disappears" from the negotiations. Based on the remarks of another U.S. expert, it is also plausible that the issue of intellectual property rights is hampering Du Pont's efforts in the PRC. Finally, between 1986 and 1992, production of CFCs more than doubled, from 11,540 tons to 24,941 tons.

Chinese officials have noted additional constraints to the effective implementation of the Montreal Protocol. Environmental officials predict that by the year 2010, "the lowest rate of growth in the use of ODS" by various industries will be 6 percent; the highest rate "is expected to reach 18–20 percent." Already, in this regard, the trends have been negative. Although per capita usage remains quite low, China's percentage of the world's total ODS consumption has increased dramatically, from 3 percent in 1986 to 18 percent in 1994. These officials expect that as the state attempts to control and reduce the use of ODS, some industries and consumers will resist, making implementation more difficult. In addition, they fear that because China is moving from a planned to a market economy, neither its market mechanisms nor its regulatory abilities are adequate to enforce implementation.

Finally, they suggest that while the international capability to implement technological changes to prevent ozone depletion is high, limited funds have prevented the PRC from developing a similar level of commercialization for these technologies. In this vein, they note that the mechanisms for technology transfer from the developed to the developing states are not well developed and hinder the capacity of developing countries to implement these technologies. One Chinese scientist, intimately involved in the internal Chinese negotiations over the Montreal Protocol, claims that the World Bank, UNDP, and Asian Development Bank consultancy process is not favorable to the PRC and does not support indigenous Chinese technological development. She argues that these organizations hire consultants who favor the technologies with which they are familiar (e.g., from Du Pont, ICI, etc.).

Relatedly, Chinese officials comment that the "application procedures and restrictions on the use of funds received from the Multilateral Fund are excessively elaborate and involve a long waiting period." Concerned departments and enterprises are required to assign a certain level of manpower and financial resources to conduct first-stage preparations, including the acquisition and translation of relevant materials, negotiations, and the hosting of foreign experts. All these factors have served to dampen the enthusiasm of some enterprises. As noted earlier, the Chinese estimate that they will need $1.4 billion for the 140 projects they have proposed to phase out 60 percent of ODS. Thus far, however, they have received only $30 million from external funding sources. As one scientist has complained, China consumes 70 percent of ODS among developing countries but receives only 20 percent of the funds provided.

There is also substantial variation among the responsible ministries in terms of the rate and commitment of implementation. The Ministry of Light Industry, the Ministry of Electronics, and the Ministry of Chemical Industry, which were the largest consumers of CFCs at the time China signed the Montreal Protocol, are considered to be more "advanced" in their outlook on developing and implementing ODS substitutes; the Minister of Light Industry, especially, has played a leadership role in attempting to fulfill China's commitment. In contrast, the ministries of Domestic Trade and Commerce have been less aggressive in working to reduce consumption of ODS. According to NEPA officials, the Ministry of Domestic Trade, which is responsible for refrigerators, is short of manpower and is highly decentralized.

The regionalization of the PRC has introduced problems for central authorities in attempting to implement the Protocol. The center's plans for gradual elimination of ODS will have adverse effects on local-level industries. Especially for small enterprises, such policies can be very problematic. Furthermore, the projects funded by the Multilateral Fund must be approved by local governments and have local financial support for implementation. According to Chinese officials, thus far the provinces and cities have not developed legislation to support the elimination of ODS.

Lessons Learned: The Policy Implications

We have seen that China entered into these five international agreements to secure access to hard currency and technology. While many individual Chinese understand and endorse the environmental purposes of the agreements, they were able to convince the skeptics by pointing to the agreements' short-term benefits and possible international penalties. This calculus of cost and gain not only was at work during the accession stage but also remains in effect during the implementation stage,

Table 11.3
China's compliance with treaty reporting requirements, 1991–1996: Report transmission/receipt date

Treaty	Entry into force	1991	1992	1993	1994	1995	1996
CITES	4/8/81	9/22/92	9/25/93	9/28/94	10/19/95	10/30/96	N/R
LC	12/14/85	R	9/7/93	8/23/94	6/19/97	6/19/97	N/R
MP	9/12/91	R	R	R	R	R	R

Notes: For all treaties, N/R indicates that no report was received by July 1997. R indicates that a report was filed, but the date of receipt is unavailable. Montreal Protocol reports received include those filed on time, late, or subsequently revised; in some cases a country may have met two or more years' reporting requirements with the submission of a single report.

particularly in the decisions at lower levels to comply with or disregard the regulations that Beijing has promulgated.

The implementing agencies in Beijing fulfill the procedural requirements of the accords, as is shown in the data presented in table 11.3, but in most instances are unable to monitor provincial and local compliance with the substantive obligations of the treaties. They possess neither the resources nor the manpower to enforce central directives, and the State Council and the Ministry of Finance, facing severe shortfalls in central government revenues, have been unwilling to commit the requisite funds to strengthen the implementing agencies. Seeing the priorities as being at the center, local officials for the most part have not demonstrated a commitment to the implementation of these treaties. Whether through payment for looking the other way, or active participation by the officials themselves in noncompliance, in each treaty there is a financial incentive for evasion.

Although the State Council and National People's Congress have been unwilling to sharply increase appropriations to deal with these problems, these leading bodies of the Chinese government are aware of and disturbed by the situation at the local level. Their solution has been to launch a three-year effort whereby State Council officials and National People's Congress representatives will regularly inspect the compliance of provincial and local officials with environmental protection laws. The National People's Congress is also in the midst of expanding and toughening environmental legislation; Qu Geping left his position at NEPA in 1994 to head the legislative effort at the Congress. Through drafting and promulgation of National Agenda 21 and other long-term planning documents, China's environmentalists are attempting to raise awareness of the country's ecological problems. And discussion is under way to raise NEPA to ministerial status and to transfer responsibility for many, if not all, environmental agreements to it.

We believe these are all constructive steps. Enhancing NEPA's status within the Chinese bureaucratic hierarchy would certainly be useful. But there are many other measures that both the international community and China could undertake to improve performance.

As far as the international environmental organizations are concerned, we have five recommendations.

First, we have seen that the accession stage heavily affects the implementation stage. International negotiators should understand the likely consequences of China's designation of its lead negotiating agency and seek information about the domestic dynamics shaping China's negotiating stance. Here are to be found the first clues about who in China favors the agreement, who is opposed, and what the various actors hope to gain or fear losing. In cooperation with Chinese allies, the international negotiators at this stage should develop strategies to broaden the base of domestic support. And realistic assessments need to be made about China's implementation capabilities, so that expectations on both sides can remain realistic.

Second, once the agreement goes into effect, simply throwing money at the Chinese implementing agency is not helpful. As Schaller demonstrates in the CITES case, and other international experts mention concerning the World Heritage Convention, there is significant abuse of funds. Chinese officials privately confirm that wages for Chinese personnel or the per diem charges of foreigners far exceed actual costs. Chinese personnel receive far less in yuan costs than international agencies are billed in foreign exchange. The difference is diverted to underwrite administrative costs of the agency. As a result, to a much greater extent than they realize, international organizations may support the Beijing operation, with the result that the central government's effort will atrophy once external support ceases. It is incumbent upon international agencies to understand the Chinese central government's actual budgetary and manpower commitments to sustain the agreements, and to insist that these commitments be honored. International negotiators should accurately assess their own leverage in extracting satisfactory commitments from China's leaders. While the latter are deeply committed to preserving China's sovereignty, they do respond to tough-minded international bargaining.

Third, many of the agreements require sophisticated equipment, technology, and substantial external financing. It does little good to entice China to enter an agreement and then not provide the expected, necessary support. Yet that has happened repeatedly, with the result that the Chinese partners are politically exposed and somewhat bitter. Engaging China in environmental protection entails a protracted, serious commitment by the international community that cannot be lightly abandoned. Manpower training has to be a core aspect of the commitment. Especially important in this regard is the stationing in China of experts technically proficient in the issues at stake. They should expect to remain for a protracted period, though not at outrageous cost, until Chinese experts are well prepared to take their place.

Fourth, the core challenge entails institutional development at the provincial and local levels, and among ancillary agencies whose cooperation is essential. While officials from the lead bureaucracy typically support effective implementation and possess the authority (on paper) to enforce the agreement, they lack the power and authority to ensure compliance by other agencies. Thus, for CITES, the Ministry of Forestry is at odds with both the Ministry of Foreign Trade and Economic Cooperation and, potentially, the People's Liberation Army. In the case of the London Convention, the effective supervisory power of the Ministry of Petroleum over its own dumping does not bode well for the efforts of the State Oceanic Administration. In addition, in efforts to comply with the World Heritage Convention, Chinese officials in the State Cultural Relics Bureau are challenged by officials from the national tourist agencies. These second-level agencies are powerful, and their incentive for compliance is unclear. Thus, international agencies should pay attention to developing the interest and capacity of the ancillary agencies. Although the principal agency will typically seek to monopolize the resources, international agencies should resist the temptation to lavish all their attention on the lead agency, and cultivate the secondary institutions as well. Moreover, nurturing institutions in Beijing while ignoring those at the provincial level will not suffice. Attention must be given to enhancing the implementational capacities at lower levels, especially in locales considerably affected by environmental problems.

Fifth, international agencies and foreign governments must pay more attention to the full consequences of the institutional and structural changes they urge upon the Chinese government. Privatization, administrative decentralization, and price reform without concomitant development of a legal system, an independent judiciary, and an effective government revenue system weaken the authority of the central government to enforce environmental regulations. In addition to addressing environmental issues directly, international agencies and foreign governments must recognize that China's entire institutional infrastructure needs development in order to address environmental problems effectively.

Turning to China, we note four findings.

First, the key problem is effective implementation of existing regulations, not the absence of regulations. The Chinese have proved adept at drafting laws to implement the treaties they have signed. Often, in fact, they have attempted to build up a legal infrastructure based on the treaty prior to the ratification of the treaty. In a few cases, like the London Convention, Chinese officials feel that the Convention's guidance for legal statutes is insufficient. For the most part, the Chinese have developed a strong legal system on paper. Moreover, while in some cases implementational efficacy might be improved by stiffer penalties, the fact that poachers are willing to risk death suggests that improved compliance cannot be obtained through assigning further responsibility to China's already awesome coercive institutions.

Second, we strongly recommend that the Chinese government significantly increase the percentage of GNP it devotes to environmental protection, both to address the core problems that threaten to slow China's economic growth—loss of arable land, water shortages, inefficient use of energy—and to fulfill hopes for an improved quality of life.

Third, we believe that the media have a major role to play in improving public understanding of environmental issues. Most Chinese with whom we spoke believed that the public was not aware of the dangers posed by eroding air, water, and soil quality, and the public knew even less about such issues as global climatic change, depletion of the ozone layer, or preservation of biodiversity. Hence, the public was not receptive to closing factories that manufacture or use ODS or to inspecting shops that use tropical timber. Improvement of environmental journalism in China, encouragement of more vigorous environmental reporting on television, and launching of county-level newspapers would be appropriate measures the Chinese government could undertake. At present, most of the 2000 counties in China lack their own newspapers. Counties (*xian*) are still the basic building block of the Chinese political system. Enterprises and activities under their control account for a substantial portion of the violations of the five agreements, yet these violations go unreported and unnoticed.

Fourth, we believe that adherence to environmental regulations would be greatly enhanced by the presence of environmental associations. We know that under the current Chinese system, purely voluntary associations face severe constraints. They must secure governmental approval and register with the Ministry of Civil Affairs. Clearly, the Chinese government is not prepared to welcome the formation of NGOs and interest groups. But even within the constraints it sets for itself, the government could allow and even encourage the development of "GONGOs" that are typically found in other Asian authoritarian systems: government-organized nongovernment organizations. We believe such organizations, semiofficial but with support and membership drawn from the private sector, could do much in the environmental sphere. Their absence, indeed, sets China apart from all other major countries, where environmental activists, organized together, have played roles in raising public awareness and monitoring governmental performance.

In the final analysis, the underlying issue for both the international community and China in the implementation of the five agreements is finances. How will the costs be allocated? What portion will be borne by international organizations and individual foreign governments? And within China, how will the costs be allocated among the central, provincial, and local governments? We do not find in any of these cases a readiness within China or at the international level to grapple with this problem in concerted, sustained fashion. If difficulties exist on these relatively inexpensive treaties, the problems regarding the larger, more complex issues of global climatic change and biodiversity can only be more vexing. In our five cases, agreements could be reached because the finances either were not directly confronted or pious and unrealistic pledges were made. Before even more agreements are reached without adequate attention being paid to implementational and financial issues, the Chinese experience suggests that attention must be given to these issues at the international and national levels. The five agreements seem like a good place to begin.

Notes

1. Although the official did not mention it, Daya Bay is also the site of a nuclear reactor. It is plausible that radioactive materials were being dumped into the bay.
2. As of 1997, 30 percent of the CFCs consumed in China are imported.

12

Embedded Capacities: India's Compliance with International Environmental Accords

Ronald J. Herring and Erach Bharucha

Domestic Landscape: Structure and Actors

There are good reasons for skepticism concerning the independent contribution of international accords to domestic environmental protection. Signing often represents symbolic politics among state elites problematically claiming to represent national societies and removed from the constraints of domestic politics. Extrapolating from Anthony Downs's work on "issue-attention cycles" (1972), the real test of even soft international law comes when domestic costs become a matter of political process; obstacles to compliance that were invisible under the rarified negotiation process emerge as it becomes clear whose ox is to be gored. Capacity to comply is embedded in the state's relation to the society it claims to represent and in positional imperatives of the international system.

This chapter considers India's compliance with global environmental treaties as part of a comparative analysis across a broad sample of nations and treaties: the Convention on International Trade in Endangered Species of Flora and Fauna (CITES); the International Tropical Timber Agreement (ITTA); the World Heritage Convention (WHC); and the Montreal Protocol (MP). India has not signed the London Convention on the Prevention of Marine Pollution by Dumping of Wastes and Other Matter, which is part of the comparative frame, but does not formally object to the treaty.[1]

We conclude that there is a divergence in effects between those treaties that are congruent with national policy evolution and those that are not. International accords for preservation of nature and natural resources—CITES, ITTA, WHC—are consonant treaties; these reinforced India's strong de jure domestic commitments. In contrast, the Montreal Protocol represents a dissonant accord; extensively criticized by India, and altered as a result, the Protocol deflected the trajectory of domestic policy significantly. Table 12.1 shows when India became a party to these treaties.

All generalizations about India are suspect; it is a nation of continental diversity in biological and social characteristics, political economy, and administrative

Table 12.1
Treaty membership of India

Treaty	Year treaty came into effect	Signed by India	Ratified by India
WHC	1977	1977	1977
CITES	1975	1974	1976
ITTA	1985	1986	1986
LC	1975	not a member	not a member
MP	1989	1992	1992

capacity. Its high biological diversity in terms of biomes, species, and indigenous cultivars warrants rank as a megadiversity nation.[2] India's Constitution provides, among the Directive Principles of State Policy, the basis for environmental protection: "The State shall endeavor to protect and improve the environment and to safeguard the forests and wildlife in the country" (article 48A). Moreover, "It shall be the duty of every citizen of India to protect and improve the natural environment ... and to have compassion for living creatures" (article 51A[g]).

These constitutional directives are supported by considerable sophistication of legal instruments and institutions, and by one prominent strand of a rich cultural heritage: what Samar Singh as CITES Standing Committee chairman, called "the Gandhian ethic of restraint and frugal use" (1986:210).

The political structure of India is federal; the parameters of federalism are under continuous stress and renegotiation. Most issues of environmental importance—water, land, forests, fisheries, public health, agriculture, wildlife—were subjects originally reserved to states in list III of the Constitution.[3] Administrative capacity and political will for conservation in the states are unevenly developed; the center has had a ministry of environment only since 1985. Yet environmental policy has come from the top and center, to frequently resistant state governments. An appendix to the Tiwari Report, which recommended creation of a separate Ministry of Environment, commented, for example, that "State governments are not interested in preserving wildlife" (GOI, Dept. of Science and Techrology 1980:91). The report of the Ministry of Environment and Forests (MOEF) for the World Bank, *Environmental Action Programmes*, was more sweeping:

> There is a reluctance in most states to strengthen their departments of environment, despite the ever-increasing role that such a department is being called upon to play. Many states also want the decentralization of environmental powers from the Central Government ... but are not willing to correspondingly strengthen their departments. This reluctance symptomizes the low priority that many states give to the conservation of the environment. Even the outlays suggested in the annual and five year plans, for environment, by the Planning Commission are invariably cut down. (1992a:24)

More recently, a committee of the Ministry of Environment concluded:

> On innumerable occasions the Central Government has opposed the action proposed to be taken or taken by the State Governments in relation to National Parks and Sanctuaries be it denotification, delimitation, grant of leases for mining, industrial development, utilisation of funds, etc. Invariably such objections have gone unheeded and the State Governments have gone ahead with their plans, presumably on the basis that wildlife is a common legislative subject. (MOEF 1996:80)

Even if states are cooperative, federalism introduces administrative difficulties. One recent report concluded: "A major problem has been the lack of exchange of wildlife crime related intelligence between states. In one instance, an offender with a standing arrest warrant in one state operates openly in another" (MOEF 1994:13).

The Centre's political complexion has undergone significant alteration, from single-party dominance in the first two decades of independence to instability in recent decades. An unstable Centre necessitates complex bargaining with the states in which Delhi reserves its political capital for urgent matters (such as maintaining a fragile majority in Parliament).[4] Budgetary allocations reflect increasing state responsibility: in the seventh Five-Year Plan, the Centre was to spend more than three-fourths of the total allocation for the environment; in the eighth Plan (1992–1997), allocation to the states was more than three-fourths of the total (MOEF *Annual Report*: annual). Nations sign international agreements as unitary representatives of nations; both planned devolution of powers and political conflicts over authority between Delhi and the states dilute the capacity of the central state to speak for, monitor, or direct the political system. The Ahmed Committee concluded:

> With a great deal of fluctuation and instability in politics, there has been a deterioration in the relationships between the centre and the state governments. Wildlife management related issues have become the target of political conflicts and party rivalry and endless recommendations for wildlife have been paralysed in the field. (MOEF 1996:26)

Global environmental policy has been dominated by the ministries; neither political parties nor conventional interest groups in the pluralist sense have been significant actors.[5] The Ministry of Environment and Forests is the lead ministry for international accords, though others are crucial to implementation. Environment has relatively low status and power among central ministries; it has often lacked a cabinet-rank Minister, but has been headed by a Minister of State. The portfolio has comparatively few opportunities for power or patronage at the Centre or in the states. The Ministry has a pool of personnel with great experience and skill, but bureaucratic turnover limits institutionalization of expertise and coordination within the Ministry. The pathologies of ordinary bureaucracy are very much in evidence. Beyond the MOEF, effective coordination among ministries on environmental matters is problematic.

Parliament has not been a major forum for international environmental accords, though environmental issues in general became more prominent there in the 1980s.

In 1980, there were 387 environmental questions in Parliament; the number jumped to 633 the following year. The year 1980 was pivotal for environmental policy: Prime Minister Indira Gandhi reintroduced environmental activism from the Centre; the controversial imposition of Delhi's authority over states' developmental interests stopped the Silent Valley dam project in Kerala; the Forest Conservation Act put new restrictions on land use for development; and the Tiwari Report recommended creation of a separate Department of the Environment (GOI, Dept. of Science and Technology 1980). By this measure, a peak of interest was reached in 1986, when 1,552 environmental questions were asked in Parliament.[6]

Parliamentary questions have been effective tools for focusing attention of ministries, revealing lacunae, and forcing action. Moreover, parliamentary questions allow NGOs to target members of Parliament for lobbying; the World Wide Fund for Nature–India (WWF-I) uses the data to advise other NGOs on effective activism. The Ministry of Environment and Forests, which was created in 1985, has become institutionalized by becoming a focal point for calling-attention motions in Parliament. Answers reveal the paucity of hard data available on important issues of compliance, often because state governments will not or cannot supply data to the Ministry—another consequence of federalism and lack of monitoring capacity.

In India, as elsewhere, NGOs increasingly energize, shame, or pressure governments on environmental issues. These oppositional arms of society stiffen or induce states' will to act. They represent crucial cybernetic loops, feeding back information the state may not want to know or be known. The term "NGO" itself evoked positive images as states fell into disfavor under the global liberalizing zeitgeist of the 1980s: local, participatory, activist organizations of high principle.

This state–NGO oppositional dynamic certainly applies to India, but the image from popular discourse requires qualification. Many NGOs are salariat organizations that more closely resemble a subcontractor to the state and international agencies than voluntaristic social movements. High principles compete with opportunism and new elite niches. Moreover, the NGO sector is by no means monolithic, nor is it automatically a force for international environmental accords.

NGOs can be divided along a political dimension: those that self-consciously put "people first" (the social ecology perspective) and those primarily concerned with ecological integrity. There is a roughly coterminous divide between social activist and scientific organizations. The traditional preservationist NGOs have organized much of the scientific research on environmental issues in India, at times in collaboration with the government and international NGOs. The Bombay Natural History Society is perhaps the best known example.

Environmental NGOs number at least 7,000; their local activism influences most environmental issues. NGOs with connections to Delhi and interest in international accords are few. WWF-I in Delhi increasingly assumes the role of clearinghouse and lead center. Located symbolically in the same complex as the Ford Foundation, World Bank, and other international agencies, it has the broadest ambit and most resources. The Centre for Environmental Law was moved to the WWF-I building to support conservation causes in court and study legal implications of international accords for existing Indian law. WWF-I is the center for coordinating a network of Environmental Information System (ENVIS) centers around India. The ENVIS Center 07 in Delhi has been effective in providing information to NGOs around the country (and even around the world) *about* NGOs in India, particular research scientists, educational resources, and environmental activities in Parliament.[7]

On many issues, state–NGO fusion, not opposition, constitutes a new structure for environmental policy dynamics. Official recognition was formalized by the opening of an NGO Cell in the Ministry of Environment on June 5, 1992, to coincide with World Environment Day (MOEF *Annual Report* 1992–1993:100). WWF-I is integrated with the Ministry through multiple linkages of personnel, educational activities, and some funding and research. A TRAFFIC office for CITES monitoring was set up within WWF with a director concurrently working as a consultant to the Ministry. Likewise, collaboration between the NGO Tata Energy Research Institute (TERI) and the Ozone Cell of the Ministry formulated plans for complying with the Montreal Protocol.

It is impossible to list all important NGOs, but the Centre for Science and Environment (CSE) deserves special mention; intellectual and policy agendas of Indian environmentalists and officials have been deeply influenced by its work. The Centre has been a leader in developing a position for the South on the global environmental crisis as a product of the North, which now seeks to impose costs on the South to remedy its effects.[8] Given the depth of this perception, the international origin of accords may become an independent cause for suspicion or opposition, rather than support, within India. Conflicts over sovereignty find some NGOs on the internationalist side of the fence; others, on the traditional nationalist side.

Poverty, dualism, and land scarcity dominate the domestic political economy: very high levels of absolute poverty coexist with an emerging middle class of large size in absolute terms—larger than in Britain and France combined, the government often claims. A small Europe competes with a poor majority for resources and space (Gadgil and Guha 1995). Transnational consumer aspirations powerfully constrain any agenda of environmentalism that threatens growth, modernization, and international integration. With 16 percent of the world's people on 2 percent of the world's land—much of it quite degraded—the option of sequestering nature is not viable in India. State capacity for environmental protection presupposes the participation and consent of people whose subsistence routines depend on natural resources—governance rather than rule. Internationally, India's size and diversified economy create less dependence than in most poor countries, and thus independence in the politics of global environmental regimes. Nevertheless, aggregate poverty and remaining strands of structural dependence are facts of national economic life, limiting policy choices and shaping strategies.

Consonant Treaties: Nature and Natural Resources

CITES, ITTA, and WHC all deal with natural systems; at question is the state's capacity to intervene in a society where livelihoods and normative claims of access resist state incursions. India's domestic legislation has moved in parallel to or preceded global accords in this area, and is often more stringent than international treaties. Effectiveness of domestic policy is mixed, and difficult to assess. India's *Environmental Action Programmes* (GOI 1992a:25) acknowledged that shortages of manpower and data severely restrict the state's knowledge of what effect its policies have. Data on monetary allocations and personnel in the MOEF show sharp increases over time; protected areas have increased significantly since 1980.[9] Nevertheless, poaching of endangered species and environmental degradation of protected areas continue, often at an accelerated pace (MOEF 1994, 1996; GOI 1996).

Convention on International Trade in Endangered Species of Flora and Fauna

India has been especially important to CITES, ratifying the treaty early (July 20, 1976), and elected Chairman of the Standing Committee for an unprecedented three consecutive terms, beginning in 1983. The CITES logo was designed in India (Samar Singh 1986:199).

India's legal framework for implementing CITES is unusually comprehensive; compliance is difficult to monitor and enforce. Primary mechanisms are the Wildlife (Protection) Act of 1972, the Customs Act of 1962, and the Import–Export Policy; the Director of Wildlife Conservation is the Management Authority. Wildlife was moved from the State List to the Concurrent List by constitutional amendment in 1976, giving the Centre more control. Chapter XVI of the Import–Export Policy (1993), under the Foreign Trade (Development and Regulation) Act of 1992, prohibits exports of "all forms of wildlife including parts and products." Wildlife is defined as "all plants and animals." India's CITES *Annual Report* (1993:1) notes: "With the domestic trade in wild animals already banned and virtually a complete ban of export of all forms of wildlife, there is hardly any room left for unscrupulous elements to export wildlife ... in the garb of permissible items."

Import of both animals and plants is allowed on recommendation of the Chief Wildlife Warden of a state government, subject to the provisions of CITES.[10] Permission is essentially limited to zoological or scientific purposes. India is not a major importer; in 1993 the total number of animals, mostly for zoos, was seventy-one. Imports of flora are likewise minimal.

Legal exports of animals have for years been very small, mostly to zoos. CITES-certified exports of flora are increasing rapidly; in 1993, listing of exports took thirty-seven pages of the *Annual Report*. All exports were certified "cultivated," and without exception were for purposes of trade (as opposed to scientific uses). This increase seems to be evidence both of increased international trade in exotic plants and of increased awareness of CITES regulations. Not all "cultivated" certifications are authentic, but traders are increasingly aware that they must have them. Only a legal procurement certificate (or No-Objection Certificate) is required, and is obtained from Divisional Forest Officers; verification is not always easy. Distinguishing species and origins of plants is even more demanding than for animals. The threat to endangered flora from trade seems to be growing but is little researched.[11] Moreover, the community censure that often attaches to killing of animals—or some trees—is not so easily activated for plants in general.

Illegal export of plants is a relatively new concern; animals have dominated compliance efforts. The range is great: ivory, rhino horn, tiger bone, furs of lesser cats, musk, peacock feathers, reptile skins, tortoises and many others; the scale is impossible to estimate.[12] International trade indicates the existence of an internal market for animal products that is illegal under the Wildlife Act.

The short history of regulation indicates progressive tightening of the de jure net. Exports of wild animal skins and garments were banned in 1979, along with indigenous ivory. Strengthened efforts against poaching were triggered by evidence of increased killing of the flagship species, tigers, in the 1980s. India's wildlife conservation efforts were symbolized by Project Tiger from 1973 onward. Though it is now fashionable to downplay glamorous megafauna, the world has lost probably 95 percent of its tiger population during the twentieth century; India remains the best hope for avoiding extinction of this majestic animal. The central government's response—the Control of Poaching and Illegal Trade in Wildlife Scheme (1986)—included improvement of telecommunications through a network of wireless stations and walkie-talkie sets, more vehicles for enhanced mobility, arms for protection staff, establishing check-posts and rewards for information or apprehension of offenders—in effect, creating slush funds for encouraging intelligence and diligence (*LSQ* 4776, 4781, 26-8-87). Subsequently, alarm at the escalation of trade in tiger bone symbolized widespread ineffectiveness in protecting even the flagship species.[13]

Believing that smuggling would be diminished if internal trade were more strictly controlled, on November 20, 1986, the government prohibited carrying on business or trade in articles made from listed animals under chapter V-A of the Wildlife Act. Enforcement was disrupted by a legal challenge. Traders of furs and skins argued that chapter V-A violated their rights under article 19g of the Constitution, the right to earn a livelihood. Traders tried to sell legal stocks in the interim; the government claimed that "stocks" were a cover for continual purchases.

The Delhi High Court ruled on January 23, 1987, that the government must buy the stocks at market value, which the government argued it could not do except by legitimizing the commoditization of illegal commodities.

The Ivory Traders and Manufacturers Association raised a similar challenge. A stay granted by the High Court was used by nearly 300 petitioners along with the Cottage Industries Association. The result was mobilization of ivory craftsmen and dealers and a long court battle (Chengappa 1993; Panjwani 1994).

Effective May 23, 1992, a total ban on import and domestic sale, transfer, or display, in any commercial place, of African elephant ivory was enacted. The intent was to protect endangered Asian elephants by preventing "laundering" of Asian ivory as African ivory. By 1992, both stays had been vacated and the right to livelihood protection had been finally dismissed by the Supreme Court's refusal to hear the case. WWF-I was prominent in these, and other, legal battles to enforce both CITES and Indian law.[14]

Smuggling routes are difficult to monitor because of India's very long borders, many in inaccessible areas. Major CITES enforcement efforts are concentrated in the four metropoles; smugglers avoid these points, favoring small ports such as Tuticorin, though a surprising number of seizures are made in Delhi (TRAFFIC-India 1994). Tiger bones and parts move through Bhutan and Bangladesh to Southeast and East Asia; tiger pelts and birds move west to the Persian Gulf. Monkeys and a variety of pets go to the United States. Furs and snake skins move through Nepal to Europe. Kashmir, where military conflict reduces the reach of the state, is a major fur-trading center. Calcutta has been the major bird export nexus. Traders use the posts, rails, pack animals into Tibet and Bangladesh, ships to the Persian Gulf, luggage aboard commercial aircraft, and many other vectors. The variety of routes and techniques creates severe jurisdictional and tactical difficulties for enforcers.[15]

Though illegal consignments are regularly seized, penalties for kingpins are extremely rare. Cases drag on for years; seized stocks remain the subject of long legal battles. It is difficult to know whether enforcement efforts are reducing smuggling. Certainly the seizures are increasing in volume and value—especially in 1994—which may indicate more traffic or better enforcement. The most knowledgeable people think it is both: more traffic, because of international price movements, and somewhat better enforcement. Enforcement is improved by the small sums available to officers to buy intelligence; an improvement would be to match the customs' practice of 10 percent of the value of the haul as a reward, but seized stocks are assigned no monetary value and are destroyed. It is now harder to forge CITES certificates, but not impossible. Coordination among police, customs, wildlife wardens, and at least nine other agencies involved in enforcement is improving, in part because of efforts initiated by WWF-I and the MOEF, with support from the CITES Secretariat.[16]

CITES enforcement in India is a continuous struggle against formidable odds. High-value products will always find markets; borders are porous; enforcement personnel are spread thin, often outnumbered and/or outgunned; and corruption is always possible in a high-value game.[17] Better administration will not stop trade when stakes are high, as we have learned universally from narcotics. The importance of CITES is the recognition that simultaneous work on both demand and supply sides of the equation, through international cooperation, is necessary for improving survival chances of species threatened by commoditization and protected by domestic law.

International Tropical Timber Agreement

As with CITES, Indian law is stricter than the conservationist exhortations of ITTA; Indian policy rejects its commodity-promoting intent. Chapter XVI of the Import–Export Policy (1993), section 7, prohibits export of "wood and wood products in the form of logs, timber, stumps, roots, barks, chips, powder, flakes, dust, pulp and charcoal." The only explicit exception to this very thorough cataloging is sandalwood handicrafts (section 9). Restriction of exports and a liberal policy toward timber imports are meant to prevent further deforestation.[18]

Signing the ITTA did not directly influence afforestation or protection policies.[19] The treaty imposes no environmental claims on India, not only because of its commodity-promoting nature but also because India does not allow export of timber, with rare exceptions,[20] but only of value-added products. Nevertheless, objectives of India's forest policy have long been consonant with those of ITTA, article 1, paragraph h: "sustainable utilization and conservation" and "ecological balance." India now has 639,182 square kilometers, or 75.01 million hectares, of "forests"—19.44 percent of its geographical area.[21] The Dutta Committee Report stated in 1996: "Vast tracts of forests have been rendered treeless within the last twenty years and the pace is accelerating from year to year" (GOI 1996:11). The *National Conservation Strategy* lists major threats: "Our forest wealth is dwindling due to over-grazing, over-exploitation both for commercial and household needs, encroachments, unsustainable practices including certain practices of shifting cultivation and developmental activities such as roads, buildings, irrigation and power projects" (GOI, MOEF 1992b:3).

Relative weighting of these causes evokes intense controversy, as it has historically. In the precolonial period, forests were not conceptualized as limited or threatened, but rather as a resource to be converted to human use values; forest-clearing for settlement and agriculture was honored (Raghunandan 1987). The modern policy logic of forest conservation began in the late nineteenth century as an outgrowth of European "scientific forestry," organized to obtain sustained yield for extraction (which increased dramatically with the demands of empire (see, e.g.,

Guha 1985, 1989). Forests essentially became state property with various functions. Reserved Forests were used by the government for timber production; Protected Forests permitted limited extraction by local people for their traditional subsistence needs: fuel, fodder, and nonwood products. Continuation of this colonial system is increasingly attacked as ineffective, unjust, and undemocratic (Hiremath et al. 1994).

India's National Forest Policy of 1952 set a goal of 33 percent forest cover for the country; states enacted their own legislation. In 1976, forests were moved from the State List of the Constitution to the Concurrent List, giving Delhi more control. Social forestry programs for afforestation began in 1976, on recommendations from the National Commission on Agriculture.[22] In 1980, the Forest (Conservation) Act was passed to prevent deforestation and protect habitat for wildlife conservation. This controversial act prohibited conversion of any forest area to nonforest use without Delhi's approval. Approved diversions for development had to be compensated by an equal area of afforestation. The Act spawned substantial center–state conflict. A typical representation in Parliament taps difficult issues:

> There is an urgent need to provide employment to the people in Maharashtra, especially in the district Nasik, but due to the Forest Conservation Act, 1980 ... all the development works in the area have been held up. Many development works which were started earlier have become stalled. Crores of rupees have already been spent on them ... I would request that a little relaxation would be provided ... so that employment could be generated and the people may not have to starve. (*LSQ* M/R 377, 26-3-90)

Amendment of the Forest (Conservation) Act in 1988 coincided with a new National Forest Policy. Noting that the 1952 Forest Policy's goal of restoring forest coverage to 33 percent had failed, and that destruction of "genetic diversity" had been extensive, the policy envisioned joint management, power-sharing between villagers and forestry officials, compensatory afforestation for developmental diversions of forests, and ecorestoration with joint usufruct. In theory, the inadequacy of the command-and-control logic of colonial forestry was appreciated, though working out new institutional arrangements in the face of suspicious citizens and recalcitrant officials will require commitment and creativity (on which, see Guha 1994).

Despite official gestures toward participatory forest management, policy to ensure "sustainable use" remains controversial. Politically, the contradiction is between centralized bureaucratic control and devolution to states and communities. Normatively, there is conflict over conceptualization of forest dwellers' daily practices as "concessions and privileges" (granted by the state), as opposed to rights inherently vested in local people. Environmentally, the conflict is between preservationist "deep ecology" and the social ecology of development favored by most activist NGOs (Herring 1991). Empirically, in terms of forest conservation, there are no easy conclusions and many deep disagreements.[23] Conflicting claims to resource stewardship, conservation values, employment, and social justice are no easier to resolve in India than in the old-growth forests of the United States.

World Heritage Convention

India's World Heritage sites are listed in table 12.2.

India implements protection of WHC Natural Sites through its system of national parks; World Heritage Sites receive no more legal and administrative protection than other national parks.[24] Of India's five Natural Heritage Sites, two have been threatened in recent years—Manas seriously, and Kaziranga less so.

Manas is one of India's richest protected areas, containing twenty-two known endangered species: some animals are almost endemic; The Manas Tiger Reserve was established in 1928; the park now covers 2,837 square kilometers. Designation

Table 12.2
World Heritage Sites in India

Year	Natural site	Cultural site
1983		Ajanta caves
		Ellora caves
		Agra Fort
		Taj Mahal
1984		Sun Temple, Konorak
		group of monuments at Mahabalipuram
1985	Kaziranga National Park	
	Manas Wildlife Sanctuary	
	Keoladeo National Park	
1986		Churches and convents of Goa
		Group of monuments at Khajuraho
		Group of monuments at Hampi
		Fatehpur Sikri
1987	Sundarbans National Park	Group of monuments at Pattadakal
		Elephanta caves
		Brihadisvara temple, Thanjavur
1988	Nanda Devi National Park	
1989		Buddhist monastery at Sanchi
1993		Humuyun's Tomb Delhi
		Qutb Minar and its monuments, Delhi
Total	5	16

as a WHC Site came in 1986. Manas harbors the capped langur, golden langur, slow loris, clouded leopard, pygmy hog, and other rare species. Among the most serious threats is extensive poaching of the greater one-horned rhinoceros. As in much of India's northeast, ethno-secessionist conflict has affected state control in the area.

Agitation for a separate state by Bodos who live in the vicinity of Manas began March 2, 1987; in addition to autonomy, their demands included an end to plantation monoculture, prohibition of foreign liquor, preventing exploitation by middlemen in forest products, expulsion of Assamese (the majority community in the state of Assam), and withdrawal of the paramilitary Central Reserve Police. Armed guerrillas of the Bodo Security Force found sanctuary in the park. Before 1988, the core area was relatively unaffected; 120 square kilometers of core area was eventually occupied by militants and their supporters.

Weapons were removed from anti-poaching forces' camps for fear that Bodos might raid them for weapons; in other areas, patrolling was stopped for fear of violence. Structures were burned and field officers were evacuated after a number were killed; much of the area was in effect surrendered to guerrillas. Burnings of anti-poaching forces' camps in joint attacks by rhino poachers and the Bodo militants were reported. The Ahmed Committee reported that the "agitation is being financed, at least in part, by the illegal trade in endangered species" (GOI, MOEF 1996:127). As staff departed, the "wildlife and timber mafia" moved in without restraint.

In a statement on the situation to Parliament, Shri Udhab Barman (Barpeta) urged the government to act because Manas "has been in the grip of armed poachers and militants, between whom the dividing line is very thin" (*LSQ* M/R 377 (IV), 4-26-93). The Dutta Committee described "determined and organized gangs of poachers equipped with sophisticated weapons and links with international smugglers" (GOI 1996:25). In March 1994, a bomb was thrown into the house of the Divisional Forest Officer. By early 1996, nineteen of forty-three camps were manned by officials; threat of militants prevented manning the others. Wildlife were destroyed, including threatened species such as the swamp deer. The grasslands, which are the last refuge of the floricans, hispid hare, pygmy hog, and other species, were extensively burned; timber was felled.[25]

Poaching in the Northeast is facilitated by the widespread availability of arms, linked to insurgency and counterinsurgency. Horns are traded by extremist groups for arms, and arms are rented out by ex-soldiers and farmers; regular arms bazaars exist (Menon 1996:68).

An early response in 1990, funded by WWF-I, was the conventional provision-of-opportunity model: development of a buffer zone with cooperatives, apiculture, pisciculture, and so on. A Memorandum of Settlement creating a Bodo Autonomous Council, covering 2,000 villages, was signed in 1993. Nevertheless, the conflict is not resolved; guerrillas remain active, now demanding that more villages be included in their domain, and ethnic violence in the area has escalated. Accusations that the state government misused Delhi's allocations to protect the area are persistent (e.g., GOI, MOEF 1996:29). By 1996, the government of Assam was advertising the reopening of Manas, but explicitly warned tourists that entry was at one's own risk; extremist activities have declined but represent a potential threat.[26]

Kaziranga, also in Assam, is perhaps best known as the most important protected area for the endangered greater one-horned rhinoceros; it is home to more than half the world population of the species (Menon 1996:97). Kaziranga has been threatened by both large-scale poaching and development. The Subramanian Committee Report (GOI, MOEF 1994:4) includes Kaziranga on its list of "sensitive and disturbed areas ... [where] presence of militant elements hamper[s] the work of wildlife protection staff." Their recommendation is that "special measures should be taken by the Ministry of Home Affairs to clear these areas of these elements."

As in Manas, gangs of poachers have periodically outgunned forest guards, though officials are better armed than in Manas (Menon 1996:11–16). The state government blamed the Centre for inadequate funding; the Centre countered that allocated funds were underutilized. In 1990 Kaziranga was placed on the Threatened List of the International Union for Conservation of Nature; Manas was already on the list. Kaziranga had been listed previously because of poaching and plans for rail connections; it was removed from the list and then placed on it again as plans for a state-subsidized oil refinery (Numaligarh) nearby were announced.[27] Efforts to expand the park are hampered by disputes over property claims of what the government considers "squatters." As in Manas, there have been center–state conflicts over responsibility, and speculation among wildlife activists that timber and poaching gangs could not thrive without political connections.[28]

All protected areas are susceptible to people–wildlife conflicts; in the Sundarbans WHC Site, these involve tigers that kill people. The widely publicized "tiger widows" of the area demand compensation from the state for depredations of animals that the state has in effect declared its property.[29] In the Bharatpur WHC Site (Keoladeo National Park), a ban on grazing of domestic buffaloes in 1982 led to conflict between officials and villagers, shooting deaths, and an unanticipated biological outcome. The ungrazed wetland became choked with weeds and paspallum grass previously kept in check by grazing, creating suboptimal habitat for avifauna (many migratory, and thus a global concern) for which the area had been made a WHC Site. Compromises with villagers have allowed human removal of grasses. Threats from the WHC Secretariat to delist Bharatpur have been resented and resisted by the government of India.[30]

Cultural WHC Sites seek to conserve the human landscape rather than the biological. India's sites are protected under the Ancient Monuments and Archeological Sites and Remains Act of 1958, administered by the Archeological Survey of India. Sites reflect the rich diversity of India's heritage: Hindu, Buddhist, Jain, Muslim, and Christian (see table 12.2 for List of World Heritage Sites). Protection of sites—for archaeological significance, as places of worship, and as tourist centers—symbolically reaffirms the socio-cultural diversity of the nation. Inclusion on the World Heritage List does not, however, seem to have been stressed as a means of increasing vigilance, though state governments seeking tourist revenues periodically lobby for WHC designation of their sites. The primary NGO involved, INTACH (Indian National Trust for Archeological and Cultural Heritage), provides energy for monitoring and fund-raising. Though preservation of cultural heritage is more easily appreciated within the country than values of biodiversity, maintenance has not been effective. Threats come from encroachment, neglect, development, and ambient air pollution, a situation most prominently illustrated by the Taj Mahal.[31]

Though of symbolic importance, the WHC has had minimal environmental impact in India. When steps to protect the Sundarbans were listed in Parliament (*LSQ* 1227, 3-8-93), they included measures under the Project Tiger and Biosphere Reserve programmes, but *not* the WHC. Financial assistance through the WHC has been largely restricted to cultural sites; the MOEF does not even list the WHC among its international responsibilities (MOEF *Annual Report* 1992–1993:94), though it is the nodal agency for WHC Natural Sites. Relations with the Secretariat have been troubled on issues such as Manas and Bharatpur, arising from the classic constraint on international cooperation—national sovereignty.

Evolution of an Ozone Regime: The Montreal Protocol

The ozone regime manifests a genuine impact of international politics and treaty obligations on domestic process: perception of the problem, formulation of solutions, and technical means to ends were largely external to India. Signing conservationist accords was consonant with Indian policy; signing Montreal required alteration of domestic policy for global ends. India's initial reluctance to sign the protocol, particularly when reinforced by China, threatened to scuttle the accord (Benedick 1991:183–186). India was then influential in changing the ozone regime to address concerns of the South, but national compliance remains explicitly conditional and problematic for reasons identified in the earliest debates.

The process through which an agreement is forged may be as important as the law that results. Negotiations framed ozone protection as a North–South conflict, not as a global commons dilemma. India was not a major producer or consumer of ozone-depleting substances (ODS), nor were its scientific efforts regarding the stratosphere concerned with ODS. The policy elite considered ozone to be a "Western" problem[32] that neither affected India nor was a consequence of her actions. India entered negotiations as a result of external stimulus and with a model that projected serious developmental costs for poor countries.

A task force combining representatives from the NGO sector, industry, and the MOEF was established in 1988.[33] Its report, *The Economic Implications for the Developing Countries of the Montreal Protocol*, conceptualized a broad range of transition costs, including "retardation" of user industries, consequent losses to the national economy, and loss of export markets (Development Alternatives 1990:13). Retarded growth—the "hiccup cost"—was projected to continue for ten years in chlorofluorocarbon (CFC)-dependent activities. More significantly, given India's critical power situation, an energy penalty of up to 500 megawatts of installed capacity was projected (Development Alternatives 1990:26–27). A study funded by the British government, the Touche Ross Report, estimated transition costs at $1.2 billion. Compensation for these developmental costs then became a necessary condition for cooperation under the "polluter pays" principle. Mahesh Prasad, as Secretary of Environment, told the industrial countries: "We are willing to cooperate but we will not take your burden on our shoulders."[34]

As it became clear that without India and China, the protocol would be meaningless, both nations were widely accused of "international blackmail" and "implied threat." A report from the influential Indian Centre for Science and Environment stated: "As these countries could not conceivably be blamed for what had happened in the past—using just 2 percent of the world's CFCs—they were being blamed for what could happen in the future" (Agarwal and Narain 1992:7). Minister of Environment Z. R. Ansari explained India's reluctance:

> The technology of substitutes, conservation, recycling and equipment modification will be the monopoly of a few countries in the developed world.... The question that haunts us is the extent of resources required to get the technology as well as the products from the companies in the developed world. (Ibid.)

The Protocol could well create a global market rigged to India's national disadvantage; language urging industrial countries to "facilitate" access and provide subsidies was "delightfully vague." There was notable asymmetry: vagueness on issues of concern to India and clear and specific provisions for restrictions (Agarwal and Narain 1992:9; *TOI* 4-6-91; *LSQ* 7583, 14-5-90).

India therefore argued in London in 1990 for mandatory technology transfers to poor (article 5) nations. When Western governments argued that technology was privately owned, and thus beyond their control, Maneka Gandhi retorted that the entire Montreal process was about government restrictions on private firms; this type of intervention was universal.[35] The Secretary of Environment argued that if ozone truly constituted a global commons dilemma, then the technology to solve the problem should be available as common property. Delegations from Brazil, China, Malaysia, and other nations worried that concentrated ownership of replacement technology would permit oligopsonistic price gouging—what Malaysia's representative called "environmental colonialism" (Benedick 1991:189).

Amendments in London in 1990 made concessions on these points, but India's support for the treaty was both politically thin and substantively conditional. The government officially understands the modified article 10 to make national compliance contingent on implementation of mechanisms for financial assistance and transfer of technology (*CP*:2, 35).

Calculating Costs

Given the conflict preceding the London Amendments, and their ambiguous language,[36] calculating costs of phaseout becomes crucial for compliance. India established the Task Force on National Strategy of Phasing Out Ozone Depleting Substances on May 29, 1991, almost a year after the London Amendments but a year before India became a party (on June 16, 1992). Some costs of conversion are straightforward, but others raise difficult conceptual and political issues. By 1994, disagreements between India and the Executive Committee of the Fund over compensation had become serious.

The lead institution planning compliance was the Ozone Cell of the MOEF, with technical support from the Tata Energy Research Institute and the U.N. Development Program. Representatives of industry and officials from seven ministries were involved. India's current per capita consumption of controlled substances was then 8.8 grams and unlikely to exceed 20 grams—a very small number relative to the global problem and to the 300 grams allowed under the Protocol.[37] Plans for phaseout of even this low level are complicated because use is concentrated in areas of rapid growth. The National Program for Phase-out, financed partly by the Multilateral Fund, noted that sales of consumer goods of the type that involve the use of ODS (air conditioners, refrigeration, consumer electronics, aerosol sprays, etc.) have been increasing at a rate of about 15 percent per annum, whereas real per capita expenditures have been increasing at 2.4 percent (*NP*:16). ODS are also expanding rapidly beyond middle-class consumption into critical fields such as health delivery, fire fighting, food preservation and shipping, and pharmaceuticals.

Inherent uncertainties of cost projections make all estimates approximate. The Task Force estimated the cost at between $1.4 and $2.45 billion (*RTF*:20), including costs to producers, intermediate users, and consumers, as well as R&D programs, but *not* including costs of recovery and recycling, destroying substances and equipment that cannot be recycled or converted, substitutes for cleaning applications of ODS, any halon substitution, or social costs of plant closings and induced "economic drag." The Country Program (hereafter *CP*) submitted to the Executive Committee of the Fund in 1993 estimated $1.964 billion, including $15 million for strengthening institutional capacity. Large costs for technology development, transfer, and adaptation to "local production and circumstances" are included. The cost per kilo phased out is $9.85 (*CP*:64, table 13).

Refrigerators may not be the most difficult issue in India's compliance (at an estimated cost of $620 million), but the political metonymy of refrigerators has been significant. India argued that conversion costs should not be borne by consumers. Minister of Environment Maneka Gandhi said publicly, "If I sign this treaty, even my father will throw me out of the house" (interview 8-2-93). Montreal was believed to risk making old refrigerators obsolete and new ones prohibitively expensive. Refrigerators (and air-conditioning) were considered more necessities than luxuries in the Indian climate; national poverty required that refrigerators have a very long life. Symbolically, even moderate middle-class aspirations in the poor world were threatened by solutions for a problem created in countries where refrigerators were taken for granted (even thrown away periodically).[38] More important, refrigerators were a lead connection to developing export markets; CFC regulation threatened not simply privileged national consumers but also India's position in the global economy.

Planning Contingent Compliance

Compliance imposes a complex planning task, even at low levels of aggregate ODS use. The working group established by MOEF includes seven ministries and seven departments (*CP*:4). There are serious trade-offs, both economic and environmental. Some substitutes have high global warming potential, flammability, toxicity, or other undesirable externalities (particularly the volatile organic compounds). Private firms are undecided, facing considerable uncertainty and risk exacerbated by economic liberalization. The government envisions a mix of policy instruments to deploy once there is a discernible pattern of behavior: regulation, taxation, subsidies for R&D, low-interest loans for switchovers. Monitoring for reporting to the Secretariat is coordinated by the Ozone Cell. Its data will come from the state Pollution Control Boards (PCB) and central PCB (*CP*:56). At the state level, "the Small Industries Service Institutes which are State Offices of the Office of the Development Commissioner, State Offices of National Small Industries Corporation (NSIC), Directorates of Industries of the State Governments, State Pollution Control Boards, will be identified as the nodal agencies. . . ."

Planning thus is technically and institutionally complex; the Task Force urged caution: "India need not take any hasty decisions on the switch-over as they may prove to be costly. In any case our contribution to ozone depletion is negligible" (*RTF*:17). Quotas on production and consumption are not to be enforced until

1998. Decisions will ultimately have to be communicated to myriad unorganized-sector operators, and repair and maintenance people; no representatives of the "unorganized sector" have been major players in the planning process. In part this is inevitable, since the unorganized sector is, after all, unorganized. Yet much of the discussion of recycling, recapture, storage, and substitution of ODS will be irrelevant unless it accords with realities on the ground. Power is in the hands of the organized sector; once supplies are shut off or switched over, the unorganized sector will have no alternative to compliance and little chance for compensation from the Multilateral Fund. It is here that transition pain will be concentrated.

By order of the Chief Controller of Imports and Exports, Ministry of Commerce, India banned exports and trade in controlled substances with nonparty countries in June 1993; trade with countries that are parties to the Protocol is permitted with a license (*IED* 7-1-93). A steering committee to evaluate fundable projects was constituted in September 1993.[39] Compliance now depends on the action of firms.

Central planning is currently in disgrace, but the license-permit-quota system, along with the infancy of ODS industries in India, produced a concentrated production structure amenable to regulation. There are only five CFC-producing firms, four carbon tetrachloride (CTC) firms, and one MCF (metheyl chloroform) firm. Halons are produced by two government-initiated enterprises. India is now self-sufficient in production of CFCs and their applications, and an exporter to West and Southeast Asia. The relatively new plants entailed heavy investment and will not reach their break-even point for some time. Though production is concentrated, about 65 percent of the *use* of ODS is in the small-scale sector, the accounting of which is incomplete and difficult (see Appendix II, Table 2). Aerosols are produced by 200 medium-scale firms, but there are many times that number of small fillers in the unorganized sector. Foams likewise span 300 medium-scale firms that are identifiable and countless tiny enterprises.

The largest ODS in India is carbon tetrachloride, virtually all domestically produced, accounting for nearly 50 percent of use (in electronics, rubber manufacturing, pharmaceuticals, pesticides, and other applications); relatively little is known about its distribution and possible substitutes. Although the national program seeks to ensure that small enterprises are fully compensated for phaseout, including appropriate retraining and provision for worker severance, this large financial burden will have to be borne internally—if it is borne at all.

India's conditionality in compliance is clear and public. The Country Program forwarded to the Executive Committee of the Multilateral Fund said (*CP*:35):

> India is committed to adhere to the provisions of the Montreal Protocol ... provided sufficient funds are made available and required technologies are transferred in accordance with the provisions of the protocol. In view of the shortage of funds available to industry and the non-

availability of appropriate technology, ODS phaseout progress in India will be determined by the availability of assistance from the Multilateral Fund. This assistance is expected to include cost of conversion of existing facilities, costs arising from premature retirement or enforced idleness, and costs of patent and designs and investment costs for establishing new production facilities.

The Executive Committee challenged India's Country Program for "double counting" and inclusion of nonfundable categories (Dasgupta 1994). More seriously, the Multilateral Fund seems to be rationing resources, which will inevitably be less than India's claims. A Negative List (of costs not eligible for compensation) has been established, expanded, and altered, often after a proposal has begun its journey toward funding. Internal delays in choosing proposals to forward, as well as uncertainty about criteria of the Executive Committee and implementing agencies, have discouraged firms from formulating projects, thus slowing compliance. Technicians at the World Bank and UNDP, which are implementing agencies, assumed de facto control of the allocation process rather than the parties; administrative costs are very high.

As a result, Minister of State for Environment Kamal Nath, previously Chairman of the Executive Committee, suggested at the sixth Meeting of the Parties that article 5 countries stop the switch-over until developed countries gave full assurances of timely and predictable replenishment of the Multilateral Fund. Noting the large arrears in contributions, he argued that the Protocol was being "whittled away at the implementation level," creating a mood of "distrust and discontent."[40]

Compliance at this stage is inevitably slow because private firms and the nation face difficult choices. (see table 12.3a–e). To adopt intermediate technologies is risky, as HCFCs themselves become proscribed under the protocol. Industry worries that "if we go through two cycles, they will not compensate the second time—maybe not the first" (CII Interview, 2-3-95). Waiting for resolution of the technology flux creates two negative effects: dependency on whoever develops new technology and the "hiccup" effect of suppressing investment during the waiting period. Both dangers, long feared by Indian officials, are now real. The energy penalty of switchover is not part of the amended compensation guidelines, nor is research and development for adaptation to local conditions; both are critical industrial issues. For example, industry has delayed conversion to HCFC134a in refrigeration because the substance will become illegal and has negative environmental effects, publicized by European NGOs. Alternative hydrocarbon-based systems being developed in Europe present safety problems under Indian conditions (TERI 1994); amelioration of these problems is on the Negative List for compensation.

Assistance from the Multilateral Fund was initially small—about $11.5 million, around 0.5 percent of the roughly $2 billion estimated in the Country Programme. Minister Kamal Nath said it was "disappointing" that much smaller countries have

India: ozone-depleting substance

Table 12.3a
ODS production, imports, exports, and consumption in 1991 (metric tonnes and percent)

Substance/source	Domestic products MT	Imports MT	Exports MT	ODS consumption MT	ODS consumption %	ODP value	ODP consumption MT	ODP consumption %
Annex A. Group I								
CFC-11	1450	510	60	1900	18.3	1.0	1900	14.5
CFC-12	3280	0	430	2850	27.5	1.0	2850	21.7
CFC-113	40	280	–	320	3.1	0.8	256	2.0
CFC total	4770	790	490	5070	48.9		5006	38.2
Annex A. Group II								
Halon 1211	50	500	–	550	5.3	3	1650	12.6
Halon 1301	–	200	–	200	1.9	10	2000	5.2
Halon total	50	700	–	750	7.2		3650	27.8
Annex B. Group II								
CTC (used as solvent)	3920	80	–	4000	38.6	1.1	4400	33.5
Annex B. Group III								
Methyl chloroform	540	10	–	550	5.3	0.1	55	0.5
Total	9280	1580	490	10370	100		13111	100

1. Imports of CFC-114 and CFC-115 and exports of CTC were negligible in 1991
Source: *India Country Programme*, September 1993, Table 1.

Table 12.3b
1991 ODS consumption by sector (metric tonnes and percent)

	Refrigeration and air-conditioning	Foams	Solvents	Aerosols	Fire ext.	Total
Tonnes	1990	1580	4950	1100	750	10370
%	(19.2)	(15.2)	(47.7)	(10.6)	(7.2)	(100)
ODP tonnes	1990	1576	4795	1100	3650	13111
%	(15.2)	(12.0)	(36.6)	(8.4)	(27.8)	(100)

Source: *India Country Programme*, September 1993, Table 2.

Table 12.3c
Estimated unconstrained ODS demand by substance (in 1996, 2000, 2005, and 2010), in metric tonnes

	1991 actuals		1996		2000		2005		2010	
Subs	Tons	ODP tons	Tons	ODP tons	Tons	ODP tons	Tons	ODP tons	Tons	ODP tons
CFC-11	1900	1900	4350	4350	7800	7800	15500	15500	27200	27200
CFC-12	2850	2850	5960	5960	10560	10560	20900	20900	41200	41200
CFC-113	320	256	750	600	1580	1264	4240	3392	10300	8240
H-1211	550	1650	1050	3150	1350	4050	1750	5250	2100	6300
H-130	200	2000	340	3400	500	5000	800	8000	1000	10000
CTC	4000	4400	10600	11660	23800	26180	48900	53790	96000	105600
MCF	550	55	1120	135	2000	240	3560	427	6400	768
Total	10370	13111	24170	29255	47590	55094	95650	107259	184200	199308

Source: *India Country Programme*, September 1993, Table 4.

Table 12.3d
Estimated unconstrained ODS demand by sector (in 1996, 2000, 2005, and 2010), in metric tonnes

	1991 actuals		1996		2000		2005		2010	
Sec	Tons	ODP tons	Tons	ODP tons	Tons	ODP tons	Tons	ODP tons	Tons	ODP tons
R&AC	1990	1990	3780	3780	6330	6330	11860	11860	23760	23760
Foams	1580	1576	3570	3563	5910	5897	10700	10684	16580	16555
Solvents	4950	4795	12700	12632	27820	28137	57540	58465	114640	116573
Aerosols	1100	1100	2730	2730	5680	5680	13000	13000	26120	26120
Fire ext.	750	3650	1390	6550	1850	9050	2550	13250	3100	16300
Total	10370	13111	24170	29255	47590	55094	95650	107259	184200	199308

Source: *India Country Programme*, September 1993, Table 5.

Table 12.3e
CFC substitutes under consideration for R & AC sector

Sl. No.	Product	Present	Currently evolving alternatives/substitutes
1.	Domestic refrigerators	CFC-12 (refrigerant)	HFC-134a, HFC-152a, mixtures and blends, hydrocarbons (propane/ butane, propane/iso-butane mixtures).
		CFC-11 (foam blowing)	HCFC-22, HCFC-22/HCFC-142b, cyclo-pentane, HCFC-141b, HFC-134a
2.	Refrigerated cabinets (deep freezer, iee cream cabinets, bottle coolers, visi-coolers)	CFC-12 (refrigerant)	HFC-134a, HFC-152a, mixtures, hydrocarbons (propane/butane, propane/ isobutane mixtures)
		CFC-11 (foam blowing)	HCFC-22, HCFC-22/ HCFC-142b, cyclo-pentane, HCFC-141b, HFC-134a
3.	Water coolers	CFC-12 (refrigerant)	HCFC-22, HFC-134a, HFC-152a, hydrocarbons and mixtures
		HCFC-22 (refrigerant)	HCFC-22
4.	Auto (car) A/C	CFC-12 (refrigerant)	HFC-134a, mixtures
5.	Bus & van A/C and refrigerated road transport	CFC-12 (refrigerant)	HCFC-22, HFC-134a, mixtures
6.	Train A/C	CFC-12 (refrigerant)	HCFC-22, HFC-134a
7.	Central A/C plants	HCFC-22 (refrigerant)	HCFC-22
		CFC-11 (refrigerant)	HCFC-123
		CFC-12 (refrigerant)	HFC-134a
		Ammonia (refrigerant)	Ammonia
8.	Processed chillers	CFC-12 (refrigerant)	HCFC-22, HFC-134a
9.	Ice candy M/C	CFC-12 (refrigerant)	HCFC-22, HFC-134a

Source: *India Country Programme*, September 1993, Table 6.

received so much more than India (*PIO* 3-10-94). By his figures, of the $200 million approved by late 1994, China got more than $37 million, or 19 percent of the total; India, only 6 percent. Even the Philippines, which joined after India, got more funds, as did Egypt and Thailand. The Ministry of Environment has three major concerns: first, phaseout resources lag behind what the West promised; second, after the ozone hole is mended, there is no incentive for the Multilateral Fund to be replenished; finally, the South will be left behind in the global economy—after the North eliminates controlled substances, it will ban products made with them, stranding those at the bottom of the product cycle.

Meeting with representatives of industry in October 1994, the MOEF agreed to pursue the following objectives at the international level: avoiding enforcement of the Negative List and frequent changes in funding mechanism; facilitating transfer of technology for ODS phaseout and ODS production in India without any preconditions (technology bundling, demands for equity position, etc.); avoiding efforts to accelerate the phaseout schedule of CFCs, halons, and HCFCs; retraction of the 15 percent production capacity allowed to developed countries (to allow India to meet that residual market demand).[41] Success in these negotiations will determine the pace and extent of compliance.

Opponents of the Montreal Protocol now seem prophetic; the assurances of London, problematic. After seemingly successful international bargaining, India agreed to help solve a problem it did not create. Nevertheless, negotiations were restricted almost entirely to a thin stratum of officials, elite NGOs, and private capital; costs were putative; compensation was assumed. The real costs of Montreal have not been—arguably could not have been—debated in anything like a plebiscitary fashion. Neither consumers nor small enterprises—the bulk of ODS users—were engaged. The government's stated contingency of compliance leaves room for retreat if anyone with power notices the costs.

Frequent celebration of Montreal as a success in resolving global commons dilemmas obscures both the problematic sense in which state actors represent their societies and the persistent feelings of injustice emanating from lower tiers of the international political economy. Maneka Gandhi said:

> The resolution of these issues in respect of the Montreal Protocol to enable the developing countries to be partners in saving the ozone layer is an acid test of the willingness of the developed countries to promote a true partnership among all the countries of the world for managing global change. (*HT* 19-4-90)

Antinomies of Institutional Capacity: Power and Governance

Public policy comes alive or perishes on the ground, in concrete relations between social actors and claimants to authority. Characteristics of international treaties matter greatly in terms of what institutional and relational capacity is required for

Table 12.4
India's compliance with treaty reporting requirements, 1991–1996: Report transmission/receipt date

Treaty	Entry into force	1991	1992	1993	1994	1995	1996
CITES	10/18/76	12/9/92	10/12/93	10/26/94	10/20/95	11/29/96	N/R
LC	Not a party						
MP	9/27/92	—	R	R	R	R	R

Notes: For all treaties, N/R indicates that no report was received by July 1997. R indicates that a report was filed, but the date of receipt is unavailable. Montreal Protocol reports received include those filed on time, late, or subsequently revised; in some cases a country may have met two or more years' reporting requirements with the submission of a single report.

compliance. As a large state with a well-trained bureaucracy, India has little difficulty meeting prescribed reporting requirements, as table 12.4 shows. Compliance with the substantive provisions of treaties is more complex. CITES, ITTA, and WHC confront issues in which local populations have strong normative and political claims against state restrictions; livelihoods are at stake. Here state institutional capacity hinges on governance—that elusive mix of authority, transparency, representation, participation, and cooperation that complements power with legitimacy.

In contrast, Montreal links will and capacity. Implementation is a centralized, technocratic operation for which the state is well prepared and endowed, and to which the major players are resigned; residues of the command economy and concentrated production facilitate administration. Institutional capacity for sophisticated and timely planning, aided by international cooperation, has been demonstrated. Negative effects on the public will be diffused, probably unnoticed in the general inflation, and difficult to link concretely to the Protocol. Small-scale users of ODS will be hurt, but the state has the capacity to ensure compliance by choking off supplies at the producer level. Will and capacity to comply could be compromised, however, if the Multilateral Fund has too little money or flexibility on cost definitions for India to remain a full participant. Compliance would then be slower and more erratic, but structural pressures in the global economy will drive ODS phase-out in any case.[42]

Regulating nature tests state capacity more severely. Here capacity is a complex interaction of institutional capability variables (how many officers, with how many vehicles, walkie-talkies, and weapons, with how much organizational coherence, etc.) that constitute necessary but not sufficient conditions for enforcement, and relational variables that are far more important after a threshold capability is reached. These variable relationships between bearers of state power and civil society constitute a capacity for governance that command-and-control systems often undermine.

The Ahmed Committee, charged by the High Court of Delhi to make recommendations on protected areas (GIO, MOEF 1996:30), concluded starkly: recent surveys of protected areas, sanctuaries, and national parks show "an alarming and depressing scenario [characterized by] .... general apathy and lack of motivations, interest, competence among the staff, at different levels." Further, "the situation has deteriorated to such an extreme, that nowadays to be posted as Director of a National Park is regarded as a punishment!" The Committee's report cites as causes "lack of motivation and low morale ... inability to control poaching ... increasing tensions between local communities and Park management ... special problems of encroachment, illegal mining, illicit fishing, ganja [cannabis] cultivation, prawn farming and intensive tourism of pilgrims" (GOI, MOEF 1996:31).

Incapacity in a physical or staffing sense is widely recognized and easily understood, and is remedied with resources. This is an area where international support is useful. Protection staff lack sufficient vehicles, communication devices, arms, and, at the local level, such amenities as medicine, weather gear, boots, and the like. Their morale is undermined not only by these matters but also by lack of educational and employment opportunities for their families (GOI, MOEF 1994:7; GOI, MOEF 1996:32 et passim). Moreover, "[v]acancies in Wildlife Wings are not filled. At the time of writing of this report, there were 450 vacancies to be filled in Tiger Reserves. This is particularly unfortunate because these are high profile conservation areas. The status of other Protected Areas is much worse" (GOI, MOEF 1996:34).

Relational capacity for governance is more difficult to create. Protection of natural areas creates conflict with the use rights, property, and often safety of surrounding populations in a land-scarce economy; belief in state capacity is a necessary condition for working relationships, but is often missing on the ground. One illustration might be useful: Jagan, *sarpanch* (headman) of Kailashpuri, a village removed from Ranthambore National Park, where there was resistance to "resettlement," captured the core of oppositional perception: Our sacrifice to the cause of the tiger was a joke. The Forest Department could not take care of us, the animals or the forest, so what right do they have to be employed? They left the tiger in the hands of the poacher and left us in the hands of God. (*IT* 15-8-92)

Simultaneously, wildlife professionals believe that "concessions" to local populations have gone too far, compromising authority with accommodation. In a section of the Ahmed Committee Report entitled "Wildlife Conservation: An Issue Too Sensitive for Anyone to Touch," we find:

> All politicians and leaders of political parties seem to be unwilling to stand up for wildlife and take the risk of formulating a "pro-wildlife policy." Wildlife conservation, which has been implemented mainly throught the Protected Areas system and the Wildlife (Protection) Act, is currently under atttack as symptomatic of a power system which is undemocratic, authoritarian and contemptuous of the rights and the needs of the local communities affected by the imposition of the protective measures which favour wildlife! (GOI, MOEF 1996:37)

In the Report's view, conservation objectives "seem to *have lost their intrinsic and specific value*, as people are fumbling and groping through a maze of all other issues except those concerning wildlife and wild habitats. The end sufferer is the wilderness of India." There has to be balance, but the balance that used to be there "has collapsed under the encroachment upon the wild habitat by a growing population to answer to their need for more land to cultivate" (GOI, MOEF 1996:37).

These dilemmas are serious in a political democracy; the member of Parliament from Almora said in Parliament:

> Due to the increase in ... tigers in the Corbett National Park ... and non-availability of adequate food for them, the tigers come in the areas adjacent to this park and it creates a terror in the neighboring villages.... In my constituency about 200 people have been either killed or injured by tigers during the last three years. There is terror in the villages ... Sillour and Silt villages are gripped in terror.

He asked for more guards, barbed wire fence, and victim compensation, "as it is being paid to the persons killed or injured in rail or air accidents (*LSQ* M/R 377 (I), 27-4-90). Furious at the state's callousness in compensating loss, and its incapacity to control its marauding property, villagers predictably defend themselves by killing predators. When the state is more enemy than public trust, confrontation and evasion are more likely than cooperation; governance gives way to coercion.

State incapacity to control smugglers, socially ensconced bandits, kidnappers, insurgents, and drug dealers weakens compliance with international accords but is difficult to remedy.[43] This incapacity reflects the larger phenomenon of parallel power beyond the reach of state authority and pervasive antagonistic relations of rural people to the state. For more than a decade, the famous poacher and smuggler Veerappan killed elephants for ivory in South India; when tuskers became rare, he turned to sandalwood. Efforts to track and capture Veerappan led to police personnel being killed in ambushes and encounters—six individuals in 1993 alone, including senior officers. He knew the terrain, had kinship connections among the Padiyachi Gounder community, and moved easily among the villages, where police are feared and distrusted. In 1993, the job of catching Veerappan was transferred from the police to the paramilitary Border Security Force, which treated all villagers as probable criminals.[44] Veerappan could be construed as a folk hero pursuing traditional subsistence routines—or as a murderer of people and elephants, as well as a local bully who terrorizes villagers, thus representing a microcosm of the contesting ideologies of conservation.

More generally, the Subramanian Report concludes: "No poaching can occur without the passive knowledge or active help of villagers living in and around the protected areas." But villagers often refuse to share this information with the state: "the most important reason being a widespread hostility in the rural areas against the protected areas which have put severe restrictions on the use of the forests for their basic needs—grazing, firewood and small timber and collection of NTFPs [nontraditional forest products] for a livelihood" (GOI, MOEF 1994:17). The very remoteness of reserves, necessary for their biodiversity function, makes them attractive to anyone (such as poachers) who wishes to avoid the state, thereby undermining their biodiversity function. The crux of governance is that the state must show that law matters—elephants cannot be killed with impunity—but simultaneously that it is a servant of the people, not a bully or terrorist.

Consequences of hostility to the state on protection issues extend beyond conservation outcomes. Just as rural rebellion often begins in "banditry" in remote areas, so challenges to state authority often begin in, or are legitimated by, disputes over claims to natural resources, frequently connected to demands for autonomous political space. M. S. Pal, member of Parliament for Nanital, argued that the Forest Conservation Act (1980) had stopped development work in the Uttarakhand region of Uttar Pradesh. As a result, "resentment among the people is growing and the feeling of a separate state is gaining momentum" (*LSQ* 066 M/R 377, 22-3-90). The government announced the creation of a separate state on August 15, 1996. This connection between subnationalism and natural-resource use, one step more radical than the "sagebrush rebellion" over federal control of land use in the western United States, may be used instrumentally by politicians, but is a predictable outcome of intrusive state power that is not locally sanctioned or leavened with participation.

More mundane capacity variables include the incentive structure that glues the state together. An appendix to the Tiwari Report (GOI, Dept. of Science and Technology 1980:90) noted:

> The Wildlife Guards while performing the same protective duties in more difficult situations than their counterparts in the Police Service, receive considerably lower emoluments due to the fact that their unions cannot exert the same pressure on Government. As a result guards rather tend to associate with poachers than resist them.

Politicization of public service—often abetted by democratization and decentralization—likewise affects capacity. When even the lowest level of Forest Guard is considered a patronage resource, one expects a relative decline in institutional capacity. For relational capacity, individual officers matter greatly; trusting relations between the state and the local society, though rare, must be built on a very micro scale. The forest policy of 1988 implicitly recognized the limits to and counterproductive consequences of a coercive and conflictual relation between state and local society. Natural-resource conservation requires breaking the antinomy of state–society in favor of institutionalized cooperation. Joint forest management, involving experiments in participatory cooperation between people and the Forest Department, new concepts of buffer management through ecodevelopment, and sensitive appreciation of people–wildlife conflicts—all indicate potential for institutional reform conducive to governance, but certainly are no panacea.[45]

There is a cognitive shift necessary for governance as well. Policies highlighting buffer zones and subsistence job creation implicitly assume that it is only the most desperate who prey on public natural resources, and that the poor have income ceilings that are easily satisfied (unlike the rest of us).[46] But greed is a more destructive force than desperation; much environmental degradation ultimately results from forces that are not local and rural, but urban and international—as rural people understand but policy elites often do not (Gadgil and Guha 1995).

Lessons

In explaining the effectiveness of India's environmental policy, structural characteristics discussed throughout provide an explanatory rough cut: contested federalism; growth imperatives driven by democracy, mass poverty, and subordinate position in the global economy; and relations between state and society at the point of contact. The final category is the most important and complex; variables are fluid, situationally specific, and difficult to generalize, particularly in a continental political economy. Unfortunately for grand theory, individuals make large differences. During the debates surrounding the controversial London Amendments, the "government of India" was de facto a few individuals whose differing perceptions mattered fundamentally.

More generally, the structure of roles and incentives, as well as intrastate relations (e.g., among ministries), may set parameters, but a great deal of variance in particular tests of capacity is determined by individuals specific to the time and place. Individuals are not perfectly substitutable; they differ in worldview, values, energy, and connections—hence in personalistic capabilities and constraints. Courageous officers resist threats—of transfer or worse—when implementing unpopular policy; some die as a result. Others, being more rational, compromise. Some are brilliant innovators able to follow and lead, inspire and learn simultaneously. Governance ultimately resides in these analytically messy micro relations between the local state and its subjects.

Policy analysis inevitably confronts the Archimedes problem: given a place to stand and a lever long enough, one can move the world. Real politics and real resources constrain where the state can stand and the length of its levers. Priorities compete; macropolitical capacity is constrained by multiple demands on limited resources. Effective performance on policy X constrains performance on Y. Governance involves compromises; the politically rational may not be policy-friendly. With vigorous political competition, the room for maneuver is often small.

At the deepest level, as pressed by some NGOs, compliance with the spirit of conservationist accords would require changes in the operative conceptualization of development itself (CSE 1986; Raghunandan 1987; Herring 1991). "Sustainable development," for all its trendy ambiguities, is debated from positions of great sophistication and experience in India, but that discourse lacks political power and confronts powerful interests. Moreover, international competitive pressures make it difficult for any nation, poor or rich, to sacrifice growth for environmental values. For poorer nations, the choice is more difficult, normatively and practically.[47]

Nevertheless, several themes within the realm of the possible recur. Cybernetic capacity and interagency coordination are particularly important. Knowledge is a necessary condition for administration, but data collection and analysis present difficult choices when so many pressing issues clamor for attention and resources. Building cybernetic capacity is often hindered by the state's unwillingness to admit what it does not know. Some of the most vital knowledge-building is being done by NGOs, often with international support and, increasingly, with government's blessings. The workshops organized by TRAFFIC-India for CITES serve as a model on both scores: disparate parts of the state meet with NGO activists and technical people, providing a unique forum for sharing information and coordinating strategies. International support could be profitably increased. Enhanced institutional capacity for CITES and WHC—more infrastructure and personnel—would improve compliance, but the character of state–society relations at a very local and situation-specific level is ultimately more important, and less amenable to amelioration.

Were the state more like the unitary actor that international lawyers and academics assume it to be, some problems would prove more tractable. Interministerial —and sometimes intraministerial—coordination is a major problem. The relative lack of power of the Ministry of Environment and Forests among ministries at the center aggravates inevitable coordination problems across ministries. Functional missions of ministries assigned to serve as nodal authorities also make a difference.[48] The Environmental Law Centre, established to resolve legal tangles between international accords and domestic law, is a useful model for improving coordination and compliance.

Myrdal's famous (1968) characterization of "soft states" in South Asia is apt; (selectively) permeable membranes contribute to regulatory incapacity. But just as the state is less unitary, and less an agent, than much theory presupposes, so it is also less stable, manifesting continual reformations in response to political dynamics. Popular protest and environmental policy produce new relationships between state functions and nonstate institutions—particularly NGOs. Cooperation between WWF-India and the MOEF, for example, aids in enforcing CITES, rationalizing domestic law in accordance with international agreements, improving the knowledge base through which policy is adjusted, and supporting litigation and environmental education without which policy proves ineffectual. The integration of TERI and the Ozone Cell created new planning capacity. The addition of an NGO Cell in the MOEF in 1992 was a new and dynamic phenomenon.

Compliance with international accords and domestic law is hindered by obstacles that resist quick fixes, many of them going to the heart of governance itself. These difficulties are exacerbated by national poverty, difficult center–state relationships, an unstable center, and bureaucratic inertia. The last reflects a familiar, perhaps universal, pattern. Inaction on new initiatives occurs for lack of precedent. Reports and files are sent down the administrative ladder for "looking into" and sent back up for "'advice," "opinion," or "comments." The loops can repeat indefinitely until someone attempts to act. A certain amount of noncompliance is not willful, but rooted in common bureaucratic pathologies: lethargy, conservatism, rigidity, and arrogance, inter alia. There may be no cure for bureaucracy. Nevertheless, experiments in joint forest management and ecorestoration schemes, NGO–state cooperation, and local assertions of authority are reinventing state–society relations in tentative episodes.

Nevertheless, some problems are inherently intractable in liberal democracies. Compliance with CITES and WHC in India would be aided by suppression of armed banditry and secessionist movements, but such capacity is difficult to reconcile with limited government, even assuming someone would know how to do it.

International Justice: Fairness, Will, and Capacity

Compliance presupposes capacity and will. In the nature-conserving treaties under consideration, India's domestic legislation is strict; effectiveness is constrained by divided will and incapacity. Montreal makes significant demands that are beyond the technical and financial capacity of India absent international assistance promised under the London Amendments. Failure to implement those transfers jeopardizes both capacity and will to comply.

Inferior position in a decidedly hierarchical world system constrains both will and capacity for environmental solutions. At the White House conference on global warming in 1990, India's greenest Environment Minister, Ms. Maneka Gandhi, argued that poor nations necessarily put development before environment; democratic institutions ensured that "legitimate aspirations" of the poor to catch up as soon as possible to rich-nation standards could not be denied:

> When we cannot afford to set apart resources even to replace the unacceptable pesticides which affect the health of the people today, can we afford to make investments to avert a global problem which will be manifested 50 years hence, particularly when we are not contributing to the creation of the global problem? (*HT* 19-4-90)

For consequential treaties, perceptions of fair play in an unequal world matter. There is concern in the South that new environmental regimes will make their industrialization slower and more costly in comparative historical terms precisely because of the noninternalized costs of the North's industrialization (Agarwal and

Narain 1992). Moreover, international environmental regimes create a difficult tension. The international system simultaneously urges economic liberalization and environmental management; the magic of the market and the specter of externalities both require urgent attention and sacrifices in the periphery. The former necessitates disengaging an intrusive state; the latter requires an activist, interventionist state, not only monitoring technologies and production processes but also actively planning to intervene and guide them. It is not difficult to deduce hypocrisy from the North's admonition to "do as we say, not as we did."

The degeneration of global commons discourse on ozone to compensation discourse followed naturally from the global history of the threat, the distribution of developmental costs, and subsequent North–South framing of negotiations. One prominent indigenous view of the Montreal Protocol is that India sold out, settling for a petty bribe.[49]

Yet in the framework of structural political economy, there was little choice. Coincident in time with Montreal, Indian industry was moving into international markets sensitive to the Protocol. Exports of refrigerators (largely to the USSR and West Asia) had reached the level of 4,000 annually and were climbing. Navin Fluorine won a contract from Iraq to supply a turnkey CFC plant worth $28 million, causing considerable friction between India and the United States.[50] The post-1991 development strategy of attracting foreign investment and promoting globalization ruled out defiance of global economic powers and policies. As Indian capital achieved the maturity to compete internationally, potential internal resistance was transformed into a force for international engagement; industry recognized that the international market was being irrevocably restructured by Montreal. Global firms with deep pockets, strong research and development budgets, and economies of scale cannot be matched in India, much less in Cameroon; joint, and asymmetric, ventures—or deepened import dependence—are necessitated by the ozone regime.

Montreal reinforces global integration of Indian firms to allow technical tie-ins that are Protocol-compliant but indigenously unavailable (e.g., Seshan 1995:180). Transfers of technology are proving Mahesh Prasad and Z. R. Ansari to have been prescient: multinational corporations (MNCs) do not want to sell or license new technology—partly for fear that Indian industry would then become low-cost competition—instead, they hold out for majority equity participation in joint ventures.[51] The global market-structuring effects of Montreal are more important for the ozone layer than either will or capacity to comply with the Protocol in the poor world.

International effects on India's environmental policy thus operate through channels more powerful than treaty compliance. International information flows and some technical assistance have improved India's capacity to enforce CITES, but world market forces spur the killing of endangered species. Market control in neighboring countries and end-consumer nations is critical for enforcing India's prohibitions on trade-driven decimation of species. Liberalizing reforms in the former socialist bloc had perverse effects on CITES in India, creating new international market players operating with hard currency and fewer constraints.[52] Liberalization in India, much influenced by international actors, has produced a competition among states to court capital, foreign and domestic, often at the expense of "denotifying" (legally removing from special status) protected areas or relaxing standards.[53]

Particularly since Rio, sovereignty in management of domestic nature is politically sensitive. Disputes between India and the WHC Secretariat over issues of access and monitoring have not aided compliance on the ground, and risk national resentment at international interference.[54] India's representative to the 1993 ITTO meeting in Kuala Lumpur had a distinctly critical view of the ITTA: "the countries with no forests left are telling us how to manage ours." He was frustrated by the disproportionate allocation of voting rights, the assumptions embedded in "tropical timber"—as if destruction of temperate forests were no problem—and refusal of the North to agree to a proposal to generate resources for forest protection and regeneration through a levy on internationally traded timber.

For treaties other than Montreal, the specifically international component of India's commitments is of more symbolic than environmental importance: reinforcing norms of international cooperation and conservation norms that are clearly under counterattack (e.g., Rowell 1996). Reciprocal spread of conservationist ideas and innovations through international communities has greatly influenced environmental policy in India. New thinking and research were spurred by the Rio conference; institutional innovations came immediately afterward. Drafting of the Wildlife Act was influenced by global conservationist communities, just as India's wildlife community influenced CITES. International science and technical practice were central to formation of ozone policy. But specifically normative discourse on the justice of international environmental regimes remains both disputed and potentially inimical to compliance.

Skepticism in India about the fairness of outcomes and the good faith of major players in the Montreal process, for example, has not been resolved. India's task force questioned the distributive outcome:

> The allocation of per capita consumption limits is disproportionate ... and appears to be arbitrary.... The USA after implementing control measures ... would consume about 2.2 million tonnes of CFC-11 and 12 up to the year 2000 [compared with India's limit of about 0.1 million tonnes]. On a per capita basis [this] works out to be 97 grams and 8461 grams for India and USA respectively. (*RTF*:27)

The Secretary of Environment, who bargained hard for the London Amendments, said in 1993: "The West had known about this ozone hole since 1974. They waited to do anything until they had developed substitutes they could sell as monopolists."[55]

Appeals to international justice are certainly in part politically instrumental, deployed in a complex dramaturgy of relatively wealthy elites who claim to speak for abstractions of national interest—and for the poor. Nonpoor individuals in both North and South derive benefits from global environmental crises: consultancies, travel, appointments, employment; adjustment costs are disproportionately borne by the poor. Nevertheless, fairness of international structures and processes is a necessary condition for willing compliance. The Montreal Protocol nearly failed on grounds of unfairness raised by poorer nations; widely hailed as a global commons success story, it bears an onerous responsibility in laying the groundwork for more politically difficult and complex accords. As chairman of the Multilateral Fund, India's Environment Minister Kamal Nath chided parties for overdues and "cautioned the world community that the failure to implement the protocol would threaten the conventions on climate change [and] biodiversity ... finally leading to the negation of the Rio process" (*HT* 9-10-94). Compliance in governance of the global commons becomes more thinkable to the extent that the rhetoric about global community and shared sacrifices resonates with practice.

Abbreviations

AST: Assam Tribune (Guwahati)

BNHS: Bombay Natural History Society

CP: Government of India, Ministry of Environment and Forests, *Country Programme: Phaseout of Ozone Depleting Substances under the Montreal Protocol* (New Delhi: MOEF, September 1993)

ET: Economic Times (Delhi)

DH: Deccan Herald (Pune)

HT: Hindustan Times (Delhi)

IED: Indian Express (New Delhi)

IT: India Today

IPS: InterPress Service

LSQ: Lok Sabha Questions

MH: Maharashtra Herald (Pune)

NEO: North East Observer (Guahati)

NP: Government of India, Ministry of Environment and Forests, *India: National Programme: Phaseout of Ozone Depleting Substances Under the Montreal Protocol* (New Delhi: MOEF, 1993)

NWT: Newstime (Hyderabad)

PAT: Patriot (New Delhi)

PIO: Pioneer (New Delhi)

STM: Statesman (Calcutta)

RSQ: Rajya Sabha Questions

RTF: Government of India, *Report of the Task Force on National Strategy of Phasing Out Ozone Depleting Substances* (New Delhi: MOEF, 1993)

TEL: Telegraph (Calcutta)

TOI: Times of India

Notes

1. It is handled by the Ministry of Surface Transport, not the Ministry of Environment. The government's position is that there are unresolved substantive questions: Is there overlap with the Basel Convention? with the Law of the Sea Convention (which India has signed but not ratified)? Some officials in the Ministry of Surface Transport believe that the Merchant Shipping Act of 1958, as recently amended, covers the same ground as the London Convention and conforms to international standards. Penalties under that act are severe; nevertheless, marine pollution is recognized as a severe and understudied problem (e.g., *LSQ* 1174, 27-3-85; *LSQ* 354, 25-2-87).

2. There are ten major biogeographic regions and sixteen major forest types, including 14,500 species of angiosperms. Biodiversity "hot spots" are also areas of great endemism. India's ecosystem diversity is of special importance because it ranges from cold alpine conditions in the Himalayas to extremely hot and arid deserts to wet evergreen rainforests in the Northeast and the South. For an overview of taxonomic studies, see MOEF, *Annual Report 1992–1993*, pp. 15–26.

3. Forests and wildlife were added to the list of joint center–state responsibility by amendment 42 to the Constitution in 1976 (GOI, Dept. of Science and Technooogy 1980:entry 17A).

4. On differential (and often weak) capacity and interest in the states, see GOI, MOEF 1994:13 et passim; GOI, MOEF 1996:ii, 34, 38, passim; *LSQ* 3320, 4-17-85. Narasimha Rao's vulnerable government capitulated to state demands for weakening environmental impact statement requirements and delegating more power to the states. No-development zones (NDZ) around national parks and reserve forests were compressed (by 80 percent), and concessions to state demands that "non-polluting" industries be allowed within the NDZs were made (*Financial Express* 2-5-93). The modal form of center–state interaction is for the center to formulate policy guidelines, or "schemes," and announce some formula for cost-sharing with the states. (For examples, see MOEF, *Annual Report 1992–1993*:38–42 passim.)

5. Interest groups have traditionally pursued their interests through cultivation of special connections with the bureaucracy rather than overt campaigns to influence legislators or public opinion. One notable exception—the Confederation of Indian Industry (CII)—projects an overtly modernist, internationalist profile. CII's Environmental Cell was involved in formulating the Country Programme for the Montreal Protocol. It has also produced a book on environmental law for business and has started a periodical, *Green Business Opportunities*.

6. The number fell to 924 in 1989, still well above the beginning of the decade, but less than the peak (ENVIS Center 07 1993:table 2, annexures I and II). Although a short time series, and not the best of all indicators, the data support Downs's (1972) notion of an environmental issue-attention cycle. In the first two years of its existence, the Ministry received

around 45 percent of the total environmenal questions in the two houses of Parliament (Lok Sabha and Rajya Sabha); afterward, the percentage going to the Ministry exceeded 60 percent. Data from ENVIS (annual) 1980–1990.

7. During the year 1992–1993, the Delhi ENVIS center responded to 2139 written requests for information from within India, more than one-fourth of which were requests for information about other NGOs (ENVIS 1993 [np]). In the same period, the center responded to 143 written inquiries from abroad. See also MOEF, *Annual Report 1992–1993*:90.

8. See, e.g., Agarwal and Narain 1992 on the Montreal Protocol. More recently, CSE disputed the estimates of costs of transition emanating from the Ozone Cell as the basis for India's contingent compliance with Montreal.

9. The Ministry personnel increased from 520 in 1985–1986, the first year there was a separate ministry, to 1,056 in 1989. On aggregate area under protection, see MOEF, *Annual Report 1992–1993*:90. Budgetary allocations since 1980 have exploded in nominal terms. Shekar Singh (1993:15) notes, however, that following Rio the rhetorical commitment to the environment was not matched by allocations in the annual plans, which declined from 0.563 percent of the budget in 1992–1993 to 0.497 percent in 1993–1994.

10. Since 1980, the import of wild animals as pets is subject to the provisions of CITES, regulated by the Ministry of Commerce's Public Notice no. 27-ITC (PN)/80. See also Bajaj 1996.

11. Only six species of highly endangered flora are protected by the Wildlife (Protection) Act; very little is known about threats to specific species. It was a TRAFFIC-India study of trade in agarwood that facilitated CITES listing of *Aquilaria malaccensis*. Agar, worth Rs 3000 per kilo, is used for incense, perfume, medicines, writing material, occult/ceremonial purposes, timber, fumigation, and many other purposes (Chakrabarty et al. 1994; Chengappa 1993). *Taxus baccata*, from which taxol is made; *Clochicum luteum*, a medicinal plant; *Paphio paedilium* orchids; and *Gloreosa superba*, for allopathic drugs are among other illegal exports. WWF-India has produced a small handbook (WWF-I 1994) for enforcement personnel, and a more extensive looseleaf reference guide with indigenous names of controlled species in local languages as well as Hindi and English.

12. When asked in Parliament, the Finance Minister underlined the obvious: "Smuggling being a clandestine activity, it is not feasible to estimate the value of snake skin smuggled out of the country" *LSQ* 3764, 19-8-87. One does not know whether increasingly large seizures indicate increased activity or better enforcement, but seizures are increasing. In the first week of November 1994, the Indian government announced the largest seizure ever of skins and hides (in Kashmir), totaling over 1,000, including the largest tiger skin ever seized, fourteen feet from nose to tail. A listing that a TRAFFIC researcher says may represent 10 percent of the seizures in the country in 1993–1994 covers 303 instances (TRAFFIC-India 1994). The one exception to strict trade prohibitions—shed antlers and molted feathers—also permits undetectable abuse.

13. The *New York Times* of (November 20, 1994) reported new agreements of Asian nations to curb tiger trade at the meeting of CITES parties in Florida. China, though invited, chose not to attend the Global Tiger Forum in India in 1994. The most comprehensive recent assessment of success in tiger protection is the Dutta Committee Report (GOI 1996).

14. See *TRAFFIC Bulletin* 13 (May/June 1993). On the cases, *PAT* 5-24-93; Panjwani 1994:26–42. Some residual problems with stocks remain. Ron Herring was at the Ministry in February 1995 when a public-sector unit (Bharat Leather Corporation) was still petitioning the MOEF for a special dispensation to sell stocks of reptile skins, pleading financial hardship. The (Delhi) *Financial Express* (1-26-97) reported that the corporation was still seeking per-

mission for a one-time sale of snake skins and snakeskin products—they had 156,707 pieces of reptile skin alone. Though the stock represents a potential loss of Rs 10 million to a cash-starved public-sector unit, the Minister ruled that nothing can be done without going through CITES.

15. For example, "One dealer of Western Utttar Pradesh has mailed to foreign buyers a price list of live animals including leopards which can be shipped abroad by 'arrangement' with airline staff" (GOI, MOEF 1996:131).

16. The Customs, Coast Guard, Navy, state wildlife officials, police, Border Security Force, Indo-Tibetan Police, Railway Protection Force, Department of Revenue Intelligence, Commerce Intelligence, Controller of Import and Export, and others have come together with NGOs and activist individuals. The multiplicity of agencies renders coordination and training more difficult and complicates data collection: reports on offenders do not go to any central monitoring center, but rise through different administrative hierarchies.

17. Elephant tusks worth Rs 4,000–6,000 a kilo in India—200 days of employment for a landless laborer—are sold in Bangkok for $60,000 each. Leopard skins sell for $10,000 each. Tiger bones costing $90 per kilo in Delhi bring $250 per kilo in China and East Asia. Tiger skins sell for $15,000; a single bowl of tiger penis soup, for $320. Reports of payments to local people to kill tigers range from the equivalent of $5 to $300 per animal; rural wages in tiger areas are less than $1 per day. Between 1988 and 1992, tiger bones increased in price from $150 per kilo to more than $250 per kilo. Rhino horn, which in the 1970s was worth $3,000 per kilo, was up to $10,000 per kilo by 1985 and $17,500 in 1993. There is a saying in the Nepal border areas that you take a tiger to Nepal and come back on a motorbike. Price data are only representative, culled from a number of press accounts and from TRAFFIC-India and CAT News, among others. See also GOI, MOEF 1994, 1996; GOI 1996: passim. Allegations that big traders are protected are common; one official investigation report remarked, "[w]hen an important wildlife trader of Delhi was being pursued vigorously, all the forest officials of Uttar Pradesh dealing with the case were summarily transferred. This could send signals to other officials that it is not worthwhile pursuing such cases" (GOI, MOEF 1996:131).

18. For example, *LSQ* 1763, 6-3-87. The Finance Minister said in a typical response: "Import of timber has been allowed with a view to conserve the country's depleting forest resources," on open general license at a concessional duty of 10 percent. Significant imports of timber (largely from Southeast Asia) began in the early 1980s; in 1982–1983, the value was Rs 308.47 crores; in 1983–1984, Rs 162.07 crores; in 1984–1985, Rs 293.70 crores (data from Directorate General of Commercial Intelligence and Statistics). Though we do not have full confidence in the data, imports for 1989–1991 were somewhat higher, averaging Rs 411.95 crores annually, less than U.S.$137 million.

19. The government of India for years had little to do with the ITTO, partly because the nodal authority was the Ministry of Commerce, which took no interest. Negligible impact on policy in India does not mean that ITTA has no impact at the normative level globally. A former member of the Planning Commission told us a remarkable story of an official in London calling him to ask about a particular individual log shipped from India. Such concern would have been inconceivable a few decades ago.

20. One hears unconfirmed reports of smuggling, primarily of sandalwood. Debra Callister (1992) reports that "species banned from export from India in an unworked state are alleged to be available in retail trade in the United Kingdom." Some exceptions are made for reasons of foreign policy—for example, teak to the Gulf states for "luxury consumption," according to the official who had to sanction the export. The value-added concession is legitimated as a spur to handicrafts, which are produced by "weaker sectors." A timber exporter explained to

us that the loophole was used by timber contractors to reduce sandalwood logs to dust and chips, claim these as by-products of handicraft production, then export them for extraction of oil. As the regulatory regime has been tightened, timber-exporting firms have virtually disappeared; the last active exporter spends much time in Delhi lobbying for exemptions.

21. Ministry surveys establish three categories: 385,000 square kilometers of "dense forest," 239,930 of "open forest" and 59,640 of "scrub" (GOI, MOEF, *Annual Report 1992–1993*: 35). As to ITTA classification, officials are puzzled that oaks and pines growing in India are more "tropical" than those growing in England or the United States. It was believed through the 1980s that India's forests were being destroyed at the rate of 0.3 percent per annum. Though recent satellite images show a small increase in forest cover, this could reflect a change in imaging techniques. (See GOI 1991). Much of the dispute about the extent of forest cover is biologically meaningless; threats to biodiversity are masked as monoculture biomass replaces endemic forest systems.

22. Social forestry schemes in the sixth Five-Year Plan (1980–1985) covered 4.65 million hectares; the seventh Plan (1985–1990) planned for 9 million hectares, and covered 7.14 million (MOEF data). On issues of forest policy generally, see Samar Singh 1986: chapter 4, 27ff.

23. In 1994, the government proposed the Conservation of Forests and Natural Ecosystems Bill, which became known to NGOs through leaks from the Ministry. This bill, meant to replace the 1927 colonial Forest Act, is widely perceived to represent a strengthening of the centralized command-and-control logic. NGO resistance prevented the bill from going immediately to Parliament; an alternative NGO bill was circulated and discussed. For commentaries by leading activists and scholars, see Hiremath et al. 1994 (the government's draft bill is reproduced pp. 91–222); see also Guha 1994; *TOI* 1-2-95.

24. There are now 75 national parks, 421 wildlife sanctuaries, and 19 tiger reserves in India. The aggregate protected area is similar to that of the United States, just under 5 percent—one-fifth that of Costa Rica but more than many nations. Much "protection" is tenuous or bogus. Current research seeks to identify "biodiversity hot spots" for special protection, given limited resources and intense competition for space.

25. "Mafia" designation by Sanjay Deb Roy (1994:2), the lead conservation official in the area. This section is also based on Menon 1996; WWF-I 1992; GOI 1996:25–26, 52; *TEL* 18-12-89; *AST* 21-2-93; *STM* 21-2-93; *EI* 11-6-90; *IT* 31-8-92; *STM* 18-10-90; *TOI* 12-8-89; *PAT* 22-1-90; *AST* 11-1-93; *IED* 9-5-89; *Hindu* 2-23-96, 5-24-96, 12-19-96; *LSQ* 5062, 19-4-93; *LSQ* 1924, 10-3-93; GOI, MOEF 1992b:4 et passim; and interviews in the MOEF and wildlife community, as well as a study that cannot be quoted or cited.

26. *TOI* 12-1-96; *North East Times* (Guwahati) 7-16-96.

27. Accounts in *TOI* 11-30-1990; *AST* 4-1-90; *TEL* 12-12-92; *NEO* 4-6-93; *DH* 8-6-89; *NWT* 5-27-9l; *Hindu* 9-23-96, 10-17-96, 10-26-96; *North East Times* 7-16-96; and interviews. On rhino poaching specifically, see Menon 1996. The central government's position was explained in 1991 in *RSQ* 394 May 1991: a centrally sponsored scheme for rhino protection, with Rs 50 million in the seventh Five-year Plan, for more protection staff, vehicles, arms, and wireless sets in addition to the centrally sponsored Scheme for the Control of Poaching and Illegal Trade in Wildlife. These schemes allow 50 percent cost-sharing for the state government of Assam.

28. Threats to Kaziranga have elicited rapid and vigorous responses from the global wildlife community to reports that staff have run out of basics like batteries and water filters; starving elephants; malnutrition; courage of staff fighting poachers, dying for want of ammunition, and so forth (much of which is confirmed in official reports from Delhi). An appeal from the

London-based Environment Investigation Agency (The Political Wilderness: India's Tiger Crisis) stressed dedication of the forest guards despite meager rations of food and medicines, but posited a larger problem: one Mr. Currey, who collected $35,000 in four days, says wildlifers worldwide are wary of sending money to India: "Where is all the money for Kaziranga going? If the Centre is routing it to the state, what is the state doing with it?" (*TEL* 10-25-96.

29. For example, *IPS* 25-1-90. Wildlife officials who work in the area argue that press reports overstate the killings, but recognize that frequent deaths of people and livestock from wildlife present intractable problems.

30. See BNHS 1988; Madsen 1995. *PI* (11-2-95) carried a front-page story on the threatened delisting, citing WHC Secretariat concern that the endangered Siberian cranes no longer visit Bharatpur. Cranes returned in small numbers in 1996, after a two-year absence (*Hindu* 2-2-96).

31. See the August 1993 issue of *Down to Earth*; the WWF-I library in Delhi has a large file documenting deterioration of archaeological sites.

32. Former Minister of State for Environment and Forests Maneka Gandhi, who negotiated the London amendments, told us "they tried to scare us with skin cancer; Indians do not fear death. This is really all about money." See also the Minister's comments in Parliament, *LSQ* 6061, 23-4-90. In India, as in much of the Third World, the ozone-hole scare was often dismissed as a problem for people with light skin. *MH* (7-7-92) quoted A. P. Mishra, an Indian atmospheric scientist, to the effect that there will be no ozone holes over the subcontinent. Anil Agarwal and Sunita Narain, of the influential Centre for Science and Environment, published the telling article "MNC Cons the Third World with Ozone Hole Scare" (1991). An academic treatment that reflects India's dominant position and explains the international justice issues is Makhijani, Bickel, and Makhijani 1990.

33. Navin Fluorine and Sri Ram Compressors represented the industrial (ODS-producing and -consuming) sector. Pressure came from External Affairs; the Ministry of Environment did not play a very active role.

34. Personal communication. This account of India's positions is based on interviews with the principals in New Delhi in July–August 1993, press accounts (especially *TOI* 1-7-90, 8-7-90), and Benedick's first-hand account in *Ozone Diplomacy* (1991). The earliest estimate of costs is Touche Ross Management Consultants, assisted by Cremer and Werner and S. B. Billimoria and Company, "Reducing the Consumption of Ozone Depleting Substances in India—Complying with the Montreal Protocol" (New Delhi, 1991), unpublished document.

35. "Either you [sell us] the technology or you change your laws or you change your patent rights.... Start working on it." This section draws on interviews with Ms. Gandhi and Mr. Mahesh Prasad in New Delhi in 1993, supplemented by press coverage, particularly that of Usha Rai for the *Times of India*; quotation from Benedick (1991:189).

36. Countries of the South that had not signed the Protocol attempted to link availability and affordability of technologies to compliance. In their proposed draft amendment, "the obligation to comply" would have been made subject to adequate financial compensation *and* "preferential and non-commercial" transfer of technology (Benedick 1991:189). Industrial nations opposed making compliance formally contingent on the size of the compensation fund, worrying that disputed cost calculations could provide poor countries with an exit from treaty obligations. The final amendment is a compromise. Annex II, article 5, states that "developing the capacity to fulfil the obligations ... will depend upon the effective implementation" of transfer of technology and financing thereof. Benedick argues (1991:196) that this formulation merely recognizes reality, but does "not go so far as to release parties from their treaty obligations."

37. *RTF*:15. All references to the text are to the revised *Report* (hereafter *RTF*) of March 31, 1993. Projections of use by substance and sector under an unconstrained demand scenario are given in appendix II, tables 3 and 4.

38. For example, "In India, most products have a long life span (as opposed to the throw away practices commonly adopted in some developed countries), and consequently the servicing of products to meet local consumer demands has to be borne in mind so as to ensure that the consumer is not affected by the ODS phaseout strategy" (*CP*:35).

39. With a maximum of four industrial representatives, none with a direct interest, and separate representation for CII and ICMA. GOI F. No. 4/34/92 IC-I, order dated 20 September 1993 (MOEF Ozone Cell).

40. As a result, India was awarded $4 million in anticipation of projects to be approved later. *PIO* 03-10-94, 08-10-94; *HT* 08-10-94; *TOI* 09-10-94; and interviews in Delhi, February 1995.

41. Confederation of Indian Industry, WS4/SKJ/ACTION (New Delhi, n.d. [Oct. 1994]), unpublished memorandum; see also the Minister's comments, *HT* 9-10-94; also *TOI* 8-10-94. Concerns about compensation to smaller firms are expressed in *Comments of Government of India on the Document UNEP/OxL.Pro/ExCom/19/54 as Requested in ExCom Decision 19/31* (New Delhi: MOEF, Ozone Cell, n.d. [presumably 1996]).

42. This sanguine assessment applies with more certainty to CFCs and halons than to CTC (methyl bromide is not a large issue). CTC abatement is not a priority in India, in part because CFC deadlines are earlier and in part because, once listed as an ODS under the Montreal Protocol, there is little incentive for the government to regulate use through domestic legislation, since no compensation would be forthcoming.

43. Press reports (e.g., *TOI* 16-7-90) periodically mention drug production in protected areas. *Cannabis sativa* (marijuana) grows naturally in Corbett and Dudwa Tiger Reserves and is both cultivated and harvested wild for smuggling. When this issue was raised in Parliament (*LSQ* 2896, 27-8-90), the government responded with a central scheme to assist the state in eradication, including the use of tractors. Drug smugglers also poach animals.

44. Police in Karnataka and Tamil Nadu had killed twenty people in "encounters" while pursuing Veerappan as of 1993 (*IT* 31-7-93:24–26). This section is based on press reports (e.g., *TOI* 30-7-93; *HT* 5-8-93, 14-8-93) and interviews with wildlife officials. There is talk of amnesty for Veerappan, now that the famous brigand Phoolan Devi has been elected to Parliament (*Hindu* 5-25-96, 6-6-96).

45. For an example of state–society relations that do seem to work in forestry under Indian conditions, see Guha (1994). On the pervasive ill effects of coercive state intervention in conservation, Bromley and Chapagain (1984); Guha (1985, 1989); Schenk-Sandbergen (1988). On participatory conservation strategies, Agarwal (1993); Gadgil et al. (1993).

46. The World Bank's ambitious plan for ecodevelopment to protect fragile ecosystems in India, opposed widely by NGOs and the environmental intelligentsia, makes exactly these assumptions. The Bank's plan does not target any of the WHC Natural Sites already designated as of global importance.

47. For example, in 1989, Minister of Environment V. R. Ansari said: "The developing countries muster the resources to meet the minimum needs of their citizens at great sacrifice. These countries will be unable to spare further resources for the substitutes to CFCs. The poor of the developing countries will look askance at a government that spends resources on substitutes to CFCs to prevent depletion of the ozone layer ... while they continue to wallow in poverty, hunger, disease and ignorance" (Rosencranz and Milligan 1990:313).

48. For example, ITTA was given to the Ministry of Commerce; the Ministry of Environment was essentially uninvolved, despite its responsibility for forests. As a result, India did not apply for technical or financial support for sustainable forestry projects. The first MOEF representative to attend an ITTO meeting, at Kuala Lumpur in 1993, not only was uninformed about India's participation in the ITTO but found the treaty framework objectionable. He questioned India's coding as a producer nation and as an exporter of tropical timber; India is an *importer* and has banned exports. The privileging and definition of "tropical" seemed odd; much of India's timbering, like that of the United States, is of temperate hardwoods. And the weighted system of voting rights seemed unfair to him.

49. For example, Bidwai 1991. Though Minister Routray clearly explained compensation provisions in Parliament (*LSQ* 1098, 16-8-90), statements by Maneka Gandhi and the contemporary press treated the compensation fund as a fixed transfer payment. When asked by the Prime Minister in a cabinet meeting if Montreal was good for India, Ms. Gandhi said, "Yes, we got $40 million" (interview). An article in *Business Standard* (2-6-95) repeated the common inaccurate assertion that "India has been allocated Rs 6,500 crores" [$2.2 billion] by the Fund."

50. The American ambassador sent a nasty note to the government of India about the plant. India responded that neither Iraq nor India had signed the Montreal Protocol; the deal was perfectly legal. The American position later appeared hypocritical when documents from Iraq (obtained by us from the former Secretary of MOEF) surfaced. It turns out that an American firm had bid on the same contract and lost out to India on cost grounds, reducing the American position from principle to sour grapes.

51. Interviews, New Delhi, January–February 1995; see also Dasgupta 1994. The Allwyn–Hitachi joint venture to build non-CFC compressors in response to the entry of Whirlpool (USA) and Samsung (Korea) into Indian production is a recent case in point (*Hindu* 10-23-95, 7-26-96).

52. English-speaking infiltrators for TRAFFIC-India were stopped just short of the inner circle while investigating smuggling in Nepal; their accents were not Slavic (or Italian, Italy being the entry point for Central European illegal trade). See Van Gruisen and Sinclair 1992.

53. Kothari (1996) argues the point more generally: structural adjustment threatens more rapid environmental degradation. Shekar Singh (1993:14) argues that liberalization of export policy, "reportedly necessitated by the World Bank," has greatly increased the danger to plants, sandalwood forests, and coastal areas, among other resources. The question is too complex for treatment here; it is clear that liberalization in pure form poses increased environmental risks. The balance in India is difficult to discern at this stage. For an instructive debate among Indian economists, see *Down to Earth* 1992.

54. When the WHC Secretariat wrote to the MOEF, asking whether India was aware of civil conflict in Manas, the nodal authority was stunned; forest officials were being killed, and the government was engaged in negotiations to cede authority to the insurgents. Neither the MOEF nor the Secretariat had the capacity to quell insurgency. Threats to delist Bharatpur angered wildlife professionals: there are many possible reasons for the absence of Siberian cranes, most of which are beyond India's borders and control.

55. Interview in New Delhi, August 3, 1993.

13

Cameroon's Environmental Accords: Signed, Sealed, but Undelivered

Piers Blaikie and John Mope Simo

This chapter is an account of the extent of implementation of and compliance with environmental accords by a relatively small African country. Nonetheless, Cameroon is an important party when viewed within the context of the African continent as a whole. The country still has significant tropical timber reserves, areas of natural beauty and cultural significance of global quality, outstanding biodiversity, manufacturing plants using and exporting substances harmful to the ozone layer, and, finally, an offshore oil industry with high inherent risks of marine pollution.

Cameroon has signed all the accords except the Dumping Convention. It has had considerable policy experience in many of the areas of environmental planning and law, and in practical, field-level implementation. It is also far from being one of Africa's poorest countries. Yet in spite of these factors, this is a rather bleak account of nonimplementation and noncompliance. The explanation of this state of affairs lies partly in the profound economic decline of the country, especially since the mid-1980s; partly in the persistence of exploitative neocolonial relationships involving the use of natural resources (in this case, tropical timber and oil); and partly in a crumbling government apparatus beset by public expenditure cuts and a decline in standards of public service. The emphasis therefore moves from the details of the accords themselves, and their particular legal and political aspects, to a more general malaise that pervades *all* such accords and the capability of the state in general. While the specifics of the accords are important in the narratives of their implementation and compliance by the state and civil society, this case study also highlights the issues of underdevelopment, foreign debt, neocolonialism, and the growing incapacities of the state as they affect all treaties in Cameroon—all of which have wider relevance to other countries of the African continent.

Characteristics of the Country

The Republic of Cameroon lies between 3° and 13° north latitude and 8° and 16° west longitude. It is bordered by Nigeria to the west and northwest, Chad to the northeast, the Central African Republic to the east, the People's Republic of Congo to the southeast, and Gabon and Equatorial Guinea to the south. The country stretches about 1,250 kilometers along the northeast-to-southwest axis, and about 800 kilometers from east to west. The borders are characterized by virtually no roads and very poor communications in general, so they are easy to cross without encountering official controls. This is of particular importance in the illegal movement of low-bulk and high-value commodities, such as ivory, animal trophies, and smaller live animals. Its area is 475,442 square kilometers, and its population is estimated at about 12.5 million (1994), of which 55 percent are under twenty years, and 6 percent over sixty. About 64 percent live in rural areas, and 36 percent in urban areas. Like other African countries, Cameroon is rapidly urbanizing, from 16 percent of the population living in urban areas in 1965 to 46 percent in 1987. (For a general description of Cameroon, see Ngwa 1988.)

The Economy

Officially Cameroon is a "middle-income country" (World Bank 1989), although this classification is no longer relevant, in view of the precipitous decline in the country's economic fortunes since 1987. During the period 1977–1987, real GDP per capita rose from $410 to $920 (World Bank 1992). This was mainly due to oil exports, but also to timber and other agricultural products. Since 1987, the economy has deteriorated sharply. GDP has declined by 18 percent, and GDP per capita fell by 50 percent between 1985 and 1991. The implications of the economic crisis are examined in more detail below.

The economy is dominated by the production of primary products and has little capacity for processing or adding value. Small farms (less than two hectares) occupy 90 percent of the cultivated area, and supply 90 percent of agricultural production and 80 percent of marketed products (but only 3 percent of GDP). The dominant farming system is of the slash-and-burn type (which creates strong pressures to clear forest lands that are directly related to population growth).

Cameroon is a tropical country endowed with outstanding natural resources of global significance. It has within its national territory 188,000 square kilometers of dense, humid evergreen forest, and therefore possesses highly valuable timber resources, as well as great biodiversity (a characteristic of this type of forest). There is also a wide range of other ecological zones, some of which are of limited extent and rare within Africa as a whole. The management of timber, biodiversity, and areas of outstanding natural beauty and scientific interest is interlinked, as is the implementation of the international treaties that deal with them. Indeed, in many senses, these resources are different ways of defining the same natural objects. Thus the conservation of a rare animal species may frustrate the exploitation of another resource (e.g., timber for export), and managing of both involves trade-offs and complementarities. For the purposes of examining each of the accords and resources in Cameroon, the resources are discussed first.

Timber

Cameroon is the only West African country that still has large reserves of tropical and subtropical forest. They cover over 70 percent of the national territory, and rank third in area in Africa after Zaire and Gabon. The moist, dense evergreen and semideciduous forest is the best known because of its commercial value, and is relevant to the ITTA, although it comprises only part of the forests of Cameroon.

The commercially important humid zones lie in the south of the country, particularly in the southeast; the west–central part has already been heavily logged for commercially exploitable species. However, the savanna forests of the north–central part of the country provide fuelwood for the local and urban populations, as well as habitat for wildlife, and promote soil and water conservation. The dense forest in the moister south provides soil nutrients and land for cultivation by slash-and-burn techniques for most of the rural population.

Therefore, although ITTA concerns itself only with the commercial value of the moist tropical evergreen and semideciduous forests in the south and east of the country, the other functions of forests in Cameroon are much wider and affect the concerns of CITES, as well as the livelihoods of the majority of rural people. In the medium term, therefore, future population growth (from over an estimated 12.5 million in 1994 to possibly 17 million by the year 2000) will substantially increase the demand for both land to cultivate and fuelwood, thereby placing additional pressure on the forest. In the main commercial areas in the semideciduous and evergreen areas, population densities are lower than in the north and west, so commercial sustainability will become the dominant issue there. In the north, however, sustainability is threatened by a fuelwood problem; use is currently well above the rate of maximum sustainable cut in most areas (Ndjatsana 1993). Overall, it is estimated that by 2010 fuelwood will become the major forest product of Cameroon (UNDP/FAO 1988:3). Fuelwood consumption, at present 7 million cubic meters per year, is predicted to reach 12.4 million cubic meters by the year 2000 (Zama 1995:264).

It is estimated that deforestation is running at between 80,000 and 150,000 hectares per year (UNEP 1992c:18), mainly as a result of shifting cultivation (often following partial clearance by logging), although elsewhere (UNEP 1992:6) it is stated that 200,000 hectares of forest is being cleared per year by agriculturalists alone. It is also estimated that 1 million hectares of forest were cleared between 1976 and 1986 (UNDP/FAO 1988). Commercial exploitation has increased since about 1985, but harvesting methods are frequently wasteful and primitive, rates being often less than 5 cubic meters per hectare. Many species are felled and left on the ground, since research and marketing have not been carried out to exploit more than a small number of species. In addition, other vegetation and the topsoil are frequently damaged, so that natural regeneration of the forest is slow. Only 15 percent of logging is carried out by national companies. Levels of processing in the country remain low.

Apart from two modern plywood factories and sawmills, SOFIBEL and ALPICAM, Cameroonian wood processing suffers from antiquated and inappropriate equipment, shortage of skills, and a lack of product standardization. Also, the domestic demand for fuelwood dwarfs that for export. In this context, rapid population growth in the country will increase the demand for both fuelwood and land for cultivation, making the prospect of successful conservation and planned exploitation of forests in the future even more remote. In turn, this will have profound implications for implementation of and compliance with the ITTA.

The timber sector employs 20,000 people officially, and many more unofficially who are involved in illegal, small-scale felling. Commercial harvesting covers half of the exploitable closed forest, and exploitation rates are typically very low (often three–five cubic meters per hectare and involving the cutting of 10–20 percent of the trees). Methods of extraction are usually wasteful and highly damaging to remaining forest stock. Thus the maintenance of a sustainable forest estate, as envisaged in the ITTA, faces formidable problems.

Biodiversity

"Biodiversity" is a term that has been interpreted in a number of ways. Defined simply, it refers to the number of different kinds of life-forms. Three aspects are important: the number of different life-forms (species abundance); the number of forms that occur nowhere else (endemism); and the number of different types of habitat (ecosystem diversity). The humid tropics in general is the principal global repository of biodiversity, and it has been estimated that over half the number of species on earth exist in the rain forest (Wilson 1988). Cameroon is one of the most important countries in the world in terms of all three components of biodiversity (Alpert 1993). The country has been identified by the U.S. National Academy of Sciences as a conservation priority area demanding special attention on the basis of high biodiversity, high endemism, and rapid conversion of forests for other uses (World Bank 1992:168). To date, 297 species of mammals, 848 species of birds, 300 species of anurans, and 9,000 species of plants have been identified; there are at least 156 endemic species, 45 on Mount Cameroon alone. Therefore the CITES accord is of utmost relevance to conserving biodiversity in Cameroon.

World Heritage Sites

World Heritage Sites are defined by criteria in paragraphs 23 and 35 of the Convention, the first concerning constructed forms, paintings, or association with events and beliefs, and the second with physical and biological formations of universal value. Cameroon applied for four sites to be recognized under the second group of criteria (Waza, Bénoué, Mount Cameroon and Dja; only the last has been recognized).

Production and Use of CFCs, Halons, and Other Ozone-Damaging Substances
Cameroon both imports and manufactures products harmful to the ozone layer (Nemba 1993). Importation of CFCs 11, 12, and 22, principally in the compressors of refrigerators, freezers, and air conditioners, and for the manufacture of rigid (polystyrene) foams, constitutes the major share of products in annex A of the Convention. In 1991 Cameroon imported about 300 tonnes ozone-harming substances through two major import agents (SATICAM and Hoechst-Cameroon). PLANTICAM, SONOPOL, and SAPCAM also import solvents. This amounted to 121 Ozone-Depleting Potential (ODP) units. It has been estimated that the average life of refrigeration equipment in Cameroon is only five years, and therefore this was the length of time set for the elimination of these substances in equipment already imported into the country (estimated in 1995 300,000 refrigerators and 280,000 freezers). From 1987 to 1995 a total of 542 ODP units was imported. However, Cameroon manufactures refrigerators, cigarettes, and foams, all of which involve the use of proscribed substances. The firms involved here are FAEM, SICAMEC and SITABAC, all located in the Douala-Edéa area, and British American Tobacco, near Yaoundé. FAEM (Fabrication d'Appareil Electro-Ménager), much the largest company, produces refrigerators and freezers for the West–Central African market, with an estimated annual demand of 100,000 units.

Political Institutions

Cameroon's constitution was drawn up in 1961 (the year in which the English-speaking state joined the new federated State of Cameroon), and amended in 1984 and 1991. The Head of State, who is also Commander in Chief of the armed forces and President of the Republic, has a five-year term of office, and is elected through direct and secret universal suffrage. The President appoints the Prime Minister, who is Head of Government and in turn appoints ministers and secretaries of state on the nomination of the President. Thus there is a high degree of concentration of power in the hands of the President, who is a typical "strongman" in the contemporary African state, and is "the centrifugal force around which all else revolves" (Sandbrook 1986:323). The constitution stipulates the powers of the National Assembly, which is composed of 180 members. The main legislative and executive body in the socioeconomic domain is the Economic and Social Council, although the discretionary powers of the President often allow him to overrule it.

On September 1, 1966, multiparty democracy (which ceased to exist in French-speaking East Cameroon in 1962) was officially suspended in English-speaking West Cameroon. The two leading political parties—Ahmadou Ahidjo's Union Camerounaise in East Cameroon and John Ngu Foncha's Kamerun National Democratic Party in West Cameroon—decided to merge. The outcome was a single-party structure throughout the former federated state, the Union Nationale Camerounaise. The experiment with federalism ended in 1966 when Cameroonians voted in a referendum for what some observers described as an imposed unitary state. This change resulted in the formation of the first government of the United Republic of Cameroon with Ahmadou Ahidjo as President (Ngwa 1988:2–6). This political framework was abandoned by Paul Biya in 1983 with a return to the nomenclature Republic. At the same time, the social and economic system called "planned liberalism," introduced by Ahidjo, was scrapped in favor of what the present President, Paul Biya, described as "communal liberalism." The difference is symbolic only, and does not alter the fundamental principles and actions of politicians and policy makers.

A multiparty democracy was officially restored in 1990, and presidential elections by universal suffrage were held in October 1992 that were, according to official sources, narrowly won by the incumbent President, Paul Biya, who ran against John Fru of the Social Democratic Front (SDF) and a few other candidates. The SDF claimed fraud, and these claims were supported by independent international observers. This was mainly due to the ruling party's vote-buying and the manipulation of the results before they were released. In spite of both international pressure and considerable unrest at home, the President has refused to allow another election. Although he has accepted popular demands for amendments to the constitution, the date for the beginning of such talks has not been announced, to the disappointment of the vast majority of Cameroonians.

While all the "democratization processes" can be described as cosmetic at best, human rights issues and freedom of the press have not fared any better, and the political and economic crisis has continued. In short, the move toward multiparty democracy has stalled. This has had a detrimental impact upon transparency and accountability in government, and hence upon the effectiveness of the accords in question.

This brief account of recent political history of Cameroon resembles a stereotype of economic and political decline elsewhere in Africa. The more general implications of this decline for the prospects of compliance with international agreements are discussed later; here we confine ourselves to an explanation of this decline in Cameroon. After independence, the anti-colonial front broke up along African linguistic and anglophone–francophone cleavages, creating a crisis of legitimacy. The political requirements of regime and personal survival encouraged both authoritarian rule and the growth of clientelism and prebendalism (the distribution of state offices to consolidate political authority).

Because of the lack of a large and organized business class in Cameroon, the state has had to promote economic development. However, the administration has increasingly failed to act in a disinterested and rational manner to that end, riven as it has become by corruption and political favoritism. Cameroon has not experienced efficient economic management, particularly since about 1980, and its government suffers from a profound lack of legitimacy expressed by protest rallies, strikes, and military crackdown against popular protest. This decline of the country is undoubtedly a major explanatory factor in the failure to implement and comply with the environmental accords to any substantial degree, and is elaborated upon later in this chapter.

Policy History

Cameroon has signed and ratified five of the six treaties under study. The (London) Dumping Convention has neither been signed nor been ratified. The dates of signature and ratification of the other treaties are shown in table 13.1.

The policy history of ITTA, CITES, WHC, and the Montreal Protocol in Cameroon is scanty, since the country was only peripherally involved in the international negotiations. They were signed and then ratified, often after a considerable period, but very little has been done by Cameroon specifically to legislate for or to implement the treaty obligations. This rather meager history is discussed first. However, the history of policies relating to forestry, national parks, wildlife, and industrial pollution in more general terms is both much longer and more eventful. It is instructive to examine this more general policy history, and to discuss why on the one hand there was, at least on paper, a fairly comprehensive set of regulations and enforcement structures in place before the signing of the accords—and, on the other

Table 13.1
Treaty membership of Cameroon

Treaty	Year treaty came into effect	Signed by Cameroon	Ratified by Cameroon
WHC	1977	1982	1982
CITES	1975	1981	1981
ITTA	1985	1985	1985
LC	1975	not a member	not a member
MP	1989	1989	1989

hand, very little evidence of specific activities and changes in behavior as a direct result of having signed the treaties.

Policy History with Reference to the Accords

Cameroon sent a delegation to the conferences that negotiated and drafted the Tropical Timber Convention, but failed to do so for CITES, the Vienna Convention, and the Montreal Protocol. In the latter cases, the Cameroon ambassador to the country in which the conferences took place attended the meetings and signed on Cameroon's behalf. After signing the CITES accord, the ambassador's report was lost, or at least not acknowledged for a number of years (accounts differ). It was only at the prompting of the Wildlife Officer that it was unearthed. In the meantime, the CITES Secretariat investigated complaints regarding forged signatures of the Minister and Director of Wildlife in 1982 and the export of gorillas (described elsewhere in this chapter). Cameroon finally signed the accord under personal pressures put on Cameroon delegates who attended subsequent meetings.

Regarding the World Heritage Convention, the promulgation of three sites was undertaken, but there has been no subsequent provision for increased protection in terms of personnel, transport, research facilities, and so on. Their designation therefore remains a symbolic and purely cartographic exercise at the present time.

The Montreal Protocol was signed by Cameroon on August 30, 1989. The first meeting was on May 1, 1989, in Helsinki, and Cameroon did not send a delegation. The second meeting was in London in June 1989, but Cameroon canceled the planned visit by its delegation on short notice because there were no travel funds, it being close to the end of the financial year. Instead, the ambassador in London attended the meeting, signed, and reported back. The government did not have the funds to commission its own study to provide the necessary information to the Secretariat, and the World Bank granted $30,000 for this purpose. Apparently the Secretariat had to press for the report a number of times. A draft report was submitted in March 1993, and final editing and certain recalculations were undertaken by the authors during that summer. The report was then accepted. A National Coordinator for the Global Environmental Facility (GEF) was appointed in late 1994. There has also been a U.N. Environment Programme (UNEP)-sponsored workshop on the implementation of the Montreal Protocol.

Policy History in Related Areas

Since the late 1960s, Cameroon has participated in multiple conventions, and has signed a considerable number of global and pan-African treaties (Kamto 1992). It has signed twelve pan-African conventions that deal with the regional environment, and twenty-six global conventions, including two on climate change and biodiversity. However, Kamto (1992) notes that two treaties on marine pollution (the Prevention of Pollution of the Sea by Oil and the wider-scope Convention on Wetlands of International Importance Especially as Waterfowl Habitat) were not signed by Cameroon, although it would have been in its interest to do so, especially in light of the development of the oil industry on and near the Cameroon coast. Furthermore, UNEP (1992c:40) advises that it is necessary to establish a systematic and exhaustive list of all regional and international conventions signed by Cameroon but not yet ratified. Verbal inquiry about this list to senior staff in the Ministry of Environment and Forests (MINEF) was almost universally met with uncertainty about exactly which conventions or treaties had been signed and ratified.

However, the reasons for this collective uncertainty clearly do not lie in the country's inexperience in these policy areas. Therefore we have to look elsewhere for reasons for noncompliance with these international treaties. In a number of areas of environmental policy, particularly forests and conservation of flora and fauna, Cameroon has had a long history, both as a colony and as an independent nation. Therefore the signing of international environmental accords did not demand establishment of completely novel legal and administrative structures in a policy void. A brief review of this experience is given below.

Timber and Forest Management. Timber exploitation had been going on for many years, and especially the last thirty, prior to ITTA ratification in 1985. Currently, Cameroon is the sixth largest world exporter of tropical timber, and the third largest in Africa.

Apart from official production there is much illegal production, estimated to be on the order of half of official production. A little over 150 timber exploitation licenses have been issued, of which only 23 were to Cameroon nationals (who were responsible for 10 percent of total production). On the other hand, fifteen of the seventeen international companies produce more than 40,000 cubic meters per year each, and dominate production, though trends since the mid-1990s have shown a marked increase in the share of total annual area licensed to Cameroon nationals (Ndjatsana 1993:36).

The history of forest policy in Cameroon is marked by the absence of effective legislation and regulation to ensure a sustainable forest estate in the way envisaged in general terms by the ITTA. Licenses for felling were granted without any consideration for land-use planning until their suspension in February 1992. It has been the intention of the government since then to replace licensing with concessions given to the highest bidder and subject to land-use planning.

The raising of revenue from forestry has been uneven. There are currently twelve different taxes, and official statistics from MINEF show that since about 1990, their collection has been subject to large variations from year to year. Officials privately admit that if tax collection were better administered, revenues would be about ten times the sums officially recorded. Persistent rumors circulate that certain of the largest logging companies have long evaded taxes on a regular basis.

At the ministerial level, a number of ministries have been involved with the forestry sector: the Ministry of Agriculture, which until April 1992 supervised the Department of Forestry; the National Centre for Forest Development (CENADEFOR); the National Office of Forest Regeneration (ONAREF); the Department of Agricultural Education (Ecole Nationale des Eaux et Forets at M'Balmayo); the Ministry for Mines and Energy, for wood used as a source of energy; the Secretary of State for Tourism, in the area of wildlife management; the Ministry of Higher Education and Scientific and Technical Research, for academic research in forestry matters; and the Ministry for Economic Planning and Regional Development, for national projects and collaboration with international aid organizations.

There has been a long record of overmanning and dissipation of related duties over widely dispersed and noncommunicating groups of officials. A list of such problems is given in the UNDP/FAO *Tropical Forestry Action Plan* (1988). This report's verdict on the sixth [National] Plan is that "reading the Plan one sees it has not accorded a very high priority to the forestry sector ... [and] the part of the plan concerned with the forestry sector is not integrated into an overall policy for environmental protection and the development of rural areas nor within a coordinated framework along with plans for agricultural and livestock development" (UNDP/FAO 1988:6). The Plan was suspended following the introduction of structural adjustment policies in 1988. Thus the ongoing context of policy and planning into which ITTA was inserted was one of a lack of efficient planning, disorganization among a multitude of different organizations, and a fundamental lack of monitoring and control in the field.

The new forestry, wildlife, and fisheries law was signed by the Minister of Environment and Forests in the second half of 1994. But so far action has been held up for two reasons. First, there are ambiguities in the relationship between the land law and the traditional land-tenure systems, the latter being understood and accepted by the people. Second, the new law does not go very far in ensuring sustainable use and management of the resources expected by the World Bank/IMF as one of the conditions for providing more money to handle the third phase of the country's structural adjustment program.

The Protection of Wildlife and Special Sites (CITES and WHC). The notion of World Heritage Sites as specified in the WHC has existed only since 1976. However, the protection of a World Heritage Site involves many measures similar to those for national parks and reserves, of which a considerable number have been established in Cameroon for over half a century. The government has taken an active role in the establishment and management of national parks that, in most cases, it inherited

Table 13.2
National parks and reserves in Cameroon

Name	Date reserved	Size (hectares)	Habitat (region)
Korup National Park (1982)*	1962	125,900	forest
Pangar-Djerem National Park (1982)*	1968	300,000	forest-savanna
Dja National Park (1982)*	1950	526,000	forest
Kalamaloue National Park (1972)*	1947	4,500	savanna
Waza National Park (1980)*	1934	165,000	sahel savanna
Faro National Park (1980)*	1930	330,000	savanna
Benoue National Park (1968)*	1932	180,000	savanna
Bouba-Njida(h) National Park(1968)*	1947	220,000	savanna
Mozogo-Gokoro National Park (1968)*	1932	1,400	savanna
Kimbi River Wildlife Reserve	1964	5,625	montane
Mbi Crater Wildlife Reserve	1964	370	montane
Douala Edéa Forest/Wildlife Reserve	1932	160,000	forest
Campo Forest & Wildlife Reserve	1932	300,000	forest
Kalgou Wildlife Reserve	1932	3,000	savanna
Lake Ossa Natural Reserve	1968	4,000	forest
Lobeke Lake Forest Reserve	1974	43,000	forest

* Year of national park designation.

in the form of reserves from its colonial rulers, the French and British. It has since added considerably to this inheritance. The parks within the savanna biome, with their outstanding population of ungulates and their predators, can support a modest tourist industry and thus can play a part in economic development, while the recently established parks in the tropical forest, such as Korup, Dja, and Pangar-Djerem, do not have this potential but represent a major scientific contribution of world importance.

Cameroon has nine national parks and seven reserves (see table 13.2).

Most of the national parks are located in the savanna and sahel regions. Virtually all reserves except the Kimbi River and Mbi Crater were created by order of the Haut Commissaire de la République Française between 1932 and 1950. The two others were created in 1963, after independence. It was also after independence that the five northernmost and largest reserves were transformed into national parks by order of the Secretary of State for Rural Development and provided with the basic tourist facilities.

Among the forest reserves established by the French in 1950 are Waza, Faro, Benoue, Bouba-Njida, Kalamaloue, Dja, Campo, and Douala-Edéa. At independence in 1961, only East Cameroon had any form of service to manage wildlife. In

1968, the Ministry of Agriculture in East Cameroon transformed a number of the game reserves into national parks; Waza, Benoue, Mozogo-Gokoro, Bouba-Njida and Kalamaloue were established in 1972. These were administered as national parks by Pierre Flizot, a French *coopérant* (technical advisor). The national parks eventually gained eminence in West Africa, although management left much to be desired. However, according to a prominent wildlife officer, credit must be given to expatriates who, in spite of their limited professional knowledge of wildlife management, were able to establish large conservation areas, a feat considered impossible today. It took two decades for Korup and Faro national parks to be created.

By 1971, Cameroon, responding to international pressure to elevate the Douala-Edéa Reserve in the central coastal area and Korup Reserve on the border with Nigeria to national park status, designated these two areas as wildlife parks for scientific purposes. Wildlife exploitation was prohibited, timber exploiters were evicted, and local residents economically dependent upon exploiters were "reeducated." By 1974, the Douala-Edéa Reserve had an appointed conservator and a guard post. Full park status has, however, been frustrated by discovery of oil in Cameroon's coastal areas and by the possibility that the Douala-Edéa area may hold important oil reserves.

The Korup Reserve was awarded full national park status in 1982, as were two new tropical parks: Dja National Park, which sits astride the transition zone from the coastal forest to the Congo forest, and Pangar-Djerem National Park, which extends from the Guinea savanna into tropical high forests, thus transecting the floristically interesting and scientifically important savanna–forest transition. At present, however, this park is isolated, undeveloped, and full of poachers.

Implementation and Compliance with the Treaties

Actions Taken over Time

One of the most visible actions taken by the Cameroon government was the creation of the Ministry of Environment and Forests (MINEF) in April 1992, following a Presidential decree (no. 92/069). The implementation of the sustainability clauses in ITTA, CITES, and World Heritage requires that a more unified policy and implementation structure be created. Coordinated policy-making is vital, bearing in mind the linked nature of the management of forests, wildlife, biodiversity, and World Heritage Sites. Staff members were appointed and took up their positions in April 1993. However, UNEP (1992e, 11) states that the present structure still "lacks a clear focus, and ... issues such as bio-diversity, pollution control and monitoring, and desertification seem to have been overlooked in the function descriptions [of the various suborganizations within it]."

The impetus for this reorganization, however, did not come from the Cameroon government in response to the need to comply with and implement the accords. Undoubtedly the IMF structural adjustment program was instrumental in this reorganization, and UNEP played the leading organizational role in creating MINEF. However, the new forest and wildlife law, although highly relevant to providing the prerequisite framework for implementation of CITES, ITTA, and WHC, was not initiated as a result of the signing of these international accords. While the law may be a necessary though far from sufficient measure, its impetus derives from structural adjustment thinking, linked to the World Bank's newfound environmentally friendly criteria for conditionality of loans, and to pressures from the leading bilateral aid agencies (particularly the Gesellschaft für Technische Zusammerastreit and the Canadian International Development Agency), and from individuals within UNEP.

A second broadly based initiative is the official adoption of UNEP's National Environmental Action Plan (NEAP). This has many of the same objectives as the accords, although they are stated in less specific terms. It is an ambitious document with an annex of eighteen pages of recommendations, but many of the pervasive problems of economic decline—such as low and falling standards of public service, and impoverishment of large sections of the population—are not addressed. Nonetheless, the document has been written and officially accepted; some senior officials have read it; and there is, in theory, a unified ministry that can start to plan and implement its recommendations. Many of them, if implemented, would fulfill many clauses in the four treaties under study. The significance of the document in terms of changing the practice of environmental management on the ground is discussed below.

ITTA. In circumstances similar to the setting up of MINEF, the Office Nationale de Développement des Forêts (ONADEF) was established as a result of the IMF structural adjustment program in February 1990, and therefore was unrelated to the signing of the accords, although it should certainly facilitate the reforms implied in ITTA. This new organization combined the largely defunct CENADEFOR with ONADEF. It has started to implement pilot projects, but no other work has been done.

It is the stated intention of ONADEF to give logging concessions on a more rational basis, with some overall land-use planning criteria in mind. However, it is widely reported that concessions are still being given in contravention of the government's land-use criteria (e.g., they are given inside national parks and forest reserves, such as the Campo Reserve). It has also been the intention to make larger concessions to fewer and better-capitalized entrepreneurs. The number of smaller concessions under 25,000 hectares, reserved for Cameroon nationals rose from 60 to 230 between 1993 and 1995; it is these that do the most damage. The concession holders are local people who are undercapitalized, and often the logs are sold to neighboring larger (and foreign-owned) companies.

There are rumors (repeated by USAID officials on national television during April 1993) that an immense concession (800,000 hectares) has been granted to a French company (Société Forestière Industrielle de Dimako) in the southeast. This and other French-controlled companies have been accused of paying no attention whatsoever to good forestry practice and maintaining a sustainable forest estate. This implies that certain guidelines, such as maintaining the uneven age characteristics of the forest, minimizing the damage to remaining stands, protecting key species, employing dry-skid techniques, and replanting quality species should be followed. The researchers were unable to check that these guidelines were being followed directly, but informed opinion of senior forestry officials and expatriates in forestry projects was that virtually none of them were. Furthermore, MINEF and ONADEF simply do not have the personnel or the transport to monitor the activities of concession holders. They have to catch rides on the transport of the larger companies and have to rely on their data. Given the political connections of the biggest companies, it is highly unlikely that infringements, even if they were discovered, would prompt any action by the government.

ONADEF is initiating research on lesser-known species found in Cameroon forests. Fifteen out of forty marketed species constitute 90 percent of current production, but there are about 350 species, and very little is known about their utility. This is one of the main marketing objectives of ITTA.

Under the proposed new forest, wildlife, and fisheries legislation (promoted by structural adjustment), it is still difficult to give local people a share of the proceeds of wildlife exploitation. The legislation is now in its fifth draft. This version of the Forestry Bill (initially supported by World Bank and CIDA), which covers fiscal policy, forest taxes, and land-use planning, was enacted in January 1994, but the institutions that put pressure on the Cameroon government to initiate it have felt that the law does not make enough concessions. There is now a standoff between the aid community and the World Bank, on the one hand, and the government, on the other. Again, these initiatives may serve the policy objectives of ITTA incidentally, but have been made only by international and bilateral aid organizations, not by the Cameroon government in direct response to ITTA.

There have been a number of other projects funded by the ITTA or managed through it. These include a pre-project concerning silvicultural treatment through selective felling; a project for forest resource accounting; the preparation of a master land-use plan for forest areas; three small training projects; the development of logging infrastructure; and the So'o Lala and Sud-Bakundu projects, the latter of which was visited by the authors of this chapter. Both are pilot projects limited in terms of finance and area. An examination of progress so far through visual inspection of the site and progress reports of the latter project attest to the preliminary and small-scale nature of the Sud-Bakundu Project.

The funding for all these projects from 1990 to 1994 was on the order of $11 million. None of them is *directly* concerned with the sustainability clauses of the ITTA, although their objectives clearly address them in a general manner. While these projects are not evidence of compliance with the ITTA, they *are* evidence of intentions in the same general direction. Nonetheless, there remains the question of the present and future effectiveness of such projects, in terms of reaching their stated objectives as well as in terms of their impact relative to the present massive depredations on Cameroon's forest estate.

CITES. During the period between signing and ratification of CITES, the management of wildlife was divided among at least four ministries. Principally it was the Subdepartment of Wildlife in the Ministry of Planning and Regional Development that had a coordinating role. The Ministry of Tourism implemented the existing laws and ordinances, the Ministry of External Affairs was supposed to allocate funds for the international meetings associated with CITES, and the Ministry of Scientific and Technical Research was to provide technical information outside the expertise of the Ministry of Tourism. Overall administration of wildlife in general and the overview of implementation of the CITES convention in particular, therefore, received very little concerted attention. Except among a handful of senior Wildlife Officers, knowledge of CITES in the four ministries concerned was virtually nil, as shown by the fact that the absence of a report about the outcome of the Washington Convention was scarcely noticed for years.

This delay between signing and ratification was noted for all environmental accords by UNEP (1992c:40). In 1992, the administration of CITES under the Directorate of Wildlife was moved from the Ministry of Tourism to MINEF. No one was given specific overall responsibility of implementation of CITES, and it was only at the prompting and strenuous efforts of one or two senior civil servants that any official action was taken at all. UNEP (1992:40) noted:

> Cameroon's adhesion [*sic*] to these various [international environmental] conventions has not necessarily translated into domestic implementation measures. Therefore it is not surprising that these conventions, once they are signed go unheeded. This is due to the small number of national intermediaries who are responsible for transforming the international obligations into domestic laws, which eventually need to be translated into concrete measures.

By far the main issue for CITES in Cameroon is ivory smuggling and trophy hunting, although the export of birds is also lucrative. For example, one gray African parrot fetches $2500 in the United States, and there have been well-attested instances of airline officials being *forced* to load crates of parrots onto aircraft for illegal export. The most notorious ivory smugglers are currently Korean (and Chinese) road workers, who buy tusks from local poachers, saw them up, place them in containers, and smuggle them out by sea. Smuggling across the open border with Nigeria also occurs. Ironically, with the ivory ban, the price of ivory has declined, and Nigerian operators can now afford to purchase Cameroonian ivory. Recently 1.6 tonnes of ivory were seized in Hong Kong and traced to sources in Cameroon. The authorities were alerted on three occasions, citations were issued, were not answered by the accused, and were not followed up. Other cases of smuggling in contravention of CITES are discussed below.

Reform of wildlife legislation has taken a long time to effect. It is clear that laws and regulations concerning the environment were sectoral in nature, and governed exploitation of resources rather than their conservation (Gibson and Laurent 1992). Ordinance no. 73/18, the first modern forestry law of Cameroon, was replaced in 1981 by a more comprehensive law (No. 81/13, Pourtant Régime des Forêts, de la Faune et de la Pêche). There have also been changes specifically in the Wildlife Law (1971, 1981, and 1993), but none of these were in response to the signing of CITES. Rather, they were prompted by loopholes and logical inconsistencies in the law (for example, while the ostrich was a protected species, its eggs could legally be sold under the Wildlife Law of 1981).

Again, one or two senior members of the (then) Wildlife Subdepartment took the initiative to update the wildlife part of the Forestry Law, and to include additional protective measures (e.g., the legal creation of a buffer zone around protected areas and the creation of integral nature reserves and national parks). This was the only example of changes in regulations undertaken as a result of the signing of CITES. However, legal means by which local communities are entitled to wildlife, and thereby provide incentives for its protection by them, are still lacking.

Other efforts to promote wildlife conservation and enforce CITES by institutions outside those of the state have been in the form of bilateral and multilateral projects, as in the case of sustainable forestry. The objectives of most of these projects tend to concern the institutional and scientific preconditions of wildlife protection (inventories, mapping, research, training, creation of awareness, socioeconomic studies, further project preparation). There are currently seventeen elephant projects, of which only four are partially or wholly funded, including the Nyassoso Project, the Mount Kilum Forest Project, and Korup National Park.

Turning now to the question of whether the objectives of CITES are being achieved in Cameroon, the evidence to the contrary makes depressing reading. It is reasonable to assume that neither the signing nor the eventual ratification of CITES caused any diminution in the rate of poaching, nor in an increase in the resources made available to the Department of Wildlife. Alpert (1993:45) states, for example, that five of the nine faunal reserves are massively degraded, with two "totally degraded." The budget of the department is only 1 million CFA, and its director does not have a vehicle—the U.N. Development Program (UNDP) once gave him an old one, but it caught fire soon after. More germane, the low density of wildlife and protection officers on the ground (one for every 1,000 square kilometers) and local corruption at all levels of the service have made enforcement ineffective. Current reports of hunting and its implications for wildlife populations in protected areas all point to increased trapping and shooting both on a commercial and on a subsistence basis. The rate of kills is without exception judged to be nonsustainable in the areas studied. Large-scale ivory smuggling continues, and the behavior of target groups remains virtually unchanged.

Elephant populations continue to fall. Forest clearance has limited elephants to remaining forest areas (particularly in the southeast and in protected areas such as forest reserves). It is probably true that the pressures to hunt elephants for ivory and to eliminate what is considered a feared pest have risen since about 1980, but there has been no concomitant increase in resources allocated for their protection throughout Cameroon. More recently, the national press carried an article about the killing of eleven elephants in Yakadouma in 1992. Dublin et al. (1995) chronicles the steady decline in elephant populations in the country. Seizures of ivory originating in Cameroon have numbered fifty-nine since 1993, with no discernible trend (Dublin et al. 1995:47).

Hunting for bush meat (so-called beef) also has remained at a high level. More sophisticated firearms are now being used (sometimes by soldiers and gendarmes or loaned by them to others). More frequently nonlocal hunters pay the village chief a small fee to poach on a commercial scale. Cross-border smuggling of ivory and bush meat to Nigeria and the Central African Republic is acknowledged as a growing problem.

Studies of the distribution of a wide range of other endangered species on Schedule A (e.g., the African elephant, Preuss monkey, chimpanzee, gorilla, mandrill, red-eared monkey) around Mount Cameroon (Gadsby and Jenkins 1992) indicate a serious depletion of many, and the danger of extinction of a smaller number (for studies of particular species or reserves in Cameroon, see Infield 1988). However, there are simply no data for present populations, let alone historical ones, on which to make a credible quantitative estimate of the changes in populations of endangered species before and after signing and ratification of CITES. All that can be said is that many experienced ecologists believe that most of the species on Schedules A and B are continuing to decline, but with considerable variation between species and areas.

World Cultural and Natural Heritage. Waza, Benoue, and Dja have been put forward as sites under WHC, but only Dja has been accepted as a World Heritage Site.

Table 13.3
World Heritage Sites in Cameroon

Year	Natural site	Cultural site
1987	Dja Faunal Reserve	
Total	1	0

Table 13.3 shows this. The GEF is currently being used to prepare Mount Cameroon as a fourth site for either a bioreserve or a World Heritage Site. All these except Dja are situated in the savanna zone, and have received very little extra protection from the pressures of poaching, fires, tree-felling, and collection of fuelwood. The Dja Reserve (covering 526,000 hectares) has been subject to two management plans (by two sets of foreign consultants, one dated 1987 and the other undated). However, little else in the way of the provision of infrastructure and personnel has been done (e.g., drawing up an adequate budget, transformation of Dja into a national park, the creation of clearly marked buffer zones, building of housing for staff, etc.). Fifteen staff members have been appointed, though the Conservator and five of these are based sixty kilometers outside the reserve, and there is no reliable vehicle available. Nor are guards adequately armed, compared with poachers, who have modern arms and plenty of ammunition (most often weapons of war).

There are a number of multilateral projects at various early stages; all of them aim in different (and, one suspects, overlapping) ways to protect the Dja Reserve. There is a seven-country project (FED Project), at present in the fund-raising stage, to implement the necessary means to protect the site; a UNESCO Dja Project with very similar aims, but in danger of being canceled because of its high projected cost ($2.17 million); and a road project (the Kribi–Yokadouma Highway Project, financed by the African Development Bank and the World Bank), which will have very mixed and possibly threatening implications for the site.

Finally, one vital precondition for successful conservation of the site is the diversion of economic benefits of conservation to local people through an alteration of existing property rights. Therefore, a good indicator of real progress toward fulfillment of the treaty's objectives is the degree of protection-with-integration. This has not occurred, and the Dja Reserve remains the property of the state (Law 81-13 of 27.11.81, title II, chapter I).

The Montreal Protocol. The actions taken by Cameroon so far are the signing of the treaty; the provision of a report to the Secretariat, paid for by the World Bank; subsequent revision of some of the calculations, following comments from the Secretariat; and the appointment of a National Coordinator for GEF. Cameroon's

Ministry of Industry and Trade is banning importation of all refrigerators and freezers that use chlorofluorocarbons (CFCs). The Ozone Office said the ban was necessary because the stockpiles of equipment using CFCs that cannot be sold in developed countries were being dumped in Africa.

Cameroon also plans to introduce regulatory initiatives to protect the ozone layer, and is already working to increase public awareness of the ozone issue. For example, the National Ozone Office of the Ministry of Environment and Forests is giving away free bumper stickers that state "I Love Life, so I Protect the Ozone Layer."

Changes in Behavior of Target Groups

Target groups involved in activities covered by ITTA, CITES, WHC, and the Montreal Protocol include politicians, bureaucrats in the appropriate ministries (in three cases, MINEF), functionaries both in ministries and in the enforcement of laws and regulations (e.g., custom officials, the police, forest rangers), and especially the populations who live in and around the natural and cultural resources (in the case of the first three treaties).

ITTA. Although ONADEF was created for reasons other than the promotion of the objectives of ITTA, many of its policies share the latter's objectives. However, there is little evidence that the performance of this new organization has been significantly different from that of its predecessors. Three pilot projects have been started since 1990, and most of the training, research, inventory, and mapping is being undertaken by international and bilateral aid organizations. Some research continues at the Centre de Recherche Forestière (CRF) and at the Département Forestière du Centre Universitaire de Dschang (CUDS), but their research programs only incidentally cover areas necessary for the development of the sustainable forest estate envisaged by ITTA.

Environmental awareness among the forest administration is low except for a few senior people, but this situation has started to change with the reorganization of MINEF and the UNEP initiative in publishing its multidisciplinary and multi-institutional report on the environment in Cameroon. Rural people see nature as a free and nonscarce good—until the resources are at the point of disappearing; it is only then that grassroots organizations come into being to protect what remains. An example can be seen in the Kilum Mountain Forest area in North-West Province, where soil erosion and the drinking water problem have become acute after two decades of continuous logging, leading to environmental awareness among local people and the creation of institutions to express their views. Clearly, the issue of property rights is central here, and the lack of effective action by the state, in terms of encouraging responsibility for forest and wildlife resources through the development of clear property rights, has resulted in a failure to bring about a change in behavior of local people. Also, there has been very little environmental education of the public (through schools, for example), and it is confined to small numbers of forest dwellers who find themselves inside designated parks and reserves.

In these cases, more sustainable alternative agricultural technologies, which minimize the use of shifting cultivation and wildlife hunting, are being promoted. In more general terms, the structural conditions for continuing expansion of agricultural land into the forest are still in place for virtually all the rural population outside forestry and national park project areas. There has been no change in the practices of the foreign-owned timber companies—indeed, there are popular reports that illegal felling and unsustainable forestry exploitation have been getting worse. In a political speech at Doula in mid-1993, the opposition leader, Fru Ndi, made a political issue of the volume of timber being felled and exported illegally.

CITES. As we have already shown, on paper there is considerable evidence of policy initiatives to conserve wildlife and their habitats. However, although the central objective of CITES (the conservation of endangered species) is shared by many of the current wildlife projects, CITES's means to reach them center on the restriction of *trade* in these animals, whereas projects seek to conserve through protection, education, and use of alternative resources. Therefore, attention should be confined to enforcement officials and those who hunt endangered species.

Implementation of wildlife regulations has always been problematic, and those of CITES are no exception. Certificates of origin and export permits can be issued at Yaoundé and Douala airports since the appointment of Wildlife Officers there. Obam (1992:215) states that Cameroon's forests and savannas are being exploited in such a chaotic and disorganized manner as to threaten their very existence, and cites evidence of massive poaching that has become increasingly commercialized and is promoted with sophisticated weapons on a large scale. The new Forests, Wildlife and Fisheries Act (Law no. 94/01) took effect in January 1994, but the enabling legislation has yet to be formalized (Dublin et al. 1995). This act provides for the classification of wildlife into three categories, with species lists established through ministerial order.

However, Cameroon remains the only central African country offering legal elephant trophy hunts. There is also a long history of venality on the part of officials. Guns are provided to poachers by the very enforcement officials who are supposed to police wildlife. Permits sold for large sums allow the export of 5000 kilos of ivory at a time, whereas the standard *permis spéciale de grande chasse* allows a maximum of twenty-five kilos of "found" ivory (as opposed to ivory from a hunted elephant) (IUCN 1979). Other examples of actions of concerned people, both officials and the public, discussed below indicate that positive changes in behavior are difficult to find.

WHC. There is no change of behavior of target groups in the sphere of heritage sites. The recent crisis in cocoa farming has provoked a transfer of income-earning activities to poaching, particularly in the northern and western sectors of Cameroon. However, the vast size of the site at Dja, and its remoteness and lack of roads, means that most of the habitat is safe from human use at the present time.

Montreal Protocol. Since ratification of the Protocol in 1989, a plan for reducing emissions has been produced, and is under negotiation. Thus it is still too early to look for changes in behavior in target groups. However, a number of projects focusing upon training of customs officers to identify ODS imports, phasing out of ODS at specific factories, and the establishment of a National Ozone Monitoring Commission have been initiated since 1993.

A central question in this book is whether there is evidence for improvements in implementation and compliance through time. Implicit is the assumption that the signing of the accords brought consequential and focused attention and action concerning the specific issues addressed in each of the environmental accords. Such a question gives primacy to the accords themselves as the trigger for actions by the state and other institutions toward compliance and implementation. In other chapters of this book, this undoubtedly elicits fruitful answers, because it tests the hypothesis that the very existence of an accord creates an increasingly favorable climate for compliance with it.

However, in Cameroon it is very difficult to identify any chains of action by government that are directly linked to the signing of the accords. While it is possible to list related but not directly consequential actions by various parties, which may conceivably facilitate future implementation and compliance (this is done near the end of this chapter), it is impossible to attribute even these preliminary and tentative actions to the signing of the treaties. There is a cascade of institutional reorganizations, general statements of intent, shopping lists of reforms—all of them on paper and the vast majority of them instigated by multilateral and bilateral institutions. There are also conditions imposed by the World Bank that have prompted some organizational reform in government; these will be discussed later. However, very few of these changes were prompted by the signing of the accords. Therefore, in this case the issue of improvements in compliance and implementation through time cannot be resolved by a categorical denial that *any* improvements have taken place—there are some, although they may be more symbolic and rhetorical than practical—but it can be clarified in the sense that the accords were seldom the prime instigators of the changes. Indeed, Cameroon was faithful in submitting required reports only when it received external assistance to do so, as table 13.4 shows.

Table 13.4
Cameroon's compliance with treaty reporting requirements, 1991–1996: Report transmission/receipt date

Treaty	Entry into force	1991	1992	1993	1994	1995	1996
CITES	9/3/81	9/12/94	7/26/94	N/R	N/R	N/R	N/R
LC	not a party						
MP	11/28/89	R	R	R	R	R	R

Note: For all treaties, N/R indicates that no report was received by July 1997. R indicates that a report was filed but the date of receipt is unavailable. Montreal Protocol reports received include those filed on time, late, or subsequently revised; in some cases a country may have met two or more years' reporting requirements with the submission of a single report.

Factors Explaining Implementation and Compliance

Characteristics of the Country

The Pervasive Economic Crisis. Since about 1985, Cameroon, in common with many less-developed countries in sub-Saharan Africa, has experienced a deepening economic crisis. As figure 13.1 shows, Cameroon has suffered the most marked degree of economic deterioration in Africa.

The reasons are manifold and open to varying interpretations. First, there has been a precipitous decline in the real value of exports. For example, while the quantity of agricultural exports (cocoa, coffee, cotton, rubber, etc.) rose by 25 percent between 1983/1984 and 1992, their value fell by 47 percent (UNEP 1992c:5). The combined drop in oil prices and the (CFA)/U.S. dollar exchange rate meant that oil lost 42 percent of its value in one year (1984–1985), and the decline has continued. Timber is the only exception to this trend, with a production peak of 625,000 tonnes being reached in 1983/1984, and more or less maintained. Prices rose from 49,000 French francs/ton to 73,000FF/ton in the same period, resulting in an increase in export earnings of 42 percent. However, this was not enough to offset an overall decline in the value of exports in the natural resources sector from 265,000 million CFA in 1984/1985 to 178,000 million CFA in 1989/1990 (a decline of 33 percent).

Second, and partly related to the decline in the terms of trade, state revenues declined by 45 percent in the period 1984/1985–1991/1992 and, coupled with a rise in international interest rates and the negotiation of large international borrowing, Cameroon has experienced acute difficulties in repaying foreign loans, and faces a rapidly mounting foreign debt. By June 1991, its external debt stood at 1,300,000 million CFA—practically 2.5 times the present national budget.

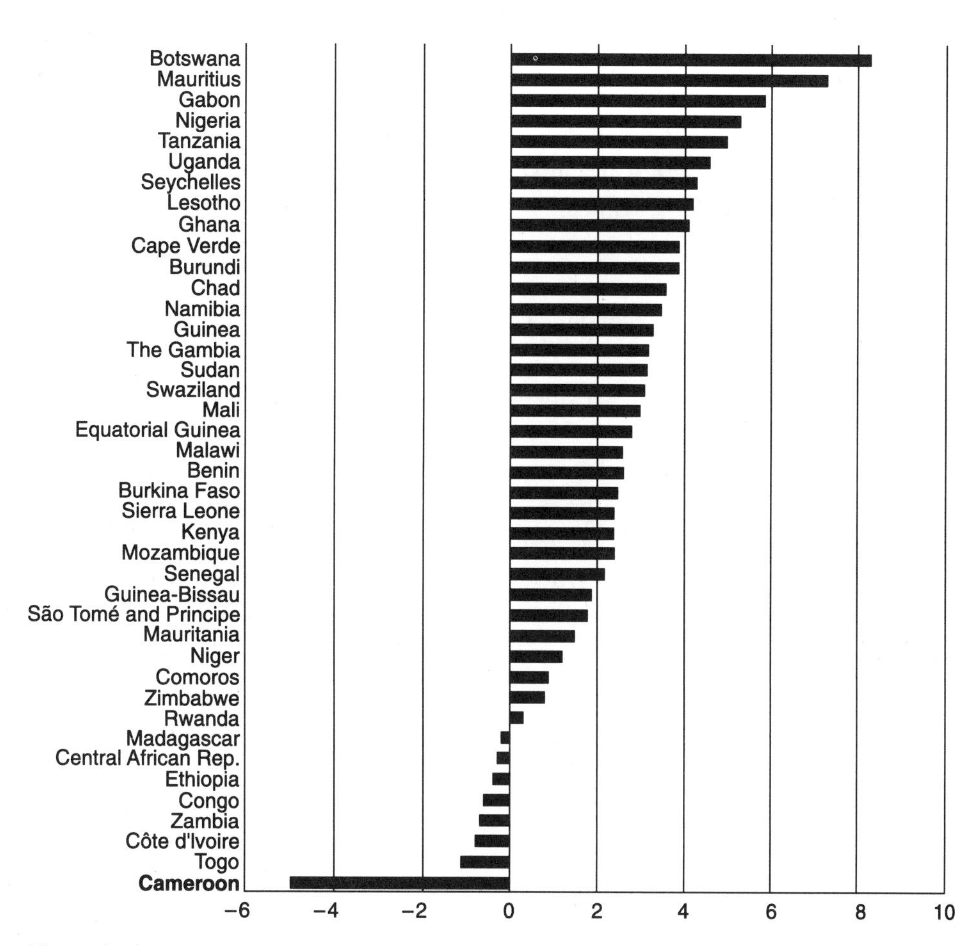

Figure 13.1
GDP growth rate in Africa

The Caisse Française de Développement decided to come to Cameroon's assistance with a bilateral loan of 30 million CFA in September 1992, three months before the World Bank required repayment of its own loan in a situation where it was obvious that Cameroon would default. The political implications for Cameroon's ability to insist upon compliance of the largest (French) logging companies with ITTA and national regulations are discussed below. However, Cameroon withheld 4.3 million CFA and paid the World Bank only 25.7 million CFA. As a result the Bank suspended all loans in April 1993 and declared that if Cameroon was still unable to pay by July, it would then be asked to repay *all* arrears without exception.

Third, there are longer-term reasons for the pronounced economic decline. Poor economic management, institutional structures of government that discourage market-efficient and competitive behavior, excessive government expenditure, and poor state revenue generation are some of the reasons given by the World Bank (1992) and by authors of a collection of essays on African development and the World Bank (Commins 1988). The first structural adjustment policies were imposed by the IMF and the World Bank in 1988. Their short-term impacts were a sharp reduction in government expenditure, especially for unproductive purposes, and a contraction in state employment.

The implications of the economic crisis for implementation and compliance are discussed below.

First, there has been acute difficulty in paying the salaries of public servants even after salary reductions of between 20 and 37 percent in January 1993 and a further 50 percent in November of the same year. Sometimes salary payments are suspended for three months at a time, and for even longer periods in more remote centers. For some entire units of the civil service, payments have been suspended altogether. It is reported that during late 1993, growing unrest and lawlessness by civil servants, particularly in the north, occurred. This retrenchment has further squeezed investment and development expenditure. Equipment, particularly transport, remains poorly maintained, which curtails local monitoring of environmental accords (e.g., forest protection, anti-poaching). Also, it is frequently stated by civil servants at all levels that morale has been seriously affected.

Certainly it is a reasonable hypothesis that very low salaries, from which no one in government service could derive anywhere near a decent standard of living, encourage corruption to augment earnings at all levels. This applies to the local gendarmes who police the traffic of logs on the roads, to Wildlife Officers issuing certificates and permits for the export of specimens, as well as to senior officials. There is a Cameroonian saying, "Le salaire c'est pour respirer et les indemnités c'est pour travailler," meaning that the salary is just enough to survive and to induce the employee to turn up at the office, but in most government agencies it is the allowances and other pecuniary advantages that provide the incentives to work.

Second, Cameroon finds itself beholden to foreign (particularly French) interests. For example, the government is unable to compel foreign logging companies to comply either with the spirit of ITTA or with its own regulations. External actors, in this case foreign, especially French-controlled, logging companies, are widely reputed to pay little heed to guidelines for sustainable forestry. Of the seventeen international logging companies, nine are French.

France is by far Cameroon's most important trading partner, and maintains close personal and cultural relations with its former colony. Therefore French companies enjoy a considerable degree of political protection at the embassy level, and at least two of the biggest logging companies have family connections at the very highest political levels in metropolitan France. Also, it was France that stepped in with loans to prevent Cameroon from defaulting on an IMF loan. Any insistence by Cameroonian officials that international or national guidelines be adhered to, or that local taxes be paid, simply cannot be pressed home. These assertions are difficult to substantiate, since the communications on which they are based are informal, confidential, and often delivered verbally or implicitly.

Third, the undermining of the capability of the state to provide good governance is both a cause and an effect of increasing dominance of foreign finance, expertise, and implementation in many aspects of government. This dominance is exercised through a large number of institutions. While bilateral aid program staff members like to think that they maintain good relations of cooperation, or at least coordination, between themselves, their diverse political and developmental objectives do not make for coherent national policy. It also means that there are many foreign institutions that provide aid within the project framework. Their outputs are therefore dispersed, often are uncoordinated, and do little to strengthen the institutional capability of the government. They also may, or may not, serve to fulfill the objectives of *nationally* signed treaties, under which the nation-state is expected to take responsibility for implementation and compliance.

Fourth, falling incomes in the countryside have led to increased pressures to pursue the commercial exploitation of wildlife, to cut timber illegally, and to clear land for agriculture. For example, the closing of the Palmol oil-palm plantations near Korup National Park has led the plantation workers to cut timber for sale, to clear land for agricultural subsistence, to poach wildlife, and thereby to put increasing pressures on the habitats in the park. While the illegal export of specimens and trophies is carried out by wealthy entrepreneurs, they rely upon large numbers of poor people to trap or hunt animals.

Fifth, chronic foreign debt and the policies put forward to reduce it in the Structural Adjustment Plan of 1988 have stimulated exports, particularly timber (but also petroleum products). Therefore, there is a policy environment in which any impediment to logging is liable not to be taken seriously.

Characteristics of the Treaties

ITTA. This treaty was negotiated under the auspices of UNCTAD to promote trade in tropical timber. The major producing and consuming countries are parties to the agreement. Most of the agreement concerns pure trade matters, but article 1, paragraph h, identifies objectives concerning sustainability, such as "the development of national policies aimed at sustainable utilisation and conservation of tropical forests and their genetic resources, and at maintaining the ecological balance in the regions concerned." This goal is to be achieved in producing countries by encouraging the expansion and diversification of trade in timber, improved forest management and wood utilization, and reforestation. The main problems with the implementation of and compliance with this agreement are listed below.

1. It is basically a trade agreement with *vague* clauses concerning sustainability, balanced ecology, and environmental concerns. There are no quantified targets for these clauses, and therefore no means of evaluating whether they are being implemented. Also, it has been claimed that ITTO is not interested in environmental concerns and disregards evidence of biodiversity degradation in tropical forests (P. Brown 1993).
2. In any case it would be very difficult to produce such targets because the concept of sustainability is riven with complexity and technical uncertainty. Environmental concerns about the soil and water conservation functions of the forest are even less well understood. However, that is not to say that it is an impossible task to derive locally some indicators of a sustainable forest estate, species diversity, and rates of cut, and to draw up a basic land-use plan. The agreement lays down no guidelines, let alone clauses for identifying, measuring, and implementing the environmental concerns of the agreement.
3. The forest in a developing country is not a commodity that is exploited by a small and identifiable group of commercial operators who can easily be monitored. Instead, it is the focus of virtually all the rural population directly and of many urban dwellers. The central object of the agreement is assumed to concern only export, but forest use in reality affects most of the population.

CITES. This convention is designed to control the export and import of animal species that are rare and/or threatened with extinction. At present it concerns only fauna, although the addition of plant species is being considered. The main problems with implementation of this convention, which derive from the nature of its terms, are as follows:

1. Articles IV and V call for the identification of animal species, which is an unavoidable and central means of implementing the convention. However, it requires considerable expertise on the part of Wildlife Officers at the point of export to be able to tell some species apart. Since there are four points of export in Cameroon (Garoua, Maroua, Ndoundere, and Douala), there have to be at least four such officers at the airport or port at all times. There are clearly opportunities for exporters to mislabel animals.
2. It requires detailed ecological knowledge to judge whether the export of animal species listed in appendices II and III put them in danger of extinction. If this information is not available, there are no other criteria on which to deny an export license, and the tendency has been for licenses to be granted for the export of a considerable number of animal species that, in the view of knowledgeable wildlife personnel, are acutely threatened with extinction (e.g., the giant frog and the crocodile).
3. Since wildlife forms an important part of the livelihoods of rural people as subsistence or a means of earning cash through sales, there is an acute enforcement problem. Poaching, through trapping or shooting, is indiscriminate. Even if the suc-

cessful trapper does not care to eat the unfortunate animal, he can find somebody who will. "Beef" is widely sold in the countryside and towns throughout Cameroon. Strictly speaking, this erosion of biodiversity in animal species cannot be explained in terms of the failure of the implementation of and compliance with CITES, since CITES deals only with the international trade of animal species. However, the purposes of CITES in protecting biodiversity are being undermined because many species are killed but not traded or, if they are traded, only intranationally.

4. Animal species listed in all appendices are often illegally imported and exported along the unguarded frontiers. For example, the current ban on the sale of ivory has reduced its price, and a number of Nigerian dealers are now buying Cameroonian ivory across the border. One of the most important species (the elephant, now transferred to appendix I) is killed for a very high-value, low-bulk commodity (ivory) that can easily be smuggled. There is also the illegal importation of ivory, from the Central African Republic, that can then be smuggled out through the port of Douala.

5. This point does not refer to CITES as presently constituted but to the possibility of including endangered tree species. Trade names for different types of timber in Cameroon often refer to up to twelve different tree species, and it would become a practical impossibility to check every unsawn log between the site of felling and the sawmill or point of export. The suggestion to extend the ambit of CITES to tree species would merely underline items 1–3 above.

World Heritage. This treaty puts constraints on the use of designated sites within a country. According to article 5, in order to conserve designated cultural and natural heritage areas in any country, parties should adopt a general policy giving a function to the site in the "life of the community" and integrate site protection into "comprehensive planning programmes"; set up services to conserve sites; engage in research and development of operating methods for counteracting dangers to the sites; and take appropriate legal, scientific, technical, administrative, and financial measures to identify, conserve, and rehabilitate the sites. The only sanction that can be applied by the Secretariat is deletion of sites from the World Heritage List when they have deteriorated to the extent that they have lost their outstanding universal value. There are two main problems with this convention:

1. There are not, and probably could not be, any quantitative measures of "outstanding universal value." In any case this is a perceptual matter and could be independent of physical changes in the site.

2. The "life of the community" may entail the exploitation and destruction of the universal values of the site. Poaching and felling of trees in Cameroon highlight this essential contradiction. Participatory management is clearly implied but in no way resolves these contradictions.

The Montreal and Vienna Ozone Protocols. According to their terms, these protocols contain the most arduous obligations of any of the treaties. The Montreal

Protocol was negotiated within the broad framework of the Vienna Convention (1985), which commits parties to take appropriate measures to protect human health and the environment from the adverse effects of human activities that modify the stratospheric ozone layer. The Montreal Protocol requires states that are parties to reduce their consumption of chlorofluorocarbons (CFCs) and to freeze that of halons. The Vienna Convention provides for monitoring and the dissemination of information and research. The duties under the treaties are quite precise.

The fact that the Montreal Protocol makes provision for less stringent target emissions for the developing countries and the lack of modern environmental laws in Cameroon may explain the delay in ratification. Also, given the social, political, and economic crisis in the country, this laxity in the obligations may make countries like Cameroon less committed to reducing their consumption levels of CFCs and halons by the year 2000 and of carbon tetrachloride by the year 2005. Another important point that makes the Montreal Protocol different from other international environmental treaties is that it goes beyond the more general encouragement to cooperate and establishes the interim Multilateral Fund to help developing countries reduce their consumption of the substances that deplete the ozone layer. Finally, the GEF provides additional money through a system of low-interest loans to poorer countries. Perhaps it is access to such resources that will eventually induce Cameroon to implement the accord quickly.

Actions of External and Internal Actors

ITTA. It is difficult to identify any direct actions that amount to altered behavior and have improved the sustainability of the forest estate in Cameroon. The most important external actors are the Secretariat of ITTO; international nongovernmental organizations (INGOs) such as CARE, the World Wildlife Fund (WWF), and the Association Bois de Feu; multilateral agencies such as the World Bank and UNDP; bilateral agencies such as USAID, CIDA, and ODA; and commercial logging companies.

There is evidence that since Cameroon signed the treaty, the World Bank has had a large input in the formulation of a new forestry policy and a modern environmental jurisprudence for the country and/or cooperated to achieve certain tasks that are in conformity with the obligations of this convention. The World Bank was also the prime mover in the formation of MINEF and ONADEF. Other external actors are important in terms of individual forestry projects. For example, CIDA has been involved in making an inventory of the tropical rain forests and their rich biodiversity, and has financed ONADEF's cartographic center at Nkolbison (Yaoundé) with teledetection equipment that will be used in West and Central Africa. Following the creation of the Ministry of Environment and Forests in 1992, CIDA submitted a zonal plan consisting of nine thematic maps aimed at establishing permanent forest areas in the country.

More recently, ODA has started a three-year Forest Management and Regeneration Project (FMRP) based in the M'balmayo Forest Reserve. The ODA is supporting the FMRP with almost £2 million. The main objective of the project is to help the government of Cameroon improve the management of its rain forest. This will be done by building on research experience through the establishment of demonstration plots for five different techniques for forest regeneration after logging and farming. These plots will provide data on establishment and management costs and on growth and yield. In addition, biological research, socioeconomic surveys, and monitoring will be carried out by staff from two European institutions, the Institute for Terrestrial Ecology (Enschede, Netherlands) and the Natural Resources Institute (United Kingdom).

Another example of INGO participation in carrying out an inventory of Cameroon's forests is that of the Dutch consortium Tropenbos. It operates in the Centre and Littoral provinces. There are also a number of ITTA-funded projects, discussed earlier. However, many of the training programs, well executed though the majority are, do not reach their ultimate objectives because most of the trainees cannot find jobs in the forestry sector, particularly following the downsizing of the staffing in this sector.

Thus there are many donors who may simultaneously be doing similar things in various parts of the country. A Canadian diplomat told the authors: "Cameroon says yes to everybody. But this creates a lot of problems when it comes to implementation of its policies or, for that matter, international environmental laws. It also leads to confusion and contradiction because different donors ... have the permission to do the same thing." For different reasons, Cameroonian forestry officials accede to requests for special concessions and tax-free exploitation permits by French lumbering companies and other foreign firms involved in the lucrative timber industry.

The most significant internal actors are high-ranking politicians, civil servants, entrepreneurs, local chief-level authorities, and members of the public. As one official in MINEF remarked, "It is not necessarily the man who owns the chain saw who plunders the forests, but those who do business with him." The implication here is that besides the legal exploiters with licenses, there are numerous others who engage in illegal trade. The key internal actors are those senior officials who organize logging concessions.

CITES. Cameroon signed the Washington Convention in 1973. There had been prior discussion of Cameroon's position among a number of senior administrators and scientists, but there were no funds for a delegation to attend, and the protocol was signed by the Cameroon ambassador to the United States. Apparently the ambassador failed to write a report to the Subdepartment of Wildlife, or it was lost (accounts differ); ratification did not occur until 1981. Very little action was taken after the Washington Convention, and Cameroon was not party to, and did not attend, the Bonn Convention in 1979. The CITES Secretariat issued a report in late 1977 regarding large exports of endangered species, and a CITES official visited Cameroon when it was discovered that a member of the Wildlife Department had been forging the signatures of the Minister and the Director of the department on a large scale, enriching himself in the process, and that a considerable proportion of export of specimens from Cameroon did not originate there but in neighboring countries, especially the Central African Republic. It was only then that an enterprising senior member of the Wildlife Department felt he had to make the case for ratification to the government, which he did in June 1981.

In the period between the signing and ratification of CITES, there were a few significant attempts to enforce the regulations that now exist by both international and national actors. The well-known case *Lépere* v. *the State* is reported below.

In September 1977, the Chief of Service for Wildlife and Forest Environment, the Conservator of Forests for the Central South Province, and the Provincial Judicial Police for the Central South discovered three and a half tonnes of ivory in a store in Yaoundé belonging to a French businessman named Lepère. The tusks were seized and Lepère was taken to court. However, the Magistrate Court and the Court of Appeal ruled in his favor, and he was eventually acquitted. The Ministry of Agriculture then appealed the decision of the two courts to the Minister of Justice and Keeper of the Seal, citing in support the signing of the Washington Convention by the Cameroon Head of State and the subsequent adherence to it by Cameroon. The matter has been referred to the Supreme Court and remains unsettled.

The Minister of Justice wrote to the Conservator of Forests (who had originally made the seizure and had insisted upon the application of the law) to say that "a higher authority" had dissuaded him from taking that course. When the Conservator persisted, he was advised that his position was not safe and that he should take a year's absence from his post for further training in Europe. In the same year, the head of the Wildlife Subdepartment, who was never informed of the time and place of the hearings, appealed to the Head of State with a copy of the letter sent to the Ministry of Justice, but no further action was taken. Six months later, all the ivory had disappeared from the Yaoundé store. To the best knowledge of the Wildlife Subdepartment head, Lepère never got the tusks back. The tusks were not, and could not have been, auctioned as stipulated in section 65 of the law because they were court exhibits. Auctioning them would have required a court clearance, which was never requested or granted. What happened to the ivory remains a mystery.

A major incident connected to this case worth mentioning is the calling up in 1982 of all the officers who were in Forestry at the time of the Lepère case to testify about the disappearance of the tusks. Since then nothing has been said to any of them.

Another illustration may be given. In 1983–1984 (two years after ratification), Cameroon authorized the exportation of seven gorillas belonging to Mr. and Mrs. Roy of Sangmelima to Holland. Although the Department of Wildlife and National Parks claimed that the gorillas were captured before the signing of the Washington Convention, various bodies, including the Animal Protection League of the United States, claimed that the CITES Bureau and the Department of Wildlife and National Parks had abetted the illegal exportation of endangered species. At the fifth CITES conference in Buenos Aires on April 13, 1985, about thirty delegates, in a letter addressed to the only Cameroon delegate there, expressed their hope that representatives of the Management and Scientific Authorities would be present at future conferences. They also highlighted their concern at recent exports of gorillas from Cameroon and their hope that Cameroon would adopt and enforce a policy of total legal protection of gorillas and a total prohibition on exports.

Two other instances of corruption may indicate the general lack of enforcement, although these are based on hearsay. The Governor of a northern province asked the Conservator of Forests to provide him with some bush meat from a reserve illegally, so that he could serve it to his friends. When the Conservator refused, the Governor wrote to the Minister of Tourism, complaining of his insubordination and asking that he be removed from his post. Another case, dating from the mid-1980s, involved the use of confiscated firearms by Wildlife Officers for the organized poaching of wildlife from their own reserves. Since these cases, we have been unable to discover further attempts at prosecution or to pursue the law.

WHC. The main external actors in this case are bilateral aid agencies and INGOs that have offered finance and expertise. A number of country representatives of WWF, ODA, and USAID used their personal friendships with key Cameroonian conservation officials (some three names are mentioned repeatedly) in order to encourage the application to WHC for consideration of three sites.

It is worth asking how unusual these actions in Cameroon are when viewed from a pan-African viewpoint. A case may be made that the use of the environment and the political-economic structures within which it occurs in Africa, especially West and Central Africa, can be generalized without undue violence to country specifics. Colchester (1993:15) states:

> The close convergence of interests between the post-independence bureaucracies, the ruling indigenous elites and foreign capital opened a deep gulf between the rulers and the ruled. The

result was that though "development" was certainly promoted in all three countries [referring to Gabon, the Congo, and the Central African Republic], it was a kind that bypassed the rural poor. Prestige projects in construction, mining, railways, oil, forestry and agribusiness served the interests of the urban elite and foreign companies, perpetuating the enclave economies of the colonial era.

There follows an account of the "mayhem" (Colchester 1993:17) caused by uncontrolled logging, graft, force, and coercion in the three countries referred to in the quotation above, as well as in Equatorial Guinea, Zaire and Cameroon. In all of them, although they have different colonial histories and present political-economic configurations, the "close convergence of interests" referred to above prevails. It smothers accountability and transparency, and any form of political representation that favors them. It is hardly surprising that most environmental accords will tend to run counter to these interests, and while there are countervailing forces in each of these countries, they will tend to be frustrated.

Dynamic Models of How Implementation and Compliance Work

Evidence for Implementation and Compliance

It is useful to list the evidence for implementation and compliance for each accord, although the caveat must be borne in mind that such evidence is seldom intentionally linked to the accords themselves.

ITTA. Redrafting of the Forest Law (now in its fifth draft and before Parliament, although this was prompted not by the ITTO but by the World Bank and its Structural Adjustment Programme); the reorganization of environmental policy-making and implementation by the creation of MINEF; the creation of a number of indigenous environmental NGOs, although their capacity to act as watchdogs is very limited; the UNEP initiative to move toward a National Environmental Action Plan (NEAP); the creation of the Forestry School at Mbalmayo and the training of forestry staff; and a variety of bilateral and multilateral projects of an exploratory, information-gathering, research, and preliminary nature to improve forest management.

CITES. The redrafting of the Wildlife Law in 1981 and again as part of the Forest Law (see above); the creation of the Wildlife School at Marua and training of wildlife staff; the creation of national parks, faunal reserves, and reserve forests by cartographic and legal means, some of which are supported by bilateral and multilateral projects; and the installation of the machinery to issue certificates of origin and export permits at points of exit, and Wildlife Officers at Douala and Yaoundé airports.

Montreal Protocol. The provision of a report to the Secretariat on the amount of ozone-depleting substances used in Cameroon, and the estimated cost of phasing them out.

World Heritage Convention. The identification and promulgation of three areas (Waza, Benue, and Dja), and the preliminary planning of the successful application (Dja).

Nevertheless, it is clear that very modest steps toward implementation, but virtually none toward compliance, have been taken. These are dwarfed by the overall continuation of the practices the accords seek to eliminate.

Generalizations

Notions of the implementation of and compliance with international environmental accords are conventionally predicated on notions of the liberal-democratic state that is supposed to intervene in order to manage the environment in a "rational" manner —above the competing interests of different parties—and whose decisions are carried out by apolitical and impartial state functionaries. In this model, impartial and objective scientific information acts as the arbiter between private interests. While this view has been under sustained criticism for some time (see chapter 4 in this volume), any progress toward the goal of improving compliance with international agreements rests upon modernist assumptions. These include the application of science to policy and a rational legislative and administrative program. In turn, these activities assume both the Weberian necessity of an efficient, specialized, rational (i.e., nonarbitrary) administration (Weber 1947), and a state with the necessary legitimacy for the program to be carried out. To what extent are these assumptions valid, specifically for Cameroon and, in more general terms, for other states in Africa?

In this chapter there have been claims that the largest logging companies evade taxes, that planning is inefficient, that ministries are overstaffed, that field monitoring of the accords is almost wholly lacking, that corruption in the Ministry of Wildlife is endemic, and that former colonial powers exercise strong political bargaining power in favor of their transnational companies. Thus, in short, it is suggested that Cameroon simply does not have a state that can deliver in substantial measure the means by which environmental accords can reach their objectives.

In the matter of resource use and environmental planning, we do not even claim that present actions are the result of rational scientific analysis. On the one hand there are the scientific papers and texts and the international treaties on which they are based, and on the other hand there are the actions of state officials at all levels, which seldom are justified or judged by the criterion of fulfilling these treaties (or any other environmental regulations). There are certainly formal structures and institutions (e.g., MINEF or ONADEF) and routinized actions that may be put in train. Planning documents prepared by foreign consultants are adopted officially, but actual implementation and enforcement in the field are meager. This outcome can be likened to a control panel of electric switches—they are in place, labeled, and visible. Throwing switches may give the impression of choice and decisive action, but the panel is not wired to the power source.

There are important general issues concerning pervasive political and economic reasons for this state of affairs in Cameroon and other African states. It is tempting in any brief review of the problems faced by African states like Cameroon to deemphasize the diversity of African states, and to employ development narratives that treat "Africa" as an exception to policy and strategy success stories elsewhere in the world (see the "Except Africa" thesis of Roe 1995). Nonetheless, it is widely agreed that a syndrome of a lack of legitimacy of the state, authoritarianism and neopatrimonial rule, political instability, and corruption and clientelism exists in many African states and can be explained by common historical elements.

In addition to this dismal set of related symptoms, exogenous factors such as the current harsh economic climate, the power of transnational companies, and the continuation of (neo)colonial relations have combined to create a downward spiral of interrelated political and economic decline (see Dumont 1969; Bates 1981; Hart 1982; Hyden 1983; Sandbrook 1986; Abernethy 1988). Here, we are not primarily concerned to explain the interrelations between these symptoms, but rather to show that they exist in Cameroon, and how they are obstacles to compliance with environmental accords there and perhaps elsewhere in Africa.

The downward spiral is in motion in many African countries, but its form and seriousness vary. Cameroon has perhaps experienced a sharper decline from a more advantaged position than many other countries. It was considered a middle-income country and also had a comparatively strong state (Sandbrook 1986:327). Today, the evidence for its economic and political decline is everywhere. The state's control over its natural resources is constrained by the leverage exercised by French neocolonial interests, especially in forestry, and by agreements signed with transnational corporations concerning oil exploitation on terms very advantageous to them.

As Colchester (1993) has shown for other countries in West Africa, the combination of exogenous economic and political interests and the weakness of the state makes substantial compliance with international accords a remote prospect. Zama (1995) attests to the way the Cameroonian government has virtually handed over control of the forest to foreign companies by the drafting of a bill introduced in Parliament in 1993. The bill directly contradicts the spirit of the Cameroonian Constitution, which stipulates that natural resources should be exploited for the benefit of all. Local people living in the forest have been denied use rights, and as one deputy said, "The Bill was not drafted by Cameroonians" (Zama 1995:264). Furthermore, the method of awarding concessions is shrouded in secrecy, and the terms of the concessions strongly encourage nonsustainable harvesting and short-term maximization of profits.

There are two other important elements that link the state and economic decline in Africa, and that have serious implications for the prospects of compliance with the international accords. These are a lack of legitimacy of the state and the size and capability of the bureaucracy.

The low level of legitimacy of the Cameroon state was mentioned earlier. It adversely affects the degree to which local laws and regulations can be applied, particularly in relation to CITES, the sustainability clauses of the ITTA, and the WHC. Local rural people have lost their customary rights of access to the forest and its fauna, and they see in their daily lives how foreign companies, corrupt officials, and local "big men" exploit these resources. They feel there is no reason to heed any restrictions imposed by the state in the name of international agreements. Poaching, corruption, and illegal felling are the result, and the state does not have the capacity to enforce policy or to monitor the conservation and use of these resources.

The second generalization concerns the size and capability of the bureaucracy. While Abernethy (1988) and Commins (1988) lay considerable stress on the bloated size and costs of the bureaucracy, Sandbrook (1986) emphasizes bureaucratic behaviors that are inimical to rational decision-making, leadership, and economic growth. Thus the recent downsizing of the government apparatus in Cameroon as a result of structural adjustment policies has probably made only a small difference in its capacity to implement and comply with the international agreements. Until the early 1980s, Cameroon had one of the better administrations in tropical Africa, but it has undoubtedly suffered from increasing levels of clientelism, prebendalism, and corruption since about 1980. In other words, if the size of the bureaucracy had not been reduced drastically, its performance and abilities to bring about implementation and compliance would probably have been little better. This conclusion may well apply to other countries in Africa.

There are a number of usually senior administrators who are well aware of the scientific rationality of the treaties and the necessary implementation and compliance measures. However, they are *few* in number and represent the "old guard" in a rapidly changing and deteriorating administration. They have often become isolated and frustrated in their jobs, and are absolutely unable to form an epistemic community. Partly this is due to their inevitable criticism of existing practice, and partly to the fact that any corruption that may exist requires complicity by large numbers of people. It must also be borne in mind that they were, for obvious reasons, the main informants to the authors of this chapter. Their view is pessimistic and somewhat cynical, and they come from a beleaguered minority.

Enforcement of all the accords (except the Montreal Protocol) works against the short-term economic interests of very large numbers of people. Timber is used as fuelwood on a day-to-day basis; the land on which it stands is required for agriculture under present technologies and population growth; and certain species offer opportunities for immense and rapid enrichment to foreign firms as well as small-scale loggers. Wildlife provides meat for the majority of the population, both urban and rural; some species fetch very high prices as live specimens or as trophies, or yield ivory. Simply, most people who are part of the state apparatus or in civil society make money from continuing to exploit these resources or from selling access to them. Since the aggregate rate of exploitation is nonsustainable, the purpose of the accords will tend to be frustrated.

Since the state has made very little legal provision to enable income to flow to those local people who engage in conservation activities (particularly of wildlife), there is no incentive for them to obey conservation laws and ordinances that claim wildlife and other natural resources as property of the state.

ITTA, CITES, and WHC are difficult to enforce because, without participatory management and enforcement by local people, they involve the policing of large territories, often in areas of low population density, by centralized, understaffed, and underpaid state agencies.

Prescriptions for Improving Compliance

First, since the general problem of nonimplementation and meager compliance in Cameroon lies in the inability of the state, any prescription must address this issue first. It is tempting to believe that the provision of better planning tools (e.g., forest and fauna inventories, mapping exercises, training, better information storage and retrieval) will substantially improve implementation of these accords. While these certainly are a necessary prerequisite, they are far from sufficient. Also, there is undoubtedly a growing cynicism toward the "development industry" in the 1990s. Endless reports, institutional strengthenings, international workshops, project evaluations, training programs—in short, a breathless flow of expertise and technical transfer—are all received by African countries like Cameroon in a socioeconomic environment of visible decline and disintegration.

Second, there must be the political will and institutional capability to manage resources in a sustainable and nonpolluting way, but pervasive failures of government to provide fair and certain enforcement of the law prevent this from coming about. Rather than assume that there will be such a change (and there are various models available, both neoclassical and radical), more pragmatic and modest proposals, often at the local and project level, may be more appropriate.

Third, maintaining a sustainable forest estate, particularly in the moist tropical forest, is even more problematic. Here, there are very large sums of money to be made from commercial exploitation without regard to sound ecological principles. It is rarely feasible to trust that sustainable commercial forestry practices will be followed by those who are not involved in cutting down the trees. Sound commercial practice requires careful, intensive, and expert monitoring. The mobilization of NGOs (both international and local) and political parties may be helpful in alerting international opinion, particularly aid agencies and lending institutions, to continued flouting of the environmental clauses of ITTA and the frequent bypassing of CITES. ITTA may be too much of a trading organization, and too involved in the shorter-term commercial interests of the trade, to be able to act energetically for environmental concerns. Bold criticism and advocacy are difficult within the country, and foreign commentators have a role to play (e.g., the two American senators whose views on the practices of some logging companies were aired on national television). It may also be possible to expose some foreign-owned timber companies in various embarrassing international arenas.

Fourth, in the case of CITES and WHC, the state should step back and let NGOs and local communities manage the forests in a variety of comanagement regimes, and get foreign aid to help in providing the infrastructure to make ecotourism possible. Structural adjustment and the pervasive economic crisis have obliged the state to retreat in any case. The problem then arises that appropriate property rights, the flow of benefits from environmental conservation to those who use the natural resources involved, and an effective regulatory policy are not in place. Without them, local compliance with international accords will remain a chimera. Attracting funds through the marketing of these resources and the sharing of those funds with local communities has been tried successfully elsewhere, in a wide variety of different institutional forms.

Examples of what can be done can be found in Costa Rica, where the government, NGOs, universities, local people, and fund-raising establishments work together for the conservation of wildlife and forests. Other examples exist in Zimbabwe and Zambia with regard to wildlife, and with regard to national parks in Annapurna and Everest national parks in Nepal. While these and other examples tend to be reified and idealized in the development literature, they at the very least provide some indications of how incentives can be provided through participatory management institutions. In some cases these examples have raised contradictions between the wishes of (some) local people to use a resource unsustainably and sound, long-term management. But in all cases, it has required an incompetent state to step back and renegotiate fundamental property rights away from state monopoly and in favor of local people. In Cameroon, the politics for such a radical shift is not yet in place.

In the context of this book as a whole, the conclusion of this chapter may seem pessimistic and negative. After all, one of the main purposes of the research effort on which this book is based is to improve implementation of and compliance with international environmental accords in the future. The case of Cameroon, together with several others in the book, throws a rather ominous light on environmental conservation and sustainability. It is an extreme case, although unfortunately there are others like it. Together they demonstrate how the conditions, both external and internal, for the reproduction of underdevelopment compromise the international project of effective environmental management.

14

Brazil: Regional Inequalities and Ecological Diversity in a Federal System

Murillo de Aragão and Stephen Bunker

Brazil's compliance with international environmental accords varies greatly from treaty to treaty and is not consistent nationally. Most of this variation emerges from the diverse geographic and political conditions in different regions of the country. Compliance is greatest where national authority is strongest. The national government generally wields the most power in the more affluent urban areas, and is least effective in the rural hinterlands. Given Brazil's settlement patterns, compliance is more likely in the densely populated coastal cities than in Amazonia and other interior zones.

Brazil is party to the five environmental accords on which this work focuses (see table 14.1).

Brazil, with an area of 8.5 million square kilometers and inhabited by 150 million people as of 1997, is the largest country of South America. It borders every country in South America except Chile and Ecuador. In its territory lie most of the Amazon forest and the Amazon River, the largest river in the world in volume of water. Its long north–south extension, stretching from north of the equator to south of the tropics, gives Brazil a wide range of climates and a variety of ecosystems.

Discovered in 1500 and colonized by Portugal, Brazil became independent as a kingdom in 1822. In 1889 it became a federal republic. Like other countries in Latin America, it has alternated between democracy and authoritarianism. In 1985, after twenty-one years of a military regime, Brazil became a democracy again and elected its president by direct vote.

Brazil is a federal republic with three administrative levels: federal, state, and municipal. The states, headed by governors and state deputies who are directly elected, are divided into more than 5,000 municipal districts. Each district has a mayor and a city council, also elected by direct vote. Brazil's economic crisis and its incomplete transition from a centralized military regime to a more decentralized constitutional democracy largely determine the political role each level of government plays.

Brazil has suffered from a public administration crisis created by low salaries, mismanagement, and an unbalanced allocation of financial and human resources.

Table 14.1
Treaty membership of Brazil

Treaty	Year treaty came into effect	Signed by Brazil	Ratified by Brazil
WHC	1977	1977	1977
CITES	1975	1973	1975
ITTA	1985	1985	1985
LC	1975	1982	1982
MP	1989	1990	1990

The redemocratization of Brazil in 1985 and the administrative reform promoted by the Collor administration in 1990 aggravated this problem. The severe shortage of trained government personnel in the poorer regions exacerbated the regional differences in administrative capacity.

The lack of administrative control, as well as regulatory conflicts between the federal, state, and municipal governments, has made uniform national compliance with environmental accords nearly impossible.

The pace of economic development in the different Brazilian regions has been as uneven as the climate is diverse. Population densities, per capita income, the quality of communication networks and local administrative capacities, and other factors that affect not only the implementation of the treaties under study, but also the economic activities the treaties seek to regulate, vary enormously from region to region.

The five agreements studied relate to different geographical areas in Brazil, which makes the regional differences essential to an understanding of Brazil's compliance with the agreements.

For example, CITES and ITTA are most relevant to the more sparsely populated, least institutionally developed, and most remote areas of the north and west, while the ozone-related accords and dumping treaties are most applicable in the densely populated industrialized areas farther south. CITES and ITTA aim to regulate extractive economies in areas with a long history of unregulated appropriation from nature, while the other treaties regulate industrial activities in areas with a long history of state intervention in the economy.

Brazil's economic and social problems, in particular the gross disparity in income distribution, necessarily relegate environmental concerns to a lower level of priority. Sociologist Herbert de Souza described the relation between citizenship, democracy, poverty, and the environment thus:

> In Brazil the degradation of the environment and of society, of people and of nature constitute [*sic*] two sides of the same coin, in the same style of development and of the absence of democracy. Ecology here is people. People here are the primary ecological question. But here it is also clear that the two survive or destroy themselves together. (de Souza 1992, p. 48)

Absent a consistent popular demand for environmental preservation, and given the general lack of revenues, the government has paid scant attention to environmental problems. The uneven distribution of political and economic power has made the situation worse.

The Paradoxes of Brazil

Brazil is a country of great contrasts. These affect Brazil's implementation of and compliance with international accords.

Paradoxes of the Economy

While Brazil has experienced significant economic growth—the World Bank records it, as of mid-1997, as a middle income country—the pace of economic development has been very uneven among the different regions.

Brazil achieved an average annual GNP growth rate of 6 percent during the period 1945–1988 that was reflected in patterns of land settlement and use. By the 1990s, 78 percent of the Brazilian population lived in cities, compared with 37 percent in the 1960s. Industrialization accompanied this massive urbanization: in the 1960s, only 8.5 percent of the labor force was employed in industry; that figure grew to 22 percent by 1992. During the same period, the market share of industrial products increased from less than 10 percent of total exports to 70 percent.

The positive effects of Brazil's economic growth were countered by the country's high external debt. Brazil's debt reached approximately U.S.$120 billion in 1994, and as of mid-1997 was being renegotiated. External public debt of more than U.S.$60 billion causes serious imbalances in public accounts, with interest rates reaching 40 percent in real terms per year. This high cost consumes resources that could otherwise be allocated to improving administrative efficiency and compliance with environmental accords.

Industrialization has created an enormous income disparity in Brazil. According to Roberto Guimaraes, "in the period of most rapid growth, between the 1960s and 1980s, the richest 10% of the labor force increased its share of income from 40% to 50%, while the poorest 50% saw its share reduced from a modest 17% to only 12%" (Guimaraes 1992, p. 59).

The 1995 United Nations Development Program's (UNDP) *Report on Human Development* indicated that Brazil had one of the highest rates of inequality, in terms of income distribution, in the world. While the effort to combat inflation since the start of the Real Plan in July 1994 had brought some economic benefits, as of mid-1997 Brazil remained a country of great contrasts. The 1996 UNDP study coordinated by the Institute of Applied Economic Research (IPEA, an agency connected to the Ministry of Planning) concluded that Brazil was comprised of three subcountries with different levels of economic and social development: the state of Rio Grande do Sul, near Argentina and Uruguay, had the best level of human development, comparable with that of the Czech Republic, whereas Paraíba's human development index was less than Kenya's and only a little better than India's.

The Brazilian disparity in per capita income was also very large. The Federal District, where the capital is located, had the highest per capita income in the country, with rates comparable to those of the First World, around U.S.$10,200, while in Piauí, 60 percent of the population had an income below the poverty line. São Paulo, the richest state in the country, had the highest absolute number of poor people.

The Environmental Legislation Paradox

Brazil has enacted wide-ranging and comprehensive legislation at all three administrative levels: national, state, and municipal. But enforcement of the legislation has been very weak.

Environmental legislation has a long history in Brazil, dating back to the Ordenagoes Manuelinas, the Code promulgated by King Dom Manuel in 1516 when Brazil was still a colony, which outlined a reforestation policy (Wainer 1991; *Revista Forense* 1992). Despite this policy, however, the Portuguese devastated the forests for brazilwood, a source of easy profit from the new and distant colony.

During the 1970s Brazil adapted to the environmental rules adopted by the international community, and became concerned with the protection of fauna and flora, water, and historic heritage even before the environmental boom. In 1981, while the military regime was still in power, Brazil implemented a national environmental policy for the first time.

In 1988, Brazil adopted a new constitution, which to many Brazilian jurists is one of the most advanced in its environmental protection provisions; twenty-four articles deal directly with the environment. The most significant constitutional provisions relevant to environmental protection are the following: (a) treating environmental protection as an essential component of Brazil's industrial and agricultural activities; (b) requiring environmental impact assessments for obtaining approval of any project that may cause environmental harm; (c) giving the Attorney General, who is independent of the other federal powers, responsibility for initiating inquiries and judicial actions to protect the environment; and (d) giving any citizen standing, without need to prove personal injury, to propose action to annul any act that damages the environment.

A complex framework of juridical norms makes these provisions operational. As of mid-1997, Brazil had sixty-nine important laws dealing with the environment, (twenty-nine of these focusing specifically on the environment); five decree-laws dealing with the status of laws concerned with fishing, mining, and industrial pollution; and fifty-three acts or international treaties of which Brazil is a signatory that directly or indirectly touch upon environmental concerns. In 1985 the legislature enacted Law 7.347, authorizing civic organizations to initiate suits, which has served as the legal basis for judicial actions enjoining acts threatening the environment. Thus even before Brazil became a party to the five environmental accords under study, it possessed broad and detailed environmental legislation. However, despite this and the adoption of the 1988 Constitution, the latter's lack of internal coherence, the absence of political commitment, and the public administration crisis have resulted in incomplete enforcement of the constitutional provisions and environmental laws.

Evidence of failure to comply abounds. For example, as of mid-1997 there have been no reports of judicial actions against the illegal trade in fauna, in spite of the nonbailable nature of the crime. A social conscience for implementing legislation is lacking. Sometimes, acts considered criminal are not socially condemned, such as owning wild animals in the northern region of the country. At other times, the law is rigorous but the means for its implementation are fragile and inefficient.

Congressional action on environmental issues also is inconsistent. For example, while Congress, in response to media attention, created a commission to investigate the activities of the multinational timber companies in the Amazon forest, it proceeded slowly in looking at legislation to strengthen environmental crimes, a subject that did not preoccupy the media but that is important to compliance with certain treaties.[1] However, at the beginning of 1998, Brazil finally approved a new environmental law that makes certain actions a crime and strengthens enforcement.

Responsibility for environmental protection at the federal level has changed over time. Since the 1930s a number of ministries have been involved in implementing and complying with policies related to forests, fishing, hunting, water, cultural heritage, and national parks. Until the 1980s, the Brazilian Institute for Forest Development (IBDF) and a number of other agencies scattered among the various ministries were responsible for the direct application of environmental regulations. IBDF's main role was to promote the development and exploitation of forests. The government centralized the administration of environmental policy when it created the Brazilian Institute for the Environment and Renewable Natural Resources (IBAMA) to replace the IBDF and other federal agencies in 1989. The Ministry of the Interior was in charge of public policy efforts aimed at environmental protection from 1898 to 1992. In 1992, the Ministry of the Environment assumed responsibility for this task. IBAMA operates under the Ministry of the Environment.

Paradoxes of the Media

Contradictions are evident in the state of the press. Brazil has newspapers and magazines with graphic and editorial quality similar to those in Europe and North

America. Two weekly magazines—*VEJA* and *Istoé*—are among the ten best-sellers in the world. There is a high degree of press freedom; Brazilian press investigations into public scandals led to the impeachment of former President Fernando Collor in 1992 and to the annulment of the political rights of deputies in 1993. Articles on the environment appear in the press almost every day.

However, in spite of an abundance of publications, communication through the media is hampered by a high rate of illiteracy among the people and the lack of purchasing power in the vast majority of households. These factors impede the adequate flow of information necessary for public participation in environmental protection. The reading public is small in relation to the population. (Less than three copies of magazines are sold per person per year. In the United States, the number is more than thirty copies per year.) The reading public does not exceed 40 million, out of 150 million Brazilians.

Television is the most popular mode of communication, reaching 95 percent of the country and over 60 percent of the population. In the 1990s, 73 percent of households own at least one television set. However, television in Brazil serves more as a means of entertainment than as a means of education. The television informs but, in general, does not investigate environmental issues. Hence, the majority of the population is not sufficiently exposed to the problem through this media.

Paradoxes of the Private Sector

Despite suffering from highly unequal income distribution and extreme unevenness in social and economic development, Brazil has achieved a remarkable level of social organization. According to Sergio Abranches, the number of NGOs created in Rio de Janeiro between 1970 and 1990 exceeded the total number created since the beginning of the century (Abranches 1993). *VEJA*, the leading popular journal, reported that in 1993 approximately 5,000 NGOs operated in Brazil with a total budget of U.S.$700 million (Bernardes and Nanne 1994). Forty percent of these NGOs focused on environmental issues. Other sources indicated that there were 1,533 NGOs in the National Environmental Institutions Cadastre in 1992. In 1996, a new census published by Mater Natura and World Wildlife Fund put the number of environmental NGOs at 725, still an impressive number.

During the 1990s the effectiveness of the environmental NGOs has been questionable. The activities of the majority are at the local or regional level. There is no coordinated or permanent NGO activity at the federal level that can pressure the Executive and the legislature to give priority to environmental issues on the national agenda.

While financial resources have been made available for environmental protection projects in Brazil, they have often been wasted. From 1994 to 1997, the

Brazilian Bank for Economic and Social Development (BNDES) allocated U.S.$150 million to finance projects to fight the pollution of Guanabara Bay. However, until mid-1997, not one cent had been used for this project. There was a lack of conscience, of interest, of involvement of the federal and state authorities, and of society's mobilization to stimulate the action of official agents and of the business community, who argued that the loan interest rates were very high.

While resources have been available to the business community for environmental protection, with the exception of the large companies, businesses have been largely uninterested. According to BNDES, during the 1990s over U.S.$2 billion was loaned to environmental projects of private companies, but at least half of the companies that benefited from these resources were large exporters who needed to follow international environmental standards. Many big enterprises—Banco do Brasil, Exxon, Alcan, Vale do Rio Doce, Shell, among others—invested millions of dollars in environmental projects.[2] The World Bank estimated that between 1992 and 1995, international funding sources invested U.S.$1,447 million in environmental projects in Brazil; these loans go to the government for disbursement.

The federal budgetary resources available to protect the environment also led to paradoxes. The Ministry of Culture (AC), responsible for all official cultural, policy, including the preservation of the historic heritage, had a very small budget, approximately U.S.$150 million in 1995.[3] Even so, because of the involvement of society and the authorities, among other factors, the AC achieved the best performance on treaties. The significant involvement of the various groups affected by the conservation of the historic heritage compensated for the financial limitations.

The federal government contributes little to Brazil's environmental budget. State and municipal governments and other public sources fund approximately 40 percent of Brazil's environmental costs. In 1992 and 1993 the federal budget allocated only U.S.$203 million and U.S.$242 million, respectively, for the implementation of public environmental policies. In 1995, the federal government allocated approximately U.S.$1,250 million to the Ministry of the Environment,[4] a little over 1 percent of the total national budget,[5] for environmental affairs, but this still has not provided the resources necessary to ensure compliance with environmental laws and policies. Consequently, implementation of treaties like CITES has been below expectations. The resources have been insufficient to maintain officials to monitor the forest for smugglers and fires. Even worse, these resources have not been used efficiently, nor have the activities of various units been coordinated.

The Brazilian paradoxes are interrelated. They permeate a country where the First World and the Third World meet. Brazilian paradoxes explain the inconsistency of the environmental policies in the country, as revealed in the analysis of the treaties below.

Implementation and Compliance

Before analyzing compliance with the five agreements, it is important to note that for the World Heritage Convention and CITES, it was possible to obtain extensive research materials at IBAMA, UNESCO, and the Instituto do Patrimônio Histórico (IPHAN). Furthermore, the press has covered these treaties amply since the early 1980s. However, the history of Brazil's activities prior to joining any of the agreements is more elusive, and the types of sources and data available for the treaties are varied. Another barrier to consistent analysis across treaties is that four of the treaties were considered under the military regime, when there was little public access to data and little or no participation by NGOs.

The World Heritage Convention

Brazil joined the World Heritage Convention (WHC) because of its desire to protect numerous natural and historical sites. As of December 1997, it had nine sites on the World Heritage List, including two natural sites. The listing of sites by year is shown in table 14.2.

As of mid-1997, UNESCO was examining whether the city of São Luis met the criteria for listing as a cultural site.

During the debates that led to the WHC's creation, the Brazilian delegation successfully argued that the agreement should not create international custody of areas on the WHC list, because it would impinge on state sovereignty and constitute a dangerous precedent in international law. It insisted that the treaty be voluntary

Table 14.2
World Heritage Sites in Brazil: Years Listed

Year	Natural Site	Cultural Site
1980		Historic Town of Ouro Preto
1982		Historic Centre of the Town of Olinda
1984		Jesuit Missions of the Guaranis*
1985		Historic Centre of Salvador de Bahia
		Sanctuary of Bom Jesus do Congonhas
1986	Iguaçu National Park	
1987		Brasilia
1991	Serra da Capivara National Park	
1997		Historic Center of Sao Luis
Total	2	7

* Site shared by Argentina and Brazil. Ruins of Sao Migueldas das Missoes in Brazil are part of this cultural site.

and cooperative, expressing an international effort to preserve historical sites that would be supplementary to similar national efforts.

Brazil also argued that the states, not the treaty organization, should be in charge of nominating sites to the World Heritage List. Brazil opposed U.S. efforts to apply the WHC to conserve natural areas, and pointed out that the 1972 Stockholm conference would be a better multilateral forum for the discussion of environmental issues.

Brazil joined the WHC in 1977 with reservations, because it disagreed with the financial contributions required. Its tardy adherence was mostly due to the military government's reluctance to publicly recognize an area under strong influence of the intellectual Left. Since the WHC did not contain mandatory clauses, it did not require the approval of the Brazilian Congress.

Actions over Time. In December 1977, after Brazil joined the WHC, Congress approved a law mandating the creation of areas of special tourist interest. In 1980, Brazil had its first site included on the World Heritage List: the historic city of Ouro Prêto, built during the colonial period.

Most of the cultural sites on the World Heritage List have been regulated by preexisting federal legislation, such as the Capanema Law (1935) on historic and cultural heritage and the 1961 Law 3.924 relating to archaeological sites. However, state and municipal regulations were introduced after the listing of Olinda, Brasília, and Salvador.

The cultural sites are under the administrative responsibility of IPHAN, which operates under the Ministry of Culture. In September 1994, President Itamar Franco promoted the transformation of the Instituto Brasileiro de Patrimônio Cultural (IBPC) into IPHAN as part of a reform aimed at making administration of conservation of historic sites and monuments more efficient. IPHAN is responsible for the preservation of all areas and sites registered by the Brazilian federal legislation, including 3,500 estates, monuments, museums, and parks, and six of the eight sites on the WHC list. While IPHAN has no police power of its own, it may call upon the Attorney General and the Federal Police to enforce its directives. It has a decentralized structure consisting of regional agencies that have had varying levels of success because of differences in their capacity to mobilize public and private initiatives.

The two natural sites are under the authority of a different federal body, Brazilian Institute for Environment and Renewable Natural Resources (IBAMA), which operates under the Ministry of the Environment.

Iguaçu Park, covering almost 170,000 hectares, is one of the oldest parks in the country and a major tourist attraction. It borders on a natural park of the same name in Argentina. In 1986, Iguaçu Park was included on the WHC list in the wake of an intense, four-year mobilization of NGOs aimed at preventing the paving of a road that crosses the preserved area. A public suit launched in federal court against the state of Paraná succeeded in preventing the paving. The NGOs initiating the suit maintained that the road would facilitate illegal timber activities in the park. The road remained unpaved as of mid-1997.

The other natural site on the Brazilian WHC list is the 100,000-hectare Serra da Capivara National Park, located in Piauí, one of the poorest states of the country. This park is treasured as the only remaining area where *caatinga* (scrub savanna) vegetation survives in its natural state, and was declared a national park in 1979 (decree 83.548).

The inclusion of the Serra da Capivara on the World Heritage List in 1991 grew out of the personal efforts of Professor Niede Guidon, a Brazilian archaeologist of international renown. Professor Guidon, who has carried out research in the area since the 1970s, created the Fundação do Homem Americano (Foundation of the American Man), a private entity that administers the park with IBAMA's authorization, in 1986.

Changes in the Behavior of Target Groups. In the early 1980s, coinciding with the political breakdown of the military regime, the Brazilian government began to pay more attention to the WHC. Architect and designer Aloisio Magalhaes, of the Secretariat of Culture in the Ministry of Education, was largely responsible for this change in behavior. During this period, Brazil made its first financial contribution to the WHC organization and joined the Committee on World Heritage. Because of the personal efforts of Professor Magalhaes, Brazil initiated procedures to include the city of Olinda on the World Heritage List.

At the same time that the national government increased its participation in the WHC, some state and local governments began to express great interest in environmental concerns. Their interest can be attributed to several factors: (1) the federal government's commitment to establishing the National Environmental Policy in 1981; (2) increasing international concern about environmental and historical questions; (3) pressure from residents of Brazil's more developed areas and the increasing interest of local communities; and (4) media focus on environmental and national history issues, which was boosted by the newly reestablished freedom of the press.

Parallel to the change in the Brazilian government's attitude, redemocratization permitted citizens to question government policies and to form public-interest groups, as reflected in the litigation to prevent the paving of the road in Iguaçu Park.

Is Brazil Achieving the Purposes of the WHC? According to the Ministry of Foreign Relations and UNESCO, Brazil was up to date on its financial obligations as of early 1997. However, not all WHC sites have been equally conserved, a reflection of Brazil's internal contradictions and the disparate administrative competences of the agencies involved. Several organizations, including at least two at the federal level, and state and municipal governments are responsible for implementing the treaty. A lack of resources, inconsistent public governance, and varying levels of social commitment to preserve the listed sites exacerbate the decentralization of command.

Brazil's fulfillment of the purposes of the WHC varies from site to site. This subsection considers Brazil's performance in relation to six of its eight sites and concludes that implementation varies widely among regions, but that the listing of the sites has had some notable effect.

Although Iguaçu Park had been a natural park for over sixty years before Brazil joined the WHC, its registration on the World Heritage List was crucial in aiding the community's long struggle to prevent paving of a road that crosses the park. After the site was listed, 400 families who illegally occupied 12,000 hectares of the park were transferred. In subsequent years, however, new tourism projects threatened to disfigure the landscape.

While local IBAMA officials affirmed that the park was in good condition despite the shortage of resources, the local community argued that the park was poorly maintained and that it lacked clearly marked borders and buffer zones. In 1996, the Minister of the Environment announced investments of U.S.$20 million during the next five years to improve tourist facilities and at the same time to provide adequate environmental controls and create a buffer zone.

Inclusion of the Ruins of São Miguel das Missoes on the World Heritage List (as part of the Jesuit Missions of the Guarinis) crowned the efforts the state of Rio Grande do Sul began in 1925 to restore the ruins. Local officials in charge of preserving the site have benefited from specialized training under the WHC. A 1990 administrative crisis, in which the thirteen officials responsible for the site were threatened with dismissal, was in part averted because the area was recognized as a World Heritage Site. Despite the lack of resources, the ruins of São Miguel were, as of mid-1997, in a good state of conservation.

Brasília, built in the 1950s, is the only contemporary architectural site on the WHC list. UNESCO officials visited the site in late 1994 and indicated that it was well preserved. This is due to the government's implementation of relevant legislation and to the fact that growth—in terms of population and economic activities—is mostly limited to the outskirts. Curiously, Brasília became a World Heritage Site before Brazil considered it a part of its national heritage. One individual, Governor José Aparecido de Oliveira, former Minister of Culture, was responsible, through his personal, direct efforts, for nominating Brasília to the World Heritage List.

The city of Salvador was also in fairly good condition, although restoration activities have been criticized. While it was listed as a World Heritage Site in 1985, a systematic effort to restore the area began only with the election of Governor Antonio Carlos Magalhaes in 1991. In 1994 the state government, with the support of private enterprises, finished a U.S.$30 million restoration of 350 old houses in the historic center of the city. Technical support from the WHC has led to specialized training in restoring monuments.

Other sites have had less success. Despite recent federal financial contributions, Serra da Capivara Park cannot count on a regular flow of government, private, or official international funds for infrastructure projects. In 1993 the park administration petitioned the WHC Secretariat to include the site on the list of threatened sites. UNESCO officials visited the site in October 1994, and reported that it was in good condition. In 1996, the park received a U.S.$1.2 million grant from the Inter-American Development Bank to establish occupation programs for the population living in the surrounding areas, and UNESCO has financed the registration of all the rock paintings in the parks.

Ouro Prêto, the first Brazilian site on the World Heritage List, is in the process of being restored with assistance from UNESCO, the city government, the Federal Development Bank, the Inter-American Development Bank, the Fundação do Museu de Arte Sacra da Paróquia do Pilar, businesses, private entities, and the community. While it would be unfair to conclude that the inclusion of the site on the World Heritage List is responsible for mobilizing all of these resources to preserve the site, it is fair to note that it became easier to gain international aid and support from private companies after it appeared on the World Heritage List.

Considering that the WHC is not an agreement obligating states to nominate sites and does not have financial resources commensurate with its requirements, Brazil's adherence to the convention has represented the development of a national consciousness for the preservation of historic and cultural sites and, in a sense, the environment. At the time of this study, it may have been too early to conclude that Brazil would continue to achieve the WHC's purposes. However, there is little doubt that the listed sites would be in worse condition had Brazil not joined.

Factors Explaining Implementation and Compliance. The size of Brazil and the economic and social disparities among the regions affect all of the agreements under study. Because of the significant inequalities among the regions in which the WHC sites are located, the sites occupy vastly different positions in the hierarchy of social and governmental concerns. Who is motivated to preserve the sites largely determines the extent and quality of the preservation efforts. Factors that have had a positive effect on implementation of and compliance with the treaty include the interest of the community in environmental and historic issues; the consciousness of the ruling elite in some of the sites listed; the efforts of some Brazilian personalities; the support of the business community; the positive intervention of the judiciary in investigating and condemning acts that threaten world heritages; and the support of international organizations.

Local communities have been critical to preserving sites. For example, the state of Rio Grande do Sul has been concerned with preserving the Ruins of São Miguel das Missoes since 1925. In 1990, the municipal government and the community, with the support of the local branch of the Brazilian Bar Association, took legal action to guarantee the continued employment of the officials who care for the site. In addition, the local government, in conjunction with the local community, successfully lobbied the Ministry of Culture and obtained special funding of U.S.$270,000 in 1993.

Community action also played an important role in Paraná in preserving Iguaçu Park. Community activists criticized the park's conditions and attempts to install ecologically destructive tourist projects, and conveyed their concerns to UNESCO. Iguaçu Park had only eighteen civil servants in charge of its care in 1997.

Meanwhile, in Piauí, one of the poorest states in the country, there is little local concern about the Serra da Capivara Park. Besides providing irregular funding for maintenance, the federal administration had hired only four officials to work there in 1993. What little has been achieved, including its placement on the World Heritage List, is due to the personal efforts of one scientist.

In Brasília, the local political environment should be given credit for the preservationist effort. The local government has been concerned with preserving the city's innovative character since the city's founding in the late 1950s.

The judicial system has begun to protect national historical and environmental sites, particularly since Congress enacted the Law of Diffuse Rights of 1985, which gave public-interest groups and attorneys standing to defend the public interest, especially in matters of the environment and consumer rights. Some legal actions have helped further the WHC's interests. SODECA, a community group in Olinda, obtained an injunction to impede the construction of a branch of a federal bank in the preserved area. The courts were involved with the WHC in legal procedures related to Iguaçu Park and the Ruins of São Miguel das Missoes. In contrast, a ten-year-old suit against landowners with false titles situated within the limits of the Serra da Capivara Park continued, as of mid-1997, without resolution.

The business community has also become increasingly involved in the preservation effort, although it would be able to participate more effectively if the government instituted tax incentives. Even without a more effective tax policy, however, significant examples exist of the business community's involvement in the preservation of historic and environmental sites: Fundação Roberto Marinho, connected to Globo Television Network, sponsored restoration works in Ouro Prêto and Salvador, and Pirelli, the Italian tire firm, sponsored the illumination of historical monuments in Ouro Prêto. Other companies, such as Alcan, Exxon, Banco Real, and Banco do Brasil, sponsor the conservation of sites not on the World Heritage List.

Many companies support the preservation of historic and natural sites in an effort to improve their image; supporting preservation efforts is recognized as an important marketing tool. The national press generally provides good coverage to preservation issues and emphasizes the fact that some of the sites are of world heritage status. Undoubtedly, the inclusion of a site on the World Heritage List brings it increased attention from the Brazilian press.

Support by UNESCO and organizations such as the Inter-American Development Bank has also helped Brazil to implement and comply with the treaty. UNESCO's education and training programs and its support for specific programs have been particularly helpful. In 1997, there were at least three cooperative education projects involving UNESCO and Brazilian universities that aimed at restoring and preserving the historic heritage of Brazil.

Some factors that have impeded better implementation and compliance with the WHC are not directly tied to the treaty. Most important, World Heritage Sites are located in seven Brazilian states, each with different social and economic realities and uneven standards of public administration. For example, in Piauí, local law enforcement has not been as efficient as that in Paraná in preventing environmental damage to the Serra da Capivara Park.

Implementation also depends on the commitment of many public organizations, such as IBAMA, IPHAN, Commission for the Advancement of Higher Education Staff, the federal universities, and the Federal Police, among others. While administrative decentralization could have a positive effect if there were adequate budgets and reasonable efficiency, this is not the case. The federal administration and the states have few resources for the proper management of the historic and environmental World Heritage Sites in Brazil.

Except for a few sites, administrative and legal incompetence and insufficient financial resources continue to plague Brazil's WHC sites. Iguaçu Park, for example, turns the visitors' fees it collects over to the federal government. However, the federal government fails to return the 50 percent share earmarked for conserving the park. IPHAN is dependent on the goodwill of local governments and the Federal Police, which have other priorities and are ill-equipped to exercise their administrative and supervisory powers of historical and cultural preservation. IBAMA's situation is analogous: it, too, lacks the officials and resources it needs to supervise Brazil's extensive environmental resources.

Despite the examples of community participation given above, the lack of a general preservationist conscience still poses a problem. Acts suggesting a lack of public awareness or interest, such as littering, graffiti, and theft of historic pieces, still occur. In Brazil, the preservation of the historical heritage is still not an important social concern.

CITES

CITES is especially important for Brazil, which, because of its rich natural resources and wilderness areas, is particularly vulnerable to poachers, smugglers, and others who trade in endangered species. Implementation is problematic, however, since the protected species occur in poor, remote areas where government presence is thin and local economies have long depended on extracting natural resources.

Interviews indicated that Brazil did not participate in the initial CITES meetings. Brazil ratified the treaty in 1975, the same year it took effect internationally. Because of its binding nature, the Brazilian Congress formally approved the treaty as part of the implementation process. At the time, the CITES regulations appeared to be less rigorous than the Brazilian legislation already in force, including the Forest Code of 1965 and the 1967 Hunting and Fishing Codes. The Brazilian authorities' perception that they were already adequately regulating commerce in rare species, and their lack of awareness of serious threats to the Brazilian flora and fauna, facilitated the treaty's reception but retarded its implementation.

Actions over Time. Brazil's political situation at the time it joined CITES affected its initial implementation efforts. In 1975, the military regime promoted "developmentalism," signified by the expression "Greater Brazil," which paid no heed to environmental concerns. There was no national agency to supervise implementation of the treaty.

Only with the government's establishment of the National Environmental Program in 1981 did it begin to pay greater attention to CITES. The Brazilian Institute of Forest Development (IBDF) assumed administrative authority for the treaty and became the scientific authority on fauna. Jardim Botánico do Rio de Janeiro (Botanical Garden of Rio de Janeiro) became the scientific authority on flora. In 1982, the IBDF established an export and import licensing system for fauna and flora covered by CITES. Seven years later, IBAMA was created, and inherited most of IBDF's responsibilities, including the administration of CITES.

Other agencies have been involved with CITES, although there have been no official coordinating mechanisms between IBAMA and the other agencies. These agencies include (a) the International Trade Department, which issues permits for exports and imports of CITES species; (b) the Ministry of Agriculture, which operates through the Division of Sanitary Control, and inspects plants and animals in ports and airports; (c) the Federal Revenue Service, which collects taxes and supervises import and export documentation; and (d) the Federal Police, which is in charge of preventing the smuggling of endangered species.

Brazil has fulfilled its procedural obligations under the treaty. It filed the required annual reports on export and import of endangered species through 1995 (the 1996 report was not due until October 3, 1997). The reports were filed within two months of the date they were due. In 1992, Brazil developed and offered software for controlling CITES information to treaty parties. (See table 14.4, at the end of this chapter, for Brazil's compliance with reporting requirements of CITES.)

Changes in the Behavior of Target Groups. In 1986, an important shift occurred that facilitated compliance with CITES: neighboring countries started to collaborate in implementing CITES. Several adjacent states, such as Bolivia and Paraguay, started to repress the smuggling of endangered species coming from Brazil. It was suspected that these countries had previously legalized these species and then exported them to Europe and North America. Exporters who once legally traded in CITES species accepted and adapted to the treaty's rules. However, smugglers who took advantage of Brazil's fragile supervisory structure continued to challenge the government's efforts to comply with CITES.

NGO interest in CITES activities has grown. Although in 1997 there were no NGOs dedicated exclusively to monitoring and controlling the commerce in wild fauna and flora, during the 1980s NGOS showed increasing interest in projects to preserve endangered species. Several initiatives supported by NGOs stand out: the Tamar Project, to conserve sea turtles and their ecosystems; the Mico-Leão Dourado Project, to protect the remains of the Mata Atlântica, the natural habitat of the gold lion monkey and the work of the SOS-Mata Atlântica Foundation, for the protection of the coastal forest in the southeast region of Brazil.

Some NGOs, like the Defensores da Terra and the SOS Mata-Atlântica Foundation, with the support of the Brazilian Bar Association (Mato Grosso state), participated in the debates on a congressional bill to make the legislation pertaining to the commerce in endangered species more efficient. Unfortunately, by mid-1997, the bill had not yet been approved by the Congress.

International NGOs such as World Wildlife Fund and TRAFFIC International have also become active. TRAFFIC, which in 1997 was about to open an office in Brazil, disseminated information about the illegal commerce in species listed by the treaty. World Wildlife has been very active in Brazil in the 1990s, and in 1995 issued an important report on traffic in wild animals in Brazil. It also distributed 500 kits with videos and printed information on the commerce in illegal species to NGOs, schools, and universities to improve the public awareness of the problem. This initiative led to a petition with 32,000 signatures, delivered to the Minister of the Environment in July 1996, asking for a census of endangered species and more resources for environmental protection and education programs.

Since 1980, the implementation of CITES has stimulated more responsible behavior among those who wish to legally export and import endangered species. The increasing number of licenses reflected this change in behavior. Nevertheless, the legal commerce in species listed by CITES represents only 5 percent of the total.

Smugglers continue to operate intensively, taking advantage of the fragility of the country's system of control and repression of the illegal trade in wild species.

Is Brazil Achieving the Purposes of CITES? Brazil does not fulfill the rigorous standards for internal regulation or the precepts recommended by CITES. With respect to its financial obligations, Brazil is up to date as of 1996. It was in arrears from 1989 to 1992, but it met its financial obligations in 1992. Brazil also satisfied its procedural obligation to provide an annual report of its export and import permits. With respect to the regulation of the international trade in endangered species, Brazil's performance is very bad.

According to TRAFFIC International, Brazil was responsible for 15 percent of the world's illegal trade in fauna. This represents around 12 million animals per year, 30 percent of which were smuggled to other countries. The illegal commerce in wild animals has a value of around U.S.$1.5 billion per year. This information was not officially confirmed in 1997. Nevertheless, unofficially, Brazilian authorities have recognized the lack of success of the combat against the wild animal traffic in the country. According to a report in *Folha de São Paulo*, 7,000 birds were smuggled every year from Amazonas state (Lozano 1994). The same report confirmed that alligator skins (*pele de jacaré*), parrots, fish, and other creatures were illegally traded in the Amazon region. Comparing the information gathered about the contraband trade in wild animals and the number of export licenses issued by IBAMA, shown in table 14.3, we can see the size of the problem in Brazil.

There was no precise information as of mid-1997 on the extent of trade in wild flora. However, it was known to be extensive and to involve contraband wildflowers

Table 14.3
Brazil's issuance of export and import licenses for CITES species

Year	Export licenses	Import licenses
1986	497	—
1987	617	7
1988	700	25
1989	673	50
1990	695	67
1991	705	62
1992	650	32
1993	—	—
1994	639	92
1995	474	162
1996	688	195

and medicinal plants. There has been an intense exploitation of and illegal trade in mahogany, but in mid-1996 cutting was suspended for two years.

At the CITES Conference of Parties in November 1994, Brazil was sharply criticized for its performance in implementing CITES. It was a main target for international attempts to include mahogany on the CITES list of endangered species.

Unfortunately, smuggling of endangered species in Brazil has continued because it is still a profitable industry. The trade in CITES-protected species is probably just the tip of the iceberg of such trade, and without CITES the situation would have probably been much worse.

Factors Explaining Implementation and Compliance. While Brazilian legislation implementation of CITES is in many respects rigorous, many factors make compliance with CITES difficult. These include the country's vast size and unsupervised borders, the lack of collaboration between the federal and state agencies, the differences among the states in their willingness to enforce CITES, the absence of sufficient qualified personnel to control and repress illegal commerce, the poverty of people living in areas rich in endangered species of fauna and flora, and the absence of centers for separating and reintroducing animals that are apprehended in illegal trades.

Brazil covers a vast area with poorly supervised frontiers. In 1994, the government bought a surveillance radar system from a group of American companies led by Raytheon. The government expects to install this system in the Amazon over the next few years. However, significant obstacles remain. Brazil's ability to implement CITES is complicated by the fact that many of its environmentally rich regions are accessible from border zones. It lacks the human and financial resources needed to control these areas effectively, not only in respect to CITES but also to prevent forest burning and the indiscriminate cutting of trees.

IBAMA has assigned few civil servants to the protection of wildlife. In the vast Amazonas state, IBAMA has only twenty-eight officials working to control the illegal commerce in animals and wood. One of the reasons for the shortage of manpower is that IBAMA has difficulty recruiting skilled people. In general, IBAMA employees earn low salaries and lack a defined career track.

Eighty percent of CITES exports from Brazil consists of flora, and almost all of it comes from technically sophisticated plant nurseries in the center-south. This figure suggests that Brazil is not making rational use of the Amazon region's immense flora potential and that the region must therefore be supplying clandestine channels of commerce. This discrepancy is a product of the inaccessibility of forest areas in the north and the fact that local and state authorities rarely collaborate to restrain illegal commerce, both domestic and foreign, in threatened species. In areas with high levels of poverty and ignorance, residents depend on illegal commerce in protected species as a means of survival.

Brazil generates a significant internal demand for wild animals. An extensive report by the newspaper *O Globo* on June 21, 1991, described Rio de Janeiro as a huge center of national and international illegal trade in wildlife. It is estimated that in the metropolitan region of Rio de Janeiro alone, more than 85,000 parrots are kept captive. Police repression has done little to curb Rio's wildlife black market.

In the vast Pantanál swamplands of Mato Grosso state, public agencies do little to stop the indiscriminate hunting of alligators. Ironically, a project for the sustainable commercial exploitation of alligator leather waited more than fifteen months for a business permit. Neither government nor private entrepreneurs have shown much support for or interest in implementing the rational exploitation of flora and fauna because of the long time period involved. Long-term projects are incompatible with the frequent changes of political command in the environmental agencies and with the electoral interest of state authorities. Moreover, absent continual government repression, illegal trade is much more profitable than a legalized business.

From a scientific perspective, CITES focuses on fragile ecosystems consisting of areas with high density of fauna and flora. In Brazil, in contrast with temperate countries, the biomass contains a large number of species, although each species is relatively few in numbers. Species diffusion and diversity require sustainable management on a case-specific approach and demand specialized scientific understanding. On the other hand, these characteristics have made the extinction of particular species in Brazil gradual, in contrast with the sudden disappearance of some large herds and flocks in the United States and Europe.

At the Fort Lauderdale, Florida, Conference of the Parties, members concluded that Brazilian legislation was inadequate for controlling trade in endangered species, as were the laws of Argentina, Spain, Japan, and other countries. However, the quality of Brazil's legislation is not the paramount concern. More important is Brazil's inability to enforce its existing legislation. Brazilian legislation imposes strict punishment: for example, selling or killing a wild bird leads to arrest without possibility of posting bail. Improved implementation of existing legislation is key to ending illegal trade in endangered species, but it is almost impossible to achieve where the population is sparse, there are few or no roads, and the government presence is thin.

CITES has benefited Brazil. IBAMA's administrative stability in implementing CITES is an exception in the Brazilian administration of international trade. Brazil is gradually obtaining the support of its neighbors in the struggle against smuggling and in the "legalization" of CITES species exports through these countries; collaboration with consumer countries is improving; and authorities have been informed of seizures of illegal imports that originated in Brazil. However, the Brazilian government had yet to appropriate funds to finance training programs in 1996, and there were no programs for carrying out censuses of the endangered species. According to a list published by IBAMA in 1989, until early 1984, censuses were taken on fewer than 5 species among more than 207 species in the process of extinction. The species surveyed include the Pantanal alligator and the gold lion monkey.

NGOs have played an important role in the implementation of CITES. Some Brazilian NGOs sought to coordinate with TRAFFIC-International to monitor trade in endangered species. Other NGOs and INGOs, such as Greenpeace and the WWF, have lobbied the Brazilian authorities to improve the implementation of CITES and compliance with Brazilian legislation related to CITES.

ITTA

The ITTA is a pioneer treaty drafted to promote sustainable trade in tropical woods. Elaborated within the spirit of UNCTAD proposals, the main goal of which was to assure fair conditions of trade between industrialized and developing countries, the ITTA has assumed growing importance in recent years as a tool for promoting international forest preservation.

Brazil was active in the negotiation of the ITTA and subsequently joined the accord in order to develop its share of the international trade in tropical woods, estimated in 1985 at U.S.$7 billion. While not a large exporter of wood at the time, Brazil had 300 million hectares of forest resources (Malaysia, the largest timber exporter at the time, had only 13 million hectares). Second, Brazil expected that cooperation with the ITTA would encourage regional exploitation of natural resources and the development of wood-processing industries.

Brazil joined the treaty in March 1985. Because the agreement did not contain mandatory clauses, it did not require approval by the Brazilian Congress.

The interministerial working group created to examine Brazil's adherence to the ITTA recommended that Brazil serve as the headquarters of the ITTO, the ITTA Secretariat, because of its potential as an important exporter, its position as a natural leader among wood-producing countries in Latin America and Africa, and the benefit of holding 7 percent of the votes in the treaty (serving as headquarters would entitle the country to additional votes). However, Japan's offer of technical, bureaucratic, and financial support to the ITTO made Brazilian aspirations impractical. Brazil initially supported Holland as the headquarters country. Later, it changed its support to Japan. Unconfirmed rumors at that time suggested that Brazil's support of Japan's candidacy was connected to being able to sell Japan the wood needed for the construction of the Tucurui hydroelectric dam.

During the 1990s, the ITTA was the subject of intense debates among member countries. In 1990, at Bali, the members committed themselves to implementing sustainable forestry standards in the export of tropical wood by the year 2000, and to increased funding for sustainable forestry management projects. The 1994 successor agreement to the ITTA consolidated these developments. Brazil participated in the negotiations and, as of May 1997, has adhered to the new agreement in the hope that it will stimulate international trade in tropical timber on a sustainable basis.

Actions over Time. The Division of Basic Products of the Ministry of Foreign Relations, which represents Brazil at treaty meetings, is also in charge of implementing ITTA. IBAMA helps the Division to select projects that will receive technical and financial support from the ITTO.

In general, Brazilian law supports ITTA objectives. For example, article 15 of the Forest Code of 1965 states that the Amazon forest, from which Brazil extracts most of its timber, must be exploited in a "rational form." In 1987, the IBDF established a policy permitting sustainable logging in areas of less than fifty hectares. In areas over fifty hectares, logging activities must be connected with an IBAMA-approved forest management program. In July 1989, a new law (7.803) strengthened this policy considerably; programs for logging in primary forests and forest plantations, whether privately or publicly owned, must include plans to reforest and adopt forest management practices compatible with the local ecosystem. Furthermore, in 1994, IBAMA issued two new decrees that established more rigid requirements for the sustainable management of Brazilian forests. One requires an Environmental Impact Report (EIR) from anyone who invests in enterprises that may cause environmental damage, and the other outlines the procedures that must be followed in order to protect the environment. Although these decrees are important for preserving forests, some NGOs have complained that they do not go far enough, and note that an EIR is not required for the logging of areas of less than 2,000 hectares.

Brazil adopted measures in 1990 to limit nonsustainable extraction of mahogany, a wood of great commercial value that accounts for 70 percent of the country's wood exports and the exploitation of which is the target of great international pressure. These measures were (a) the creation of a quota system intended to reduce the volume of mahogany exports; (b) a 50 percent reduction in unprocessed exports; (c) the limitation of permitted mahogany exports for new exporting firms to one-sixth of total annual sales; and (d) the maintenance of current levels of mahogany exports for firms that carry out reforestation projects. At the 1994 CITES Conference of the Parties, international pressures against Brazilian mahogany exports culminated with the Netherlands' proposition that mahogany be considered an endangered species.

In July 1996, as a result of international pressure, Brazil imposed a two-year ban on the exploitation of mahogany.[6] It also stipulated that cutting in the Amazon forest could not exceed 20 percent of the total area. The World Wildlife Federation considered these good initiatives, according to its director in Brazil, Garo Batmanian.

However, this NGO also expressed its misgivings that the lack of control would stimulate the harvest of mahogany in neighboring countries, especially Peru and Colombia. In January 1998, the President of Brazil announced in London that Brazil would join the World Wildlife Campaign for Forests by committing to protect 10 percent of its forests to the year 2000. This would triple the forest area under conservation.

The implementation of the ITTA in Brazil occurs through government financing of projects proposed by IBAMA, other state and municipal agencies (including municipal administrations), foundations, NGOs, and other organizations. In general, these projects deal with four aspects of tropical forest exploitation: forest management, training of personnel, dissemination of information and research, and development of new technologies. Projects are presented to the Division of Base Products of the Ministry of Foreign Relations, which sends them to the Brazilian Agency for Cooperation (one of its divisions), to IBAMA, and to the Ministry of Environment.

Between 1987 and 1990, the ITTO approved only two small projects, both destined for IBAMA (at the time, IBDF), with a total value of U.S.$200,000. Between 1990 and 1992, the ITTO approved disbursements of approximately U.S.$8 million for fifteen projects, among them two projects that included counterpart financing by the Brazilian government. At the end of 1993, over six projects were in the final phase of approval and waiting for funding.

Between 1994 and 1997, only seven projects were approved and funded under the treaty, totaling approximately U.S.$5 million. Four were carried out by NGOs. Three of these four were promoted by wood trade associations. The rest were implemented by official entities, such as the Universidade Federal do Paraná Foundation and IBAMA.

Changes in the Behavior of the Target Groups. The main targets of the ITTA are the federal government, which implements it; the state governments, on whose territory large reserves of tropical wood are located; the companies that cut and sell wood; indigenous groups who live in the large natural reserves; and the NGOs and other organizations that can participate in projects and programs in the field of the treaty.

For Brazil, the main focus of the treaty has been the possible gain of a market for tropical wood. The ITTA has lost importance because of other initiatives directed at the development of sustainable forestry practices, such as the Pilot Program for the Amazon, which the G-7 funded in part, and because it was perceived to have failed in its objective of promoting the export of processed tropical wood. State governments, especially in the Amazon region, have been critical of conservationist proposals. They believe that federal authorities, NGOs, and the international community are insensitive to the social and economic realities of the region.

Brazil's failure to implement the ITTA can also be attributed in part to state government complicity with business interests in the area of forest exploitation. IBAMA estimated that wood smuggling, mostly from the Amazon region, would total U.S.$81 million in 1994, more than half of the legal mahogany exports in 1993. IBAMA officials admit that local and state governments in the Amazon region support and protect certain wood smugglers, at least by omission. For example, in Pará state, loggers built over 3,000 kilometers of illegal roads passing through protected and federal areas in order to transport wood. According to reports by Greenpeace and Núcleo de Direitos Indígenas, local authorities do not interfere with the use of these roads. A lack of personnel explains part of the government's laxness. IBAMA has only 28 officials to oversee the entire Brazilian Amazon—only one official per 42,000 square kilometers.

In general, the business community has not played a significant role in implementing the ITTA, other than to thwart its objectives. No big companies were involved in the projects supported by the ITTA. Loggers in the northern region have not been responsive to treaty proposals for sustainable development because these would reduce their profits. They have had no reason to comply, because they are aware of the government's difficulties in enforcing environmental legislation. The Association of Wood-Exporting Industries of the state of Pará (AIMEX), an organization of wood exporters mainly from Pará and Amapá states, signed an agreement with the Ministry of Environment in January 1993 that included a pledge not to extract wood from environmental preserves and/or indigenous areas. However, in spite of signing this agreement, loggers affiliated with AIMEX have continued their activities in prohibited areas.

NGO activity in relation to the ITTA falls into two categories. The first concerns the projects financed by the ITTO, and the second relates to the preservation of tropical forests in general. In spite of the participation of well-known NGOs in ITTA projects, including Fundação Pro-Natureza (FUNATURA), Núcleo de Direitos Indígenas (NDI), and Instituto de Estudos Amazônicos e Ambientais (IEA), NGOs in general have rejected the commercial aspects of the treaty and have been loath to participate in its projects. Some NGOs consider the ITTA to be a cover to protect commercial interests. The low level of funding allocated to ITTA projects also explains the low level of NGO involvement (Fearnside 1994).

However, a large number of NGOs have been active in working to preserve tropical forests. Grupo de Trabalho Amazônico (GTA) represents 240 NGOs from the Amazon states or connected with the Amazon that participate in the Pilot Program for the Conservation of Tropical Forests, supported by the G-7. These NGOs have a higher level of interest in preserving tropical forests than do NGOs from other areas.

Are the Purposes of the Treaty Being Achieved? According to data from the Ministry of Foreign Relations, Brazil is meeting its bureaucratic obligations to the treaty. Brazil participated in the thirteen meetings held by the end of 1993 and presented a description and evaluation of the Brazilian forest sector elaborated by FUNATURA and IBAMA.

Although Brazil has met the treaty's administrative requirements, it has only partially complied with ITTA objectives. It has developed few projects in the ambit of the ITTA. Brazilian exports of tropical woods have not been substantially affected by the treaty; on the contrary, some species, like mahogany, have been exported in ever greater quantities with only a low degree of processing. Despite certain projects in progress, there has not been a significant volume of technology and information transfer on the management of tropical forests. Nor have consumer countries opened their markets to wood products processed in timber-extracting countries like Brazil.

Blame for the failure to implement ITTA cannot be placed on Brazil alone. This treaty has ambitious goals and limited resources in relation to the strong commercial interests of producing and consuming countries. Brazil has viewed ITTA as an unfair commercial agreement; its environmental clauses are increasingly restrictive and reduce the competitiveness of tropical woods relative to temperate woods, which the treaty does not cover but that represent 90 percent of the international timber trade. The new ITTA agreement redresses some of these problems.

Factors Explaining Implementation and Compliance. Exports of tropical timber represented less than 0.2 percent of total national exports and therefore were not of great importance to Brazil's overall international trade. Of this total, 70 percent of wood exports originated in the northern (Amazonian) region. These exports generated few jobs or other benefits for the local population, since there is no significant processing involved. Most logging activity in northern Brazil supplied the needs of southern Brazil. As a producer member of the ITTA, Brazil is an atypical case, given that its exports of tropical timber have been such a small part of its overall trade volume. Brazil's importance as a member country stems from its potential for future timber export.

Despite Brazil's rigorous environmental legislation, including detailed conditions for the commercial cutting of wood, logging has not been carried out within reasonable sustainability parameters. For example, until recently IBAMA could authorize the logging of areas up to fifty hectares, but these authorizations, generally for agricultural purposes, did not specify the quantity of wood to be extracted. As a result, loggers could easily obtain a "legalization" of wood illegally extracted by attributing its harvest to farmers whom IBAMA authorized to log. Although forestry issues have reached scandalous proportions in Brazil—where, according to data in 1991, 2 hectares per minute were deforested, totaling 1.11 million hectares per year—neither the government nor the Brazilian elite is sufficiently committed to strengthening sustainable development practices (Fearnside. 1993).

In the 1990s, new actors have appeared: multinational timber companies from Malaysia, Portugal, the United States, and Denmark, among other countries. Supported by the local and state authorities of the Amazon region and with IBAMA's fragile surveillance, twenty-two of these companies, mainly from Asia, operated at efficiency levels never before attained in terms of wood extraction.

With little or no environmental concern, these companies have claimed to be cutting only more common kinds of wood from the Amazon forest, although they have extracted more highly valued wood. A study concluded in 1996 by the Strategic Affairs Secretariat, an intelligence agency of the Presidency of the Republic, accused these companies of smuggling contraband wood, permit falsification, and environmental devastation, and the authorities of negligence. Unfortunately there is not, except in rare cases, an environmental concern about the extraction of wood by the companies operating in the Amazon region.[7] The goals of self-sustainability recommended by the ITTA are irrelevant to them.[8]

Although Brazilian laws regarding timber extraction exceed the ITTA's provisions, their implementation is difficult. IBAMA has had few officials, and scant equipment and resources, to monitor the logging companies and to ascertain that the wood really comes from the area for which the logging authorization was given.[9] This is especially important because IBAMA can authorize the clearing of forest for agriculture. These authorizations do not specify the amount of lumber to be cut and thus can be used as cover for wood taken from other areas.

In addition, the resources available for implementation are few. Until mid-1997, only U.S.$13.3 million of the treaty's funds went to a little over twenty projects in Brazil. In part, the small number of projects reveals the lack of interest in or the mistrust of the NGOs in relation to the goals of the treaty. The ITTA has had little impact in Brazil.

Moreover, state governments, especially in the Amazon region, are critical of conservationist proposals. Contacts with state officials show their belief that the federal government and the international organizations are insensitive to the social and economic realities of the region. Consequently, the state governments are not significantly involved in implementing the treaty.

Furthermore, the communities located in the regions of tropical wood extraction are economically dependent on the exploitation of natural resources. According to the NDI, some indigenous groups have defended their right to sell timber from their reserves. Also, in general, the local communities have not possessed an adequate level of environmental consciousness. Hence, in addition to the forest management projects, Brazil needs educational projects on sustainable forestry and rational exploitation of the environment.

The statement of diplomat José Alfredo Graça Lima, head of Brazil's permanent mission in Geneva, to various international organizations synthesizes the disappointment of the Brazilian government in the 1983 ITTA: "Ten years ago, we accepted an agreement about trees imagining that it would foment international trade in that product. We were fooled."

London Dumping Convention

Marine pollution is another problem Brazil inevitably faces, given its extensive coastline. In the early 1980s, a country that was party to the London Dumping Convention (now known as the London Convention of 1972, or LC) dumped chemical wastes in Brazilian waters, thereby demonstrating Brazil's urgent need to affiliate itself with the Convention. Brazil was not involved with treaty negotiations and did not join the LC until 1982. The Ministry of Foreign Relations, the Ministry of the Navy, and the Environmental Division of the Ministry of the Interior were the forces behind Brazil's signing of the convention.

The Ministry of the Navy presented two arguments in support of joining the LC: (a) joining would not create new obligations for Brazil because it did not dump toxic waste in the sea; and (b) with its immense Atlantic coast, Brazil had a strong interest in controlling ocean dumping. On the basis of these arguments, President João Baptista Figueiredo sent a message to Congress, which approved adherence to the LC in March 1982. However, the fact that Brazil already had detailed environmental legislation addressing marine pollution in place, and that it did not dump toxic substances in the sea, led to little government interest in drafting additional legislation based on the LC.

Actions over Time. The Brazilian government's perception that Brazil did not dump wastes as defined by the treaty discouraged it from implementing the LC. For example, in 1983 the Directory of Ports and Coasts of the Ministry of the Navy, which regulated the dumping of oil and other pollutants in national waters, issued an internal directive, Portomarinst no.32/02 (eventually implemented in June 1992), in which it referred to two existing laws as legal precedents, one on maritime traffic (Decree 87.648/1982) and one on maritime and river pollution (Law 5.357/1967). The Directory dedicated five lines to the LC in its directive, in which it made regional environmental preservation agencies responsible for issuing written technical opinions about what substances could be dumped into marine waters. According to the directive, the Ministry of the Navy would play only a bureaucratic and administrative role in the issuance of these written opinions. The Division of the Environment of the Ministry of the Interior also did little to ensure that an agency assumed responsibility for implementing the LC.

No acts or administrative decrees, beyond existing legislation concerning marine and fluvial pollution, regulated the implementation of the LC in Brazil in 1997. A group of IBAMA experts, working in conjunction with the Ministries of Foreign Relations and of the Navy, scientific institutions, and the National Nuclear Energy Commission, was in the process of designing a project for the administrative regulation of the LC.

As of late July 1997, Brazil had fulfilled its procedural obligations to provide annual reports on dumping permits issued for that year to the secretariat at the International Maritime Organization. While the 1991 report was not filed until January 24, 1996, and the 1992–1995 reports were filed on March 29, 1996, Brazil filed its 1996 report on July 2, 1997, indicating a continuing intention to comply with the requirement. (See table 14.4, near the end of this chapter, for Brazil's compliance with reporting requirements of the LC.)

Changes in the Behavior of Target Groups. The Ministry of the Navy is responsible for supervising Brazilian waters, including dumping authorizations and assessment of fines. During the 1990s, the Ministry of the Environment became more involved with the LC. Since the Ministry of Foreign Affairs' recommendation that Brazil join the LC in 1982, however, the Navy has paid little attention to the treaty. In part, a lack of technical and financial resources has hindered its supervision of territorial waters. In addition, its steadfast belief that Brazil does not carry out ocean dumping has given the Navy little impetus to improve supervision. According to the Navy, the only form of ocean dumping that Brazilian actors have practiced regularly is port dredging, which has not conformed to the recommendations of the LC's New Assessment Procedures. The Navy also reported some incidental dumping, attributable to accidents at sea.

The main target group for the treaty has therefore been the federal and state administrations. However, the Brazilian port administrations, chiefly managed by federal or state agencies, have practically ignored the LC. The business sector, including the fishing sector, has not been involved in implementing the treaty because it has not dumped toxic waste in Brazilian ocean waters in forms covered by the treaty.

Brazil has modified its position on the dumping of radioactive wastes. Previously, considering the impossibility of dumping of these wastes on land because of vehement state government opposition, Brazil opposed a total ban. By 1997, however, even without a clear policy for disposing of these wastes on land, Brazil no longer considered sea dumping to be a viable option. Brazil changed its position in order to reconcile it with the position of the majority of South Atlantic countries, with the goal of transforming the South Atlantic into an area free of pollution caused by nonregional countries. The new Brazilian position is generally compatible with arguments that Greenpeace International advanced in defense of the treaty.

Is Brazil Achieving the Purposes of the Treaty? As of September 1992, Brazil had yet to inform the LC Secretariat of the name of the administrative authority charged with implementing the treaty in Brazil. IBAMA was preparing to assume this function in 1996. Brazil's lack of leadership made it impossible for it to fulfill its administrative obligations under the LC. Brazil has failed to develop a system for issuing dumping permits and has enacted no provision requiring the government to report dumping activities to the Secretariat.

Nevertheless, Brazil practices neither of the forms of dumping banned by the convention, nor those that were permitted in territorial waters, the Exclusive Economic Zone (EEZ), or international waters. Therefore, despite Brazil's failure to implement the treaty adequately, it has not violated the LC's substantive part.

Factors Explaining Implementation and Compliance. Brazil has 7,308 kilometers of coastline. Decree 1.089/1970, issued in 1970 by the military regime and enacted without any international negotiation, created a territorial sea extending 200 miles from the coast.

The Convention of the United Nations of the Law of the Sea, concluded in Montego Bay, Jamaica, in 1982, of which Brazil is a party sanctioned in practice Brazilian aspirations to create an EEZ. However, the demarcation of the limits of the territorial sea and EEZ, as well as the scientific and economic studies that the Convention of the Law of the Sea obligates treaty parties to undertake, have been delayed by the lack of resources and political will. IBAMA has struggled with enormous budget and human resource difficulties. Its Division of Environmental Protection within the Department of Environmental Quality designated four officials responsible for five international conventions, including the LC, in 1997.

In effect, the Brazilian government had only superficial knowledge of the underwater and subsoil characteristics of the Brazilian continental shelf and the EEZ. Studies of the marine currents along the Brazilian coasts were equally incomplete, especially concerning their effects on the distribution of pollutants. In light of these facts, the prospects for implementation of the LC in Brazil were poor. The protection of Brazilian waters from clandestine dumping, an argument that motivated Brazilian adherence to the treaty, has not received the required attention.

The press has given greater coverage to questions related to land-based sources of marine pollution. The LC, like the majority of treaties signed by Brazil (with the exception of the conventions on ozone and the WHC), has been largely ignored. Brazilian NGOs have shown little or no interest in the LC. The question of the demarcation and use of the territorial sea and EEZ was not a subject of national debate, as was, for example, the question of the demarcation of indigenous peoples' lands and areas of environmental protection. The business sector has not shown interest in ocean dumping, an indication that illegal dumping of toxic wastes was perhaps occurring on land. Judging by legislation existing by 1997, Brazilian authorities are not interested in addressing the dumping of toxic wastes in Brazilian waters.

The LC was not originally intended to resolve the environmental questions of developing countries, but only to regulate the dumping of toxic wastes by the developed countries. Recognizing this limitation of the LC in its original form, its parties need to reconsider their objectives.

Montreal Protocol

As an industrializing country, Brazil had to address the question of ozone-depleting gases released by industry and the manufacture of consumer goods. Brazil was not a party to the Vienna Convention at the time of the Montreal Protocol negotiations, and was therefore not involved in the formation of the Protocol. It joined both treaties in 1990, after the Montreal Protocol was in effect. Before then, the Brazilian Ministry of Health had banned the use of ozone-depleting gases, such as aerosol propellants, in almost all cosmetics, housecleaning products, and medicines in 1988. The important London Amendment of 1991 to the Protocol was integrated into national legislation in December 1992.

The Brazilian delegation participated in the negotiations of the London Amendment. It was particularly successful in drawing attention to its concerns, the result being that the adjustments of the London Amendment conformed to Brazilian external environmental policy regarding differentiated responsibilities. The policy was that all countries are responsible for the environment, but in different ways, in accordance with the degree of development, financial stability, technological progress, and contribution to global pollution. Apart from the success of the Brazilian diplomats, which encouraged Brazilian participation in the treaty, Brazil's adherence to the ozone treaties clearly reflected the desire of the Collor administration to carry out an "environmental offensive" in the international arena after repeated denunciations by the national and foreign press and NGOs of Brazil's incapacity to control environmental damage.

The adjustments adopted dealt with the main Brazilian demands: (1) maintenance of the ten-year grace period for developing countries, during which the elimination of production and consumption is not mentioned; (2) the creation of a multilateral fund administered by the rich countries and the less-developed countries (LDCs) together that assures the availability of resources to the latter for industrial conversion; (3) access to ozone-friendly technologies; and (4) arrangements to release LDCs from obligations if they are not proportionate to their financial and technological capabilities.

Actions over Time. In the final months of 1989, IBAMA, the Ministry of Foreign Relations, and business groups—Du Pont, Hoechst, Dow Chemical, Consul, Climax, and Prodoscimo—met informally, with the support of the U.S. Environmental Protection Agency, in order to carry out a national study of the use of substances that damage the ozone layer. During the political negotiations that preceded the inauguration of Fernando Collor in March 1990, it was decided that the implementation of the ozone treaties would be coordinated by the Ministry of Economy and not by IBAMA. That decision resulted from business community fears of "environmentalist" implementation without a clear understanding of the economic impact.

In October 1991, an interministerial act created the Grupo de Trabalho do Ozonio (Work Group on Ozone; GTO), which included representatives of many branches of the federal administration as well as of public and private organizations. The GTO tried to elaborate on the 1989 study, but the shortage of financial resources, the constant changing of coordinators, and the lack of business collaboration weakened the initiative. Finally, in September 1992, the study was completed. Like the first study, however, its results were not considered to be of high quality. At the same time, the organizational structure of the Executive branch was reformed, and the GTO was transferred from the Ministry of Economy to the Ministry of Industry and Commerce.

The registration of firms that produce, sell, and consume gases that can harm the ozone layer was completed in the first half of 1993. Around 800 firms were registered. This program relied on the support of NGOs in some states. In order to maintain a current database of firms related to the ozone issue, the Multilateral Fund provides the Brazilian government with financial resources. Seminars were conducted to provide small firms with information about and assistance in industrial conversion. U.S.$10.9 million is available for industrial conversion of large producers of air conditioners and refrigerators and one project related to foam production. A training program for 25,000 registered technicians who repair refrigerators and air conditioners is to be promoted by the Associação Brasileira de Ar Condicionado, Ventilação e Aquecimento, with financial support from the Multilateral Fund.

Brazil has participated in ozone-layer monitoring projects in the southern hemisphere through the radio astronomy project of the Federal University of Santa Maria in Rio Grande do Sul state with the cooperation of the Instituto Nacional de Pesquisas Espaciais, NASA, and Argentine, Canadian, Chilean, and Peruvian scientists. This projects relied on resources provided by the World Bank through the Global Environment Facility program.

As of the end of July 1997, Brazil had filed the annual reports on consumption of ozone-depleting substances as required by the Montreal Protocol for all years through 1995. The 1996 report had not yet been received. (See table 14.4.)

Changes in Behavior of Target Groups. Many producers of gases are multinational enterprises that obey the directives of their head offices. At the beginning of the implementation of the treaties in Brazil, most of the resistance by these firms to the ozone regulations had already been attenuated by pressure from NGOs and the governments of their home countries. The large Brazilian companies that consumed noxious gases were exporters and knew that their exports would be damaged by the imposition of nontariff barriers to products that were not ozone-friendly. If multinational orientation and export commitment push forward the full implementation of the treaty, Brazil—as an LDC—has until 2010 to fully adopt the requirements of the treaty. That fact undoubtedly reduced resistance to the treaties' implementation. The exclusion of IBAMA and, later, the Ministry of Environment from the control of implementation also reduced business opposition. The target groups were easily identified and were interested in the issue for economic and public image reasons.

The Brazilian Congress, as usual with environmental issues, is superficially involved with the treaty, especially because the norms related to the treaties were produced by the Executive branch in the form of acts and decrees. Only one bill, authored by former representative César Cals Neto, sought to address the issue, but it has not made conclusive progress.

The role of Brazilian NGOs is limited but positive. The Brazilian affiliate of Friends of the Earth, in Pôrto Alegre, collaborated in the efforts to register the producers, distributors, and consumers of gases harmful to the ozone layer and also lobbied for the approval of a state law banning the use of CFCs in new refrigerators, air conditioners, and packing. An example of the involvement of INGOs is Greenpeace's initiative to present samples of two German-made ozone-friendly refrigerators to Brazilian industry and officials.

Are the Purposes of the Treaties Being Achieved? In 1997, Brazil was fulfilling its financial obligations in terms of article 5 of the Convention, designated to support the administrative structure of the treaties, but was behind in its provision of information. Only recently had the national study been set in motion, and annual data, reports of exports to nonparticipating nations, and information about other provisions taken by the Brazilian government were lacking. (Table 14.4 shows Brazil's compliance with reporting requirements of the LC.)

CFCs were practically banned as propellants for cosmetics, housecleaning products, and medicines in 1988. All consumer products that use gases as propellants have a notice that the gas used is ozone-friendly. In the refrigeration industry, responsible for the use of 46.9 percent of the CFCs in Brazil, there were medium-to-long-term plans in progress in 1997 for the production of ozone-friendly products. The auto industry, notably General Motors and Fiat, initiated the production of new models with air conditioners using HFC 134 instead of CFCs. Coca-Cola, because of a decision by its head office, decided to gradually replace the CFCs in refrigeration equipment in its machines (post-mix and pre-mix) used in bars and restaurants. Both initiatives were announced to the public in 1993 to show the commitment of the companies to the environment.

Plastic foam accounted for 40.1 percent of CFC use in Brazil. At the end of the 1980s there was an important change in the industrial profile of the sector. Previously the production of foams had been concentrated in little more than ten large factories. When smaller machines became available, the production of foam was decentralized into small units owned by furniture and mattress makers, as well as the auto parts industry.

There has been no business association to represent the interests of foam producers, in part because they are in different markets but principally because foam is rarely a final product. Nevertheless, some movement in search of ozone-friendly alternatives can be verified. For example, Mercedes-Benz, which builds trucks and buses in Brazil, has tested the use of coconut derivatives to replace plastic foams as upholstery.

The majority of the Multilateral Fund industrial recycling projects are connected to refrigeration projects and do not concern foam and packaging. Also, since the foam sector does not export, it has not been subject to the pressures of consumers in nations where the implementation of the treaties is more rigorous. The end of the use of noxious gases by the foam and packaging sector perhaps will depend on the end of the sale of these gases by the large chemical enterprises.

Factors Explaining Implementation and Compliance. Target groups have followed the process closely and are well informed. The diffusion of information about industrial conversion projects for small and medium-size enterprises, which generally are not exporters, has been a special concern. In the second half of 1993, attempts were made to reach small and medium-size enterprises, but the foam industry did not receive adequate information about the treaties. Scientific and commercial issues related to ozone have been discussed in the national press, with news about Brazilian participation in the ozone layer-monitoring projects regularly published. Industry news about CFCs has emphasized the use of ozone-damaging gases in refrigeration. For the general public, the ozone question—when related to industry—means refrigeration, not packaging or foam.

The purposes of the treaties have been achieved because of (a) an initial impulse from the national government, even before the adherence of Brazil to the treaty; (b) the press; (c) the multinational characteristics of the industry that produces the gases; (d) pressure on exported products; (e) international funding for projects; and (f) the delay until 2010 of full implementation of the treaty.

Conclusions

Effective implementation probably has been most problematic in the two treaties most specifically relevant to Brazil, CITES and ITTA. Attempts to regulate the exploitation of forests and endangered species took place in large, remote, ecologically complex, sparsely populated, and underdeveloped regions of Brazil. These areas were very difficult to police, and allowed great informality of business enterprise.

Economies in these regions have long depended on the extraction of natural resources, so there has been little political or cultural sympathy for these treaties' aims. Curbing extraction by target populations has been extremely difficult, especially in the case of CITES, where individual poachers and small traders deal in commodities that are easily hidden. These problems were aggravated by extreme regional inequalities, both political and economic. The problems of transition from a military dictatorship to a democracy complicated implementation even more.

Protecting specific sites under the WHC has been a far simpler matter, depending on local political processes, both governmental and nongovernmental. Implementation of the Vienna Convention and the Montreal Protocol has been bureaucratically and politically simple as well. Target groups were easily identified and accessible, and were subject to governmental pressures to comply with the spirit of these agreements. The terms of the agreements have been phased in relatively slowly.

Compliance with all of the treaties has been reached in one area: the Ministry of Foreign Relations announced in October 1994 that Brazil was up to date with its financial obligations to the agreements. Table 14.4 compares Brazil's compliance with reporting requirements of CITES, the LC, and the Montreal Protocol over time. It reflects all reports filed by July 1997.

In general, the effects of the treaties are extremely difficult to separate from the effects of other pressures, national and international, to protect the environment. The political and economic crises that Brazil has experienced since the late 1970s further complicate analysis. In some cases, national legislation predated the relevant treaties, and the implementation of this legislation varied before and after the signing of the treaties. The existence of the treaties was presumably useful to advocacy, but overall, endogenous forces have seemed to be far more determinant of both governmental and civil actions and behavior.

Nonetheless, there are significant differences in implementation and compliance among the accords that reflect the very different ecological processes that are involved in the problems the accords are trying to address. CITES and ITTA are oriented

Table 14.4
Brazil's compliance with treaty reporting requirements, 1991–1996: Report transmission/receipt date

Treaty	Entry into force	1991	1992	1993	1994	1995	1996
CITES	11/4/75	11/30/92	11/30/93 (extension granted)	1/23/95	11/27/95	12/12/96	N/R
LC	8/25/82	1/24/96	3/29/96	3/29/96	3/29/96	3/29/96	7/2/97
MP	6/17/90	R	R	R	R	R	N/R

Notes: For all treaties, N/R indicates that no report was received by July, 1997. R indicates that a report was filed, but the date of receipt is unavailable. Montreal Protocol reports received include those filed on time, late, or subsequently revised; in some cases a country may have met two or more years' reporting requirements with the submission of a single report.

toward rural and wilderness areas. The London Convention of 1972 and the Montreal Protocol deal with industrial products and effluents. By their nature, these are associated with urban areas. Most of the World Heritage Sites in Brazil are cultural sites in urban areas, where regulation arguably is less costly and more effective. Ecological diversity and regional inequalities inherent in the federal system in Brazil are critical to explaining Brazilian compliance with these international accords.

Notes

The authors received the valuable collaboration of the economist Kenneth Nobrega and of the biologist Patricia Gonçalves Baptista de Carvalho for this chapter.

1. Bill 1164 of 1991, proposed by former President Fernando Collor, intended to provide instruments for penal and administrative sanctions against activities prejudicial to the environment. Until May 1997, this project waited for a definition from the Senate to be approved or rejected.

2. The American Chamber of Commerce of São Paulo awards a prize annually to the best business program for environmental preservation.

3. The 1995 budget for the Ministry of Culture was divided in the following manner: U.S.$79 million for personnel, U.S.$56 million for costs, U.S.$14 million for structural investments, and U.S.$1 million for financial investments.

4. The 1995 budget for the Ministry of the Environment was divided in the following manner: U.S.$455 million for personnel, U.S.$22 million for external interests, U.S.$678 million for operating costs, U.S.$60 million for investments, U.S.$26 million for financial inversion, U.S.$3 million for amortization of the internal debt, U.S.$28 million for amortization of the external debt, and U.S.$65 million for other expenses.

5. The Union budget executed in 1995 was U.S.$107,641 million.

6. The decree does not apply to exploitation in planted forests, a breach that allows logging companies to work a while longer in old concessions (*Correio Braziliens* 7/26/96). Information provided by IBAMA in May 1997 indicated that the United States made a motion, sup-

ported by Bolivia, to include mahogany in the scope of CITES. At the meeting at Harare, Zimbabwe, in June, Brazil vetoed the creation of an international certificate that would attest that the mahogany had been extracted from sustainable ecological reserves. Eighty percent of Brazilian mahogany is extracted from illegal areas (*VEJA* 1997b).

7. *VEJA* (1997a) published an extensive news report denouncing the activities of these timber companies in the Amazon region and IBAMA's lack of surveillance.

8. In October 1996, a special committee was created by the House of Representatives to investigate the Asian companies' activities. The committee was scheduled to finish its investigations at the end of June 1997.

9. Since 1994, the National Institute of Space Research has not divulged the extent of the deforestation in the Amazon (*O Globo*, 5/5/97).

15

Assessing the Record and Designing Strategies to Engage Countries

Harold K. Jacobson and Edith Brown Weiss

In a world of sovereign states, international accords are the traditional and readily available instruments for dealing with common problems. The world community has relied heavily on them in its efforts to protect and enhance the environment. Given the accelerating pace and spread of human activities that potentially threaten the environment, it is important to ask how successful this strategy has been and how promising it is for addressing future problems. Moreover, there may be ways in which it could be used more effectively.

The key issue is whether or not international environmental accords contribute to modifying the behavior of states and, through states, that of enterprises and individuals. For this to happen, states that are parties to international environmental accords must implement and comply with them.

By the time of the 1992 United Nations Conference on the Environment and Development, the implementation of and compliance with international environmental accords had become such a salient issue that a study of the enforcement of international environmental agreements was commissioned as part of the preparatory work for the conference (Sand 1992). The Conference subsequently proclaimed (*Agenda 21*, paragraph 39.3(e)) that one specific objective of the post-Rio action program was

> To ensure the effective, full and prompt implementation of legally binding instruments and to facilitate timely review and adjustment of agreements or instruments by the parties concerned, taking into account the special needs and concerns of all countries, in particular developing countries. (Robinson 1993:622)

The analyses in this book contribute to understanding what can be done to promote the achievement of this objective.

The research shows what six large and crucial, and two smaller but representative, countries and the European Union have done to implement and comply with five major international environmental treaties, all of which have been in effect for substantial periods of time. Three of the treaties were negotiated in the 1970s and two in the 1980s. By 1997, CITES and the London Convention had been in effect for more than two decades, the World Heritage Convention for twenty years, the

International Tropical Timber Agreement for more than a decade, and the Montreal Protocol for eight years. The record is thus sufficiently long for meaningful conclusions to be drawn.

What does the record show? What have these nine political units done to implement and comply with the five international environmental treaties that they have joined? What explains the extent of implementation and compliance, and what explains changes in implementation and compliance? What does the record of the past suggest about strategies for the future?

The Secular Trend Toward Strengthened Implementation and Compliance

First we must assess the extent of implementation and compliance. A secular trend toward strengthened implementation and compliance was visible by 1995, and strengthened efforts generally meant better implementation and compliance. By the mid-1990s, several of the countries in this study were in substantial compliance with many or all of the treaties, and all of the countries in the study had taken some steps to comply. Moreover, the trend over the decades was toward greater compliance.

The record of compliance with respect to these environmental treaties is consistent with the experience with international agreements generally recounted in chapter 3. And viewed against the assessment of compliance with national laws and regulations within the United States and with Community regulations and directives within the European Union, as surveyed in chapter 2, the record at the international level is comparable or better.

In no case was compliance perfect, and not all nine political units were doing a good job of implementing and complying with all of the five treaties—indeed, a few were not parties to all of them—but the overall trend was positive. Parties took an increasing number of actions to implement the treaties, and in most cases procedural and substantive compliance improved. The political units generally were acting more in ways that conformed to the spirit of the treaties, but in some cases compliance could only be termed weak or even very weak, and in some others the level of compliance had declined.

This trend toward strengthened implementation and compliance had several dimensions. It involved the treaties themselves, the international institutions charged with overseeing their operation, other international governmental organizations, national and international nongovernmental organizations, industry, expert communities, and laws, regulations, and institutions within countries.

At the international level, several things occurred. First, countries continued to become parties to the treaties, so that their membership became increasingly global and comprehensive (see chapter 5, especially figure 5.1). Countries that had been slow to join conventions eventually did so. There are several examples of this trend among the countries considered here. Russia and Japan joined the World Heritage Convention in 1988 and 1992, respectively. The WHC had entered into force in 1977. China became a party to the WHC and to the London Convention in 1985, the latter a decade after it had entered into force. And China and India joined the Montreal Protocol in 1991 and 1992, respectively, several years after it came into effect and a year or more after the Montreal Protocol Fund had been created. Japan became a party to CITES in 1980, five years after the convention came into effect. Cameroon joined CITES in 1981 and the World Heritage Convention in 1982. Hungary became a party to the London Convention in 1976 and the WHC in 1985.

Second, for all of the treaties, there was greater attention over time to implementation and compliance and to strengthening the supervisory mechanisms. The treaty budgets increased, secretariats generally grew modestly in size, and more attention was paid to monitoring and compliance. The functioning of the Montreal Protocol's Implementation Committee and the adoption of the noncompliance procedures are strong examples of this trend.

A third aspect at the international level of the trend toward strengthened implementation and compliance was the fact that the parties to three of the treaties had agreed to deepen their commitments. The London Convention was amended several times, with addition of new substances to the Prohibited List, including, in 1990, radioactive wastes. In 1990, parties to the Montreal Protocol adopted adjustments and amendments that advanced the target dates for phasing out controlled substances, brought new chemicals under control, established the Implementation Committee, and set up a new financial mechanism to provide financial and technical cooperation to developing countries. In 1994, the International Tropical Timber Agreement was renegotiated. In the new agreement, parties producing tropical timber agreed to try to export wood only from sustainably managed forests by the year 2000. Consumer countries agreed to establish a new fund to help producer countries meet the objective of sustainably managed forests, and they issued a separate formal statement pledging to follow comparable forest conservation guidelines for their own forests and committing themselves to the objective of achieving sustainable management of their forests by 2000.

Finally, in the decades following 1972, international nongovernmental organizations that played a role with respect to these international environmental treaties gained members, resources, and sophistication (Princen and Finger 1994). By the 1990s, the CFC Alliance, Greenpeace, TRAFFIC, and the Worldwide Fund for Nature were strong and extremely active. The International Union for the Conservation of Nature, a curious hybrid international nongovernmental and governmental organization, also grew dramatically.

These changes at the international level were matched by equally important changes at the national level. First there were institutional changes. At the time of the Stockholm Conference, among the nine political units in this study, only the United States had an Environmental Protection Agency (EPA). In January 1970 the United States established the Council on Environmental Quality, and it established EPA that July. By the time of the Rio Conference, all of the political units in our study had established bodies with mandates like EPA's. India created its Department of the Environment in 1980. In 1978, the Soviet Union took steps toward creating a separate agency to deal with environmental issues, steps that were completed a decade later with the establishment of Goskompriroda, which eventually became the Ministry of Environmental Protection. The Hungarian Ministry of the Environment and China's National Environmental Protection Agency were created in the 1980s. Cameroon created its Ministry for the Environment and Forests in 1992, and Brazil's Ministry of the Environment was established in the same year.

The European Union is a special case. The Community did not gain formal legal competence to deal directly with environmental issues until the passage of the Single European Act in 1987. The Community, however, adopted its first Action Programme on the Environment in 1972, and the European Commission's Directorate General XI, which is responsible for matters relating to the environment, consumer protection, and nuclear safety, was established in 1981.

Research institutes and advisory bodies also were established, such as the Brazilian Institute for the Environment and Renewable Natural Resources, which was created in 1989. All of these institutions came to play crucial roles in dealing with issues of implementation and compliance. The strength of the institutions and their centrality to policy formulation of course varied, though none gained the stature of the traditionally strong Ministries of Economics, Finance, and Foreign Affairs. By the 1990s, however, all were important focal points for policies relating to compliance.

As the countries became parties to the treaties, they took important steps to implement and comply with them. In some cases, such as that of China, which adopted the Marine Environmental Protection Act 1983 prior to joining the London Convention in 1985, they acted so as to prepare for accession. In other cases—for instance, with respect to the Montreal Protocol—the countries adopted the necessary legislation and regulations after they had signed and ratified the agreement. None of the nine political units in this study had failed to take some action to implement the treaties to which they were parties. And all had made some efforts to bring practices within their borders into compliance.

Within the United States, the European Union, Japan, Russia, India, and Brazil, national nongovernmental organizations were considerably stronger in the 1990s than they were in the 1970s (Brenton 1994; Fischer 1993). They were actively promoting environmental causes and monitoring compliance with international environmental treaties and national laws and regulations.

These broad points having been made on the positive side, there are some important qualifications. The performance of some countries with respect to CITES had sharply declined since the mid-1980s. With respect to developing countries, the substantive obligations of the Montreal Protocol were not yet severe. Thus it would be premature to make a strong judgment about their performance. And the obligations for the industrialized countries involved actions that were relatively easy to take; harder tasks would be required in the future. That is, the industrialized countries had phased out the use of ozone-depleting substances for nonessential products and in instances where substitutes were relatively easily available. Phasing out more essential uses of ozone-depleting substances, where substitutes were not easily available, would be more difficult. The signs were positive, but they provided only a modest basis for projecting a positive trend.

Comparisons Among Treaties and Political Units

Within this broad framework of all political units doing something to comply and a general secular trend toward increased compliance, there were substantial differences among the treaties and the political units. A snapshot at any particular time would show these differences, and the pace of change varied among treaties and political units. It is to these differences that we now turn. We will deal first with the differences among the treaties and then address the differences among the countries. We focus particularly on compliance with the substantive obligations imposed by the treaties.

Among the five treaties, implementation and compliance seemed to be stronger with respect to the Montreal Protocol and the London Convention. Both treaties have relatively precise procedural and substantive obligations, and in general the political units included in this study, and parties to the treaties more broadly, were complying with these obligations. Even so, there were some egregious infractions at least of the spirit of the treaties, most notably the Russian Navy's dumping of radioactive waste in the Sea of Japan in the fall of 1993.

The Convention on International Trade in Endangered Species also imposed precise obligations, but it encountered serious and widely publicized difficulties in the late 1980s and early 1990s. There was a strong impression that smuggling of species in which trade was banned had sharply increased. Since illegal activities are never totally transparent, it is difficult to know how well founded this impression was. Monitoring activity had increased, and the impression of increased illegal trade may have been in part the result of more and better data gained through monitoring. But the evidence was sufficient to provide a substantial foundation for the suspicion that there was increased illegal activity.

Because the obligations they impose are less precise, judging the extent of compliance with the World Heritage Convention and the International Tropical Timber Agreement is more difficult. Certainly the eight political units in our study that were parties to the World Heritage Convention had proposed, and were maintaining, protected sites. Most countries in our study were protecting their World Heritage Sites to a reasonable degree, but there were a few serious exceptions and problematic cases. Manas National Park in India was under the control of rebels, the Everglades National Park in the United States was seriously threatened by human development, and not enough was being done to protect the Dja Reserve in Cameroon. Many of the sites could have benefited from more resources being allocated to protection and conservation.

The International Tropical Timber Agreement contained nonbinding guidelines for the sustainable management of tropical forests, but judging the extent to which these influenced forest management practices is extremely difficult. The agreement's reporting requirements relate to international trade in tropical timber products rather than to forest management. It is possible to make inferences about forest management from trade data, but these are inferences at best. The popular impression, buttressed by reports from nongovernmental organizations and other sources, is that tropical forests were being rapidly cut, and this impression is confirmed by data gained from observation satellites. An argument can be made, however, that there is a growing consensus around the concept of sustainable management. The stronger terms of the renegotiated Agreement would be a central element of this argument. The fact that the secretariat was invited to conduct an on-site investigation in Sarawak and in other places is noteworthy. The increased size of the secretariat and budget, and the increased number of technical assistance projects were also important.

No political unit did a perfect job of implementing and complying with all of the procedural and substantive obligations imposed by the treaties, but the European Union, Japan, and the United States did better than the other units in our study. They had relatively good records of submitting the reports required by the treaties, although certain other countries also had good records (see tables 5.3, 5.5, and 5.7). Their reports were generally complete, relatively accurate, and on time.

These three political units were phasing out the production of ozone-depleting substances ahead of the required target dates. The major member states of the European Union, and Japan and the United States were fully in compliance with the London Convention. The United States was a leader in implementing the World Heritage Convention, but in the 1990s some questions were raised within the United States, and among those involved in the convention, about U.S. willingness to protect endangered sites, especially the Everglades National Park. The Park was put on the Endangered List in 1993, partly on the basis of a report prepared by the

Department of the Interior, but without the United States voting for the listing. Japan finally adhered to the World Heritage Convention in 1992. Even before it became a party to the Convention, however, Japan's standards for preservation of natural and cultural sites were high. Since becoming a party, it has been in full compliance with the Convention, as have most members of the European Union.

Given the vague quality of the obligations that the International Tropical Timber Agreement placed on consuming countries, it would be hard for the European Union, the United States, and Japan not to be in formal compliance with this treaty. Japan, however, clearly has imported tropical timber from forests that were not managed in sustainable ways. Even though it is not a party to CITES, the European Union sought to fulfill the obligations of the treaty. Although Italy had difficulty with some provisions of CITES, the overall record of the member states of the European Union with respect to the treaty was positive. The United States and Japan generally fulfilled their obligations under CITES. At the same time, it is clear that protected species somehow made their way into the United States and Japan, and were sold and purchased there. Japan had entered a number of reservations with respect to a number of species, and until the mid-1990s its administration of the requirements of the treaty—for instance, in the materials and training that were given to customs officials—was not as strong as it could have been.

Russia was officially phasing out the production of ozone-depleting substances. There was evidence, however, that some of these substances were being sold from Russia. The Soviet Union and later Russia contravened the spirit of the London Convention through the Navy's activities. The fact that after the collapse of communism, government officials revealed these transgressions and addressed the issues candidly should be seen as a positive sign. Issues relating to compliance had become more transparent, allowing public discussion and pressure. Russia adhered to the World Heritage Convention in 1989, and Russian sites appear to be properly protected. It adhered to the International Tropical Timber Agreement in 1986, and it was in compliance with the treaty (its obligations were not onerous). Although the Soviet Union's record for compliance with CITES was good, decentralization and the economic downturn made Russia's compliance with CITES problematic.

Like Russia, as its transition to a democratic and market system proceeded, Hungary had difficulty fulfilling its obligations under CITES. It met its obligations under the other three treaties to which it was a party, but these obligations were not onerous. It was not a producer of ozone-depleting substances. As a landlocked country, it was hardly in a position to engage in extensive ocean dumping. Its World Heritage sites were protected.

Brazil, China, and India are large, populous countries with economies that were growing rapidly in the 1990s. Starting in the late 1970s, China began moving away from almost total reliance on planning and governmental ownership of the means of production, and toward greater reliance on market forces and private ownership. Although the state never played as large a role in the economies of Brazil and India, those countries, too, have been moving toward greater reliance on market forces and private ownership. These broad characteristics had an impact on the compliance behavior of the three countries.

All three were in substantial compliance with their reporting obligations under the Montreal Protocol. China and India, however, had production facilities coming on-line that would need to be brought under control for them to continue to be in compliance with the substantive targets.

India was not a party to the London Convention. Brazil and China tried to comply with the Convention, but they both have long coasts that are difficult to control.

India was in substantial compliance with the International Tropical Timber Agreement. Brazil cut and exported a relatively large quantity of tropical timber, and China imported tropical timber from various sources. All three countries had sites that were listed under the World Heritage Convention, and all three tried to protect these sites. India, however, had grave problems, particularly with Manas National Park. Brazil and China had varying difficulties enforcing CITES.

Cameroon had the greatest difficulty of the nine political units in our study in implementing and complying with the treaties. It was not a party to the London Convention, and it had difficulties in fulfilling its obligations under the International Tropical Timber Agreement, the World Heritage Agreement, and CITES. It had great difficulty responding to the reporting requirements of CITES. Many of Cameroon's difficulties stemmed from its limited resources, which became even more scarce with the economic crises that the country suffered in the 1980s.

Table 15.1 gives a synoptic picture of the extent to which the nine political units were complying with the substantive requirements of the five treaties in 1995. The scoring in the table represents our subjective judgment, informed by the work of our colleagues. Since each treaty includes several substantive obligations, these judgments represent an overall assessment. We divide compliance into four levels, including no action taken to comply.

Weak compliance means that the country had taken some actions to implement and comply with the obligations accepted under the treaty, but those actions fell significantly short of what was required. Countries in weak compliance could also have taken some actions that willfully contravened the obligations. Weak compliance resulted from ineffective administration and unintentionally ignoring obligations, as well as outright defiance.

Moderate compliance means that the country had taken appropriate action to implement the treaties and was in compliance with most of the obligations. Moderate compliance was generally the result of ineffective or weak administration.

Table 15.1
Compliance of the political units with the substantive obligations of the treaties as of 1995

Level of compliance	World Heritage	CITES	Tropical Timber	London Convention	Montreal Protocol
Substantial compliance	Brazil China Hungary Japan Russia USA	India USA	EU India Russia USA	Brazil Hungary Japan USA	Brazil EU Hungary India Japan USA
Moderate compliance	India	Hungary Japan Russia	Japan	China Russia	Cameroon China Russia
Weak compliance	Cameroon	Brazil Cameroon China	Brazil Cameroon China		
No action taken to comply					
Not a party to the agreement	EU	EU	Hungary	Cameroon EU India	

Substantial compliance means basically fulfilling obligations, though there may be minor infractions. Thus the United States was judged to be in substantial compliance with its obligations under the World Heritage Convention despite the situation in the Everglades and the possible threat to Yellowstone National Park.

As can be seen, all of the political units in our study that were parties to the treaties had taken at least some actions to implement them, and all were at least to some extent in compliance with the substantive obligations they had accepted. The record is better than it would have been earlier. At various times earlier, there would have been entries in the "no action taken to comply" cells. If our study had included all of the parties to each of the treaties, there would have been entries in these cells in 1995.

Table 15.1 shows that there were discernible differences among countries and treaties with respect to the extent of compliance. It indicates that the compliance records of the European Union, Japan, and the United States were on average better than those of the other units. It also indicates that on average, compliance with the Montreal Protocol and the London Convention was better than with the other three treaties, and that compliance with the World Heritage Convention was better than with CITES and the ITTA. All of the political units in the study that were parties to the Montreal Protocol and the London Convention were in either substantial or moderate compliance with the substantive obligations. All but Cameroon were in substantial or moderate compliance with the World Heritage Convention.

Although the European Union was not a party to three of the agreements, its member states were, and they were generally in substantial compliance with the substantive obligations of these agreements. Since CITES dealt with international trade, the European Union (EU) would have liked to have been a party to it, and the European Commission acts as if the EU were a party. The EU would be judged to be in substantial accord with CITES.

As we noted in chapter 1, even strong implementation and compliance with treaties do not ensure their effectiveness either in terms of the objectives of the treaties or in dealing with the problems that led to the treaties. The record in this regard is mixed with respect to the five treaties that are included in our study.

The Montreal Protocol and the London Convention seem, respectively, to have contributed to a decline in the production and consumption of ozone-depleting substances and in the intentional dumping of wastes in the high seas. The World Heritage Convention appears to have contributed to the preservation of cultural and natural resources, though there could be some debate about whether or not the particular cultural and natural resources that were being protected were always the world's most valuable ones. The International Tropical Timber Agreement has not yet resulted in the "sustainable utilization" of forest resources and, unfortunately despite CITES there appears to have been, especially since the mid-1980s, an increase in the illicit trade in endangered species. Moreover, while some endangered species have become less critically endangered, others have become more so, but arguably the situation could have become even worse absent the treaty. The treaty may have contributed to greater attention being focused on issues of species preservation within parties to it, and to their developing and strengthening their administrative capacities to deal with such issues.

Explanations of Compliance Behavior: Revisiting the Model

Using the elements of this broad and coarse assessment as benchmarks, what explains what happened? The model that was presented in chapter 1 (figure 1.1) grouped the variables that we thought might be important into four broad categories: (1) characteristics of the activity involved; (2) characteristics of the accord; (3) the international environment; and (4) factors involving the country. Our analysis enables us to amplify the diagram by filling in aspects of each box; to see better the way that interactions occur; and to have a clearer view of the relative importance of various factors. We will take the factors one by one, and identify the components that the analyses indicate are especially important.

There is an artificiality to the discussion that follows. The model shows that each factor interacts with the others to produce a combined effect on implementation, compliance, and effectiveness. For clarity and manageability of the discussion, however, the factors must be treated individually. Thus each of the statements in the following paragraphs requires the qualification "other things being constant."

Characteristics of the Activity Involved

With reference to the characteristics of the activity involved, our study confirms the conventional wisdom that the smaller the number of actors involved in the activity, the easier it is to regulate it (Olson 1968). It is much easier to monitor an activity conducted by a small number of actors than one conducted by many. Not only is it easier, it is also less expensive, which positively affects the cost–benefit ratio of complying with environmental accords.

Because only a limited number of facilities produced ozone-depleting substances through the early 1990s, it was relatively easy to control the production of these substances as the Montreal Protocol and amendments to it required. The situation could become more difficult as more production facilities come on-line in the latter half of the 1990s. The striking contrast between the limited number of facilities that produced ozone-depleting substances and the millions of individuals who could engage in illicit trade in endangered species helps to explain why CITES was much more difficult to enforce than the Montreal Protocol.

Of course there were other important differences between the situation with respect to ozone-depleting substances and endangered species. The large companies that produced ozone-depleting substances generally found it in their economic interests to discontinue production of these substances and to start production of substances that would not deplete the ozone layer. Local officials who were charged with enforcing the trade prohibitions on endangered species often could gain sums of money that were very large in relation to their salaries for allowing or ignoring infractions of the rules. The economic incentives fitted with behavior directed toward compliance in the ozone case, and worked against compliance in the endangered species case.

Activities conducted by large multinational corporations that sold a variety of products directly to consumers in many markets were easier to deal with than those that were conducted by smaller firms that were less visible internationally. Again, the production of ozone-depleting substances provides the example. Large multinational firms were concerned about their reputation globally. They were much more subject to the pressure of public opinion and diverse consumers throughout the world than were smaller, less well-known firms, such as those that accounted for much of the timber trade. In addition, large multinational corporations have bureaucratic structures that enforce control, and they prefer to conduct their activities in stable and uniform regulatory environments.

Not all multinational corporations, however, facilitated compliance with the international environmental accords we examined. Some of those dealing with forestry products not only did not facilitate the International Tropical Timber Agreement's requirement that tropical forests be managed sustainably, they worked against it.

Location of the activities is another issue. Our analysis was based on the assumption that the actions of seven of the nine political units that we studied—Brazil, China, the European Union, India, Japan, the Russian Federation, and the United States—would be crucial to the effectiveness of any international environmental accord. Because of the importance of these large political units in the world economy, the environmental degradation that they can cause, and the contribution that they can make to ameliorating environmental problems, we thought that their compliance with environmental accords would be crucial.

Our analysis did not contradict this assumption. Although we focused on what individual countries did, the aggregate effect of the actions of these countries had a determining effect on overall compliance with the accords. This is certainly the case with respect to the Montreal Protocol and the London Convention, even though by July 1997 these conventions had 162 and 72 parties, respectively. Most of the production facilities for ozone-depleting substances have been, and will be, located in the seven countries. Because these large countries account for such a large share of world production, what they do about ocean dumping will account for a predominant share of substances that potentially could be dumped. These countries also have a large share of cultural and natural heritage sites.

Our assumption does seem not to be valid for tropical timber and endangered species, since in both cases production is scattered and many small countries have substantial production; the experience with ITTA and CITES would seem to support doubts about the validity of the assumption in these spheres. One might feel that a substantial portion of the 53 parties to ITTA and the 137 parties to CITES would have to comply to bring about overall compliance. This would be true, however, only if one focused primarily on the actions of producers and suppliers. If one focused instead on actions of importers and consumers, then again it would be the seven countries or a set of them that had the greatest impact on the outcome. If the market in the United States, the European Union, and China for endangered species contracted, the incentives for individuals in Cameroon and Hungary to evade restrictions and export would be greatly diminished.

Thus, when we ask if there is compliance with a particular accord, we must first ask about the compliance of major parties. Analyses, for instance, that conclude that fewer than half of the parties are complying with an accord because the majority of parties submit reports late or fail to submit them at all, miss the mark (USGAO 1992b). At a minimum, to give a true picture, such analyses should weight the proportion of parties complying by their contribution to the problem. The difference is the difference between reporting that only a small proportion of the parties fulfill

Table 15.2
Characteristics of the activity involved

Number of actors involved in the activity
Effect of economic incentives
Role of multinational corporations in the activity
Concentration of activity in major countries

their obligation to report and reporting that parties responsible for the largest share of the production of ozone-depleting substances fulfill their obligation to report. Put another way, it is more effective to bring one of the seven large countries included in this study into compliance than it would be to bring a large number of smaller countries into compliance.

This of course is not to argue that the actions of smaller countries that make modest contributions to environmental problems are inconsequential. The actions of smaller countries can contribute to international momentum. If large numbers of small countries were perceived to be free-riding, this could sap the will of larger countries to comply. Conversely, if large numbers of small countries do comply, this can create a climate favorable to compliance that will affect larger countries.

Table 15.2 shows factors related to characteristics of the activity involved that we have found to be important. The number of actors involved, the effect of economic incentives, the role of multinational corporations, and the location of the activity in major countries are important to how effective implementation and compliance will be.

Since the characteristics of activities that contribute to environmental degradation are more or less fixed, accords must address activities regardless of whether their characteristics facilitate implementation and compliance. To the extent that accords can decompose problems and define points of attack, however, these generalizations could be used to shape the approaches that are taken to drafting accords.

Characteristics of the Accord

The characteristics of an accord make a difference. Not surprisingly, for parties to implement and comply with accords, they must feel that the obligations imposed are equitable (Franck 1995). The differentiated obligations under the Montreal Protocol for phasing out ozone-depleting substances—imposing the strictest and most immediate deadline on members of OECD, placing countries like the Soviet Union in an intermediate position, and allowing developing countries time to increase production before beginning to cut back—was a prerequisite for adherence to the accord of a large number of countries (Benedick 1991). India and China would not become parties to the Montreal Protocol until the agreement about compensatory financing had been reached at the London meeting in 1990. Part of the difficulty with the International Tropical Timber Agreement seemed to be a sense that burdens were disproportionately imposed upon the producer countries; the consumer countries' activities with respect to their forests were unregulated. The revised Agreement attempted to address this issue.

The experience with most international accords, as recounted in chapter 3 and in many studies, has been that the more precise the obligations, the easier it is to assess and promote compliance (Chayes and Chayes 1992; Chayes and Chayes 1995; Fisher 1981). The experience of the nine political units with the five international environmental treaties tends to confirm this conventional wisdom, but it also shows that the situation is more complicated than the simple formulation "the more precise, the better" would have it.

In cases where one party will suffer if it is complying and others are not complying or are free-riding, accurate assessment of compliance becomes an important factor in promoting compliance. On the other hand, when compliance is largely in the self-interest of the party involved, regardless of what other parties do, there can be substantial compliance by parties even with vaguely stated obligations. There are also other important distinctions.

The London Convention, CITES, and the Montreal Protocol imposed relatively precise obligations. Consequently, it was relatively easy to judge whether or not parties to these accords were fulfilling their obligations. Yet there were differences among these accords. The London Convention and the Montreal Protocol dealt with limited numbers of substances; CITES, with hundreds of species. Distinctions among the species with which CITES was concerned were sometimes subtle, yet they had to be understood by thousands of customs officials. This was difficult for customs officials even in developed countries like the United States and Japan. Two generalizations emerge. One is that precise and relatively simple obligations are easier to comply with than precise and complicated ones. The other is that stating obligations precisely cannot override other factors that work against compliance.

The World Heritage Convention and the International Tropical Timber Agreement were much vaguer; assessing implementation and compliance was more difficult. The World Heritage Convention required parties "to endeavor ... to take the appropriate legal, scientific, technical, administrative and financial measures necessary for the identification, protection, conservation, presentation and rehabilitation" of their natural and cultural heritage (article 5(d)). The International Tropical Timber Agreement states that its objectives include "To encourage the development of national policies aimed at sustainable utilization and conservation of tropical forests and their genetic resources, and at maintaining the ecological balance in the regions concerned" (article 1(h)). In neither case did the accord establish a clear standard for assessing implementation and compliance, yet more of the units in our study were in substantial compliance with the World Heritage Convention than they were with the ITTA. The probable reason was that parties were more likely to see it to be in their self-interest to comply with the World Heritage Convention.

The fact that the obligations in an accord are stated in vague terms does not necessarily mean that compliance will be weaker than if the obligations were stated more precisely. The phrasing of obligations depends on the character of the problem being addressed and the state of agreed knowledge about it, as well as the willingness of potential parties to a convention to accept precise responsibilities. It is, however, difficult to assess compliance with imprecise obligations, and this by itself permits countries not to do all that some might think they should to fulfill their obligations.

Environmental accords often deal with highly technical subjects, the characteristics of which change with economic and scientific change. To be effective in dealing with such subjects, accords need to include provisions for gaining and utilizing scientific and technical advice, and for ensuring that there is a broad consensus among the parties on the scientific and technical issues. Chapter 3 demonstrated both how important and how problematic scientific and technical knowledge is.

Of the five treaties included in this study, the Montreal Protocol had the most elaborate provisions in this respect. These provisions put in motion actions that were vital to the implementation of the treaty and compliance with it. The Montreal Protocol's Technology and Economic Assessment Panel has played an especially important role. Provisions in the London Convention for obtaining scientific and technical advice made important contributions to compliance with that treaty.

Requiring the filing of regular reports is a standard feature that four of the accords under consideration here, and most others, use to monitor implementation and compliance. Accords use reporting requirements to obtain information on what policies countries have adopted and on the extent and nature of regulated activities. Requiring reports is one of the few instruments available for assessing the extent of implementation and compliance.

One way of controlling behavior is through issuing licenses, and requiring that the issuance of licenses be reported to a central authority is a way of ensuring that the procedures are conducted properly and according to the agreement. Reporting requirements enforce a discipline. The use of reporting under CITES and the London Convention illustrates this.

To know whether or not behavior has changed, it is necessary first to establish baseline data and then to have data to measure changes against the baseline. The use of reporting under the Montreal Protocol and the International Tropical Timber Agreement illustrates this.

Essential as reporting is, it has problematic aspects. As was shown in chapter 5 (see especially figure 5.17), the overall record of compliance with reporting requirements reveals problems. From 1991 through 1995, less than 50 percent of the parties fulfilled their reporting obligations under the London Convention. During this period, from just above 70 to about 40 percent of the parties fulfilled their reporting obligations under the Montreal Protocol. Only CITES had a record of 70 percent or more of the parties fulfilling their reporting obligations during this period.

Governments, particularly of smaller developing countries such as Cameroon, are extremely overburdened. Filing reports is one more burden. The locus of the responsibility for preparing the reports may be uncertain—is it with the Foreign Office or with a ministry? Even larger and richer countries sometimes have difficulty complying fully with reporting requirements, as Russia did with the London Convention. All of the other large states in this study, however, fulfilled all of their reporting obligations through those due in 1995.

What happens to the reports is also important. As chapter 5 made clear, all the bodies overseeing the environmental accords had relatively small secretariats; not one had as many as thirty people. Such small secretariats had limited capacity for systematic analyses and use of the reports that they received, although particular individuals were very conscientious in tracking and analyzing reported data. Reports that could not be analyzed thoroughly and in a timely manner, as those under CITES could not be in the early years of the treaty, lost much of their effectiveness as tools for managing implementation and compliance. Over time, the ability to monitor the reports in a timely manner has increased.

International secretariats used the reporting exercise to clarify for government officials what the obligations of accords are, and the variety of techniques that have been and might be used to fulfill them. Activities under the Montreal Protocol provide a good example of this. Used in this way, reporting became an educational process, a tool that enabled secretariats, other states that were parties to the accord, and national and international nongovernmental organizations to intervene to encourage compliance. Under the Montreal Protocol, Cameroon and China received international financial assistance to develop the national strategies for the phasing out of ozone-depleting substances that the accord required.

The two previous paragraphs underscore the importance to the operation of an accord of having an effective and efficient secretariat. In all five of the treaties, secretariat officials played important roles in furthering implementation and compliance.

In addition to reports from parties, accords often include other provisions for obtaining information about the activities that they were designed to control. Some allow and encourage reporting by nongovernmental sources.

Nongovernmental organizations and multinational corporations can play important roles in providing information about activities that are treated in international environmental accords (Sand 1991). The World Heritage Convention and CITES acknowledged this by including provisions for formal roles for NGOs.

The TRAFFIC reports on illicit trade in endangered species provided information that governments might find it difficult to gather or publish. Greenpeace was an important source of information about ocean dumping. The knowledge that monitoring goes on outside of formal governmental and treaty channels was probably an important restraining factor on governmental actions. Not surprisingly, governments were unlikely to incriminate themselves. The knowledge that nongovernmental organizations will publicize activities whether or not governments report them may have prompted governments to report more accurately.

The multinational firms that produce ozone-depleting substances may sometimes have better information than governments about their production. Also, since there are proprietary aspects to this information, they have access that governments could not easily achieve. In such cases the private sector must be engaged for monitoring to be effective.

Provisions for some form of on-site inspection can make an important contribution to compliance. The knowledge that there will be an opportunity for observation by some external party can provide a powerful stimulus to accurate reporting. The knowledge that their actions will be checked encourages local officials to fulfill the obligations imposed by the treaty. On-site inspections can provide information that is useful to everyone who is concerned about compliance: information about unanticipated problems and techniques that succeed or fail. Although on-site inspection has not figured prominently in environmental accords, the World Heritage Convention regularly relied on on-site observation by NGOs, ITTA conducted an on-site mission in Sarawak, and several treaties made use of information transmitted by members of technical assistance missions.

Incentives provide another tool. Giving countries financial or other assistance to help them comply with their obligations can be important for advancing compliance. Under the regime established by the Vienna Convention and the Montreal Protocol and amendments to it, several countries received assistance in preparing inventories of the production and consumption of ozone-depleting substances. The agreements also provide for assistance in transforming production facilities.

Sanctions or some type of coercive action is another aspect of international accords. Of the five treaties considered here, CITES's parties agreed that restrictions could be imposed in cases of infractions of its provisions, and the Montreal Protocol provides for trade restrictions against nonparties. In addition, U.S. law requires the imposition of trade restrictions against countries that engage in trade in some endangered species. The threat of a CITES-imposed ban on trade with Italy seems to

Table 15.3
Characteristics of the accord

Perceived equity of the obligations
Precision of the obligations
Provisions for obtaining scientific and technical advice
Reporting requirements
Provisions for other forms of monitoring
Secretariat
Incentives
Sanctions

have led that country to modify its practices. The threat by the United States to impose trade restrictions appears to have been a factor inducing China to modify its behavior with respect to CITES (Crawford 1995). China seems to have taken tougher action to ban internal trade in rhino products. There is a role for sanctions, though they have been used sparingly. Our study supports deductive arguments that make the case for the usefulness of accords, including the flexibility for parties to impose sanctions or take some form of coercive action (Downs, Rocke, and Barsoom 1996).

Table 15.3 lists the characteristics of the treaties that our studies have shown to be important.

Two of the characteristics, the perceived equity and the precision of the obligations, relate to how the substantive provisions of the treaty are written. One relates to the secretariat, a crucial institutional feature. The other five relate to procedures that we have found facilitate the implementation of and compliance with the accords. These findings have implications for the drafting of treaties.

The International Environment

The international environment may well have been the most important factor explaining the acceleration in the secular trend toward improved implementation and compliance that we observed in the late 1980s and early 1990s. Starting with the Stockholm Conference in 1972, international concern for the environment has increased, and it increased sharply in the mid-1980s with the publication of the report of the World Commission on Environment and Development, *Our Common Future*, in 1987 and the preparations for the U.N. Conference on Environment and Development (UNCED) meeting in Rio de Janeiro (Brenton 1994; WCED 1987).

The Rio conference was a massive event. It was the largest gathering of leaders of countries in history. It also brought together an unprecedented number of nongovernmental organizations. Significantly improving implementation and compliance with international environmental accords was specifically addressed (UN 1992; Sand 1992).

Increased salience of environmental issues was one aspect of the international momentum that developed. It mobilized the media worldwide, roused public opinion, and energized national and international nongovernmental organizations; public opinion and NGOs put increased pressure on governments to deal with environmental issues. This enhanced implementation and compliance. The media are independent actors and could produce many of these effects by themselves, and sometimes they do, particularly after major disasters, such as oil spills. They also respond to stimuli such as the Rio conference.

Another aspect of international momentum was that more accords were signed and more countries became parties to these accords. This had an effect on implementation and compliance. Governments did not want their countries to be seen as laggards. Moreover, there are practical economic consequences. Once it became apparent that the major countries would stop producing and consuming chlorofluorocarbons, other countries did not want to deal with outmoded technologies. Finally, in the case of a treaty like CITES, it is easier for a government to attempt to enforce its obligations if all of the neighboring countries are also parties.

The international momentum also resulted in other international governmental organizations' paying more attention to environmental issues. It had effects on the international financial institutions that were particularly important. The World Bank devoted its 1992 *World Development Report* to environmental issues (World Bank 1992). The international financial institutions can play important roles in encouraging compliance (Shihata 1996). In the 1990s they increasingly would not lend money for projects that ignored environmental protection obligations that countries had accepted.

The United Nations and the U.N. Environment Programme also played important roles. They were driving forces in organizing the Rio Conference, and by their actions they helped to bring environmental issues to the forefront of the public conscience.

Table 15.4 lists the factors in the international environment that contribute to international momentum.

The important elements are major international conferences, the worldwide media and public opinion, international nongovernmental organizations, the number of parties adhering to the accord, other international governmental organizations, and the international financial institutions. Several elements of the international environment are subject to policy intervention.

Factors Involving the Country

The three clusters of factors that we have discussed are important, but countries are at the center of the compliance process. Countries must take the actions that are required to fulfill their obligations under the accords.

Table 15.4
The international environment

Major international conferences
Worldwide media/public opinion
International nongovernmental organizations
Number of parties adhering to the accord
Other international organizations
International financial institutions

The performance of the eight countries and the European Union in implementing and complying with the five accords examined in this study varied substantially across the countries, the accords, and time. Characteristics of countries make an important difference. Three of the five treaties considered here had more than 100 parties. The variance among all of the parties to these and other environmental accords was even greater than the differences that are evident in our sample.

An initial point is that the compliance of the political units included in this study, and that of parties to environmental accords more broadly, must be viewed in historical context. One very important factor shaping how well a country does in complying with the obligations it has accepted is what it traditionally did with respect to the issue being dealt with, and the legislation and regulations that it already had in place at the time it became a party to the accord.

Traditional behavior is related to a country's culture. Culture provides a context and springboard for what a country does. But countries' cultures are neither fixed nor static, nor are they always uncontested. Japan found it easy to comply with the obligations of the World Heritage Convention because it had traditionally cherished its cultural and natural sites; protection of them was deeply embedded in Japanese culture. The United States had similar traditions with respect to the natural environment dating to the early 1900s.

Our analyses confirmed our expectation that richer and more democratic countries would in general do better than poorer and less democratic countries. The European Union and its members, Japan, and the United States had the better records among the nine political units included in this analysis. The reasons for this are discussed in the following paragraphs.

A crucial factor contributing to the variance among the performances of countries is administrative capacity. Countries that have stronger administrative capacities can do a better job. Hungary's compliance, especially with the Montreal Protocol, demonstrates this point. Administrative capacity is the result of several factors. Knowledge is a fundamental ingredient in implementation and compliance. Having educated and trained personnel is important. But such individuals must have adequate financial support and an appropriate legal mandate in order to be effective.

The Indian administrative service had numerous and well-trained personnel, but its financial resources were extremely limited, and thus its effectiveness was restricted. Administrative capacity depends upon having authority. Administrators whose mandate is narrower than their assigned responsibilities, or who are subject to capricious interference, cannot do as well as their training and skills would make possible. Administrative capacity also depends on having access to relevant information. However much administrative capacity a country has, corruption can blunt its effectiveness.

Administrative capacity also is relative; it must be measured against the demands placed upon it, a point emphasized in the analysis in chapter 2 of compliance within political units. Even the European Union, Japan, and the United States, which had strong administrative capacity, had difficulty enforcing the obligations of CITES. Given the many points of entry to these political units, and the vast number of people and goods moving in and out of their territories, it would be beyond the capacity of these units to enforce complete control at their borders.

Administrative capacity correlates with total GNP and GNP per capita. Countries that have many resources in relative terms, either because they are large (e.g., Brazil, China, and India) or because they have high per capita incomes, have the resources to develop strong administrative capacities. Outside agencies also can help countries develop administrative capacity (Haas et al. 1993).

Beyond relative richness, economic factors are important, but rather indirectly. The political units in this study have widely varying per capita GNPs that have grown or declined at substantially different rates. Changes in GNP or the rate of growth of GNP had little discernible effect on implementation and compliance. Economic collapse and chaos, however, can and did have a profound effect. In Cameroon and Russia, compliance with CITES seems to have declined since the mid-1980s, and this seems to be directly attributable to the economic collapse and chaos. Limited government resources and rapid rates of inflation had an impact on the incentive structure of the individuals who must enforce the provisions of CITES, the customs inspectors. In some instances they were not paid. In others, they saw the value of their salaries decline precipitously. Conversely, the value of illicit trade in endangered species has increased. Under the circumstances, the apparent increase in illicit trade in endangered species is perhaps understandable.

How production is organized makes a difference. Under the Soviet Union, when the state was responsible for production, state enterprises both produced goods and monitored the environmental consequences of their activities. This system did not result in effective environmental protection. The state enterprises were more concerned with production goals than with environmental protection. In 1986, under the Gorbachev regime, an independent ministry was created with a mandate to be concerned about environmental issues, and this began to have an effect. But enforcing compliance became even more effective after more production was privatized. Governments seem to be better at regulating the activities of nongovernmental entities than they are at regulating activities under their own control. Separating responsibility for regulation and production appears to have advantages in terms of promoting compliance with international environmental accord obligations.

The extent to which and the ways in which a country is engaged in international trade are also important. The more a country engages in international trade that is not controlled by state authorities, the more opportunities there are for illicit trade. Russia's and Hungary's shifting from controlled international trade directed by a central plan to free international trade provides evidence of this. At the same time, the more a country is engaged in international trade and the more it relies on investment from abroad, the more subject it is to international pressures. China's engagement in the world economy in the 1980s provides evidence of how a country can feel that it has to conform to international environmental standards as part of belonging to the international economy.

Political systems and institutions have an effect on implementation and compliance, but again the effect is mixed and complex. Large countries have a much more complicated task of complying with the obligations of accords than do smaller ones. Federalism also causes complications. There are several levels of political authority in Brazil, China, the European Union, Russia, and the United States. The European Union, of course, is a special case because its authority was still in the process of being defined in the 1990s. In cases where activities with which the accord deals are widely dispersed, as in the World Heritage Convention, CITES, and the International Tropical Timber Agreement, multiple levels of political authority must be coordinated, which is not always an easy task.

Sometimes the authority of the central government, which accepts international obligations, does not reach deeply into local areas. The government in Beijing, for instance, had great difficulty controlling activities in southern and western China. The central authorities in New Delhi had little ability to control events in the Manas National Park. Brasília had difficulty securing compliance in Amazonia, where subjects of CITES and ITTA are found. Moreover, these large countries contain within their borders widely different ecological regions that require variation in the way administration is conducted. The extent of a country's borders also makes a difference. Countries that have long borders touching many countries have more difficulty controlling smuggling than those that do not.

As part of its reform, Russia attempted to decentralize authority. In the process of decentralization, the authority of Moscow over localities was weakened. This shift appears to have resulted in a decline in Russia's compliance with CITES. Whether this decline was the temporary result of an administrative restructuring or a longer-term change will be known only in the future.

Political stalemate and chaos can bring about a noticeable decline in implementation and compliance. This seems to have been the case in Brazil, Cameroon, and Russia at various times.

There are many features of democratic governments that contribute to improved implementation and compliance. Democratic governments are normally more transparent than authoritarian governments, so interested citizens can more easily monitor what their governments are doing to implement and comply with accords. In democratic governments, it is possible for citizens to bring pressure to bear for improved implementation and compliance. Also, nongovernmental organizations generally have more freedom to operate under democratic governments. In addition, fully independent courts can be used by nongovernmental organizations and citizens to force governmental action.

At the same time, however, democratic governments are normally more responsive to public opinion than authoritarian governments. Public opinion is not always supportive of environmental concerns—indeed, the economy is usually the public's greatest concern. Democratic governments allow conflicts about environmental issues to flare.

Because of the balance of factors mentioned in the preceding paragraphs, democratic governments are more likely to do a better job of implementing and complying with international environmental accords than nondemocratic governments, and the country chapters in this volume document this. This generalization does not always hold, however, and democratization does not necessarily lead automatically or quickly to improved compliance.

Democratization in Brazil and in Russia, however, seems to have contributed to improved compliance. Brazil's 1988 Constitution mentioned environmental issues, stated goals, and made commitments. Improved compliance in the Soviet Union could be attributed to the greater transparency in governmental processes that started with the reforms under Gorbachev and were continued after the collapse of the Soviet Union. The increased activities of such nongovernmental organizations as Greenpeace were particularly important to strengthening Russia's compliance.

The importance of nongovernmental organizations—TRAFFIC, Greenpeace, the Worldwide Fund for Nature, for example—has already been mentioned. They play a crucial role in implementation and compliance. They mobilize public opinion and set political agendas. They make information about problems available, sometimes information that governments do not have or would prefer to keep confidential. Often the information they make available is essential to monitoring. They bring pressure on governments directly and indirectly. Because many local and national nongovernmental organizations have connections with NGOs in other countries and with international NGOs, NGOs are a means of ensuring a uniformity of concern throughout the world. There are also significant transfers of funds among

NGOs, so those in poorer countries may have surprisingly extensive resources at their disposal. NGOs have become an instrument for universalizing concern.

However, not all nongovernmental organizations necessarily assist compliance. Some NGOs have purposes that are anathema to enhanced compliance with environmental treaties.

Individuals make a crucial difference in the implementation of and compliance with accords. Who the head of state is, counts. Brazilian President Fernando Collor played a major role in having Rio de Janeiro selected as the site for UNCED, and he advanced environmental causes within the country. Brazil's compliance with the five accords improved during his presidency. The Clinton administration in the United States appears to have been more committed to environmental goals than the administrations that immediately preceded it.

Individuals in less exalted positions can play important roles. Alexei Yablokov, the principal environmental adviser to Boris Yeltsin, insisted on revealing the Soviet Union's past violations of the London Convention and sought to bring the Russian Navy's activities into compliance with the terms of the treaty. Russell Train, as Chairman of the U.S. Council on Environmental Quality and head of the Environmental Protection Agency, initiated actions within the United States and extended them abroad. He played a crucial role in starting the international momentum in the 1970s. Maneka Gandhi, as Minister of the Environment, played an important role in bringing India into the Montreal Protocol.

Other individuals, through their knowledge, skills, and persistence, have played important roles in nongovernmental organizations. Tang Xiaoyan, through her work in atmospheric chemistry, played an important role in convincing the Chinese government that it should accede to the Montreal Protocol. The designation of some heritage sites, especially in Brazil, should clearly be attributed to individuals. Individuals are also important as members of epistemic communities.

Figure 15.1 shows how factors within countries affect implementation and compliance. It provides details that should be included in the country cluster.

A country's physical conditions, its history, its culture, and its behavior (tradition, legislation, and regulations) with respect to the activity involved prior to adhering to the treaty establish basic parameters that affect implementation and compliance. The economy, political institutions, and attitudes and values have an effect, but this is generally indirect. These factors operate through proximate variables. In our view the most important proximate variables are administrative capacity, leadership, nongovernmental organizations, and knowledge and information.

Figure 15.2 presents a comprehensive model of what we believe are the most important factors that affect compliance. They are grouped in the four major clusters that we have used for analysis. Figure 15.2 combines all of the elements that

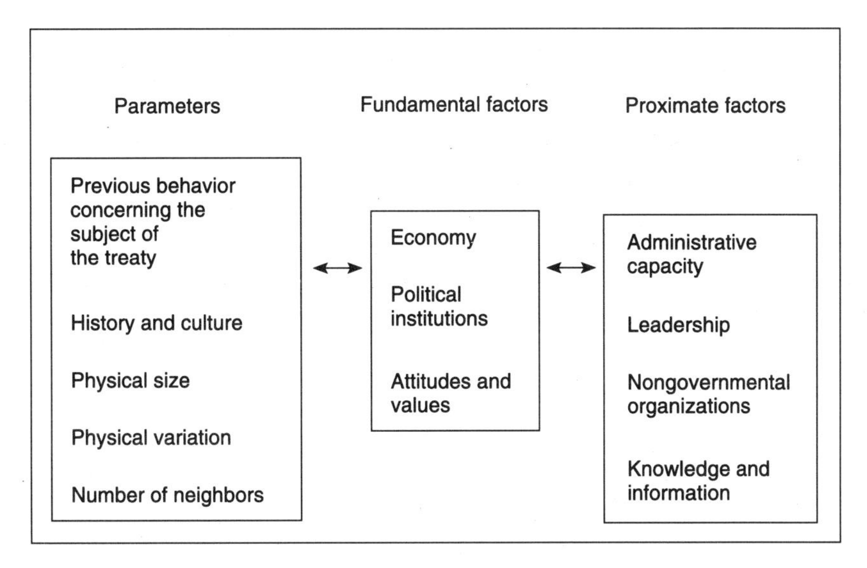

Figure 15.1
Factors involving the country

were included in tables 15.2–15.4 and figure 15.1. The most important factors within each cluster are indicated. The model is interactive: each factor affects all others, and there is feedback. Three clusters of variables—Characteristics of the Activity Involved, Characteristics of the Accord, and the International Environment—are important insofar as they affect what countries do.

The rate at which the variables in the clusters can be changed, varies. In general, those included within Characteristics of the Activity Involved are fairly fixed. Once negotiated, accords can be altered, but often only with difficulty. The international environment, in contrast, can be significantly altered relatively quickly, as it was in the run-up to the Rio conference. Within the cluster Factors Involving the Country, the proximate factors are those that have the most immediate effect and generally are those that are most susceptible to change in the short run. Thus, in the short run, the clusters of variables that are most susceptible to change and manipulation for policy purposes are the International Environment and the proximate factors within the cluster Factors Involving the Country.

Explanations of Compliance Behavior: Dynamic Processes

This comprehensive model takes into account the several factors that were introduced in chapter 1. It is a useful tool for analyzing differences among countries and treaties, but it has a static quality. It is more useful for synchronic than for diachronic analyses.

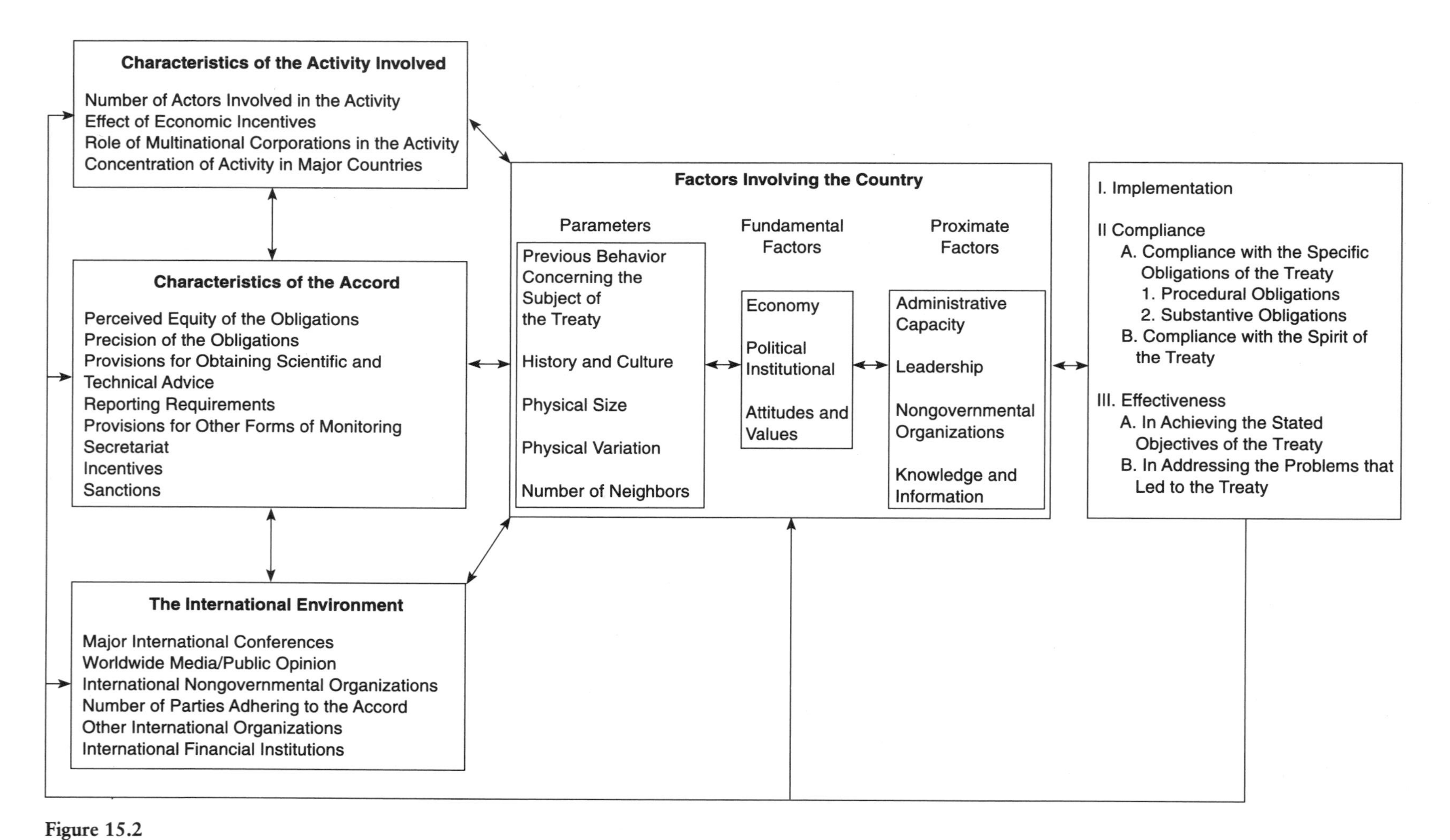

Figure 15.2
A comprehensive model of factors that affect implementation, compliance, and effectiveness

However, our research demonstrated more than differences in compliance behavior across treaties and political units; it also demonstrated change in compliance behavior over time. The model portrayed in figure 15.2 suggests how change occurs—by changing variables—and it indicates that change in one variable will affect others, but it does not show how dynamic processes occur. The history of the treaties that we have studied, and the interaction of the nine political units with these accords and with each other, clearly show that there are dynamic processes related to compliance that occur among and within countries. It is to these dynamic processes that we now turn.

We need first to consider interactions among countries. When we look comprehensively at the history of the five treaties and the nine political units, we find that frequently a large, powerful country enacted strong environmental laws domestically and then moved to establish comparable international standards. The United States enacted laws dealing with its heritage, with endangered species, and with ozone-depleting substances, and then took the lead in pressing for international treaties that would deal with the same issues. The United States played a leading role in the negotiation of the World Heritage Convention, the Convention on International Trade in Endangered Species, the London Convention, and the Montreal Protocol.

Later, as the conventions came into force, the United States and the European Union continued to play leadership roles, and at the London meeting on the Montreal Protocol, the European Union and its member countries played a more prominent role than did the United States. Japan played a leadership role with respect to the creation of the International Tropical Timber Agreement, and waged a successful campaign to have the secretariat for the agreement sited in Japan. The United States, the European Union and its members, and Japan had experience and resources that were essential to good implementation and compliance, and these resources could be shared with other countries that were less well endowed.

An important lesson of this history is that having what may be termed a "leader" is crucial to the negotiation of environmental accords and to the promotion of compliance with them. In fact, in the cases studied here, it is hard to see how effective progress would have been made without the efforts of leader countries. This finding is descriptive, not judgmental. The United States, the European Union and its member states, and Japan simply had more resources than any other political unit, and the United States had a larger economy than any other single state. Because of their economic strength, these political units were in a position to play a key role in the formation of environmental accords and in promoting compliance with them.

Now let us turn to changes within countries. Our studies make it clear that countries were in quite different positions on two dimensions at the time they joined an accord, and that their position on these dimensions changed during the life of the accord. These two dimensions are intention to comply and capacity to comply. They are related to the proximate factors within the cluster Factors Involving the Country.

Some countries clearly intended to comply with the obligations they accepted. They had thought about obligations of compliance, and either believed that they were already in substantial compliance or that they had a clear idea about the steps they needed to take to bring their practices into compliance. Other countries accepted the obligations of an accord without having thought through issues relating to bringing their practices into compliance. Some may have been blissfully unaware of the obligations they accepted; the government had simply not considered the issue. Still others may have been more cynical; they may have signed with full knowledge that they would not comply. Or a government may have been divided; the Foreign Ministry may have intended to comply while other branches of government may have had no intention of abandoning practices that contravened the accord. Countries signed accords for a variety of reasons beyond a simple intention to comply. They might sign, for instance, simply as a consequence of succumbing to international or domestic pressure.

Ability to comply is also important, as we noted at length above. Some countries had the resources that would enable them to comply; others found themselves falling short. Many assets are important for effective compliance, such as an effective and honest bureaucracy, economic resources, and public support. At the time of signing an accord, countries had different endowments of these resources and had allocated or were prepared to allocate different amounts to compliance; these endowments and allocations changed over time. Some countries were better endowed than others, but this changed. Bureaucracies that were effective and honest became ineffective and corrupt. Surpluses in government budgets disappeared and were replaced by deficits. Public support for leadership or particular policies increased or diminished. Because of economic turmoil and political and administrative changes, Russia became much less capable of complying with CITES than the Soviet Union had been. Cameroon's and Hungary's capacity for compliance in certain respects declined as well.

Figure 15.3 is based on these two dimensions: intention and capacity to comply. The countries analyzed in this study that were parties to the London Convention are placed—according to our judgment—in relationship to the axes defining the two dimensions in terms of their position at the time they signed the accord. The purpose of placing these countries in relationship to the axes is to show that historical reality supports the conceptual framework involved in the categorization.

At the time they signed the London Convention, the United States and Japan were in substantial compliance. Each had a strong intention to comply and a strong capacity to do so. They, along with several European countries, played a leading role

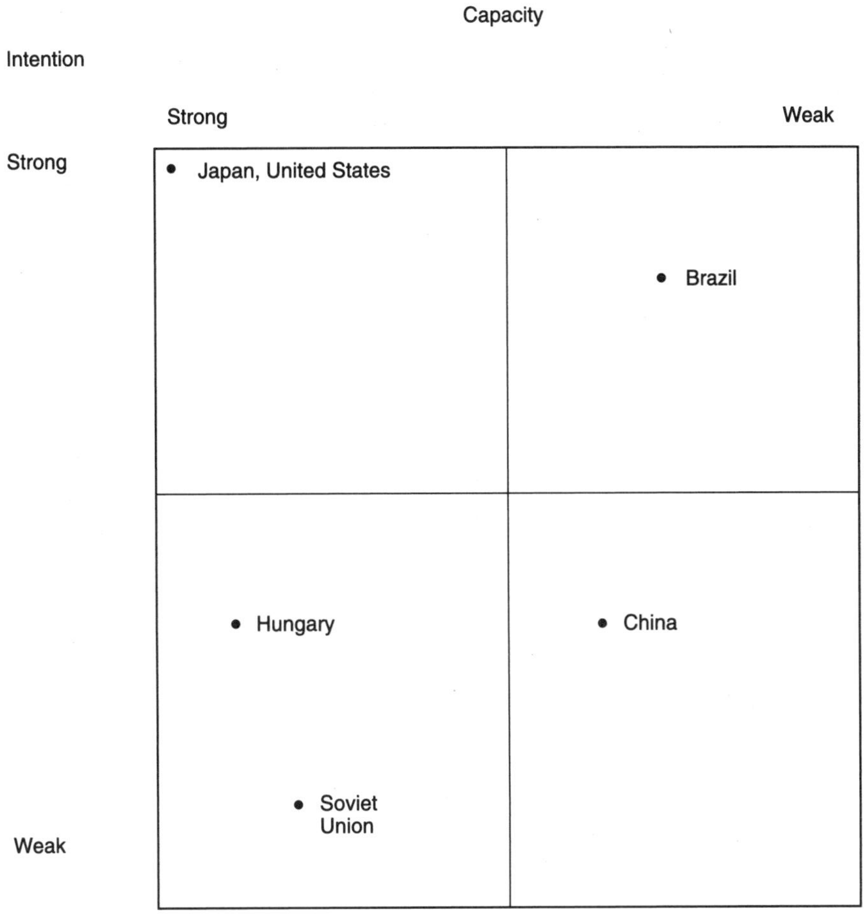

Intention \ Capacity	Strong	Weak
Strong	• Japan, United States	• Brazil
Weak	• Hungary • Soviet Union	• China

Figure 15.3
Intent and capacity to comply: Parties to the London Convention at the time they signed the accord

in creating the convention. At the time that it signed, Hungary wanted to be a good citizen, but as a landlocked country it had no need to think through the obligations. Its capacity to comply was modest, but probably adequate to the responsibilities it assumed, given its landlocked status. The Soviet Foreign Ministry may have intended to comply, but the Soviet Navy was clearly unwilling to abandon its practices that contravened the spirit of the Convention. The Soviet Union had a strong capacity to comply, and if its intention were modified, as it subsequently was, it could comply. At least some elements of the Chinese government did not want to comply with the London Convention, and in any case the central government found it difficult to police activities along China's long coast. Brazil's intention to comply was stronger, but its capacity was weak.

Countries' positions change over time. We have already noted how the Soviet Union's/Russia's intention with respect to the London Convention changed. In the run-up to the Rio conference, many countries became more interested in strengthening their compliance. The case of Brazil was particularly notable. President Collor wanted the conference to be held in Rio and wanted his country to have a good record when delegates to the conference arrived. Because of economic turmoil and political and administrative changes, Russia became much less able to comply with CITES than the Soviet Union had been. Cameroon's and Hungary's capacity for compliance in certain respects declined as well.

How can change in intention and capacity over time be explained? As the illustrations cited above indicate, many of the factors that produce change are endogenous to the countries concerned and are caused primarily by factors that have little to do with environmental issues; for example, Collor's election as President of Brazil and Russia's abandonment of communism and moves to establish democracy and a free-market economy. But exogenous factors can also affect intention and capacity. External assistance and pressure can affect a country's capacity and intention. Financial and technical assistance can help strengthen a country's capacity. A country's intention could be affected by external blandishments or pressures.

Figure 15.4 presents a hypothetical example of how these processes might work over time. The graphical representation of a country's record with respect to compliance over time illustrates several issues. Year 1 is when the country became a party to the accord. The top line in the diagram shows the level of the country's compliance gradually increasing over time, except for the decline in compliance between years 4 and 5. In this example, the country moved from moderate compliance at the time it joined to the accord to substantial compliance in year 7. Two factors contributed most to the level of compliance: the country's intention to comply and its capacity to comply. As those increased, so did the country's level of compliance, and when they declined, so did the level of compliance.

External assistance can help, in that it can contribute to building up a country's capacity, and it can compensate for temporary setbacks in capacity, as occurred in this example between years 4 and 7, and, in the cases in this book, in Russia, Hungary, and Cameroon. External assistance also contributed to building up Brazil's capacity to comply with the World Heritage Convention.

External pressure may contribute to a country's resolve to comply, but its role is limited. In figure 15.4, external pressure was in effect at the time the country acceded to the accord, slacked off for a year, and was reapplied in year 3. It increased in year 6, which contributed to the increase in intention and capacity to comply in year 7.

Intention to comply is the foundation for compliance. Capacity is essential, but a country will utilize its capacity only if it has the intention to comply. We chose to title this book *Engaging Countries* because of the crucial role of countries' intention

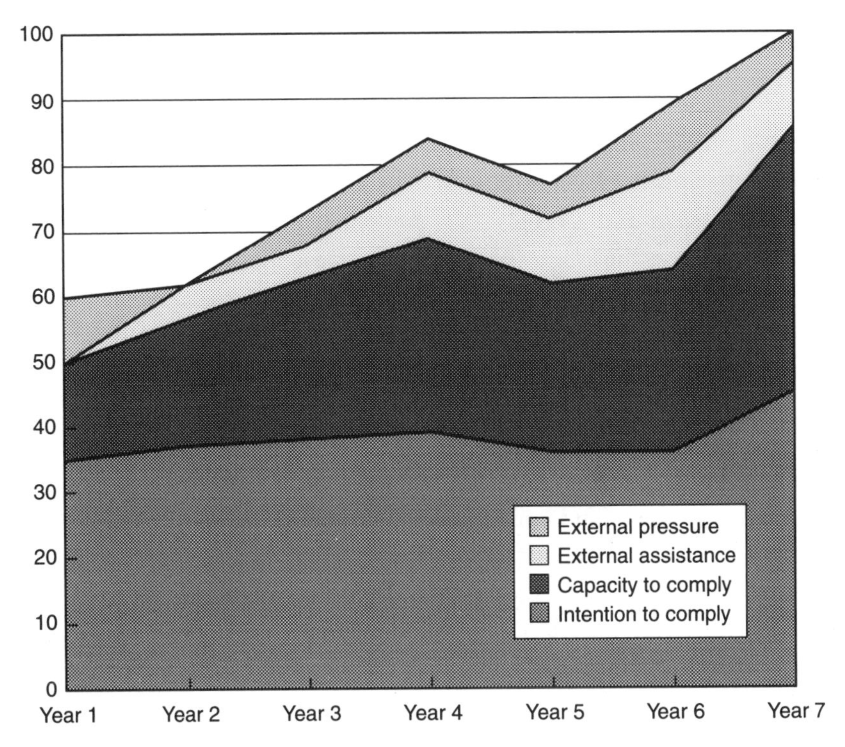

Figure 15.4
The interaction of intention and capacity to comply, and external assistance and pressure

to comply in establishing the level of compliance. The level of a country's compliance with international environmental accords depends crucially on the leaders and citizens of the country understanding that it is in their self-interest to comply, and then acting on this belief. It is they who must devote the resources to creating and maintaining a capacity for compliance, and it is they who must decide to utilize this capacity. External assistance and pressure can aid and nudge countries, but there is no substitute for the engagement of self-interest.

Stressing that countries must become engaged in the compliance effort returns our attention to figure 15.2, which portrays Factors Involving the Country that affect the level of compliance. In the short run, engaging national leaders in the effort to protect and improve the global environment made a difference, as the case of Brazil demonstrated. This strengthened Brazil's intention to comply. Supporting international and national nongovernmental organizations that focus on environmental issues, and building epistemic communities also helped to engage several countries. Providing assistance to national bureaucracies charged with responsibilities for environmental management strengthened the capacity to comply of several countries.

Our analyses underscore the longer-term importance to efforts to protect the global environment of the underlying strength and health of national political-economic systems, or what we termed the Fundamental Factors. In general, richer countries and those with democratic governments had a higher level of compliance with the international environmental accords. The strength and health of national political-economic systems and a deep public commitment are the most important ingredients in compliance; thus long-term strategies must focus squarely on these issues, as does *Agenda 21* (Robinson 1993). The process of strengthening compliance can be accelerated by concentrating on the Proximate Factors, but it is the Fundamental Factors that are essential to strengthened compliance in the longer run.

Strategies for Strengthening Compliance

What prescriptions do these findings suggest? Our prescriptions for improving compliance flow from our analyses of factors that have contributed to compliance and from the analysis of the treaties in chapter 5. The prescriptions are for strategies that deal primarily with what can be done through international law and institutions to engage countries in the substantive issues that are involved in international environmental accords, and to stimulate and encourage them to take actions to comply with these accords.

Broad Strategies

International legal and institutional strategies to encourage compliance with international agreements may be grouped into three categories: (1) sunshine methods, such as monitoring, reporting, on-site inspections, access to information, and NGO participation; (2) positive incentives, such as special funds for financial or technical assistance, training programs, or access to technology; and (3) coercive measures in the form of penalties, sanctions, and withdrawal of membership privileges. These three broad categories group the various institutional features of the treaties that were analyzed in detail in chapter 5. In terms of the diagram presented in figure 15.4, positive incentives would be placed in the category of external assistance, and sunshine methods and coercive measures would be placed in the category of external pressures. We used only two categories to simplify the diagram; now it is important to make finer distinctions. In addition to these three broad strategies, there are traditional public international law remedies available to parties for breach of an agreement, as set forth in the 1969 Vienna Convention on Treaties and in customary international law.

The three strategies are almost always employed in combination. Because of the differences among accords and countries, different mixes of strategies will work better in different circumstances. What mix will be most effective for a particular accord or country will likely change over time.

Table 15.5 shows the suite of measures contained in the five treaties that we have studied, that are attached to each of the three strategies: sunshine methods, positive incentives, and coercive measures.

Sunshine Methods. Sunshine methods are intended to bring the behavior of parties and targeted actors into the open for appropriate scrutiny, and thereby to encourage compliance. These include regular national reporting; peer scrutiny of reports; on-site monitoring; access to information by nongovernmental organizations and participation of NGOs in monitoring compliance; media access and coverage to provide public awareness; monitoring of behavior through regional workshops, corporate or private-sector networks, or consultants working on site; and informal pressures by parties and by secretariats to comply. The sunshine strategy relies on what has been termed the "reputation factor" to induce compliance. It works most effectively when there is a culture of compliance with norms (Hart 1994; Koh 1997).

Monitoring is essential to any effective program to increase compliance with treaty obligations. Monitoring may take many forms: off-site monitoring through advanced technologies that track scientific baselines or other criteria; reports by parties or nongovernmental organizations; on-site monitoring by parties, secretariat staff, or consultants; and international review of materials submitted by parties or gathered from other sources. On-site monitoring provides opportunities to learn about what is happening on the ground, including uncovering corrupt behavior by national and local officials, and by relevant nonstate actors.

National reports have served as a primary means for monitoring compliance in four of the five treaties studied. Reporting is useful in that it engages countries in implementing the agreement. Some official(s) in the country must be responsible for filing the report. Moreover, reports are an important tool for educating countries about their commitments under the agreement and potentially serve to build local capacity to comply with the agreement.

But there are also problems with relying on reporting as the key monitoring tool. Countries may be unwilling to publicize their own shortcomings in compliance through the reporting process. The brief effort in the 1980s by the World Heritage Convention to obtain national reports failed in part for this reason. Moreover, officials required to file reports under the burgeoning number of international agreements may find that their time is mostly occupied by preparing reports rather than by taking the actions called for under the agreement or addressing other high priority environmental actions. Indeed, in countries with a scarcity of skilled professional staff, the government may need to devote much of its time to meeting reporting requirements.

Table 15.5
Provisions in the five treaties that fit with sunshine methods, positive incentives, and coercive measures

	World Heritage	CITES	Tropical Timber	London Convention	Montreal Protocol
Sunshine methods					
Reports		Annual	Annual	Annual	Annual
On-site monitoring	Yes	Yes	Yes—limited		Yes
Publication of violations		Yes			
NGOs, nonstate actors monitor	Yes	Yes			
Industry monitoring			Yes		Yes
Local community involvement	Yes				
Persuasion: parties and Secretariat	Yes	Yes		Yes	Yes
Information initiatives	Yes	Yes	Yes		Yes
Positive incentives					
Technical training and assistance	Yes	Yes	Yes	(1996 Protocol)	Yes
Industry program					Yes
GEF	Yes	Yes			Yes
Special Fund	Yes		(New ITTA)		Montreal Protocol Fund
Coercive measures					
Membership status sanctions					Yes (art. V status)
Trade sanctions		Yes			Yes
Other enforcement					Yes

This raises a problem of congestion in treaty reporting. Moreover, unless the reporting requirements are standardized, the data may be difficult to compare and to evaluate. Great variations in content in national reports for the same accord may occur. In other cases, national reports may begin to look alike across different agreements, as information technology makes it easy to transfer written material from one document to another. Standardized forms can ease the burdens on governments of member states and on secretariats alike, and on nongovernmental actors reviewing the data.

Reports are useful as monitoring tools if they convey information that others can review and verify. This requires either an international process of review, or access to the reports by nongovernmental bodies, or both, as well as means to alert parties to violations by member countries. However, both secretariats and parties can be reluctant to call particular countries to account for inaccurate reports.

As indicated earlier in this chapter, for reporting to be an effective monitoring tool, the major players in a particular agreement must comply and file reports. It is arguably much less important that all parties comply fully with the requirement for this purpose, although their compliance is essential for building capacity to comply and for building a culture of compliance with the agreement among all parties.

While most agreements have focused on reporting, the World Heritage Convention has always relied primarily on site monitoring. In the future, parties to international environmental accords may turn increasingly to on-site visits to monitor compliance with treaty obligations and to verify accuracy of reports. The International Tropical Timber Organization uses many consultants to carry out country projects. On-site monitoring in some cases may give a more detailed picture of compliance than annual reports.

Sunshine methods rely on nongovernmental organizations, expert communities, and corporate actors to encourage compliance with international agreements. Nongovernmental organizations bring violations to the attention of government and of secretariats servicing the agreement. International nongovernmental organizations may work with local nongovernmental organizations to pressure countries to comply, or may exert pressure directly on governments or on corporate actors. Nonstate actors, including individuals, bring suspected violations to the attention of secretariats, parties, and, importantly, the media. While these organizations have been much more active in industrialized countries than in developing countries, their numbers are growing worldwide. Agreements like the World Heritage Convention and CITES explicitly use nongovernmental organizations in implementing the agreements.

Sometimes the most effective monitoring of compliance takes place in the private sector. Corporations have a keen interest in maintaining a level playing field by ensuring that competitors abide by the agreement. They also have the resources to monitor compliance, albeit quietly.

Sunshine methods incorporate measures that build public awareness of the agreement and the measures required, provide media access to measures of compliance and to cases of noncompliance, and use emerging information technology to build pressures upon governments and other relevant actors to comply.

Positive Incentives. International environmental accords increasingly incorporate positive incentives to induce countries to join the accord and to comply with it. The incentive approach assumes that many compliance problems exist because countries lack the capacity to comply. Incentives can provide assistance to build the capacity to comply (Bothe 1996). Incentives can also be effective in shaping the intention of the country to comply.

Incentives may take the form of special funds for financial or technical assistance, training programs and materials, access to technology, or bilateral or multilateral assistance outside the framework of the convention, from governments, multilateral development banks, or, in some cases, the private sector.

Many international environmental agreements provide for special funds to assist parties in complying. The World Heritage Convention established a special fund that, although small, has assisted countries in conserving sites on the World Heritage List. Parties to the Montreal Protocol agreed in 1990 to establish the Montreal Protocol Fund to help article V developing countries comply with their Protocol obligations. The new International Tropical Timber Agreement provides for the Bali Partnership Fund to assist countries in managing forests sustainably. Newer international agreements, such as the Framework Convention on Climate Change and the Biological Diversity Convention, also provide for funding to countries to meet the global incremental costs of complying with the agreements.

Funds are important because they build local capacity to comply, and contribute to the perceived equity of the treaties (and hence their acceptability).

Training programs and technical assistance efforts provide important incentives to comply. Training programs build local capacity to comply with treaty obligations and engage local actors in the accord. However, care must be taken to ensure that the proper people are trained and that at least some of those trained will continue working with the accords beyond six months. Various strategies can be employed: moving training programs to take place in the regions or within countries; training slightly larger numbers of people than might be needed, so that some may still be on the job in a year; and ensuring that the group to be trained includes the relevant officials in addition to any persons selected for political reasons. Other techniques include training the trainers who can then conduct regular seminars within the country, providing clear training manuals, and scrutinizing training applicants for quality.

Coercive Measures. Under the traditional framework for addressing compliance, parties rely on coercive measures: sanctions, penalties, and measures such as the withdrawal of privileges under the convention. In some areas of international law, such as trade law or national security, sanctions have been regarded as essential to achieving compliance. By contrast, in international environmental law, sanctions are rarely used. Some have suggested that sanctions are largely irrelevant to and ineffective for environmental agreements.

Sanctions have been available to enforce three of the treaties in the study: the Convention on International Trade in Endangered Species, the Montreal Protocol, and the World Heritage Convention. Both CITES and the Montreal Protocol include the possibility of trade restrictions. Sanctions may be authorized by parties to the agreement. They also have been imposed unilaterally by the United States under domestic legislation. While the trade restrictions in the Montreal Protocol under article 4 are not sanctions, but inducements to join and remain in the Protocol, they nonetheless may function as sanctions to discourage countries from withdrawing from the obligations of the Protocol. Moreover, the noncompliance procedures developed to implement the Protocol include the possibility of sanctions imposed upon the parties.

The use of trade sanctions to enforce multilateral environmental agreements raises the question of whether sanctions are consistent with obligations that countries have accepted under the General Agreement on Tariffs and Trade and the World Trade Organization (WTO) (W. Lang 1996). In the late 1990s the compatibility of trade provisions in multilateral environmental agreements with the WTO was under discussion in the WTO's Environment Committee.

Withdrawing the privileges of membership in a convention is another traditional enforcement measure. The World Heritage Convention and the Montreal Protocol provide for variants of this. While the text of the World Heritage Convention provides only for countries to nominate sites on the World Heritage List, the guidelines provide for delisting a World Heritage Site if the country violates its obligation to conserve the site. The language is ambiguous as to whether the host country must consent to the delisting. The sanction has not been used, for no site has yet been removed from the List, although the integrity of several sites is severely threatened.

The Montreal Protocol provides for withdrawing some membership privileges as a sanction for noncompliance. Countries qualifying as article V countries, and thus eligible to receive assistance from the Montreal Protocol Fund, can lose that status, and thus access to the Fund, if they do not provide baseline data on consumption levels of controlled substances within a specified time after joining the agreement.

While sanctions have not played a significant role in promoting compliance with the treaties we studied, they do have value as a weapon of last resort. Sanctions may still be needed as a latent threat to make other methods of achieving compliance effective, as outlined below. Sanctions are crucial in agreements where free-riding is possible and could carry significant rewards.

Some of the measures grouped under the sunshine strategy discussed above may, under certain circumstances, be viewed as akin to a sanction; for instance, publication of violations of an agreement, as in the case of CITES's biennial report. Or they may assist in identifying instances where sanctions are warranted, perhaps through on-site visits or regular national reports to secretariats.

Traditional Public International Law Remedies. Public international law provides remedies for breaches of treaty obligations and of customary international law. The Vienna Convention on Treaties indicates when parties may withdraw from an agreement. Rules of customary international law set forth rights of countermeasures. Some of these remedies arguably contemplate measures that withstand the provisions of the U.N. Charter. These remedies have not been used for international environmental problems. Nonetheless, such remedies are available to states, providing the legal requirements are satisfied.

Linking Compliance Strategies with Countries

Joining Strategies with Intention and Capacity

As discussed previously, individual countries have different profiles regarding their intention and capacity to comply for each international agreement. Some countries have a strong intention to comply and a strong capacity; some have a weak intention but a strong capacity; others have a weak capacity but a strong intention; still others have a weak capacity and a weak intention. Within each of these categories, states have varying degrees of intention and capacity to comply.

Strategies for compliance must be targeted to ensuring that all parties to an accord have both a strong intention and a strong capacity to comply. Different strategies need to be emphasized for different countries, depending upon their profile with respect to intention and capacity; a mix of strategies should always be available. The most appropriate strategies to induce compliance by a particular country will change over time if the country's intention and/or capacity change. All of the strategies must be geared toward engaging the country in implementing and complying with the agreement.

For countries that are strongly inclined to comply and have the capacity, sunshine methods are especially appropriate and effective. Other governments, secretariats, nonstate actors, and individuals can use them to monitor behavior and to urge parties to comply. They can build public support for compliance with the agreement. However, care must be taken to ensure that the measures are not used by unscrupulous parties to impose enormous administrative burdens (as by scurrilous claims of violations that must be investigated or that tarnish the agreement) or to devise ways to circumvent the agreement (as by filing false reports and using publicly available information to circumvent customs officials). While our research did not uncover widespread abuse of these strategies, the possibility remains.

To prevent countries from lapsing or regressing once they have undertaken to comply with the agreement and have the capacity to do so, sanctions need to be available. Sanctions are of secondary importance in ensuring that countries are not tempted to change their intention to comply and in discouraging rogue actors from undermining compliance. Sanctions are the instrument in reserve that makes sunshine methods effective.

Sometimes sunshine methods serve as a mild form of coercion by shaming would-be violators into conforming to the agreement. This effect may be observed in some cultures, such as Japan, more than in others. In this sense, sunshine methods act as indirect sanctions against parties that have strong intention and strong capacity to comply but, for whatever reason, fail to comply.

Sunshine methods facilitate the operation of democratic processes. They provide vital information so that citizens and nongovernmental organizations can hold governments accountable. They also provide reassurance that other parties to an accord are not free-riding.

When countries have the intention to comply but lack the capacity, positive incentives are especially important. This category includes many low-income countries. In some instances, countries may intend to comply but the agreement is low on their list of priorities. Incentives not only build capacity but also may positively affect the priority the country gives to complying with the agreement. Sunshine methods may have an important secondary role in mobilizing nonstate actors to press for compliance and in enabling governments, industry, and nongovernmental organizations to monitor progress toward building local capacity to comply. Sunshine methods also may help because they can build public support for devoting the resources required to comply with the agreement.

Countries that do not intend to comply, or have only a weak intention to do so, but have a strong capacity to comply may be particularly suited to some form of coercive measures, whether multilateral or unilateral in pursuit of multilateral objectives. Sanctions can be targeted to making it in a country's best interests to comply. Moreover, sunshine methods can serve as an effective secondary strategy to put pressure upon the country to comply. Positive incentives may also be useful in helping to change a country's views on compliance or to strengthen the resolve to comply.

For those countries that lack both the intention and the capacity to comply, or are weak in both, all three strategies are essential. Positive incentives help build

Intention	Capacity: Strong	Capacity: Weak
Strong	Sunshine (Sanctions)	Incentives (Sunshine)
Weak	Sanctions Sunshine (Incentives)	Incentives Sanctions Sunshine

Figure 15.5
Strategies to strengthen compliance, taking account of intention and capacity to comply

capacity. Sunshine measures may foster a culture of compliance and prevent backsliding. The threat of sanctions may encourage progress toward compliance. This category of countries is the most difficult, because the countries include many of the low-income countries throughout the world. Again, the countries in this category will vary by agreement. States, multilateral development banks, other international organizations, industrial associations, and nongovernmental actors can be effective instruments in implementing one or more of these compliance strategies.

Figure 15.5 links compliance strategies with a country's intention and capacity. The strategies are listed in each quadrant of the diagram in the order of their importance for countries in that quadrant. Strategies that are useful primarily in supporting or backup roles are in parentheses.

One of the complications in linking intention and capacity, on the one hand, and compliance strategies, on the other, is that the intention and the capacity of countries to comply may change over time. As our research revealed, national compliance changes over time in response to many factors. Thus, a country weak in intention and capacity could become stronger in both, as in Cameroon's compliance with procedural obligations under the Montreal Protocol. Or a country that was strong in capacity and intention could find that either or both were diminished, as in the weakening of the Russian Federation's capacity to comply with CITES after the dissolution of the Soviet Union. Thus, the compliance strategies needed to encourage compliance for a given country will change over time. This calls for parties to provide themselves with a range of measures that can be used flexibly to engage countries in complying with the accord.

Implementing Linkages Between Countries and Strategies

The most important point that our analyses have shown is that for there to be effective implementation of and compliance with treaties, countries must become engaged. This involves a broad commitment that is deeper than just sections of governments or even whole governments. The country must see that its interests are served by complying with the treaty.

To promote compliance, states and other actors can bring external pressures on countries through sunshine methods or through coercive measures, or they can provide positive incentives in the form of external assistance. The precise mix of measures that would be appropriate and effective will vary, depending upon the country's intention to comply and its capacity to comply.

The actions should reach local communities, encourage coordination among ministries and agencies within countries, build administrative capacity within countries, promote dissemination of information and measures to inform the public, foster technical expertise and competence, promote nongovernmental organizations that can be helpful to compliance, motivate the private sector to promote compliance, and counter corruption and payoffs that hinder compliance.

The peripheries of countries as well as the center must see that their interests are being served. Decentralizing implementation of the accord within countries will not necessarily promote compliance. Local officials may have other priorities, be more susceptible to financial incentives to ignore compliance with treaty obligations, and be less informed about the requirements of the treaty. But if local communities can become engaged in implementing the agreement so that they see the clear benefits to them of complying with the agreement, then compliance may be much stronger and more reliable in the longer term. This is particularly the case for federal countries. Positive incentives through financial and technical assistance that reaches local communities, and market activities that make compliance in the interest of local communities should be stressed.

Assistance that builds national and local administrative capacity is particularly helpful. This means increased attention to training and education at the regional and local levels. Such assistance must be carefully tailored to ensure that the benefits are long-term. Measures that promote coordination among relevant ministries and agencies in the country should also be stressed. International governmental organizations and others providing assistance should bargain hard to eliminate corrupt payments, whether through the addition of new institutions or of unqualified personnel, or direct personal payments.

The media can play a role. Newsletters and electronic dissemination of information need to be expanded, and provision must be made to ensure that local organizations and communities have access to such information. The diffusion of information about problems targeted by treaties and the actions required of governments and nonstate actors can be one of the most effective instruments in promoting compliance.

National and international nongovernmental organizations can play a major role in getting countries engaged in treaties. Not all nongovernmental organizations, however, will initially want to further national compliance. Efforts should be made to win them over. If these fail, it is important that all feel that opposing views were freely aired, considered, and rejected through fair procedures. Suppressed opposition has a tendency to resurface and can seriously weaken efforts toward compliance.

The private sector needs to be mobilized to promote compliance with the accords. This means bringing the relevant actors into a dialogue with governmental and nongovernmental actors, and encouraging measures that speak to their self-interest in promoting compliance. Nongovernmental actors should be brought into the negotiating process, if only through information and consultation. Parties and secretariats should involve them in activities to implement the agreements and to promote compliance.

Engaging Countries

As we sought to explain the difference in levels of compliance across treaties and among countries, and the overall trend toward strengthened compliance, and then developed and analyzed broad legal and institutional strategies to strengthen compliance, we implicitly and explicitly offered several prescriptions. Some of the prescriptions apply to the way accords are drafted; others, to essential institutional features; and still others, to actions that should be taken after the accords come into effect. The more important of the prescriptions that we have covered can be summarized briefly in three categories.

Negotiating the Accord

- Ensure that the obligations of the accord are perceived as equitable by parties and potential parties.
- If clearly assessing compliance is a primary concern, make the obligations as precise as possible.
- Try to ensure that the obligations are reinforced rather than contradicted by economic forces.
- Craft the treaty so that the burden of compliance is placed on a manageable number of actors. Target the major actors.

• Ensure that there are leader countries in the negotiations and that early on, they take measures to implement and comply with the agreement.

Institutional Arrangements Associated with the Accord

• Provide for regular meetings of parties, so that national and international bureaucracies will be mobilized regularly.

• Ensure that secretariats are strong enough to identify cases of noncompliance, advise various actors on how to comply, propose measures (through governments) to address issues of noncompliance, and seek support from various institutions, and other actors in which parties have confidence, to help countries.

• Include means, in which parties have confidence, for ongoing scientific assessments of problems targeted by the accord.

• Involve the international financial institutions, including the Global Environmental Facility, in building local capacity to comply with the accords.

• Link access to financial support for projects to compliance with obligations that might affect the implementation of the project.

• Involve nongovernmental organizations in helping to carry out accords.

• Develop standardized forms for reporting data, which should contribute to the effectiveness of reporting. Data required should be frugal, and perceived as equitable and essential.

• Inform the public about the agreement through the media and new information technology. This can build support for implementation and compliance.

Measures Directed at Countries

• Focus strategies for strengthening compliance on individual countries, and differentiate them according to their intention and capacity to comply.

• Make it possible for low- and middle-income countries to participate meaningfully in all agreements they may be expected to join.

• Assist countries, when necessary, in drafting implementing legislation and in initiating the steps required by the agreement for compliance.

• Make available technical assistance and capacity-building programs at the national and local levels. Regional networks of relevant national officers who have responsibility for implementation and compliance may encourage learning how to address shared compliance problems.

• Strengthen coordination among relevant ministries and departments, and between national and provincial or municipal units of government. Strengthen domestic institutions concerned with compliance.

• Build a culture of compliance by engaging both the public sector and the private sector in determining domestic needs and setting priorities.

These prescriptions are in effect tactical suggestions. They can be employed singly and in combination, and to varying degrees at various times.

In the end, engaging countries means engaging all relevant actors to promote compliance. While we speak of countries as separate political entities, the international system is a very complex web of transnational networks and local actors, joined in dynamic ways. A strategy of compliance must look beyond governments to provide incentives and pressures for all relevant actors to comply with the environmental accords.

While national compliance may be easier for countries that intend, and have the capacity, to comply, even they will need external pressures upon relevant actors, if only in the form of sunshine, and incentives at least to the private sector, to ensure compliance. For countries lacking the capacity, external incentives to relevant actors are essential, and external pressures are needed to maintain or to develop the intention to comply. Engaging states and all relevant actors from the beginning and keeping them engaged are the essential steps that must be taken to strengthen national compliance with international environmental accords.

References

Abernethy, D. B. 1988. "Bureaucratic Growth and Economic Stagnation in Sub-Saharan Africa." In *Africa's Development Challenges and the World Bank*, edited by S. K. Commins, 79–214. Boulder, CO: Lynne Rienner.

Abraham, J. 1993. "Scientific Standards and Institutional Interests: Carcinogenic Risk Assessment of Benoxaprofen in the UK and US." *Social Studies of Science* 23:387–444.

Abramson, Rudy. 1991. "Japan Yields to U.S. Pressure, Will Halt Trade in Endangered Turtles." *Los Angeles Times,* May 18, A10.

Abranches, Sergio. 1993. "Homo Indexadus." *VEJA* (April).

"ACCA Explains EPA's Final Rules for CFC & HCFC Refrigerant Sales." 1994. *Environmental Information Networks, Ozone Depletion Network Online Today* (June 13).

Adler, E. 1991. "Cognitive Evolution: A Dynamic Approach for the Study of International Relations and Their Progress." In *Progress in Postwar International Relations*, edited by Emanuel Adler and Beverly Crawford, 781–843. New York: Columbia University Press.

Adler, Robert W., and Charles Lord. 1991. "Environmental Crimes: Raising the Stakes." *George Washington Law Review* 59 (April):781–843.

Advisory Council on Historic Preservation. 1986. *Section 106, Step-by-Step*. Washington, DC: ACHP.

Agence France Press (AFP). 1993. (September 20).

Agarwal, Anil. 1985. "Politics of Environment—II." In *The State of India's Environment 1984–85*. New Delhi: Centre for Science and Environment.

———. 1993. "Ecomanagement: The Best Solutions Are Home-made." *The Hindu: Survey of the Environment* 7–11.

Agarwal, Anil, and Sunita Narain. 1991. "MNC Cons the Third World with Ozone Hole Scare." *Economic Times* (November 17).

———. 1991. *Global Warming in an Unequal World.* New Delhi: Centre for Science and Environment.

———. 1992. *Toward a Green World.* New Delhi: Centre for Science and Environment.

Agarwal, Bina. 1990. "Engendering the Environment Debate: Lessons from the Indian Subcontinent." In *WIDER Conference on Environment and Development.* Helsinki: WIDER.

Agence Europe. 1992. "(EU) Environment: A Resolution Denounces the EEC for Its Substantial Participation in the Illegal Trade of Protected Animal Species." *Europe* no. 5687 (12 March):15.

———. 1995. "EU Environment: Terms of New CITES Regulation Creating a 'Schengen of Endangered Species'." *Europe* no. 6510 (28 June):9.

Agreement Establishing the Multilateral Trade Organization. 1993. Reprinted in 33 ILM 13 (1994).

Albritton, Daniel L. 1989. "Stratospheric Ozone Depletion: Global Processes (Ozone Depletion, Greenhouse Gases, and Climate Change)." In *Proceedings of a Joint Symposium by the Board on Atmospheric Sciences and Climate and the Committee on Global Change, Commission on Physical Sciences, Mathematics, and Resources, National Research Council*, 10–19. Washington, DC: National Academy Press.

Allison, Graham. 1971. *Essence of Decision: Explaining the Cuban Missile Crisis*. Glenview, IL: Scott, Foresman.

Alpert, P. 1993. "Conserving Biodiversity in Cameroon." *Ambio* 22, 1:44–53.

Anderson, B. 1991. *Imagined Communities*, 2d edition. London: Verso.

———. 1994. "Exodus." *Critical Inquiry* 20:314–327.

Antal, László. 1979. *Fejlödés—Kitérövel*. Budapest: Pénzügyi Kutatási Intézet.

Aragão, Murillo. 1994. "Grupos de Pressão no Congresso Nacional." *Maltese* (July).

Ausubel, Jesse, and David Victor. 1992. "Verification of International Environmental Agreements." *Annual Review of Energy and Environment* 17, 1:1–43.

Axelrod, Robert, and William Zimmerman. 1981. "The Soviet Press on Soviet Foreign Policy: A Usually Reliable Source." *British Journal of Political Science* 15 (April):183–200.

"Az Alkotmánybíróság Határozatai" (Resolutions of the Constitutional Court). 1994. *Magyar Közlöny* 55:1919–1926.

Badaracco, Joseph L. Jr., and Susan J. Pharr. 1986. "Coping with Crisis: Environmental Regulation." In *America Versus Japan*, edited by Thomas K. McCraw. Boston: Harvard Business School Press.

Bakalian, Allan. 1984. "Regulation and Control of United States Ocean Dumping: A Decade of Progress, an Appraisal for the Future." *Harvard Environmental Law Review* 8 (Winter): 193–205.

Baker, Susan, Kay Milton, and Steven Yearly (eds.). 1994. *Protecting the Periphery: Environmental Policy in Peripheral Regions of the European Union*. Portland, OR: Frank Cass.

Bandyopadhyay, Jayanta, and Vandana Shiva. 1988. "Political Economy of Ecology 7–11 Movements." *Economic and Political Weekly* (June 11):1223–1332.

Bajaj, Raghmi. 1996. *CITES and the Wildlife Trade in India*. New Delhi: Center for Environmental Law.

Bardach, Eugene, and Robert A. Kagan. 1982. *Going by the Book: The Problem of Regulatory Unreasonableness*. Philadelphia: Temple University Press.

Barnes, B. 1982. *T. S. Kuhn and Social Science*. London: Macmillan.

Barnes, B., and David Bloor. 1982. "Relativism, Rationalism, and the Sociology of Knowledge." In *Rationality and Relativism*, edited by Martin Hollis. Cambridge, MA: MIT Press.

Barnett, Harold C. 1993. "Crimes Against the Environment: Superfund Enforcement at Last." *Annals AAPSS* 525 (January):119–133.

Barnthouse, L. W., J. Boreman, S. W. Christensen, C. P. Goodyear, W. Van Winkle, and D. S. Vaughn. 1984. "Population Biology in the Courtroom: The Hudson River Controversy." *BioScience* 34:14–19.

Barzdo, J., and S. Broad. 1988. *Application of CITES in the European Economic Community*, vols. 1–3. Report prepared for the Commission of the European Communities by the World Conservation Monitoring Centre and the IUCN Environmental Law Centre.

Bates, R. H. 1981. *Markets and States in Tropical Africa*. Berkeley: University of California Press.

Beck, U. 1992. *The Risk Society*. London: Sage.

Begley, Ronald. 1991. "CFC Phaseout Acceleration Effort Introduced." *Chemical Week* (December 25):12.

Begley, Sharon. 1993. "Killed by Kindness." *Newsweek* (April 12):50–54.

Bell, Ruth Greenspan. 1992. "Environmental Law Drafting in Central and Eastern Europe." *Environmental Law Reporter News and Analysis* 22, 9:10597–10605.

Benedick, Richard E. 1991. *Ozone Diplomacy: New Directions in Safeguarding the Planet*. Cambridge, MA: Harvard University Press.

Bennett, G. 1993. "The Implementation of EC Environmental Directives: The Gap Between Law and Practice." *The Science of the Total Environment* 129 (February):19–28.

Bernardes, Ernesto, and Kaike Nanne. 1994. *VEJA*. (February 9).

Bernstam, Mikhail S. 1991. *The Wealth of Nations and the Environment*. London: Institute of Economic Affairs.

Bidwai, Praful. 1991. "Montreal Protocol: Has India Capitulated?" *Economic Times* (June 8).

Bilder, Richard B. 1981. *Managing the Risks of International Agreement*. Madison: University of Wisconsin Press.

Bijker, W. E., Thomas P. Hughes, and Trevor Pinch. 1987. *The Social Construction of Technological Systems*. Cambridge, MA: MIT Press.

Bimber, B., and David Guston. 1995. "Politics By the Same Means: Government and Science in the United States." In *Handbook of Science and Technology Studies*, edited by Sheila Jasanoff, Gerald Markle, James Petersen, and Trevor Pinch. Newbury Park, CA: Sage Publications.

Birnie, Patricia. 1985. *International Regulation of Whaling: From Conservation of Whaling to Conservation of Whales and Regulation of Whale Watching*. New York: Oceana.

———. 1996. "The Case of the Convention on Trade in Endangered Species." In *Enforcing Environmental Standards: Economic Mechanisms as Viable Means?* Edited by Rüdiger Wolfrum, 233–264. Heidelberg: Springer Verlag.

Block, Paula M. 1987. "The UN Focuses on Ozone Levels." *Chemical Week* (May 13):8.

Bloor, D. 1976. *Knowledge and Social Imagery*. London: Routledge and Kegan Paul.

Boisson de Chazournes, Laurence. 1993. *Analysis and Evolution of Monitoring and Compliance Procedures of International Environmental Law. Ministerial Conference, Environment for Europe*. Paper presented at Lausanne, Switzerland, April 28–30, 1993.

Bombay Natural History Society (BNHS). 1988. *Ecology of Semitropical, Monsoonal Wetland in India: The Keoladeo National Park, Bharatpur, Rajasthan*. Bombay: BNHS.

Bond, Andrew. 1992. "Some Observations on the Russian Federation Environmental Protection Law." *Post-Soviet Geography* 33:463–474.

Bothe, Michael. 1996. "The Evaluation of Enforcement Mechanisms in International Environmental Law." In *Enforcing Environmental Standards: Economic Mechanisms as a Viable Means?* Edited by Rüdiger Wolfrum, 13–38. Heidelberg: Springer Verlag.

Boulding, Kenneth E. 1966. "The Economics of Coming Spaceship Earth." In *Environmental Quality in a Growing Economy*, edited by Henry Jarrett, 3–14. Baltimore: Johns Hopkins University Press.

Bowman, Margaret, and David Hunter. 1992. "Environmental Reforms in Post-Communist Central Europe: From High Hopes to Hard Realities." *Michigan Journal of International Law* 13, 4:921–980.

Boyle, Alan E., David A. C. Freestone, K. Kummer, and David M. Ong. 1992. "Marine Environment and Marine Pollution." In *The Effectiveness of International Environmental Agreements: A Survey of Existing Legal Instruments*, edited by Peter H. Sand, 149–255. Cambridge: Grotius.

Brack, Duncan. 1996. *International Trade and the Montreal Protocol.* London: Royal Institute of International Affairs.

Brady, Gordon L., and Blair T. Bower. 1983. "Effectiveness of the U.S. Regulatory Approach to Air Quality Management: Stationary Sources." In *International Comparisons in Implementing Pollution Laws*, 29–45. Boston: Kluwer-Nijhoff.

Brand, E. C. 1993. "Hazardous Waste Management in the European Community: Implications of 1992." *The Science of the Total Environment* 129 (February):241–251.

Brenton, Tony 1994. *The Greening of Machiavelli: The Evolution of International Environmental Politics.* London: Earthscan.

Brickman, R., Sheila Jasanoff, and Thomas Ilgen. 1985. *Controlling Chemicals: The Politics of Regulation in Europe and the United States.* Ithaca, NY: Cornell University Press.

Broad, Steven, and Tom De Meulenaer. 1992. *The Need for Improved Wildlife Trade Legislation in the EC—A TRAFFIC Discussion Paper.* Brussels: TRAFFIC-Europe.

Bromley, Daniel, and Devandra P. Chapagain. 1984. "The Village Against the Center: Resource Depletion in South Asia." *American Journal of Agricultural Economics* 60 (December):868–873.

Brown, P. 1993. "Britain Backs Trade Gag on Logging Fears." *The Guardian* 002 (March 29).

Brown Weiss, Edith. 1993. "International Environmental Law: Contemporary Issues and the Emergence of a New World Order." *Georgetown Law Journal* 81, 3:675–710.

———. 1996. *The Changing Structure of International Law.* Francis Cabell Brown Inaugural Lecture, Georgetown University Law Center. Georgetown Law Res Ipsa 45 (Spring):54–62.

Brown Weiss, Edith, Daniel B. Magraw, and Paul C. Szasz. 1992. *International Environmental Law: Basic Instruments and References.* New York: Transnational Publishers.

Brundage, Steven. 1995. Interview, Washington, DC: U.S. Department of State, March 20.

Brunner, Helen J. 1992. "Criminal Enforcement of Environmental Laws." *Environmental Law* 22, 1:1315–1341.

Bryce, Robert. 1993. "Keeping Cool May Become a More Costly Proposition." *Christian Science Monitor*, October 14, 9.

Buergenthal, Thomas. 1969. *Law-making in the International Civil Aviation Organization.* Syracuse, NY: Syracuse University Press.

Burhenne-Guilmin, Françoise. 1992. Personal communication (October).

"Bush Accelerates CFC Phaseout, Citing Thinning Ozone Layer." 1992. *Bureau of National Affairs, National Environment Daily* (February 13):19–23.

"Bush Administration to Propose Fund to Help Countries Control CFC Emissions." 1990. *Bureau of National Affairs, Environment Reporter* (June 22):372.

Callen, Kate. 1992. "Proposed Law Would Require Origin Labels for Tropical Hardwoods." *San Diego Union-Tribune*, September 27, F11.

Callon, M. 1986. "Some Elements of Sociology of Translation: Domestication of the Scallops and Fishermen of St. Brieuc Bay." In *Power, Action, and Belief: A New Sociology of Knowledge?* Edited by John Law. London: Routledge.

———. 1987. "Society in the Making: The Study of Technology as a Tool for Sociological Analysis." In *The Social Construction of Technological Systems*, edited by Wiebe E. Bijker, Thomas P. Hughes, and Trevor Pinch. Cambridge, MA: MIT Press.

Callon, M., and John Law. 1982. "On Interests and Their Transformation: Enrollment and Counter-Enrollment." *Social Studies* 12:615.

Callister, Debra. 1992. "Illegal Tropical Timber Trade: Asia-Pacific." *TRAFFIC Network Report.* New York: Oceana.

Cameron, James. 1996. "Compliance, Citizens and NGOS." In *Improving Compliance with International Environmental Law*, edited by James Cameron, Jacob Wersman, and Peter Roderick, 29–47. London: Earthscan.

Campbell, Faith T. 1979. "Trade: A Continuing Threat to Survival of Species." *National Park and Conservation Magazine* 30 (June):14–20.

———. 1991. "The Appropriations History." In *Balancing on the Brink of Extinction: The Endangered Species Act and Lessons for the Future* edited by Kathryn A. Kohm, 134–146. Washington, DC: Island Press.

———. 1993. Correspondence from Faith T. Campbell to Michael Glennon, Natural Resources Defense Council, May 5.

Canada–U.S. Free Trade Agreement. 1988. Reprinted in 27 ILM 281 (1988), art. 103.

Carnegie Commission on Science, Technology, and Government. 1992. *International Research and Assessment: Proposals for Better Organization and Decisionmaking.* New York: Carnegie Commission.

Centre for Science and Environment (CSE). 1986. *The State of India's Environment, 1984–85: The Second Citizens' Report.* New Delhi: Ravi Chopra.

Chakrabarty, Kalyan, Ashok Kumar, and Vivek Menon. 1994. *Trade in Agarwood.* New Delhi: TRAFFIC-India.

Charny, David. 1990. "Nonlegal Sanctions in Commercial Relationships." *Harvard Law Review* 104 (December):373–467.

Chase, Britan F. 1993. "Tropical Forests and Trade Policy: The Legality of Unilateral Attempts to Promote Sustainable Development Under the GATT." *Third World Quarterly* 14 (November):749–774.

Chayes, Abram, and Antonia Handler Chayes. 1993. "On Compliance." *International Organization* 47 (Spring):175–205.

Chayes, Abram and Antonia H. Chayes. 1995. *The New Sovereignty: Compliance with Treaties in International Regulatory Regimes.* Cambridge, MA: Harvard University Press.

Chayes, Antonia Handler, and Abram Chayes. 1992. "Nonproliferation Regimes in the Aftermath of the Gulf War." In *After the Storm: Lessons from the Gulf War*, edited by J. S. Nye, Jr., and R. K. Smith, 49–80. Lanham, MD: Madison Books.

Chayes, Antonia H. and Abram Chayes. 1994. "The UN Register: Transparency and Cooperative Security." In *Developing the UN Register of Conventional Arms*, edited by Malcolm Chalmers, Owen Greene, Edward J. Laurance, and Herbert Wulf, 197–223. Bradford, West Yorkshire: University of Bradford Press.

"Chemicals: Hoescht Becomes First Chemical Company to Stop Production of Chlorofluorocarbons." 1994. *International Environment Reporter (BNA)* 17 (May 4):390.

Chengappa, Raj. 1993. "Poaching: A Lethal Revival." *India Today* (May 31):111–119.

Chinese Delegation. 1993. Statement to the U.N. Conference for the Negotiating of a Successor Agreement to the International Tropical Timber Agreement. Unpublished document.

CITES. 1991. Doc SC 24.7, annex 1. 26 September. Telefax sent by Izgrev Topkov, Secretary General, CITES Secretariat, to all members of the Standing Committee.

CITES Secretariat. 1991. "Main Activities of the Secretariat to Improve Implementation of CITES in Italy." Report sent with Doc. SC 24.7, annex 1.

CITES Standing Committee. 1992a. *Minutes of the 24th CITES Standing Committee Meeting, First Session.* January 20.

———. 1992b. *Minutes of the 28th Standing Committee Meeting.* June 22–25.

———. 1992c. *Italy: Recommendations of the Standing Committee.* June 30. Notification to the Parties no. 675.

———. 1993. *Italy: Suspension of the Recommendations of the Standing Committee.* February 19. Notification to the Parties no. 722.

———. 1995. *Minutes of the Thirty-fifth Meeting of the Standing Committee.* Doc. SC. 35.13. March 21–24. *Implementation of CITES in Italy.*

Clinton, William. 1993. "Message to Congress on Rhinoceros and Tiger Trade by China and Thailand." *Weekly Compilation of Presidential Documents* 29, doc. 2300 (November 8).

Coenan, Rene. 1993. Interview, International Maritime Organization, June 17.

Cohen, Mark A. 1992. "Environmental Crime and Punishment: Legal/Economic Theory and Empirical Evidence on Enforcement of Federal Environmental Statutes." *Journal of Criminal Law and Criminology* 82, 4:1054–1108.

Colchester, M. 1993. "Slave and Enclave: Towards a Political Ecology of Equatorial Africa." Paper presented to the IWGIA conference "Question of Indigenous Peoples in Africa," Copenhagen, 1–3 June.

Collins, H. M. 1985. *Changing Order: Replication and Induction in Scientific Practice.* London: Sage Publications.

Collins, Ken, and David Earnshaw. 1992. "The Implementation of European Community Environmental Legislation." *Environmental Politics* 1 (Winter):211–249.

Commins, S. K. (ed.). 1988. *Africa's Development Challenges and the World Bank.* Boulder, CO: Lynne Rienner.

Commission of the European Communities. 1985. *Communication from the Commission to the Council Concerning the Negotiations for a Global Framework Convention on the Protection of the Ozone Layer.* DOCUMENTS. COM (85) 8 final. Luxembourg: Office for Official Publications of the European Communities.

———. 1989. *Official Journal of the European Communities* C 264 (32) (October 16).

———. 1993a. *Proposal for a Council Regulation (EEC) on Operations to Promote Tropical Forests.* DOCUMENTS. COM (93) 53 final. Luxembourg: Office for Official Publications of the European Communities.

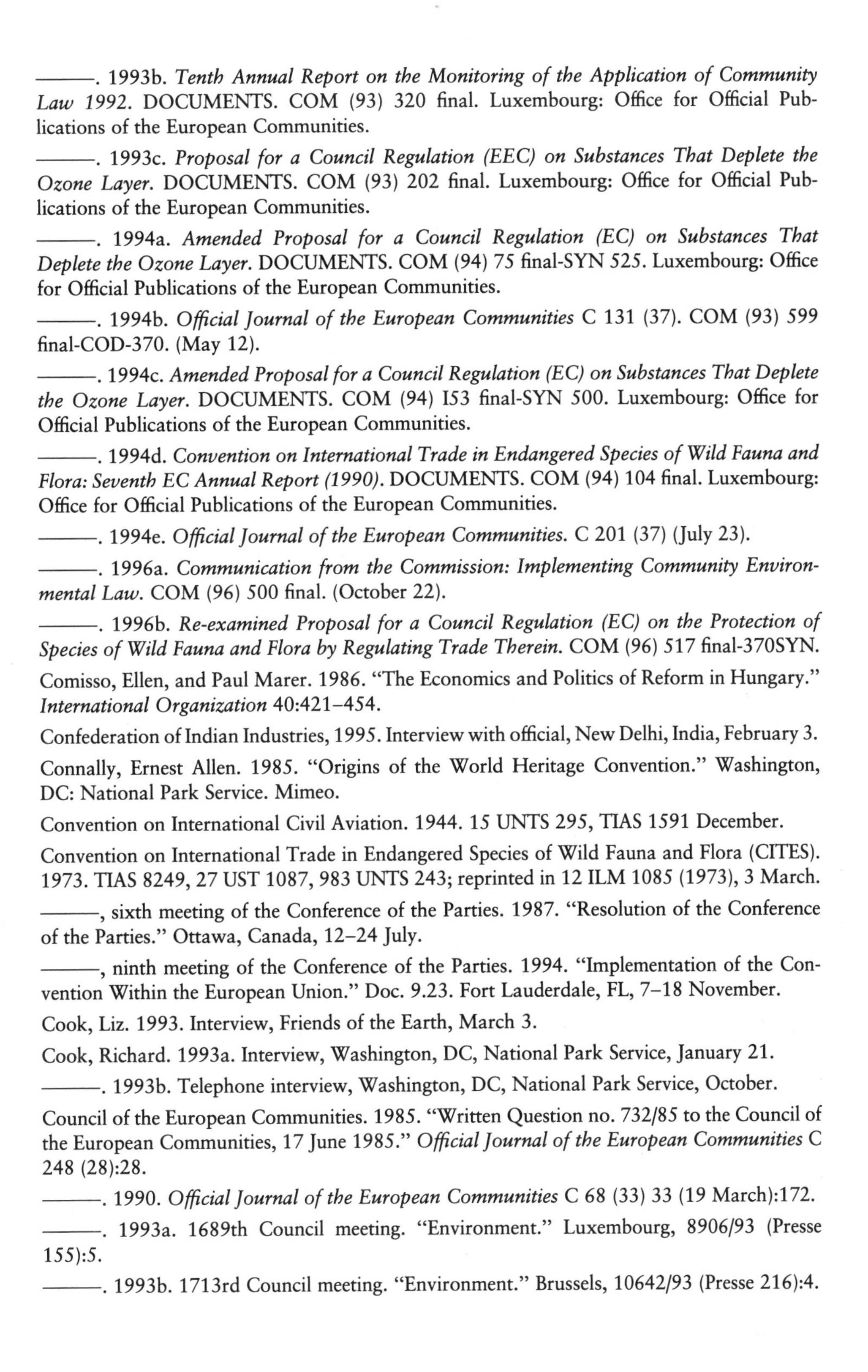

———. 1993b. *Tenth Annual Report on the Monitoring of the Application of Community Law 1992.* DOCUMENTS. COM (93) 320 final. Luxembourg: Office for Official Publications of the European Communities.

———. 1993c. *Proposal for a Council Regulation (EEC) on Substances That Deplete the Ozone Layer.* DOCUMENTS. COM (93) 202 final. Luxembourg: Office for Official Publications of the European Communities.

———. 1994a. *Amended Proposal for a Council Regulation (EC) on Substances That Deplete the Ozone Layer.* DOCUMENTS. COM (94) 75 final-SYN 525. Luxembourg: Office for Official Publications of the European Communities.

———. 1994b. *Official Journal of the European Communities* C 131 (37). COM (93) 599 final-COD-370. (May 12).

———. 1994c. *Amended Proposal for a Council Regulation (EC) on Substances That Deplete the Ozone Layer.* DOCUMENTS. COM (94) I53 final-SYN 500. Luxembourg: Office for Official Publications of the European Communities.

———. 1994d. *Convention on International Trade in Endangered Species of Wild Fauna and Flora: Seventh EC Annual Report (1990).* DOCUMENTS. COM (94) 104 final. Luxembourg: Office for Official Publications of the European Communities.

———. 1994e. *Official Journal of the European Communities.* C 201 (37) (July 23).

———. 1996a. *Communication from the Commission: Implementing Community Environmental Law.* COM (96) 500 final. (October 22).

———. 1996b. *Re-examined Proposal for a Council Regulation (EC) on the Protection of Species of Wild Fauna and Flora by Regulating Trade Therein.* COM (96) 517 final-370SYN.

Comisso, Ellen, and Paul Marer. 1986. "The Economics and Politics of Reform in Hungary." *International Organization* 40:421–454.

Confederation of Indian Industries, 1995. Interview with official, New Delhi, India, February 3.

Connally, Ernest Allen. 1985. "Origins of the World Heritage Convention." Washington, DC: National Park Service. Mimeo.

Convention on International Civil Aviation. 1944. 15 UNTS 295, TIAS 1591 December.

Convention on International Trade in Endangered Species of Wild Fauna and Flora (CITES). 1973. TIAS 8249, 27 UST 1087, 983 UNTS 243; reprinted in 12 ILM 1085 (1973), 3 March.

———, sixth meeting of the Conference of the Parties. 1987. "Resolution of the Conference of the Parties." Ottawa, Canada, 12–24 July.

———, ninth meeting of the Conference of the Parties. 1994. "Implementation of the Convention Within the European Union." Doc. 9.23. Fort Lauderdale, FL, 7–18 November.

Cook, Liz. 1993. Interview, Friends of the Earth, March 3.

Cook, Richard. 1993a. Interview, Washington, DC, National Park Service, January 21.

———. 1993b. Telephone interview, Washington, DC, National Park Service, October.

Council of the European Communities. 1985. "Written Question no. 732/85 to the Council of the European Communities, 17 June 1985." *Official Journal of the European Communities* C 248 (28):28.

———. 1990. *Official Journal of the European Communities* C 68 (33) 33 (19 March):172.

———. 1993a. 1689th Council meeting. "Environment." Luxembourg, 8906/93 (Presse 155):5.

———. 1993b. 1713rd Council meeting. "Environment." Brussels, 10642/93 (Presse 216):4.

———. 1994. "Regulation 3093/94." *Official Journal of the European Communities* L 331 (December 22):1.

Coursey, Don. 1992. "The Demand for Environmental Quality." Paper on file University of Chicago Hanrison School of Public Policy.

Cox, Robert W., and Harold K. Jacobson. 1973. *The Anatomy of Influence: Decision Making in International Organization.* New Haven: Yale University Press.

Cozzens, S. E., and Edward J. Woodhouse. 1994. "Science, Government, and the Politics of Knowledge." In *Handbook of Science and Technology Studies*, edited by Sheila Jasanoff, Gerald Markle, James Petersen, and Trevor Pinch. Newbury Park, CA: Sage Publications.

Crandall, Robert W. 1987. "Learning the Lessons." *Wilson Quarterly* (Autumn):69–80.

Crawford, Christine. 1995. "Conflicts Between the Convention on International Trade in Endangered Species and the GATT in Light of Actions to Halt the Rhinoceros and Tiger Trade." *Georgetown International Environmental Law Review* 7, 2:555–585.

Crockett, Tamara Raye, and Cynthia B. Schultz. 1991. "The Integration of Environmental Policy and the European Community: Recent Problems of Implementation and Enforcement." *Columbia Journal of Transnational Law* 29:169–191.

Curtis, Clif. 1993. Correspondence from Clif Curtis to Michael Glennon, Greenpeace International, May 3.

Cutter Information Corp. (Cutter). 1993a. "EC Proposals on HCFCs, Methyl Bromide Close to Finalization." *Environment Watch: Western Europe* 2 (4 June):11.

———. 1993b. "Report Shows Implementation of EC Environment Rules Remains Unsatisfactory." *Environment Watch: Western Europe* 2 (6 August):12–15.

———. 1993c. "Global Voluntary Label for Temperate and Tropical Timber Coming Soon, as Swiss Proposal to Mandate Label Meets with Protests." *Environment Watch: Western Europe* 2 (15 October):17.

———. 1993d. "EC Agrees on Ozone Layer Measures, Fails to Strike Deal on Packaging." *Environment Watch: Western Europe* 2 (3 December):Late News.

———. 1993e. "Main Details of EC Curbs on HCFCs, Methyl Bromide Become Clear." *Environment Watch: Western Europe* 2 (17 December):9–10.

———. 1994a. "New Tropical Timber Trade Deal Struck, but Sustainable Management Target Weakened." *Environment Watch: Western Europe* 3 (4 February):4–5.

———. 1994b. "Parliament Committee Urges Tighter Restrictions on Ozone-Eaters." *Environment Watch: Western Europe* 3 (4 February):15.

———. 1994c. "EU Angers Home Producers with Hike in Recycled CFC Import Quotas." *Environment Watch: Western Europe* 3 (4 March):1.

———. 1994d. "Dutch Environmentalists Back out of Tropical Timber Agreement." *Environment Watch: Western Europe* 3 (18 March):15.

———. 1994e. "Enforcement of Ozone-Depleter Regulations a Problem, Dutch Report Shows." *Environment Watch: Western Europe* 3 (17 June):7.

———. 1994f. "Imports Curbed as EU Ministers Complete Legislation on Ozone-Depleters." *Environment Watch: Western Europe* 3 (17 June):6.

———. 1994g. "UK Environmentalists Push for Mahogany Ban as ITTO Postpones Action on Labeling Scheme." *Environment Watch: Western Europe* 3 (17 June):12.

———. 1994h. "Dutch Government Will Not Meet Tropical Timber Deadline, Report Says." *Environment Watch: Western Europe* 3 (15 July):11.

———. 1994i. "Focus Report: Refrigeration Lags as Europe Moves Toward CFC Phaseout." *Environment Watch: Western Europe* 3 (2 September):8–9.

———. 1994j. "EU Importing 'Massive' Amounts of CFCs Illegally, Customs Data Show." *Environment Watch: Western Europe* 3 (7 October):9–10.

———. 1994k. "Enforcement, Reporting Major Problems in EU Legislation, Studies Find." *Environment Watch: Western Europe* 3 (4 November):1–3.

———. 1994l. "Dutch Drop Planned Ban on Nonsustainable Timber Imports." *Environment Watch: Western Europe* 3 (16 December):12.

———. 1995a. "Lawyers Urge EU to Reinstate Chief Environment Enforcer." *Environment Watch: Western Europe* 4 (3 February):1.

———. 1995b. "'Stop Wasting Our Time,' ECJ Adviser Tells Commission." *Environment Watch: Western Europe* 4 (7 April):2–4.

———. 1995c. "Illegal CFC Trade Detected in Spain, NGO Says." *Environment Watch: Western Europe* 4 (5 May):10.

———. 1995d. "Tropical Forest Protocol Added to Lome Convention." *Environment Watch: Western Europe* 4 (21 July):11.

———. 1995e. "EU to Push for Marked Toughening of Montreal Protocol." *Environment Watch: Western Europe* 4 (16 October):6–7.

———. 1995f. "Montreal Protocol Meeting Adopts Tighter Controls on Methyl Bromide, HCFCs." *Environment Watch: Western Europe* 4 (15 December):4.

———. 1995g. "Two Tropical Timber Labels Under Development in the Netherlands." *Environment Watch: Western Europe* 4 (15 December):13.

———. 1996a. "In Brief." *Environment Watch: Western Europe* 5 (16 February):11.

———. 1996b. "Spanish Government Drops Proposed CFC Decree." *Environment Watch: Western Europe* 5 (15 November):8.

———. 1996c. "UK Minister Calls on Europe to Fight CFC Smuggling." *Environment Watch: Western Europe* 5 (1 November):5.

Dasgupta, Sumita. 1994. "Ozone Fund." *Down to Earth* (May 31):17–19.

Dauvergne, Peter. 1997. *Shadows in the Forest: Japan and the Politics of Timber in Southeast Asia*. Cambridge, MA.: MIT Press

DeBardeleben, Joan. 1990. "Economic Reform and Environmental Protection in the USSR." *Soviet Geography* 31:237–256.

Demaret, Paul. 1993. "Environmental Policy and Commercial Policy: The Emergence of Trade-Related Environmental Measures (TREMS) in the External Relations of the European Community." In *The European Community's Commercial Policy After 1992: The Legal Dimension*, edited by Marc Maresceau, 305–386. Dordrecht, Netherlands: Martinus Nijhoff.

De Souza, Herbert. 1992. "O Papel das ONGs e a Sociedade Civil em Relação ao Meio-ambiente." *Planejarnento e Politicas, IPEA* no. 7 (June).

Development Alternatives. 1990. *The Economic Implications for Developing Countries of the Montreal Protocol*. New Delhi. Study for UNEP.

DeYoung, H. Garrett. 1990. "Du Pont Electronics Boards the Environmental Bandwagon." *Electronic Business* (August 6).

Diehl, Rita. 1993a. Correspondence from Rita Diehl to Michael Glennon, Ocean Advocates, June 3.

———. 1993b. Telephone interview, Ocean Advocates, April 12.

DiMento, Joseph F. 1993. "Criminal Enforcement of Environmental Law." *Annals AAPSS* 525 (January):134–146.

Doniger, David D. 1988. "Politics of the Ozone Layer." *Issues in Science and Technology* 4, 3:86–92.

Douglas, David. 1983. "Saving the World's Heritage—Minus the U.S.?" *Christian Science Monitor*, January 31, 23.

Douglas, M. 1966. *Purity and Danger: An Analysis of the Concepts of Pollution and Taboo*. New York: Praeger.

Douglas, Hamilton. 1979. *Ivory Trade Study*. Gland, Switzerland: IUCN.

Downs, Anthony. 1972. "Up and Down with Ecology—the 'Issue-Attention' Cycle." *The Public Interest* 28 (Summer):38–50.

Downs, George W., David M. Rocke, and Peter N. Barsoom. 1996. "Is the Good News About Compliance Good News About Cooperation?" *International Organization* 50, 3:379–406.

Drucker, Milton. 1993. Interview, U.S. Department of State, Washington, DC, November.

———. 1994. Interview, U.S. Department of State, Washington, DC, May 10.

Dublin, H. T., T. Milliken, and R. F. W. Barnes. 1995. *Four Years After the CITES Ban: Illegal Killings of Elephants, Ivory Trade and Stockpiles*. Report of the IUCN/SSC African Specialist Group, Section on Cameroon, 4–50.

Dumanoski, Dianne. 1990. "U.S. Blasted for Altering Stance on Ozone Fund." *Boston Globe*, May 10, A28.

———. 1992. "Nations Act to Speed Phaseout of Ozone-Depleting Chemicals." *Boston Globe*, November 26, A26.

Dumont, R. 1969. *False Start in Africa*, 2nd ed. New York: Praeger.

Durante, Damian. 1993. Interview, Greenpeace, November 1.

Dutch CFC Committee. 1993. *CFC Action Programme: Cooperation Between Government and Industry. Annual Report 1992*, Tilburg.

"EC Directives Need to be Backed Up With Other Incentives, Auditors Say." 1992. *International Environment Reporter (BNA)* 15, 20 (October 7):642–643.

Economy, Elizabeth. 1994. "*Domestic Reforms and International Regime Formation: China, the USSR, and Global Climate Change*." Ph.D. diss., University of Michigan.

Ehrlich, Paul R. 1975. *Island of Dreams: Environmental Crisis in Japan*. Cambridge, MA: Autumn Press.

"Endangered Species: Targets of Asia's Affluent." 1993. *Far Eastern Economic Review* (August 19):23–27.

Enders, Alice, and Amelia Porges. 1992. "Successful Conventions and Conventional Success: Saving the Ozone Layer." In *The Greening of World Trade Issues*, edited by Kym Anderson and Richard Blackhurst, 130–144. New York: Harvester Wheatsheaf.

Environmental Data Services. 1992. "Ozone Layer Left at Risk by New Global Agreement." *ENDS Report* 214 (November):13–15.

———. 1993. "Brussels Prepares Controls on HCFCs, Methyl Bromide." *ENDS Report* 217 (February):38–39.

———. 1994. "Commission Taken to Task over Reporting Record." *ENDS Report* 236 (September):36–37.

Environmental Policy in the European Community. 1990. 4th ed. Luxembourg, Office of Official Publications of the European Communities.

ENVIS Centre 07, World Wide Fund for Nature-India. 1993. *Annual Report*. New Delhi: WWF-India. Submitted to MOEF.

Epstein, Joshua, and Raj Gupta. 1990. *Controlling the Greenhouse Effect: Five Global Regimes Compared*. Washington, DC: Brookings Institution.

Esty, Daniel C. 1994. *Greening the GATT: Trade, Environment, and the Future*. Washington, DC: Institute for International Economics.

Europe Information Service. 1992. "EC Law: Community-wide Network for Better Enforcement of EC Law." *Europe Environment* no. 398 (November 17):11(I).

———. 1993a. "Ozone Layer: EC Agrees to Phase Out CFCs by January 1995." *Europe Environment* no. 401 (January 7):2(I).

———. 1993b. "Fauna and Flora: Greece Ratifies CITES Convention." *Europe Environment* no. 402 (January 19):2(I).

———. 1993c. "Tropical Forests: Commission Proposes Precise Strategy." *Europe Environment* no. 406 (March 16):1(I).

———. 1993d. "EC Council: Ozone Layer and Various Decisions." *Europe Environment* no. 407 (March 30):17–19(I).

———. 1993e. "European Parliament Urges Tropical Hardwood Import Ban." *Europe Environment* no. 411 (June 8):7(I).

———. 1993f. "Ozone Layer: Heated Debate over EC Regulation Within Council of Ministers." *Europe Environment* no. 419 (October 26):7–8(I).

———. 1993g. "Tropical Timber: EC Set to Put Forward Certification System." *Europe Environment* no. 419 (October 26):24(I).

———. 1993h. "Environment Ministers Vote to Destroy Ozone, Says SAFE." *Europe Environment* no. 422 (December 16):3(II).

———. 1994a. "European Parliament: Opinion on the Draft Regulation Concerning Ozone-Depleting Substances." *Europe Environment: Document* (supp. to *Europe Environment*) no. 426 (February 15):1–14.

———. 1994b. "Ozone Layer: Hoechst and Elf Atochem Attack Commission on Imports of CFCs." *Europe Environment* no. 427 (March 1):21–22(I).

———. 1995. "Environment Council: Decisions on Air Quality, CITES, IPPC and Seveso." *European Report* no. 2052 (June 23):24–26(IV).

European Parliament. 1988. "Report Drawn Up on Behalf of the Committee on the Environment, Public Health and Consumer Protection on the Implementation of the CITES Regulation (Council Regulation (EEC) no. 3626/82 of 3 December 1982) in the European Community (Concerning the Implementation in the Community of the Convention on International Trade in Endangered Species of Wild Fauna and Flora (the Washington Convention))." Luxembourg: Office for Official Publications of the European Communities.

———. 1990a. "Report by the Committee on Development and Cooperation on the Conservation of Tropical Forests (COM(89) 410 Final." *European Parliament Session Documents*, A3-0231/90 (26 September):2–26.

———. 1990b. "Report Drawn Up on Behalf of the Committee on the Environment, Public Health and Consumer Protection on Measures to Protect the Ecology of the Tropical Forests." *European Parliament Session Documents*, A3-0181/90 (5 July):3–20.

———. 1993a. "Report of the Committee on Development and Cooperation on the Commission Proposal for a Council Regulation (EEC) on Operations to Promote Tropical Forests (COM (93) 0053 Final)." *European Parliament Session Documents*, A3-0304/93 (15 October):2–47.

———. 1993b. "Report of the Committee on the Environment, Public Health and Consumer Protection on the Commission Proposal for a Council Regulation Laying Down Provisions with Regard to Possession of and Trade in Specimens of Species of Wild Fauna and Flora (COM (91) 0448 Final-C3-0030/92-SYN 0370)." *European Parliament Session Documents*, A3-O193/93 (15 June):52.

"European Union: Ministers Agree on Cuts in Substances Harming Ozone Layer, Fail to Move on CO_2 Tax." 1994. *International Environment Reporter (BNA)* 17 (June 15):499–500.

Everhart, William C. 1983. *The National Park Service*. Boulder, CO: Westview Press.

Ezrahi, Y. 1990. *The Descent of Icarus*. Cambridge, MA: Harvard University Press.

Fang Maoming, Zhou Jiayi, and Cui Shaozheng. 1993. "China's Ocean Waste Dumping Management and the London Convention 1972." Unpublished manuscript.

Fang Xiaoming. 1993. "China's Implementation of International Agreements." Unpublished manuscript.

Favre, David S. 1989. *International Trade in Endangered Species: A Guide to CITES*. Dordrecht, Netherlands: Martinus Nijhoff.

FBIS China Daily Report. 1993a. September 17.

———. 1993b. September 20.

———. 1994. April 7.

FBIS-SOV-89-161. 1989a. August 22.

FBIS-SOV-89-197. 1989b. October 13.

FBIS-SOV-89-239. 1989c. December 14.

FBIS-SOV-92-050. 1992a. March 12.

FBIS-SOV-92-090. 1992b. May 8.

FBIS-SOV-93-089. 1993a. May 4.

FBIS-SOV-93-122. 1993b. June 18.

FBIS-SOV-93-210. 1993c. October 20.

FBIS-SOV-94-075. 1994. April 19.

Fearnside, Philip. 1993. Interview published by *Jornal do Brasil*, May 15.

———. 1994. "Queimadas e Desmatamento I." *Jornal do Brasil*, October 10.

Finkle, Jason, and Barbara B. Crane. 1976. "The World Health Organization and the Population Issue: Organizational Values in the United Nations." *Population and Development Review* 2 (September/December):3–4, 367–393.

Fischer, Julie. 1993. *The Road from Rio: Sustainable Development and the Nongovernmental Movement in the Third World*. Westport, CT: Praeger.

Fisher, Roger. 1981. *Improving Compliance with International Law*. Charlottesville: University Press of Virginia.

Fitzgerald, Sarah. 1989. *International Wildlife Trade: Whose Business Is It?* Washington, DC: World Wildlife Fund.

Fletcher, Susan R. 1994. Memorandum from Susan R. Fletcher to Rep. Vic Fazio, January 31.

Foahom, B., and W. B. J. Jankers. 1992. "*A Programme for Tropenbos Research in Cameroon: Final Report, Tropenbos-Cameroon Programme, Phase 1.*" Wageningen, Netherlands: Tropenbos. Mimeo.

Ford, Daphne. 1994. Telephone interview, Chaco Culture National Historic Park, April 6.

Forrest, Richard. 1991. "Japanese Aid and the Environment." *The Ecologist* 21, 1:24–32.

Forster, M. J., and R. U. Osterwoldt. 1992. "Nature Conservation and Terrestrial Living Resources." In *The Effectiveness of International Environmental Agreements: A Survey of Existing Legal Instruments*, edited by Peter H. Sand, 59–122. Cambridge: Grotius.

Franck, Thomas. 1995. *Fairness in International Law and Institutions.* Oxford: Oxford University Press.

Friends of the Earth. 1992. *The International Tropical Timber Agreement: Conserving the Forests or Chainsaw Charter?* London: Friends of the Earth.

Gadgil, Madhav, and Ramachandra Guha. 1995. *Ecology and Equity: The Use and Abuse of Nature in Contemporary India.* London: Routledge.

Gadgil, Madhav, N. V. Joshi, and Suresh Patil. 1993. "Power to the People: Living Close to Nature." *The Hindu: Survey of the Environment* 58–62.

Gadsby, E. L., and P. D. Jenkins. 1992. *Report on Wildlife and Hunting in the Proposed Etinde Forest Reserve, Limbe Botanic Garden and Rainforest Conservation Project.*

Galambos, Judit. 1994. "A Decade-long Environmental Conflict on the Danube: The Gabcikovo-Nagymaros Dams." In *Environment and Democratic Transition: Policy and Politics in Central and Eastern Europe*, edited by Anna Vari and Pal Tamas, 176–226. Dordrecht, Netherlands: Kluwer.

Garner, Harold. 1993. Interview, December 8.

Gartlan, S. 1989. *La Conservation des Écosystèmes Forestières du Cameroun.* Gland, Switzerland, IUCN.

———. 1983. *Local Knowledge.* New York: Basic Books.

Geertz, C. 1973. *The Interpretation of Cultures.* New York: Basic Books.

General Affairs Agency, 1993. *The Problems of Protection Measures for Endangered Species.* Tokyo: Japan's General Affairs Agency.

George, Alexander, and Richard Smoke. 1974. *Deterrence in American Foreign Policy: Theory and Practice.* New York: Columbia University Press.

Gibson, J. E., and J. C. Laurent. 1992. *West and Central Africa Regional Environmental Law Study.* Washington, DC: USAID.

Giddens, A. 1990. *The Consequences of Modernity.* Stanford, CA: Stanford University Press.

Gillespie, B., Dave Eva, and Ron Johnston. 1979. "Carcinogenic Risk Assessment in the United States and Great Britain: The Case of Aldrin/Dieldrin." *Social Studies of Science* 9:265–301.

Glennon, Michael J. 1990a. "Has International Law Failed the Elephant?" *American Journal of International Law* 84 (January):1–43.

———. 1990b. Memorandum to Edith Brown Weiss and Harold K. Jacobson, March 1.

Gosudarstvennyi Doklad. 1993. *O Sostoyanii Okruizhauishei Prirodnoy sredy Rossiiskoi Federatsii v 1992 godu.* Moscow: Evrasia.

Gottweis, H. 1995. "German Politics of Genetic Engineering and its Deconstruction." *Social Studies of Science* 25:195–235.

Haas, P. M. 1989. "Do Regimes Matter? Epistemic Communities and Mediterranean Pollution Control." *International Organization* 43:377–403.

Government of Cameroon. 1991. "Elephant Conservation Plan." Yaoundé: Ministry of Tourism. Mimeo.

Government of India (GOI). 1991. *The State of the Forest Report.* New Delhi: MOEF.

———. 1992. *The Wildlife (Protection) Act, 1972* (as amended up to 1991). Dehra Dun: Natraj.

———. 1993. *Report of the Task Force on National Strategy of Phasing Out Ozone Depleting Substances.* New Delhi: MOEF.

———. Department of Science and Technology. 1980. *Report of the Committee for Recommending Legislative Measures and Administrative Machinery for Ensuring Environmental Protection.* New Delhi: GOI.

———. Ministry of Commerce. 1993. *Export-Import Policy 1 April 1993–31 March 1997.* New Delhi:

———. Ministry of Environment and Forests (MOEF). 1992a. *Environmental Action Programmes—India.* Interim document. New Delhi: MOEF.

———. 1992b. *National Conservation Strategy and Policy Statement on Environment and Development.* New Delhi: MOEF.

———. 1993a. *Country Programme: Phaseout of Ozone Depleting Substances Under the Montreal Protocol.* New Delhi: MOEF.

———. 1993b. *India: National Programme: Phaseout of Ozone Depleting Substances Under the Montreal Protocol.* New Delhi: MOEF.

———. 1996. MOEF (Project Tiger) *Report of the High Powered Committee Constituted on the Recommendation of the Department Related Parliamentary Committee on Environment and Forests to Undertake a Review of the Project Tiger; Carry out Evaluation and Suggest Ways and Means to Make the Project Tiger More Meaningful and Result Oriented.* New Delhi: MOEF.

Government of Japan. 1994. *Forestry White Paper* (Tokyo, Japan).

Graham, John D., (ed.). 1991. *Harnessing Science For Environmental Regulation.* New York: Praeger.

Granda, Chris. 1990. "The Montreal Protocol on Substances That Deplete the Ozone Layer." In *Nine Case Studies in International Environmental Negotiation,* edited by Lawrence E. Susskind, Esther Kiskind, and J. William Breslin, 27–48. The MIT-Harvard Public Disputes Program.

Granick, David. 1975. *Enterprise Guidance in Eastern Europe.* Princeton: Princeton University Press.

Greenpeace. 1983. *Unregulated Whaling.* London: Greenpeace.

———. 1990. *Outlaw Whalers: 1990 Report.* London: Greenpeace.

Gresser, Julian, et al. 1981. *Environmental Law in Japan.* Cambridge, MA: MIT Press.

Gross, Neil. 1989. "Charging Japan With Crimes Against the Earth." *Business Week* (October 9):108–112.

Grossman, Gene M., and Alan B. Krueger. 1992. "The Impacts of a North American Free Trade Agreement." On file with Michael Glennon, University of California, Davis.

Gruisen, Joanna Van, and Toby Sinclair. *Fur Trade in Kathmandu: Implications for India.* New Delhi: TRAFFIC-India.

Guarascio, John A. 1985. "The Regulation of Ocean Dumping After *City of New York* v. *Environmental Protection Agency*." *Environmental Affairs* 12:700–741.

Guha, Ramachandra. 1985. "Forestry and Social Protest in British Kumaun, c. 1893–1921." *Subaltern Studies* 4:54–100.

———. 1989. *The Unquiet Woods: Ecological Change and Peasant Resistance in the Himalaya*. Delhi: Oxford University Press.

———. 1994. "Switching on the Green Light." *Telegraph* (Calcutta), 25 October.

Guimaraes, Roberto. 1992. "Políticas do Meio-ambiente para o Desenvolvimeto Sutentável: Desafios Institucionais e Setorias." *Planejamento e Políticas Públicas* no. 7 (June).

Guttieres, Mario, and Aranzazu Aguilar-Lizarralde. 1994. "Italy." *European Environmental Law Review* 3, 2:38–41.

Haas, Peter. 1990. *Saving the Mediterranean: The Politics of International Environmental Cooperation*. New York: Columbia University Press.

———. 1992. "Introduction: Epistemic Communities and International Policy Coordination." *International Organization* 46 (Winter):1–35.

Haas, P. M., (ed.). 1992a. "Power, Knowledge, and International Policy Coordination." International Organization 46 (Winter).

Haas, Peter M., Robert O. Keohane, and Marc A. Levy (eds.). 1993. *Institutions for the Earth: Sources of Effective International Environment Protection*. Cambridge, MA: MIT Press.

Hahn, Robert W., and Albert M. McGartland. 1989. "The Political Economy of Instrument Choice: An Examination of the U.S. Role in Implementing the Montreal Protocol." *Northwestern University Law Review* 83:592–611.

Haigh, Nigel. 1992a. *Manual of Environmental Policy: The EC and Britain*. Essex: Longman.

———. 1992b. "The European Community and International Environmental Policy." In *The International Politics of the Environment*, edited by Andrew Hurrell and Benedict Kingsbury, 228–249. Oxford: Oxford University Press.

Haley, John O. 1978. "The Myth of the Reluctant Litigant and the Role of the Judiciary in Japan." *Journal of Japanese Studies*.

———. 1991. *Authority Without Power*. New York: Oxford University Press.

Halfon, S., and S. Jasonoff. 1993. "*Science and Compliance with International Regimes*." Paper for the International Implementation Project Young Scholars Symposium, Geneva, June 16–18.

Hammond, Allen L. (ed.). 1990. *World Resources 1990–1991: Guide to the Global Environment. A Report by the World Resources Institute in Collaboration with the United Nations Environment Programme and the United Nations Development Programme*. New York: Oxford University Press, 244–245, 254–255, 348–349.

Hammond, A. L., Eric Rodenburg, and William R. Moomaw. 1991. "Calculating National Accountability for Climate Change." *Environment* 33:11–15, 33–35.

Hanel, Peter. 1992. "Trade Liberalization in Czechoslovakia, Hungary, and Poland Through 1992: A Survey." *Comparative Economic Studies* 34:34–54.

Haraway, D. 1989. *Primate Visions: Gender, Race, and Nature in the World of Modern Science*. New York: Routledge.

Hardi, Peter. 1992. *Impediments on Environmental Policy-making and Implementation in Central and Eastern Europe*. Berkeley: UC Berkeley, Institute for International Studies.

———. 1994. *Environmental Protection in East-Central Europe. A Market-Oriented Approach*. Gütersloh, Germany: Bertelsmann Foundation.

Hart, H. L. A. *The Concept of Law*. 2d ed. New York: Oxford University Press.

Hart, K. 1982. *The Political Economy of West African Agriculture*. Cambridge: Cambridge University Press.

Hawkins, K. 1984. *Environment and Enforcement: Regulation and the Social Definition of Pollution*. Oxford: Oxford University Press.

Hayes-Renshaw, Fiona, and Helen Wallace. 1997. *The Council of Ministers*. New York: St. Martin's Press.

Hearst, David, and Paul Brown. 1994. "Soviet Union Illegally Killed Great Whales." *The Guardian*, 12 February, 3.

Henderson, Dan Fenno. 1965. *Conciliation and Japanese Law: Tokugawa and Modern*. Seattle: University of Washington Press.

Herman, Robin. 1988. "An Ecological Epiphany." *Washington Post National Weekly Review*, December 5–11, 19.

Herring, Ronald. 1991. *Politics of Nature: Common Interests, Dilemmas and the State*. Harvard Center for Population and Development Studies, Working Paper Series no. 106.

Hicks, Michael. 1994. Interview, U.S. Department of Agriculture, February 3.

Hijkoop, J., M. Abadan, and H. H. Bello. 1992. "*Les Monts Mandara: Les Hommes et l'Environnement*." Mokolo: ONADEF/CARE. Mimeo.

Hildebrand, Philipp M. 1992. "The European Community's Environmental Policy, 1957 to 1992: From Incidental Measures to an International Regime?" *Environmental Politics* 1, 4:13–44.

Hiremath, S. R., Sadanand, Kanwalli, and Sharad Kulkarni (eds.). 1994. *All About Draft Forest Bill and Forest Lands*. Dharawad: Samaj Parivartana Samudaya et al.

Hoberg, G. 1990. "Risk, Science, and Politics: Alachlor Regulation in Canada and the United States." *Canadian Journal of Political Science* 23:257–277.

Hoffman, Ellen. 1993. "Saving Our World's Heritage: World Heritage Convention's List of National and Cultural Monuments." *Omni* 16, 3:52–61.

Hood, Christopher C. 1983. *The Tools of Government*. Chatham, NJ: Chatham House.

Horowitz, Paul, and Steven Seidel. 1993. Interview, Washington, DC, EPA Global Change Division, June 13.

House of Lords, Select Committee on the European Communities. 1989. *Habitat and Species Protection*, 15th Report. Session 1988–1989. London: Her Majesty, Stationary Office.

Houston, Paul, and William J. Eaton. 1994. "Hikes in the Parks." *Los Angeles Times*, February 14, A5.

Huang Yulin. 1993. "Planning for Marine Dumping and Pollution Control." Unpublished manuscript.

Hull, Robert. 1994. "The Environmental Policy of the European Community." In *Environmental Cooperation in Europe: The Political Dimension*, edited by Otmar Holl, 145–158. Boulder, CO: Westview Press.

Humphrey, Hubert H. III, and LeRoy C. Paddock. 1990. "The Federal and State Roles in Environmental Enforcement: A Proposal for a More Effective and More Efficient Relationship." *Harvard Environmental Law Review* 14, 7:7–44.

Hungarian Interview Series. 1994. Prepared by Peter Hardi. Budapest. Mimeo.

Hunter, Susan, and Richard W. Waterman. 1992. "Determining an Agency's Regulatory Style: How Does the EPA Water Office Enforce the Law." *Western Political Quarterly* 45 (June):403–417.

Hurrell, A., and B. Kingsbury, (eds.). 1992. *The International Politics of the Environment.* Oxford: Oxford University Press.

Hutchinson, Art. 1994. Telephone interview, Mesa Verde National Park, April 7.

Huth, Hans. 1957. *Nature and the American: Three Centuries of Changing Attitudes.* Berkeley: University of California Press.

Hyden, G. 1983. *No Shortcuts to Progress.* Berkeley: University of California Press.

Infield, M. 1988. *Hunting, Trapping and Fishing in Villages Within and on the Periphery of the Korup National Park.* Report to WWF-UK. Godalming, UK.

"Interior Secretary Lujan Announces Ban on Wildlife Trade with Thailand." 1991. *P. R. Newswire,* LEXIS, News Library, Curnws file.

International Hardwood Products Association. 1989a. *Tropical Forestry Workshop: Consensus Statement on Commercial Forestry, Sustained Yield Management, and Tropical Forests.* Alexandria: IHPA.

———. 1989b. *The World's Tropical Forests: A Renewable Resource.* Alexandria: IHPA.

International Labor Conference. 1980. *Provisional Record, Sixty-sixth Session.* Geneva: International Labor Organization (ILO), 37/4–10, 19–22.

———. 1982. *Record of Proceedings, 68th Session.* Geneva: ILO.

———. 1992. *Report of the Committee of Experts on the Application of Conventions and Recommendations,* app. 2. Geneva: ILO.

International Maritime Organization (IMO). 1990. *Status of London Dumping Convention: The First Decade and Beyond, Note by the Secretariat.* London: IMO.

———. 1997. *Status of the London Convention of 1972 and the 1996 Protocol.* London: IMO.

International Monetary Fund. 1945. Articles of Agreement of the International Monetary Fund, 27 December 1945, as amended, 2 UNTS 20 (1947) 39, art. 8 sec. 5.

———. 1952. *Selected Decisions of the Executive Directors and Selected Documents,* Decision no. 102-(52/11), 13 February, 16. Washington, DC: IMF.

International Slavery Convention. (1926). 46 Stat. 2153 (1926), art. 7.

International Tropical Timber Commission. 1991a. Draft Annual Report. (XIII)/17.

———. 1991b. Draft Annual Report. (XIV)/2.

———. 1993. *Elements for the 1992 Annual Review and Assessment of the World Tropical Timber Situation.* (XIV)/3.

International Tropical Timber Organization. 1993a. *ITTO Guidelines for the Establishment and Sustainable Management of Planted Tropical Forests.* ITTO Policy Development Series no. 4.

———. 1993b. *Project Management Database Catalogue.*

———. 1996. *Annual Review and Assessment of the World Tropical Timber Situation 1996.* ITTO, GI-7/96.

International Union for the Conservation of Nature (IUCN)/UNESCO, World Heritage Committee. 1992. *The World Heritage Twenty Years Later.* Gland, Switzerland: IUCN.

Irwin, A., and B. Wynne, (eds.). 1996. *Misunderstanding Science? The Public Reconstruction of Science and Technology.* Cambridge: Cambridge University Press.

Jachtenfuchs, Markus. 1990. "The European Community and the Protection of the Ozone Layer." *Journal of Common Market Studies* 28, 3:261–278.

Jacobson, Harold, and Michel Oksenberg. 1990. *China's Participation in the IMF, the World Bank, and the GATT: Toward a Global Economic Order.* Ann Arbor: University of Michigan Press.

Jahiel, Abigail. 1994. "*Policy Implementation Under Socialist Reform: The Case of Water Pollution Management in the People's Republic of China.*" Ph.D. diss., University of Michigan.

Jasanoff, Sheila. 1986. *Risk Management and Political Culture.* New York: Russell Sage Foundation.

———. 1987. "Cultural Aspects of Risk Assessment in Britain and the United States." In *The Social and Cultural Construction of Risk*, edited by Branden B. Johnson and Vincent T. Covello. New York: Reidel.

———. 1990. *The Fifth Branch: Science Advisers as Policymakers.* Cambridge, MA: Harvard University Press.

———. 1991. "Cross National Differences in Policy Implementation." *Evaluation Review: A Journal of Applied Social Research* 15 (February):103–119.

———. 1991. "Acceptable Evidence in a Pluralistic Society." In *Acceptable Evidence* edited by Deborah G. Mayo and Rachelle D. Hollander. New York: Oxford University Press.

———. 1992. "Science, Politics, and the Renegotiation of Expertise at EPA." *Osiris* 7:195–217.

———, (ed.). 1994. *Learning from Disaster: Risk Management after Bhopal.* Philadelphia: University of Pennsylvania Press.

———. 1995. "Product, Process, or Program: Three Cultures and the Regulation of Biotechnology." In *Resistance to New Technology*, edited by Martin Bauer. Cambridge: Cambridge University Press.

———. 1996. "Science and Norms in the Global Environment." In *Social Justice and Environmental Change*, edited by Fen O. Hampson and Judith Reppy, 173–197. Ithaca, NY: Cornell University Press.

Jeanrenaud, S. 1993. "*Community-Based Conservation: Some Background Themes.*" Mimeo.

Jian Liu. 1993. "The London Convention 1972 from Chinese Perspective." Unpublished manuscript.

Japan External Trade Relations Organization (JETRO). 1991. *US and Japan in Figures.* Tokyo: JETRO.

Johns, Chris. 1994. "The Everglades: Dying for Help." *National Geographic* (April):2.

Johnson, Brian. 1991. *Expansion or Eclipse for ITTO? A Look at the First Five Years of the International Tropical Timber Organization and Its Potential—A WWF Discussion Paper.* Gland, Switzerland: World Wildlife Fund International.

Johnson, Chalmers. 1982. *MITI and the Japanese Miracle.* Tokyo: Charles E. Tuttle.

Johnson, Stanley P., and Guy Corcelle. 1989. *The Environmental Policy of the European Communities.* London: Graham & Trotman.

Jones, Clarence. 1973. "Air Pollution and Contemporary Environmental Politics." *Growth and Change* 4, 3 (July):22–28.

Kagan, Robert A. 1994. "Regulatory Enforcement." In *Handbook on Regulation and Administrative Law*, edited by David Rosenbloom and Richard D. Schwartz, 383–422. New York: Marcel Dekker.

Kamto, M. 1992. "*Élaboration d'une Stratégie Nationale de l'Environnement au Cameroun.*" Yaoundé: FAO. Mimeo.

Kawashima, T. 1963. "Dispute Resolution in Contemporary Japan." In *Law of Japan*, edited by A. von Mehren. Cambridge, MA: Harvard University Press.

Keller, E. J. 1991. "The State in Contemporary Africa: A Critical Assessment of Theory and Practice." In *Comparative Political Dynamics: Global Research Perspectives*, edited by D. A. Rustow, and K. P. Erickson, 134–159. New York: HarperCollins.

Kennedy, Paul. 1993. *Preparing for the Twenty-first Century*. New York: Random House.

Keohane, Robert O. 1984. *After Hegemony: Cooperation and Discord in the World Political Economy*. Princeton: Princeton University Press.

———. 1986. "Reciprocity in International Relations." *International Organization* 40 (Winter 1986):1–27.

———. 1988. "Reciprocity, Reputation, and Compliance with International Commitments." Paper delivered at the Annual Meeting of the American Political Science Association, Washington, DC, September 1–4.

Kim, Samuel. 1989. *China and the World*. Boulder, CO: Westview Press.

Kirkland, Richard. 1988. "Environmental Anxiety Goes Global." *Fortune* (November 21):118.

Klik, P. 1995. "Group Actions in Civil Lawsuits: The New Law in the Netherlands." *European Environmental Law Review* 4, 1:14–16.

Klingberg, Ethan. 1992. "Hungary, Safeguarding the Transition." *East European Constitutional Review* 2, 2:44–48.

Koh, Harold Hongju. 1997. "Why Do Nations Obey International Law?" *Yale Law Journal* 106:2599–2659.

Kohm, Kathryn A. (ed.). 1991. *Balancing on the Brink of Extinction: The Endangered Species Act and Lessons for the Future*. Washington, DC: Island Press.

Koplow, David A. 1992. "When Is an Amendment Not an Amendment?: Modification of Arms Control Agreements Without the Senate." *University of Chicago Law Review* 59 (Summer):981–1072.

Koskenniemi, Martti. 1992. "Breach of Treaty or Non-Compliance? Reflections on the Enforcement of the Montreal Protocol." *Yearbook of International Environmental Law* 3:123–162.

Kosloff, Laura, and Mark Trexler. 1987. "The Convention on International Trade in Endangered Species: Enforcement Theory and Practices in the United States." *Environmental Law Report* 17:10222–10227.

———. 1991. "International Implementation: The Longest Arm of the Law?" In *Balancing on the Brink of Extinction*, edited by Kathryn A. Kohm. Washington, DC: Island Press.

Kothari, Ashish. 1996. "Structural Adjustment vs. India's Environment." Paper presented at the Annual Meetings of the Association for Asian Studies, Honolulu, April.

Kothari, Ashish, et al. 1989. *Management of National Parks and Sanctuaries in India: A Status Report*. New United Press. New Delhi, India: Environmental Studies Division, Indian Institute of Public Administration

Koves, Andras. 1992. *Central and East European Economies in Transition*. Boulder, CO: Westview Press.

Kramer, Ludwig. 1995. *E. C. Treaty and Environmental Law*, 2nd ed. London: Sweet & Maxwell.

Kristof, Nicholas, and Sheryl WuDunn. 1994. *China Wakes*. New York: Random House.

Kwa, C. 1987. "Representations of Nature Mediating Between Ecology and Science Policy: The Case of the International Biological Program." *Social Studies of Science* 17:413–442.

Laky, Teréz. 1979. "Enterprises in Bargaining Position." *Acta Oeconomica* 22:227–246.

Lang, John Temple. 1986. "The Ozone Layer Convention: A New Solution to the Question of Community Participation in 'Mixed' International Agreements." *Common Market Law Review* 23 (Spring):157–176.

Lang, Winfried. 1988. "Diplomatie Zwischen Ökonomie und Ökologie. Das Beispiel des Ozonvertrags von Montreal." *Europa-Archiv* 43, 4:105–110.

———. 1993. "Diplomacy and International Environmental Law-Making: Some Observations." *Yearbook of International Environmental Law* 3:108–122.

———. 1995. "Compliance-Control in Respect of the Montreal Protocol." In *Proceedings of the 89th Annual Meeting of the American Society of International Law*, 206–210. Washington, DC: The American Society of International Law.

———. 1996. "Trade Restrictions as a Means of Enforcing Compliance with International Environmental Law." In *Enforcing Environmental Standards: Economic Mechanisms as a Viable Means?* Edited by Rüdiger Wolfrum. Heidelberg: Springer Verlag.

Lardy, Nicholas. 1994. *China in the World Economy*. Washington, DC: Institute for International Economics.

Latour, B. 1987. *Science in Action: How to Follow Scientists and Engineers Through Society*. Cambridge, MA: Harvard University Press.

Latour, B., and Steve Woolgar. 1979. *Laboratory Life: The Construction of Scientific Facts*. Princeton: Princeton University Press.

Latour, B. 1990. "Drawing Things Together." In *Representation in Scientific Practice*, edited by Michael Lynch and Steve Woolgar. Cambridge, MA: MIT Press.

———. 1993. *We Have Never Been Modern*. Cambridge, MA: Harvard University Press.

Laurance, Edward J., Simon T. Wezeman, and Herbert Wulf. 1993. *Arms Watch: SIPRI Report on the First Year of the UN Register of Conventional Arms*. London: Oxford University Press.

Lavelle, Marianne. 1992. "More Lawyers Expect to Urge Their Clients to Examine Compliance." *National Law Journal* (March 16):3.

Lawson, G. J. 1991. "*Domesticated Trees in West African Plantations: An Opportunity for Clonal Techniques*." Mbalmayo, Cameroon: ODA/ONADEF. Mimeo.

Layne, Elizabeth. 1973. "Eighty Nations Write Magna Carta for Wildlife." *Audubon* 79–102.

Lazarus, Richard J. 1991. "The Tragedy of Distrust in the Implementation of Federal Environmental Law." *Law and Contemporary Problems* 54 (Autumn):311–374.

LeDuc, Jean Patrick. 1993. Interviews, Geneva, Switzerland, June 6 and December 1.

Leitzell, Terry L. 1973. "The Ocean Dumping Convention—A Hopeful Beginning." *San Diego Law Review* 10 (May):502–513.

Leone, Robert A. 1986. *Who Profits?* New York: Basic Books.

Leopold, Aldo. 1949. *Sand County Almanac, and Sketches Here and There*. New York: Oxford University Press.

Lester, James P., and Ann O'M. Bowman. 1989. "Implementing Environmental Policy in a Federal System: A Test of the Sabatier-Mazmanian Model." *Polity* 21 (Summer):731–753.

Levy, Marc. 1991. "The Greening of the United Kingdom: An Assessment of Competing Explanations." Paper delivered at the 1991 Annual Meeting of the American Political Science Association, August 29–September 1.

———. 1993. "European Acid Rain: The Power of Tote-board Diplomacy." In *Institutions for the Earth: Sources of Effective International Environmental Protection*, edited by Peter Haas, Robert O. Keohane, and Marc Levy, 75–132. Cambridge, MA: MIT Press.

Levy, Marc A., Robert O. Keohane, and Peter M. Haas. 1993. "Improving the Effectiveness of International Environmental Institutions." In *Institutions for the Earth: Sources of Effective International Environmental Protection*, edited by Peter M. Haas, Robert O. Keohane, and Marc A. Levy, 397–426. Cambridge, MA: MIT Press.

Liberatore, Angela. 1992. "National Environmental Policies and the European Community: The Case of Italy." *European Environment* 2, 4:5–8.

Lieberson, Stanley. 1991. "Small N's and Big Conclusions: An Examination of the Reasoning in Comparative Studies Based on a Small Number of Cases." *Social Forces* 70 (December):307–320.

Lieberthal, Kenneth, and Michael D. Lampton (eds.). 1992. *Bureaucracy, Politics, and Decisionmaking in Post-Mao China*. Berkeley: University of California Press.

Limitation of Anti-Ballistic Missile Systems Treaty. 1972. 23 UST 3435, art. XII (1), (2), and (3), October 3.

Lipschutz, Ronnie. 1993. "Environmentalism in One Country: The Case of Hungary." Paper presented at the conference of the Center for German and European Studies, UC Berkeley, April 30.

Lipson, Charles. 1982. "The Transformation of Trade: The Sources and Effects of Regime Change." In *International Regimes*, edited by Stephen Krasner, 233–271. Ithaca, NY: Cornell University Press.

Lishman, John. 1994. Telephone interview, U.S. Environmental Protection Agency, March 4.

Literaturnaya Gazeta. 1989. (November 22).

Liwo, Karl Jonathan. 1991. "The Continuing Significance of the Convention on International Trade in Endangered Species of Wild Fauna and Flora During the 1990's." *Suffolk Transnational Law Journal* 15 (Fall):122–152.

Lozano, Andre. 1994. *Folha de São Paulo* (4/9/94). *O Globo* (6/21/91). *Revista Forense*. 1992. No. 317. Rio de Janeiro: Editoria Forense.

Lyster, Simon. 1985. *International Wildlife Law: An Analysis of International Treaties Concerned with the Conservation of Wildlife*. Cambridge: Grotius.

MacKenzie, D. 1990. *Inventing Accuracy: A Historical Sociology of Nuclear Missile Guidance*. Cambridge, MA: MIT Press.

MacKenzie, Debora. 1992a. "Countries 'out of Phase' Over CFC Replacements." *New Scientist* 135, 1838:6.

———. 1992b. "Large Hole in Ozone Agreement." *New Scientist* 136, 1849:5.

———. 1993. "Timber: the Beam in Europe's Eye." *New Scientist* 138, 1879:9.

MacKerron, Conrad B. 1988. "EPA Zeroes in on CFC Windfalls." *Chemical Week* (August 10):8.

Macrory, Richard. 1992. "The Enforcement of Community Environmental Laws: Some Critical Issues." *Common Market Law Review* 29, 2 (April):347–369.

Madsen, Stig Toft. 1995. "People's Rights in a Bird Sanctuary: A Case Study from Bharatpur." Unpublished manuscript, Lund University, Department of Sociology of Law.

Magraw, Daniel B. 1988. "Our Common Lands: Defending the National Parks—International Law and Park Protection." In *A Global Responsibility*, edited by D. J. Simon, 143–174. Washington, DC: Island Press.

———. 1994. Telephone interview, Environmental Protection Agency, Office of International Activities, May 10.

Makhijani, Arjun, Amanda, Bickel, and Annie Makhijani. 1990. "Ozone Depletion: Cause and Effect." *EPW* 25 (March 10):493–496.

Mankin, Bill. 1993. Interview, Global Forest Policy Project, November 2.

Mann, Dean E. 1991. "Environmental Learning in a Decentralized Political World." *Journal of International Affairs* 45, 1:67–70.

Marer, Paul. 1993. "Economic Transformation in Central and Eastern Europe." In *Making Markets*, edited by Shafiqul Islam and Michael Mandelbaum, 53–98. New York: Council on Foreign Relations.

Marsh, George P. 1864. *Man and Nature*. Reprint 1965. Cambridge: Belknap Press of Harvard University Press.

Martin, Lisa L. 1992. *Coercive Cooperation: Explaining Multilateral Economic Sanctions*. Princeton: Princeton University Press.

Maugh, Thomas. 1988. "Ozone Depletion Far Worse Than Expected." *Los Angeles Times*, March 16, Metro 1.

McCormick, John. 1989. *Reclaiming Paradise: The Global Environmental Movement*. Bloomington: Indiana University Press.

McCrea, F., and Gerald Markle. 1984. "The Estrogen Replacement Controversy in the USA and UK: Different Answers to the Same Question?" *Social Studies of Science* 14:1–26.

Menon, Vivek. 1996. *Under Siege: Poaching and Protection of the Greater One-Horned Rhinoceroses in India*. New Delhi: World Wildlife Fund International.

Meyer, F. V. 1978. *International Trade Policy*. New York: St. Martin's Press.

Misch, Ann. 1992. "Can Wildlife Traffic Be Stopped?" *World Watch* 5, 5:26–33.

Mitchell, Robert Cameroon. 1994. 2nd ed. "Public Opinion and the Green Lobby: Poised for the 1990s?" In *Environmental Policy in the 1990s*, edited by Norman J. Vig and Michael E. Kraft, 81–102. Washington, DC: Congressional Quarterly.

Mitchell, Ronald B. 1992. "Membership, Compliance, and Noncompliance in the International Convention for the Regulation of Whaling: 1946–Present." Paper presented at the 17th Annual Whaling Symposium, Sharon, MA, October.

———. 1994. *Intentional Oil Pollution at Sea: Environmental Policy and Treaty Compliance*. Cambridge, MA: MIT Press.

Molina, Mario, and Sherwood Rowland. 1974. "Stratospheric Sink for Chlorofluoromethanes: Chlorine Atom-Catalyzed Destruction of Ozone." *Nature* 249:810.

Moore, Steven. 1992. "Troubles in the High Seas: A New Era in the Regulation of U.S. Ocean Dumping." *Environmental Law* 22:913.

Morton, Terry B. 1993. Correspondence from Terry B. Morton to Michael Glennon, US/ICOMOS, April 22.

Moscow News. 1993. No. 9.

Myrdal, Gunnar. 1968. *Asian Drama: An Inquiry Into the Poverty of Nations*. New York: Random House.

National Park Service. 1962. *First World Conference on National Parks*.

National Resources Defense Council. 1991. *Saving the Ozone Layer*. Rev. ed.

Navid, Daniel. 1979. "Commission on Environmental Policy, Law and Administration (CEPLA) Investigates." *IUCN Bulletin* 10:89.

Naysnerski, Wendy, and Tom Tietenberg. 1992. "Private Enforcement of Federal Environmental Law." *Land Economics* 68 (February):28–48.

Ndjatsana, M. M. 1993. *Analyse de la Situation du Secteur Forestier du Cameroun: La Politique Forestière*.

Nelkin, Dorothy, ed. 1992. *Controversy*, 3d edition. Newbury Park: Sage.

Nelson, J. G., and E. A. Alder. 1992. *Toward Greater Understanding and Use of the World Heritage Convention*. Waterloo, Ontario: Heritage Resources Center, University of Waterloo.

Nemba, R. M. 1993. "*Programme de Pays. Mise en Oeuvre du Protocol de Montréal: Cas de Cameroun*. Ier Rapport sur la Consommation et les Usages des Substances qui Appauvrissent la Couche d'Ozone." Mimeo.

New York Times. 1991. June 11.

———. 1993a. June 6.

———. 1993b. August 31.

———. 1993c. December 21.

Ngwa, J. A. 1988. *A New Geography of Cameroon*. Burnt Mill, UK: Longman Group.

Nikitina, Elena. 1991. "New Soviet Environmental Policy: Approaches to Global Changes." *International Study Notes* 16, 1:31–37.

Noble, D. 1978. "Social Choice in Machine Design: The Case of Automatically Controlled Machine Tools, and a Challenge for Labor." *Politics and Society* 8:313–347.

Nuclear Free Seas. Trip report of Greenpeace visit to Moscow and Russian Far East, June–November 1992. 1993. Washington DC: Greenpeace.

Obam, A. 1992. *Conservation et Mise en Valeur des Forêts au Cameroun*. Yaoundé, Cameroon: National Printing Press.

"Ob Utverzhdenii Polozheniia o Mezhvedomstvennoi Kommissii po Okhrane Ozonovogo Sloia. . . ." 1993. *Sobranie Aktov* (September 6).

Oda, Hiroshi. 1992. *Japanese Law*. London: Butterworth.

OECD. 1994. OECD Environmental Performance Review: Japan, 1994 (Paris: OECD).

Okolicsányi, Károly. 1992. "Hungary: Antall's Government Proves Less Than Green." *RFE/RL Research Report* 1, 33:49–70.

Olson, Mancur. (1968). *The Logic of Collective Action*. New York: Schocken Books.

ONADEF. Introductory Public Relations Document. Yaoundé: ONADEF.

Osharenko, Gail. 1989. "Environmental Cooperation in the Arctic: Will the Soviets Cooperate." *Current Research on Peace and Violence* 3:144–157.

"O Sostoianii Okruzhaiushchei Prirodnoi sredy Rossiiskoi Federatsii v 1991 godu." 1992. *Zelenyi mir* no. 39–40.

Ostrom, Elinor. 1990. *Governing the Commons: The Evolution of Institutions for Collective Action*. Cambridge: Cambridge University Press.

Paczolay, Peter. 1993. "The New Hungarian Constitutional State: Challenges and Perspectives." In *Constitution Making in Eastern Europe*, edited by A. E. Dick Howard, 21–55. Baltimore: Johns Hopkins University Press.

Panjwani, Raj. 1994. *Courting Wildlife*. New Delhi: WWF-India.

Parson, Edward A. 1993. "Protecting the Ozone Layer." In *Institutions for the Earth*, edited by Peter M. Haas, Robert O. Keohane, and Marc A. Levy, 27–73. Cambridge, MA: MIT Press.

Peet, Gerard. 1994. "International Co-operation to Prevent Oil Spills at Sea: Not Quite the Success It Should Be." In *Green Globe Yearbook*, 41–54. Oslo: Fridtjof Nansen Institute.

Perrow, Charles. 1984. *Normal Accidents*. New York: Basic Books.

Peterson, D. J. 1993a. "Building Bureaucratic Capacity in Russia." Paper presented at the Carleton University Conference on Environmental Security After Communism, Ottawa, February 26–27.

———. 1993b. *Troubled Lands: The Legacy of Soviet Environmental Destruction*. Boulder, CO: Westview Press.

Pinch, T. 1993. "'Testing—One, Two, Three ... Testing': Toward a Sociology of Testing." *Science, Technology, and Human Values* 18:25–41.

Pope, Carl. 1993. *Testimony before House Merchant Marine and Fisheries Committee, Subcommittee on Environment and Natural Resources*. November 10.

Porter, Gareth, and Janet Walsh Brown. 1991. *Global Environmental Politics*. Boulder, CO: Westview Press.

Porter, T. M. 1992. "Objectivity as Standardization: The Rhetoric of Impersonality in Measurement, Statistics, and Cost-Benefit Analysis." In *Rethinking Objectivity, II, Annals of Scholarship*, edited by Allan Megill 9:47.

Porter, T. M. 1995. *Trust In Numbers: The Pursuit of Objectivity in Science and Public Life*. Princeton: Princeton University Press.

Prendiville, Brendan. 1994. *Environmental Politics in France*. Boulder, CO: Westview Press.

Princen, Thomas, and Matthias Finger. 1994. *Environmental NGOs in World Politics*. London: Routledge

Problemy i resheniia. 1989. Interview in JPRS-UST-89-005 (March 30):40.

Protocol to the 1979 Convention on Long-Range Transboundary Air Pollution on the Reduction of Sulphur Emissions or Their Transboundary Fluxes by at Least 30 Percent. 1985. UN Doc. ECE/EB.AIR/12. Reprinted in 27 ILM 698 (1988).

Prott, Lyndell. 1993. Interview, UNESCO, June 17.

Pryde, Philip R. 1991. *Environmental Management in the Soviet Union*. Cambridge: Cambridge University Press.

Puri, G. S., V. M. Meher-Homji, R. K. Gupta, and S. Puri. *Forest Ecology: Phytogeography and Forest Conservation*, vol. 1, 2nd ed. Delhi: Oxford/IBH Publishing.

Pursell, Carroll (ed.). 1973. *From Conservation to Ecology: The Development of Environmental Concern*. New York: Crowell.

Putnam, R. D. 1992. *Making Democracy Work: Civic Traditions in Modern Italy*. Princeton: Princeton University Press.

Quigley, Howard B. 1993. "Saving Siberia's Tigers." *National Geographic* (July):38–47.

Raclin, Linda L. 1994. "Clean Trade." *Government Executive* (January).

Raghunandan, D. 1987. "Ecology and Consciousness." *Economic and Political Weekly* 23 (March 28):545–549.

Rainforest Alliance. 1993. *Description of the Smart Wood Certification Program and the Timber Project*. New York: Rainforest Alliance.

Ramadass, M., and L. Rathakrishnan. 1993. "India's New Forest Policy 1988: An Appraisal." *Ecology* 7 (March):10.

Ramirez, Anthony. 1989. "EPA Should Clean Up Its Own Act First." *Fortune* (November 6):139–142.

Ramsar Convention Bureau. 1990. *Summary Report on the Operation of the Ramsar Bureau's Monitoring Procedure 1988–1989*. Gland, Switzerland: Ramsar Convention Bureau.

Rasmussen, Dana A. 1992. "Enforcement in the U.S. Environmental Protection Agency: Balancing the Carrots and the Sticks." *Environmental Law* 22, 1:333–348.

"Regulate Us, Please." 1994. *The Economist* (January 8):69.

Reinhold, Robert. 1993. "Hard Times Dilute Enthusiasm for Clean-Air Laws." *New York Times*, November 26, A1.

Reinicke, Wolfgang H. 1994. "Cooperative Security and the Political Economy." In *Global Engagement*, edited by Janne Nolan. Washington, DC: Brookings Institution.

Reitzes, Jody Meier. 1992. "The Inconsistent Implementation of the Environmental Laws of the European Community." *Environmental Law Reporter* 22:10523–10528.

"Republican Congress Wants to Cut Environmental Funding." 1995. *Morning Edition*, National Public Radio (January 3).

Reuters World Service. 1994. "EU Criticized over Trade in Endangered Species." September 8.

Revista Forense. 1992. No. 317 (Rio de Janeiro).

Richardson, Genevra, Anthony Ogus, and Paul Barrows. 1983. *Policing Pollution: A Study of Regulation and Enforcement*. Oxford: Oxford University Press.

Roan, Sharon. 1989. *The Ozone Crisis: The 15-Year Evolution of a Sudden Global Emergency*. New York: John Wiley & Sons.

Robbins, Chris. 1995. Telephone interview, TRAFFIC USA, April 3.

Roberts, Richard. 1990. "Tale of Birds of Prey and Their Predators: Peregrine Falcons Endangered and Some Say Highly Marketable." *Los Angeles Times*, December 22, C1.

Robinson, Nicholas A. (ed.). 1993. *Agenda 21: Earth's Action Plan*. New York: Oceana.

Roe, E. 1995. "Postcript: Except Africa." *World Development* 26 (June):1065–1069.

Rose-Ackerman, Susan. 1995. *Controlling Environmental Policy: The Limits of Public Law in Germany and the United States*. New Haven: Yale University Press.

Roselle, Mike, and Tracy Katelman. 1989. *Tropical Hardwoods: A Report on the U.S. Role in the International Hardwood Trade and the Non-Timber Economic Alternatives*.

Rosencranz, A., and Milligan, R. 1990. "CFC Abatement: The Needs of Developing Countries." *Ambio* 19 (October):312–316.

Rotham, David. 1989. "Getting CFC Substitutes to Market." *Chemical Week* (April 26): 25.

Rothman, Hal. 1989. *Preserving Different Pasts*. Urbana: University of Illinois Press.

Rowell, Andrew. 1996. *Green Backlash: Global Subversion of the Environment*. London: Routledge.

Rowlands, Ian. 1993. "The Fourth Meeting of the Parties to the Montreal Protocol: Report and Reflection." *Environment* 35, 6:25–34.

———. 1995. *The Politics of Global Atmospheric Change*. Manchester: Manchester University Press.

Roy, Deb S. 1994. "*Manas National Park: A Status Report.*" Unpublished manuscript.

Ruggie, J. G. 1975. "International Responses to Technology: Concepts and Trends." *International Organization* 29, 3:557–583.

Runte, Alfred. 1994. "Pragmatic Alliance: Western Railroads and the National Parks." *National Parks* 68 (March):30–34.

"Russia Still Dumping Nuclear Wastes at Sea." *Current Digest of the Post-Soviet Press*. 1993. 45, 14:21–22.

Sajó, András. 1994. "Legal Aspects of Environmental Protection in Hungary: Some Experiences of a Draftsman." In *Designing Institutions for Sustainable Development in Hungary: Agenda for the Future*, edited by Z. Bohniarz, R. Bolan, S. Kerekes, and J. Kindler, 33–43. Minneapolis.

Salter, L. 1988. *Mandated Science: Science and Scientists in the Making of Standards*. Dordrecht, Netherlands: Kluwer.

———. 1993. "The Housework of Capitalism." *International Journal of Political Economy* 23:105–135.

Sanada, Yoshiaki. 1989 "The Cultural Bases of the Japanese as a Key to the Myth of the Reluctant Litigant in Japan: A Prelude to the Understanding of the Japanese Legal Culture." In Japan Association of Comparative Law, *Conflict and Integration; Comparative Law in the World Today*, 105–129. Tokyo: University of Tokyo Press.

Sand, Peter H. 1985. "Protecting the Ozone Layer: The Vienna Convention Is Adopted." *Environment* 27 June:19–40.

———. 1995. "The Potential Impact of the Global Environment Facility of the World Bank, UNDP and UNEP." Mimeo.

———. (ed.). 1992. *The Effectiveness of International Environmental Agreements: A Survey of Existing Legal Agreements*. Cambridge: Grotius.

Sands, P. J. 1991. "The Role of Non-Governmental Organizations in Enforcing International Environmental Law." In *Control Over Compliance with International Law*, edited by W. E. Butler, 61–68. Dordrecht, Netherlands: Martinus Nijoff.

Sandbrook, R. 1986. "The State and Economic Stagnation in Tropical Africa." *World Development* 14, 3:319–332.

Sax, Joseph. 1991. "Ecosystems and Property Rights in the Greater Yellowstone: The Legal System in Transition." In *The Greater Yellowstone Ecosystem: Redefining America's Wilderness Heritage*, edited by Robert B. Keiter and Mark S. Boyce, 77–84. New Haven: Yale University Press.

Sbragia, Alberta M. 1992. "Environmental Policy in the European Community: The Problem of Implementation in Comparative Perspective." In *Towards a Transatlantic Environmental Policy: Conclusions from an International Roundtable Seminar*, 48–97. Washington, DC: European Institute.

———. 1993a. "Asymmetrical Integration in the European Community: The Single European Act and Institutional Development." In *The 1992 Project and the Future of Integration in Europe*, edited by Dale L. Smith and James Lee Ray, 92–109. Armonk, NY: M. E. Sharpe.

———. 1993b. "EC Environmental Policy: Atypical Ambitions and Typical Problems." In *The State of the European Community: The Maastricht Debates and Beyond*, edited by Alan W. Cafruny and Glenda G. Rosenthal. Boulder, CO: Lynne Rienner.

———. 1996. "The Push-Pull of Environmental Policy-making." In *Policy-making in the European Union*, edited by Helen and William Wallace, Oxford: Oxford University Press.

———. 1997. "Governance, Credibility, and Federalism: The European Union in Comparative Perspective." Paper presented at the ECPSR-APSA Workshop on Regional Integration and Multi-Level Governance, Bern, Switzerland, February 27–March 4.

Schaller, George B. 1993. *The Last Panda*. Chicago: University of Chicago Press.

Schenk-Sandbergen, Loes. 1988. "People, Trees and Forest in India." In *Report of the Mission of the Netherlands on the Identification of the Scope for Forestry Development Corporation in India*, annex 2. Amsterdam.

Schmidthusen, F. 1986. *La Législation Forestière dans Quelques Pays Africaines*. Étude FAO Forêts no. 65. Rome: FAO.

Schnaiberg, Allen. 1977. "*Politics, Participation and Pollution: The Environmental Movement*." In *Cities in Change: Studies in the Urban Condition*. Edited by John Walton and Donald E. Carns. Boston: Allyn and Bacon.

Schneider, Keith. 1993. "Unbending Regulations Incite Move to Alter Pollution Laws." *New York Times*, November 29, A1.

Scholtz, John T. 1984. "Voluntary Compliance and Regulatory Enforcement." *Law and Policy* 6 (October):385–404.

Schorr, David K. 1994. "Testimony before Senate Commerce Committee, Subcommittee on Foreign Commerce and Trade." *Senate Report* (February 3). LEXIS, News Library, Curnws file.

Schwartz, Herman. 1992. "The New East European Constitutional Courts." *Michigan Journal of International Law* 13, 4:741–785.

Schwartz, Seymour I., Wendy P. Cuckovich, Cecilia F. Cox, and Nancy S. Ostrom. 1989. "Improving Compliance with Hazardous Waste Regulations Among Small Businesses." *Hazardous Waste and Hazardous Materials* 6, 3:281–296.

Sclove, R. E. 1995. *Democracy and Technology*. New York: Guilford Publications.

Segerson, Kathleen, and Tom Tietenberg. 1992. "The Structure of Penalties in Environmental Enforcement: An Economic Analysis." *Journal of Environmental Economics and Management* 23 September:179–200.

Segodnia. 1994. June 18.

"Service Investigation Uncovers Ginseng Export Violations." 1992. *P.R. Newswire*, LEXIS, News Library, Curnws File.

Seshan, Sekhar. 1995. "Plugging the Hole in the Sky." *Business India* (February 13–26):180.

Seymour, Sarah. 1993. Correspondence from Sarah Seymour to Michael Glennon, Friends of Animals, April 28.

Shapin, S. 1994. *A Social History of Truth*. Chicago: University of Chicago Press.

Shapin, S., and Simon Schaffer. 1985. *Leviathan and the Air-Pump: Hobbes, Boyle, and the Experimental Life*. Princeton, NJ: Princeton University Press.

Shastri, Vanita. 1994. "The Political Economy of Policy Formation in India: The Case of Industrial Policy." Draft Ph.D. diss., Cornell University.

Shea, Cynthia Pollock. 1988. "The Chlorofluorocarbon Dispute." *New York Times*, April 10, C2.

Shihata, Ibrahim F. I. 1996. "Implementation, Enforcement, and Compliance with International Environmental Agreements—Practical Suggestions in Light of the World Bank's Experience." *Georgetown International Environmental Law Review* 9, 37 (Fall):37–51.

Shimberg, Steven J. 1991. "Stratospheric Ozone and Climate Protection: Domestic Legislation and the International Process." *Environmental Law* 21:2175–2216.

Shirk, Susan. 1994. *How China Opened Its Door*. Washington, DC: Brookings Institution.

Sikadi, V. 1991. "*Le Cameroun et les Instruments Juridiques Internationaux de Protection de l'Environnement.*" Master's thesis, University of Yaoundé.

Simon, Lyster. 1985. *International Wildlife Law: An Analysis of International Treaties Concerned with the Conservation of Wildlife*. Cambridge: Grotius.

Simon, Ron. 1992. "How to Win the Battle of the Experts." *Trial* 28 (June–December) 36–41.

Singh, Samar. 1986. *Conserving India's Natural Heritage*. Dehradun: Nataraj.

Singh, Shekhar. 1993. "Eco-Funds: Budgeting for the Environment." *The Hindu: Survey of the Environment*, 11–16.

Skolnikoff, E. 1993. *The Elusive Transformation: Science, Technology, and the Evolution of International Politics*. Princeton: Princeton University Press.

Sleeth, Peter. 1990. "Businesses Join Boycott of Exotic Wood Imports." *Denver Post*, May 29, C1.

Smil, Vaclav. 1993. *China's Environmental Crisis*. Armonk, NY: M. E. Sharpe.

Sobranie aktov. 1993. April 12.

Sostoaianie Prirodnoi Sredy i Prodoskhannaia Deiatel'nost v SSSR. 1990. Moscow: Goskompriroda.

Speart, Jessica. 1993. "War Within: Illegal Wildife Traders and the Division of Law Enforcement of the U.S. Fish and Wildlife Service." *Buzzworm* 25 (July):36.

Soroos, Marvin A. 1997. *The Endangered Atmosphere: Preserving a Global Commons*. Columbia: University of South Carolina Press.

Spencer, Leslie. 1992. "Designated Inmates." *Forbes* (October 26):100–102.

Stairs, Kevin, and Peter Taylor. 1992. "Non-Governmental Organizations and the Legal Protecti on of the Oceans: A Case Study." In *International Politics of the Environment*, edited by A. Hurrell and B. Kingsburg, 110–141. Oxford: Oxford University Press.

Stammer, Larry. 1990. "Chinese Delegates to Seek Beijing's Approval for Pact to Protect Ozone." *Los Angeles Times*, June 29, A8.

Stewart, Gwyneth. 1981. "Enforcement Problems in the Endangered Species Convention: Reservations Regarding the Reservation Clauses." *Cornell International Law Journal* 14:429–455.

Strange, Susan. 1974. "IMF: Money Managers." In *The Anatomy of Influence: Decision Making in International Organizations*, edited by Robert W. Cox and Harold K. Jacobson, 263–297. New Haven: Yale University Press.

Subak, S. 1991. "Commentary on the Greenhouse Index." *Environment* 33:2–3.

Suzuki, David, and Peter Knudtson. 1992. *Wisdom of the Elders: Sacred Native Stories of Nature*. New York: Bantam Books.

Szabó, Máté. 1991. "Changing Patterns of Mobilization in Hungary Within New Social Movements: The Case of Ecology." In *Democracy and Political Transformation: Theories and East-Central European Realities*, edited by György Szoboszlai, 170–194. Budapest: Hungarian Political Science Association.

Szell, Patrick. 1995. "The Development of Multilateral Mechanisms for Monitoring Compliance." In *Sustainable Development and International Law*, edited by Winfried Lang, 97–109. London: Graham & Trotman/Martinus Nijhoff.

Tamozhennyi Kodeks Rossiisskoi Federatsii. 1994. 4:360–366.

Tanaka, Hideo. 1985. "The Role of Law in Japanese Society." *University of British Columbia Law Review* 19:375–388.

Tata Energy Research Institute (TERI). 1994. "CFC Phaseout—Hydrocarbon Based Refrigeration." In *Proceedings of the Planning Workshop ECOFRIG Phase—IV*. New Delhi: TERI. Mimeo.

Taubes, Gary, and Allan Chen. 1987. "Made in the Shade? No Way." *Discover* 8 (August):62–71.

Tchamba, M., C. S. Wanzie, Y. Bello, and S. Gartlan. 1991. *Elephant Conservation Plan*. Yaoundé: Ministry of Tourism, Office of Fauna and Parks.

"Tensions Between Ozone Protection, Climate Change Emerging in Policy Debate." 1993. *International Environment Reporter (BNA)* 16 (November 3):811–812.

Thompson, Donald. 1994. Interview, APHIS, February 28.

Thompson, M., Michael Warburton, and T. Hatley. 1986. *Uncertainty on a Himalayan Scale*. London: Ethnographica.

Thomsen, Jorgen B., and Amie Brautigam. 1987. "CITES in the European Economic Community: Who Benefits?" *Boston University International Law Journal* 5, 2:269–287.

Thornburgh, Dick. 1991. "Criminal Enforcement of Environmental Laws—A National Priority." *George Washington Law Review* 59 (April):775–780.

Thorne-Miller, Boyce. 1993. Telephone interview, Friends of the Earth International, May 24.

Thorsell, James. 1993. Correspondence from James Thorsell to Michael Glennon, IUCN, April 6.

Tietenberg, Tom. 1991. "Innovation in Environmental Enforcement: Economic Aspects of Recent Developments in Private Enforcement and the Structure of Noncompliance Penalties." Paper presented at Annual Meeting of the European Association of Environment and Resource Economists, Stockholm, June 10–14.

Tillotson, John. 1989. "International Commodity Agreements and the European Community: Questions of Competence and Will." *Journal of World Trade* 23 (December):109–125.

TRAFFIC-Europe. 1991. *Recent Illegal Trade of Chimpanzees and Gibbons in Europe: An Example of CITES Implementation Problems in Italy*. Cambridge: TRAFFIC International.

———. 1992. *The Control of Wildlife Trade in Greece: A TRAFFIC Network Investigation*. Cambridge: TRAFFIC International.

TRAFFIC-India. 1994. "*Seizures in India*." New Delhi: TRAFFIC-India. Mimeo.

TRAFFIC-International. 1993. "CITES Law in Italy." *TRAFFIC Bulletin* 13, 3:89.

Traffic-USA. 1993. *Report*.

Treaty Banning Nuclear Weapons Tests in the Atmosphere, in Outer Space, and Under Water. 1963. 480 UNTS 43, art. I(1)(b), 5 August.

Trexler, Mark C. 1989. "The Convention on International Trade in Endangered Species of Wild Fauna and Flora: Political or Conservation Success?" Ph.D. diss., University of California, Berkeley.

Tsurikov, Stanislav. 1990. Interview. *Pravitel'svennyy Vestnik*. In JPRS-UPA-90-039. (July 7):61.

Union of International Associations (UIA). 1997. *Yearbook of International Organizations, 1997/1998*. Brussels: Union of International Associations.

United Nations (UN). 1982. International Coffee Agreement. TIAS 11095 art. 53 (16 September).

———. 1986. International Wheat Agreement. UKTS 94 (1991), art. 3.14 (March 1986).

———. 1987. International Natural Rubber Agreement. UNDoc. TD/Rubber2/EX/R.1/Add.7, UKTS 36 (1993) Cmnd. 2253, art. 45 (20 March 1987).

UNEP. 1990. *Report of the First Meeting of the Ad Hoc Group of Experts on the Reporting of Data*. UNEP/OzL.Pro/WG.2/1/4 (7 December).

———. 1992. "Survey of Existing International Agreements and Instruments and Its Follow Up." *Report by the Secretary General of the United Nations Conference on Environment and Development*. A/Conf. 151/PC/103 and addendum 1.

———. 1993a. *Report on the Situation of Human Rights in Haiti, Prepared by Marco Julio Bruni Celii, Special Rapporteur of the Commission on Human Rights, in Accordance with Commission Resolution 1992/77*. E/CN.4/1993/47.

———. 1993b. *Report on the Situation of Human Rights in Myanmar, Prepared by Yozo Kota, Special Rapporteur of the Commission on Human Rights, in Accordance with Commission Resolution 1992/58*. E/CN.4/1993/37.

United Nations Conference on Trade and Development (UNCTAD). 1982. *Report of the Sixth Preparatory Meeting on Tropical Timber, UNCTAD Trade and Development Board, Integrated Programme for Commodities*. TD/B/IPC /Timber 39.

———. 1994. *Preparation of a Successor Agreement to the International Tropical Timber Agreement, 1983, Formal Settlement by Consumer Members*. TD/Timber 2/L.6.

United Nations Development Program/Food and Agriculture Organization (UNDP/FAO). 1988. *Tropical Forestry Action Plan: Joint Inter-agency Planning and Review Mission for the Forestry Sector—Cameroon*, 3 vols. Rome: UNDP/FAO.

UNDP. 1995. *Report on Human Development*. New York: United Nations.

United Nations Environment Programme (UNEP). 1981a. *United Nations Environment Programme Governing Decision* 9/13B of May 26, 1991.

———. 1981b. *World Plan of Action on the Ozone Layer*. United Nations Environment Programme/WG/7/35/Rev. 1, annex 3.

———. 1987. *Ad Hoc Working Group of Legal and Technical Experts for the Preparation of a Protocol on Fluorocarbons*. Second Session, Vienna, 23–27 February. UNEP/WG.

———. 1991a. *Draft Report of the Second Meeting of the Implementation Committee Under the Non-compliance Procedure for the Montreal Protocol*. UNEP/OzL. Pro./Imp. Comp./2/3.

———. 1991b. *Register of International Treaties and Other Agreements in the Field of the Environment*. GC16/Inf. 4. Nairobi: UNEP.

———. 1992a. Decision IV/17C, *Application of Trade Measures Under Article 4 to Nonparties to the Protocol*. UNEP/OzL. Pro. 4/15.

———. 1992b. *Draft Report on the Fourth Meeting of the Implementation Committee Under the Non-compliance Procedure for the Montreal Protocol.* UNEP/OzL. Pro./Imp. Com./4/L.2, 9.

———. 1992c. "*Environment and Sustainable Development for Cameroon.*" Report of the Multidisciplinary and Multi-institutional Mission on the Environment. Yaoundé: UNEP. Mimeo.

———. 1992d. *Secretariat's Report on the Reporting of Data by the Parties to the Montreal Protocol.* UNEP/OzL.Pro.4/6. (August 26).

———. 1992e. *Report of the Fourth Meeting of the Parties to the Montreal Protocol on Substances That Deplete the Ozone Layer, 25 November 1992.* UNEP/OzL. Pro. 5/12.

———. 1993a. *Country Program: China, Report to the Executive Committee of the Multilateral Fund for the Implementation of the Montreal Protocol.* UNEP/OzL. Oro./ExCom/9/13.

———. 1993b. Decision V/4. In *Report of the Fifth Meeting of the Parties.* UNEP/OzL. Pro. 5/12

———. 1993c. Decision V/17. UNEP/OzL. Pro. 5/12.

———. 1993d. *Status on Ratification/Acceptance/Approval of: I. The Vienna Convention for the Protection of the Ozone Layer (1985); II. The Montreal Protocol on Substances That Deplete the Ozone Layer (1987); III. The Amendment to the Montreal Protocol (1990).* UNEP/OzL. Rat. 25.

———. 1993e. *Report of the Fifth Meeting of the Implementation Committee Under the Non-compliance Procedure for the Montreal Protocol.* UNEP/OzL Pro./Imp. Comp./5/3.

———. 1994. *Report of the Sixth Meeting of the Parties to the Montreal Protocol on Substances That Deplete the Ozone Layer.* UNEP/OzL. Pro.6/7.

———. 1995. *Report of the Seventh Meeting of the Parties to the Montreal Protocol on Substances That Deplete the Ozone Layer.* UNEP/OzL.Pro.7/12.

———. 1996a. *Final Report. Round Table Discussion and Knowledge-Sharing Networks for ODS Phase-Out.* Washington, DC. Paris: UNEP IE.

———. 1996b. *Report of the Eighth Meeting of the Parties to the Montreal Protocol on Substances that Deplete the Ozone Layer.* UNEP/Ozl. Pro. 8/12.

———. 1997. *Report of the Ninth Meeting of the Parties of the Parties to the Montreal Protocol on Substances that Deplete the Ozone Layer.* UNEP/OzL.Pro.9/12.

UNESCO. 1991. *Gestion des Resources et des Reserves de la Biosphere et Education Relative a l'Environnement (Projet Pilote de Dja).*

United Nations General Assembly. 1989. A/44/PV.42.

United Nations Security Council. 1991. Resolution no. 687. U.N.Doc. S/RES/687 (3 April).

United States, Department of State. 1993. Interview with Department of State official, Washington, DC, June 3.

United States, Department of State. 1994. "Sanctions Announced Against Taiwan for Trade in Endangered Species." *Department of State Dispatch* 5, no. 16.

United States, Fish and Wildlife Service. 1977. *Annual Report.* Washington, DC: U.S. Government Printing Office.

United States, General Accounting Office (USGAO). 1991. *Intermediate-Range Nuclear Forces Treaty Implementation.* GAO/NSIAD-91-262, appendix 4. Washington, DC: U.S. Government Printing Office.

———. 1992a. *Endangered Species Act: Types and Number of Implementing Actions.* Washington, DC: U.S. Government Printing Office.

———. 1992b. *International Environment; International Agreements Are Not Well Monitored.* GAO/RCED 92–43. Washington, DC: US Government Printing Office.

United States, International Trade Commission. 1991. *International Agreements to Protect the Environment and Wildlife.* Report to Committee on Finance, U.S. Senate.

Upham, Frank K. 1987. *Law and Social Change in Postwar Japan.* Cambridge, MA: Harvard University Press.

"U.S. Imposes Ban on Imports of Endangered Animals, Goods." 1991. *Washington Post,* July 19, A19.

Valencia, Mark. 1987. *International Conference on the Yellow Sea.* Honolulu: East-West Environment and Policy Institute.

Vandeputte, Godelieve A. 1990. "Why the European Community Should Become a Member of the Convention on International Trade in Endangered Species of Wild Fauna and Flora." *Georgetown International Environmental Law Review* 3, 2:245–264.

Van Wolferen, Karel. 1989. *The Enigma of Japanese Power.* New York: Alfred A. Knopf.

Veja. 1997a. June 18.

———. 1997b. June 27.

Victor, David G. 1995. *The Montreal Protocol's Non-compliance Procedure: Lessons for Making Other International Environmental Regimes More Effective.* Laxenburg, Austria: IIASA.

Vogel, David. 1986. *National Styles of Regulation: Environmental Policy in Great Britain and the United States.* Ithaca, NY: Cornell University Press.

———. 1987. "A Big Agenda." *Wilson Quarterly* 11 (Autumn):51–63.

———. 1989. *Fluctuating Fortunes.* New York: Basic Books.

———. 1995 *Trading Up: Consumer and Environmental Regulation in a Global Economy.* Cambridge, MA: Harvard University Press.

Waffle, Robert. 1993. Interview, International Hardwood Products Association, November 20.

Wagenbaur, Rolf. 1990. "The European Community's Policies on Implementation of Environmental Directives." *Fordham International Law Journal* 14 (February):936–950.

Wainer, Ann Helen. 1991. *Legislação Ambiental do Brasil: Subsidios para a História do Direito Ambientale.* Rio de Janeiro: Editora Forense.

Waller, Michael, and Frances Millard. 1992. "Environmental Politics in Eastern Europe." *Environmental Politics* 1, no. 2:159–185.

Walton, John, and Donald Carns (eds.). 1977. *Cities in Change: Studies in the Urban Condition.* Boston: Allyn and Bacon.

Wani, Akihro. 1988. "Introduction to Japanese Law." In *Japan Business Law Guide,* edited by Mitsuo Matsushita. Sydney: CCH International.

Wassermann, Ursula. 1984. "UNCTAD: International Timber Agreement." *Journal of World Trade Law* 18 (Jan.–Feb.):89–91

Weber, Max. 1947. *The Theory of Social and Economic Organization.* New York: The Free Press.

Westbrook, David. 1994. "Liberal Environmental Jurisprudence." *University of California at Davis Law Review* 27 (Spring):619–711.

"White House Fact Sheet on the President's Proposal for a Global Forest Convention." *Weekly Compilation of Presidential Documents* 26: Doc. 1084.

Wijnstekers, Willem. 1992. *The Evolution of CITES: A Reference to the Convention on International Trade in Endangered Species of Wild Fauna and Flora*, 3rd ed. Geneva: CITES Secretariat.

Wille, Chris. 1991. "Buy or Boycott Tropical Hardwoods?" *American Forests*. 97 (July):26–27.

Williamson, Douglas. 1994. Telephone interview, World Wildlife Fund, January 25.

Williams, Sylvia Maureen. "A Historical Background on the Chlorofluorocarbon Ozone Depletion Theory and Its Legal Implications." In *Transboundary Air Pollution*, edited by Cees Flinterman, Barbara Kwiatkowska, and John G. Lammers, 267–280. Dordrecht, Netherlands: Martinus Nijhoff.

Wilson, E. O. (ed.). 1988. *Biodiversity*. Washington, DC: National Academy Press.

Winner, L. 1977. *Autonomous Technology: Technics Out-of-Control as a Theme in Political Thought*. Cambridge, MA: MIT Press.

———. 1986. "Do Artifacts Have Politics?" *The Whale and the Reactor*. Chicago: University of Chicago Press.

Wirth, D. 1994. "Reexamining Decision-Making Process in International Environmental Law." *Iowa Law Review* 79:769–802.

World Bank. 1989. *Sub-Saharan Africa: From Crisis to Sustainable Growth. A Long-term Perspective Study*. Washington, DC: World Bank.

———. 1992. *Trends in Developing Economies*. Washington, DC: World Bank.

———. 1994. "*Cameroon: Diversity, Growth and Poverty Reduction*." Washington, DC: Human Resources and Poverty Division, Technical Department, Africa Region. Working draft.

———. 1995–1996. *Facing the Environmental Challenge*. September 1995–January 1996. Washington, DC: World Bank.

World Commission on Environment and Development (WCED). 1987. *Our Common Future*. New York: Oxford University Press.

"World Resources, 1994–1995: A Report by the World Resources Institute in Collaboration with the United Nations Environment Programme and the United Nations Development Programme." 1994. In *World Resources*. Oxford: Oxford University Press.

World Wide Fund for Nature. 1991. "Six Endangered Primates Seized by Hungarian Government." Press Release no. 53/91 (9 October).

———. 1992. *International Country Profile on China*. Gland, Switzerland: World Wide Fund International.

World Wide Fund for Nature—India. (WWF-I). *The Conservation Plan 1992–93*. New Delhi: WWF-India.

———. 1992. *Manas: The Conservation Plan 1992–93*. New Delhi: WWF-India.

———. 1994. *Wildlife Trade: A Handbook for Enforcement Staff*. New Delhi: WWF-India.

World Wildlife Fund. 1992. *Conservation and Sustainable Management of Tropical Forests: The Role of ITTO and GATT*. Gland, Switzerland: World Wildlife Fund.

Worster, D. 1977. *Nature's Economy*. Cambridge: Cambridge University Press.

Wright, S. 1994. *Molecular Politics*. Chicago: University of Chicago Press.

Wynne, B. 1987. *Risk Management and Hazardous Waste: Implementation and the Dialectics of Credibility*. Berlin: Springer Verlag.

———. 1988. "Unruly Technology." *Social Studies of Science* 18:147–167.

Wynne, B., and Sue Mayer. 1993. "How Science Fails the Environment." *New Scientist* 5 (June):33–35.

Xinhua (Beijing). 1993. (September 20).

Yablokov, A. B. 1993. "Fakty i Problemy, Sviazannye s zakhoroneniem Radioaktivnykh Otkhodov v Moriakh, Omyvaiushikh Territoriiu Rossiisiiskoi Federatsii." *Rossiiskie Vesti* (April 6):21–47.

Yearley, S. 1992. *The Green Case*. London: Routledge.

Yearley, S. 1996. *Sociology, Environmentalism, Globalization*. London: Sage Publications.

Young, Michael K. 1984. "Judicial Review of Administrative Guidance: Governmentally Encouraged Consensual Dispute Resolution in Japan." *Columbia Law Review* 84 (May):923–983.

Young, Oran. 1979. *Compliance and Public Authority: A Theory with International Applications*. Baltimore: Johns Hopkins University Press.

———. 1983. "Regime Dynamics: The Rise and Fall of International Regimes." In *International Regimes*, edited by Stephen D. Krasner. Ithaca, NY: Cornell University Press.

———. 1992. "The Effectiveness of International Institutions: Hard Cases and Critical Variables." In *Governance Without Government: Change and Order in World Politics*, edited by James N. Rosenau and Ernst-Otto Czempiel, 160–194. Cambridge, UK: Cambridge University Press.

Yann-Huei, Bill Song. 1989. "Marine Scientific Research and Marine Pollution in China." *Ocean Development and International Law* 20:601–621.

"Zakon ob Okhrane Prirody v RSFSR." 1960. *Vedomosti Verkhovnogo Soveta RSFSR* no. 40: art. 586.

Zama, I. 1995. "Achieving Sustainable Forest Management in Cameroon." *Review of European Community and International Environmental Law* 4, 3:263–270.

Zehr, S. C. 1994. "Method, Scale, and Socio-Technical Networks: Problems of Standardization in Acid Rain, Ozone Depletion, and Global Warming Research." *Science Studies* 7:47–58.

Zelenyi Mir. 1993. no. 20.

Ziegler, Charles. 1987. *Environmental Policy in the USSR*. Amherst: University of Massachusetts Press.

Zurer, Pamela. 1993. "Du Pont Accelerates CFC Phaseout." *Chemical and Engineering News* 71, 11:5–6.

Statutes Cited

Antiquities Act of 1906. 34 Stat. 225, June 8, 1906; codified at 16 U.S.C. §§ 431 *et seq.*

Archaeological Resources Protection Act of 1979. Pub. L. 96-95, 93 Stat. 721, October 31, 1979; codified at 16 U.S.C. §§ 470aa *et seq.*

Clean Air Act Amendments of 1990. Pub. L. 101-549, 104 Stat. 2399, November 15, 1990; codified at 42 U.S.C. §§ 7401 *et seq.*

Endangered Species Act of 1973. Pub. L. 93-205, 87 Stat. 884, December 28, 1973; current version at 16 U.S.C. §§ 1531–1543.

Endangered Species Act Amendments of 1978. Pub. L. 95-632, 92 Stat. 3751, November 10, 1978; codified at 16 U.S.C. §§ 1531 note, 1532–1536, 1538–1540, 1542.

Endangered Species Act Amendments of 1982. Pub. L. 97-304, 96 Stat. 1411, October 13, 1982; codified at 16 U.S.C. §§ 1531–1533, 1535–1537a, 1538–1540, 1542.

Endangered Species Act Amendments of 1988. Pub. L. 100-478, 102 Stat. 2306, Oct. 7, 1988; codified at 16 U.S.C. §§ 1531 *et seq.*

Endangered Species Conservation Act of 1969. Pub. L. 91-135, 83 Stat. 275, December 5; 1969; codified at 16 U.S.C. §§ 668aa–668cc-5.

Endangered Species Preservation Act of 1966. Pub. L. 89-669, 80 Stat. 926, October 15, 1966.

Lacey Act of 1900. 31 Stat. 187, May 25, 1900; codified at 16 U.S.C. §§ 667e, 701.

Lacey Act Amendments of 1981. Pub. L. 97-79, 95 Stat. 1073, November 16, 1981; codified at 16 U.S.C. §§ 667e, 851–856, 1540, 3371–3378; 18 U.S.C. §§ 42, 43, 44, 3054, 3112.

Marine Mammal Protection Act of 1972. Pub. L. 92-522, 86 Stat. 1027, October 21, 1972; codified at 16 U.S.C. §1361 *et seq.*

Marine Protection, Research and Sanctuaries Act of 1972. Pub. L. 92-532, 86 Stat. 1052, October 23, 1972; codified at 16 U.S.C. §§ 1431–1434; 33 U.S.C. §1401, 1401 note, 1402, 1411, 1411 notes, 1412–1421, 1441–1445.

Migratory Bird Treaty. of 1918. ch. 128, §2, 40 Stat. 755, July 18, 1918; June 20, 1936; codified at 16 U.S.C. §703 *et seq.*

National Historic Preservation Act. Pub. L. 89-665, 80 Stat. 915, October 15, 1966; codified at 16 U.S.C. §§ 470 *et seq.*

National Historic Preservation Act Amendments of 1980. Pub. L. 96-515, 94 Stat. 2987, December 12, 1980; codified at 16 U.S.C. §§ 469c–2 *et seq.*

National Historic Preservation Act Amendments of 1992. Pub. L. 102-575, 106 Stat. 4753, October 30, 1992; codified at 16 U.S.C. §§ 461 *et seq.*

National Historic Sites Act of 1935. 49 Stat. 666, August 21, 1935; codified at 16 U.S.C. §§ 461 *et seq.*

National Park Service Organic Act of 1916. Pub. L. 89-249, 39 Stat. 535, August 25, 1916; codified at 16 U.S.C. §§ 1 *et seq.*

Ocean Dumping Ban Act of 1988. Pub. L. 100-688, 102 Stat. 4139, November 18, 1988; codified at 33 U.S.C. 2267, 1414a, 1414b, 1414c.

Pelly Amendment to Fisherman's Protective Act of 1954. Pub. L. 680, 68 Stat. 883, August 27, 1954; codified at 22 U.S.C. §§ 1971 *et seq.*

Index